James William Buel

The Savage World

A Complete Natural History of the World's Creatures, Fishes, Reptiles,

Insects, Birds and Mammals

James William Buel

The Savage World
A Complete Natural History of the World's Creatures, Fishes, Reptiles, Insects, Birds and Mammals

ISBN/EAN: 9783337801748

Printed in Europe, USA, Canada, Australia, Japan

Cover: Foto ©berggeist007 / pixelio.de

More available books at **www.hansebooks.com**

FIGHT IN THE DESERT.

AVIL CO LITH. PHILA.

THE
SAVAGE WORLD:

A COMPLETE
NATURAL HISTORY

OF THE WORLD'S CREATURES,

FISHES, REPTILES, INSECTS, BIRDS AND MAMMALS.

FOUNDED UPON THE THEORY OF THE PROGRESSION OF SPECIES, AND IN ACCORDANCE
WITH GENETIC REVELATION, SCRIPTURAL TRUTHS, AND
THE HARMONY OF NATURE.

With Introduction Describing the Geological Ages, Changes in the
Earth's Crusts, Fossil Remains of Extinct Animals,
and Monsters of the Ancient Seas.

REPLETE WITH ANECDOTE, INCIDENT AND ADVENTURE. ILLUSTRATIVE OF THE HABITS
OF THE ANIMALS DESCRIBED.

Abounding with Thrilling Experiences, Wonderful Discovery, Exciting
Episodes and Descriptions of the Marvellous Curiosities
of Nature in all parts of the Globe.

BY
J. W. BUEL,

AUTHOR OF

"The Beautiful Story," "The World's Wonders," "Sea and Land," "Exile Life in Siberia," Etc., Etc.

EMBELLISHED WITH OVER 1500 ELEGANT ENGRAVINGS.

ILLUSTRATING THE ANIMALS OF THE EARTH IN THEIR NATURAL CONDITION,

AND

MAGNIFICENT COLORED PLATES.

PUBLISHED AND MANUFACTURED BY

HISTORICAL PUBLISHING COMPANY,

PHILADELPHIA, PA.

INTRODUCTION

INCE the advocacy of a theory is much like the pursuit of a material substance guided by the shadow it casts, it was not without diffidence that I undertook the preparation of a NATURAL HISTORY upon the plan herewith submitted. While the idea serving me as a basis may have been conceived by others, and is intimated by Buffon, yet I may claim for the plan an originality that is likely to invite, if not excite, the criticisms of such strict scientists as hold tenaciously to the theory that there can be no common correspondence between geology and the Genetic account of creation. While believing that there is a perfect corroboration of the testimony of Genesis, and that this fact is abundantly attested by the witness of all nature, as well as by the mute, though even more convincing evidences of geology, yet I cannot undertake an elucidation of either theory, except as an incidental explanation of the purposes of this work may involve a brief outline of the Genetic basis, upon which it has been planned and constructed. The harmony which exists between revelation and approved science—by which latter expression I mean the deductions of the world's most distinguished scientists—is briefly outlined in the introductory pages of this work, preliminary to a description of the earliest forms of life. In the belief that my readers will accept this theory, so well established by Winchell, Buckland and other great palæontologists, to whose writings I beg to offer my profound acknowledgments, I shall proceed briefly to a description of the geological ages of the world, in which will be found indisputable evidence of the progression of species, upon which demonstration THE SAVAGE WORLD has been prepared.

It is to the rocks, the strata and the fossils, that we must turn for information respecting the age of the earth and the creatures that have peopled it in the æons of the past. Though they cannot speak, yet upon each God has written, in a language which those who carefully study can read, with infinite delight, the records of the ages. Some years ago the hieroglyphics on the obelisks of Egypt and the walls of resurrected Nineveh were as so many carvings destitute of meaning, but patient study served to decipher them, and what were once unintelligible arrow-heads and crude pictures are now read as the records of a people who perished with all the monuments of their skill, learning, wealth and industry. The so-called everlasting rocks, precipitated by the mighty solvents of nature, transmuted in the alembic wherein are deposited all the ingredients of earth, eroded by waves, ground by terrific forces, vitrified by furious fires, are covered with the handwritings of Omnipotence, and are as easily decipherable as are the hewn and carved stones of buried cities, and tell the story of the ages in language no less explicit.

As nearly every school-boy now knows, the earth's crust is composed of several strata, or layers, which, as geologists have discovered, are arranged in successive series, or chronological groups, each of which is distinguishable not only by the different character of material composing it, but also by fossil remains found therein, which afford a means for proximately determining the time required in their formation. These several layers may be likened to the figures on the dial of a clock, since they serve to mark the time, or rather epochs, or eras, in the earth's existence, and to each is therefore given a name to indicate the infinitely great periods that have elapsed since God's hand began the work of fashioning the world.

These general divisions are again subdivided, just as the hours on the clock are divided into minutes, in order that the measurements of time may be more exactly reckoned. The general divisions, arranged in their regular order, are known by the following scientific names, viz. : The Archæan (meaning *beginning*), the Palæozoic (*ancient life*), the Mesozoic (*middle life*), and the Neozoic (*new life*). The subdivisions of these general groups are represented in the following table:

ERAS OF TIME—CORRESPONDING TO HOURS.	PERIODS OF TIME—CORRESPONDING TO MINUTES.
ARCHÆAN, ...	{ Laurentian Huronian
PALÆOZOIC.	{ Cambrian Lower Silurian Upper Silurian Devonian Carboniferous.
MESOZOIC, ...	{ Permian Triassic Jurassic Cretaceous.
NEOZOIC, ...	{ Tertiary Quaternary

Mr. Hinman has made another division, in which the special and characteristic fossils of each period are given, as also the relative time occupied, so to speak, by each epoch, but in which the orders are reversed; that is, on a descending scale, instead of an ascending table, as I have given above. This order has been heretofore invariably observed by writers on geology and natural history, though it is manifestly improper when applied to the latter, since all nature proceeds upon an ascending scale. His arrangement is as follows:

In the oldest, or ARCHÆAN stratum, no fossil remains have been found, though this fact does not necessarily imply that there was no animal life on our globe during that period. The inconceivably great time that has elapsed, the immense superposition of stratum on stratum, and particularly the metamorphosis that the strata have undergone by the action of fire, would have destroyed any trace of fossil remains, however numerous they may have been. But the absence of all fossils "prevents us from determining the relative ages of the different parts of this group," as Hinman observes.

The PALÆOZOIC ERA, on the other hand, is distinguished for the abundance of animal and vegetable life that then existed, though it was in this period that these evidences of abounding life first appeared. This primary animal life, however, was all of the simplest, I might say almost rudimentary, forms, so low in the scale as scarcely to be distinguishable from the vegetable. Representatives still survive in the *protozoanidæ*, as will be found explained in this work.

The CAMBRIAN AGE, or the secondary period of the Palæozoic, is remarkable for the prolific growth of sea-weeds which then flourished, but animal life was plentiful, somewhat more developed, though none of the creatures of this period were provided with a back-bone. The species most numerous were two crustaceans, viz., the trilobites (three-lobed creatures), and an animal resembling the horse-shoe crab.

The two SILURIAN AGES are distinguished as being the periods when mollusks were most numerous; of the fossils found everywhere through these strata, those species having soft bodies are most plentiful, such as the snail, nautilus, and other species of the cuttle-fish. In the Upper Silurian, trilobites begin to diminish and corals to appear, also forms of *crinoids*, or animals that were so nearly vegetable that they are called *stone-lilies*. There were also several land plants of the fern species, and a few vertebrate sea-animals of the shark kind.

In the DEVONIAN AGE fishes appeared in the greatest abundance, their fossil bones being so numerous that it has been called the "Age of Fishes." However, these primary fishes presented great simplicity of structure, as compared with those of the present period. Their bones were imperfect, and instead of scales the body was covered with shield plates. Land plants now showed greater exuberance of growth and developed into forests.

The CARBONIFEROUS PERIOD shows another remarkable change. The forests had become so rank, under the effects of warm vapors and fruitful soil, that the earth became cumbered with an excess of vegetation, which in decaying formed immense beds in marshy districts. These beds were gradually covered by new growths, and were converted into bituminous coal. This first conversion, by the action of heat, was subsequently transmuted into anthracite, graphite, and other minerals. The forests were much larger, though still retaining the fern characteristic, and new forms of vegetable growth appeared in giant canes and club mosses. It was during this period, too, that amphibians made their first appearance, forming a connecting link between fishes and reptiles. Insects also came into being now, to people the rank and lofty forests.

The MESOZOIC ERA was the period of giant reptiles, existing in great numbers and of monster size, such as the teleosaurus, ichthyosaurus, megalosaurus, pterodactyl, and other flying, swimming and creeping creatures, some of which were more than fifty feet in length, and of proportionate bulk.

In the TRIASSIC PERIOD first appeared creatures bird-like in form, the impressions of whose feet and three toes are to be seen in the sandstones of some valleys in Connecticut and New Jersey. A few mammals, of a very low type, also came into being during this era.

The JURASSIC AGE is particularly interesting for being the period when reptiles of ferocious aspect and appalling size predominated, wherein we observe a connecting link between reptiles and birds, as we shall show in the body of this work.

The CRETACEOUS PERIOD is distinguished for being the last stratum of which the fossil animal remains are wholly of extinct species. The group is subdivided into upper and lower, called the *chalk* and *greensand* formations, which are widely distributed over both continents. Thus at widely separated points in the ancient seas of four continents were similar deposits produced during the same geological period, characterized by the animal remains which they include of the same general type, and often of the same species. The ichthyornis, or fish-bird, made its appearance during this period. This curious creature had the back-bone of a fish, a keel-like breast-bone, and a long, slender beak, which was armed with socketed teeth. It was equal to a pigeon in size,

The NEOZOIC ERA represented a change no less remarkable than that of the Mesozoic, for as the latter gave birth to giant reptiles, so it was during the Neozoic period that these huge creatures disappeared to give place to higher orders of species, and were accordingly succeeded by mammals. Birds developing from the pterodactyl and the ramphorhynchus, the primitive forms, also now appeared, and the land became peopled with creatures whose forms resemble those by which we are now surrounded.

In the TERTIARY AGE, North America presented a very different aspect from that of to-day. The sea covered nearly the whole of that portion of the United States south of the junction of the Ohio river with the Mississippi. Highlands were visible in the Tennessee region, and in parts of Arkansas and also of Mexico, but a greater portion of the section named was under the sea. In Kansas, and westward to the Rocky Mountains, there were immense lakes, which, drying up, left in their beds the remains of many creatures whose species are now extinct, but which are illustrated in this work. What was equally singular is the fact that all of North America possessed a mild climate, and was covered by a luxuriant vegetation of marvellous diversity, with tropical plants growing in profusion along the shores of the Arctic sea. The giant trees of California are the representative types of the forests that distinguished the Tertiary epoch. This period is also called the "Age of Mammals" for the reasons above given. Among the numerous species, the prototypes of those now existing, and which have undergone wonderful change through constant progression, was the horse, which appeared during this era, but very dissimilar in form from that which it has since assumed. The primitive horse was scarcely larger than a fox, and had three hoofed toes on the hinder feet, and no less than four similarly hoofed toes on the fore feet. Remains have also been discovered showing the horse to have increased in size to that of a sheep, when one of the toes on each fore foot was dropped; and still later it developed into a single-hoofed animal, the size of an ass, with side toes scarcely long enough to touch the ground.

The QUATERNARY AGE is that in which we live, but it has already extended over a very long period of time, and far beyond the grasp of transmitted history. It has been within this age that some of the mightiest forces and catastrophes of nature have combined to precipitate extraordinary changes both in climate and conditions. By reason, as is supposed, of the shifting of the declination of the earth, a great deluge of ice was sent crashing down over the northern regions of both continents, converting the mild climates theretofore existing into perpetual frost, as we now find it. Not alone this, but the animals which roamed those regions were destroyed by the moving fields of ice and snow ; such monsters as the mammoth, mastodon, woolly rhinoceros, sabre-toothed tiger, cave bear, Irish elk and other species, became extinct during this period, whose bones, however, are plentifully found deeply imbedded in the earth upon the spot where they so suddenly perished, to teach us how wonderful are the ways of Providence, and how feeble is the power and understanding of man, but for whose care God is ever watchful and considerate.

J. M. Buel

CONTENTS

CONTENTS.

CONTENTS

ILLUSTRATIONS

(9)

FIERCE SNAKES OF THE JUNGLE.

THE SAVAGE WORLD.

THE BIRTH OF SPECIES.

NATURAL HISTORY, in its largest sense, treats of all things in nature the animate and inanimate, the dynamics or forces, structure of the earth, conditions, mutations, and, in short, all the operative forces of physics, as well as the habits, character, organism and species of animal life. It is therefore a most comprehensive subject, and consequently one divisible into numerous branches for distinct consideration, compassing fields for research so extensive that no single book can contain an exhaustive treatise on all the products therein. For these reasons I have chosen to confine myself in this work to a description of that particular branch of Natural History (Zoology,) which treats of the animal life on our planet, with only incidental references, as occasions seem to require, to the multitudes of plant, minute insect and infusorial life. I speak here of such references rather to acquaint the reader with a most interesting, because least understandable, fact, viz.: that the line of separation between vegetable and animal life is so indistinct that it is doubtful if the demarcation has yet been discovered. The development of microscopic life from decaying vegetation is scarcely so singular as the truth that many of the infusoria, or microscopic creatures, partake of a double nature, retaining semblances of vegetation while possessing functions of motion and digestion which characterize the animal. This subject, however, leads so directly to evolution and into metaphysics, that the general reader will feel no regrets that I have not claimed the space and patience to pursue its discussion here.

Though our subject is vast in magnitude, every step is one of extraordinary interest, unfolding new beauties and startling wonders with the introduction of each species, and raising our minds from a contemplation of these mysteries of nature to a reverential, worshipful feeling for Him who has scattered with omnipotent hand the myriads of creatures conceived and created for the peopling of our world. While our hearts are lifted up in grateful acknowledgment to the Creator, and our minds are filled with astonishment at the variety, structure and number of creatures that fill the sea, becloud the sky and make the earth a very hive of restless animates, we cannot avoid contemplating the world as it existed before the "Spirit of God moved upon the face of the waters;" before the vapors had condensed; before the great furnace (the sun) of the sky had been started, to set a myriad of worlds in motion, singing measureless praises round the circle of infinite space, and in glad hallelujahs of trembling light. With equal wonder we strive to conceive our planet as it appeared fresh from the Creator's hands, with water everywhere abounding and life in its every wave, and try vainly to imagine the condition of creation when the waters were gathered together in one place, and the dry land appeared carpeted with the Edenic flora, and everywhere, in water, land and sky, multiplied species revelled in proud ecstasy of being.

While it is not given us to know all that our longing minds perpetually try to prefigure, with similitudes by which we are surrounded, imagination is powerfully reënforced by the glimpses which we take of the inconceivably

FISHES OF THE SILURIAN AGE.

remote past through a study of comparative zoology and analogy. And in the beginning it may be well to announce, without inviting polemical discussion, that the Genetic, or scriptural, account of creation is in complete accord with the revelations of geology, a knowledge of which serves to confirm the truth of what the inspired writer recorded with an exactness that removes every doubt as to the source from whence the information emanated. Every evidence uncovered by the pick of the geologist, and every discovery made by the search of the palæonlotogist have conjunctively established the long disputed theory of the gradual unfolding and development of creation. As the flower bursts from the bud, and as the perfect creature has its growth from an egg, so has the world attained its present condition by continuous development. With God a day is as a thousand years, and a thousand years is as a day, hence who shall make so bold as to declare that the six so-called days of creation constituted a period of one hundred and forty-four hours of time as we now measure it? Even within the comparatively recent records of history the length of the day has been computed differently, and in the original Hebrew, in which Genesis was written, the term translated as *day* may also mean *period*. The order in which God spoke creation is proven by such evidence as the eternal rocks, and conforms exactly with the records of Genesis, a thing most surprising if we reject the claim of inspiration; but by every test which scientists are able to apply, this corroboration is full and complete, if we but use the term *period* where that of *day* is employed, a substitution which every theologian has agreed to.

Proceeding therefore upon the theory explained, and which every investigator heartily endorses, we may briefly consider some of the primitive forms of animal life that existed before man was created. We are able to determine what much of this life was by the discovery of fossil remains which everywhere abound as will now be explained, preliminary to the introduction of those forms now existent, and in which our chief interest must lie.

FOSSIL REMAINS OF EXTINCT ANIMALS.

The term *fossil* is used by geologists, in a restricted sense, to designate the petrified remains of animals or plants, which we find in abundance deposited, most probably, in the order of their extinction. These fossils serve to conclusively show that God must have created different species of both animals and plants at widely different periods. In many instances it has been shown that one order, or species, existed for a long space of time and then became extinct before a new order succeeded. This is most surprising, but none the less true. It is also well demonstrated that many of these successive orders appeared and passed out of existence in which vast periods of time must have elapsed, before man was created. As man is the most perfect of God's creatures, it is but the natural sequence of gradual development that he should be the last to appear, as the flower does not burst full-blown at once, but passes first through many changes and gradually opens from the bud. When the master-piece of God's work was given to have dominion over the earth and every living thing thereon, we are prepared to believe that the world had passed through very numerous and surprising changes. The igneous rocks, vitrified, or glazed by the action of fire in some cases, and left in the form of tufa, pumice or basalt in others, attest the fact that at one time the earth must have been enveloped in flames, under whose effect it underwent many changes, which now afford proof of design, being as they are, evidences of the establishment of an order of things adapted to the predetermined nature of that perfect creature about to be sent to exercise dominion over the living creatures that preceded him. By the same evidences we learn that before man was ushered into being the distribution of water on our planet was very different from the present, in that where continents now exist, there was at one time a great ocean, and where now the sea rolls in perpetual unrest, was once immense bodies of land, if not continents. The great valleys and cleavages through mountains are the imprints of ocean's fingers, or beds of what were once fresh-water lakes. Such wonderful changes of surface of course produced corresponding changes of climate; for as the mountains uprose they formed new water-sheds and affected the temperature no less signally, while the disappearance of vast bodies of water in one place and their reappearance in another would certainly be followed by pronounced changes of climate. This fact, taken in connection with the many fossil remains examined, prove that a tropical temperature once prevailed in the highest latitudes. In Greenland we find fossil remains of tropical plants and animals, while Northern Siberia yields innumerable evidences that over its now barren and ice-covered shores and plains once roamed vast herds of such equatorial animals as the elephant, rhinoceros and hippopotamus, which were probably destroyed by a sudden change of temperature during what is called the Glacial Period, about which, however, little is definitely known.

The fossil remains which we find most plentiful are those of marine species, because of their greater numbers and more regular deposition at the bottom of

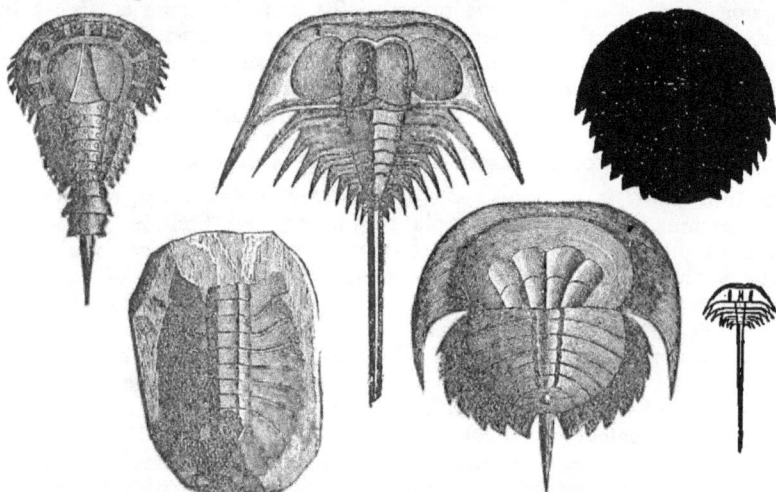

FOSSIL REMAINS OF THE EARLIEST SPECIES OF MOLLUSKS.

the sea. Those which are most commonly met with, and of which illustrations are given in this work, include

Bodies belonging to the sea.
- Shells.
- Corals and sponges.
- Radiated animals, star fishes, ammonites, sea urchins, etc.
- Reptiles, saurian species.
- Fishes of great variety.
- Whales.
- Lobsters, crabs, mollusks.
- Water plants.

Bodies belonging to the land.
- Fresh-water shells.
- Garden snail.
- Quadrupeds.
- Reptiles, several varieties, of monster size.
- Birds.
- Insects.
- Trees.

It is interesting to note some of the curious forms, of which pictures are herewith given, and observe how closely allied to these extinct species are many of the salt water creatures of our time. It is important that this analogy

AMMONITE SHELLS OF THE LOWER GREEN SANDSTONE.

STAR AMMONITE (*Olcostephanus aster*).   STRIPED SERPENT STONE (*Crioceras tabarelli*).   SHIELD STONE (*Hoplites radiatus*).

should be kept in mind in order that the connecting links between succeeding orders may be more clearly perceived.

HOW NEW FORMS CAME INTO EXISTENCE.

The fossil remains spoken of were not found in indiscriminate deposition through the secondary and tertiary stratas; on the other hand, some were obtained from the lowest beds, others from the intermediate, while several were found in the superior strata. But all, of whatever description they may be, *which occur in the secondary strata*, belong to species now wholly extinct. By far the greatest proportion of those found in the tertiary strata belong likewise to extinct species. It is only in the uppermost beds that there is any very considerable number of individuals which are identical with animals now in existence, and there they preponderate over the others.

The bones of man are not more liable to decay than those of other animals; but in no part of the earth to which the researches of geologists have extended, has there been found a single fragment of bone, belonging to the human species, incased in stone, or in any of those accumulations of gravel and loose materials which form the upper part of the series of the strata. Human bones have been occasionally met with in stones formed by petrifying processes now going on, and in caves, associated with the bones of other animals; but these are deposits possessing characteristics which prove them to have been of recent origin, as compared with even the most modern of the tertiary strata.

All the solid strata which abound in animal remains are either limestones, or contain a large proportion of lime in their composition. Many thick beds of clay also abound in them; but in that case limestone in some form or other

2

is generally associated with the clay. From this it has been inferred, and not without a strong semblance of probability, that animals have mainly contributed

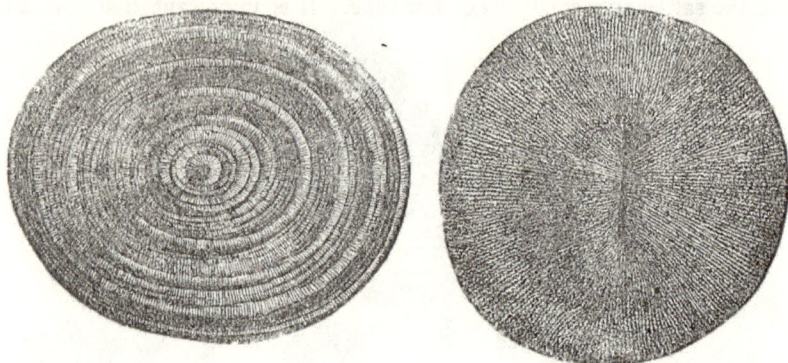

FOSSIL SEA HEDGE-HOG, FROM THE UPPER STRATA OF LIME FORMATION.

to the formation of many limestone strata, in the same way as we see them now at work forming vast limestone rocks in the coral reefs of the Pacific Ocean.

We find in the lowest beds of the series of the secondary strata that the organic remains consist chiefly of corals and shells; that is, of animals having a comparatively simple anatomical structure, and that as we ascend in the series, the proportion of animals of more complicated forms increases, the bones of land quadrupeds being almost entirely confined to the more recent members of the tertiary strata. From these circumstances, it is a received opinion among certain geologists that the first animals which were created were of an exceedingly simple structure, that they gradually became more complex in their frame,

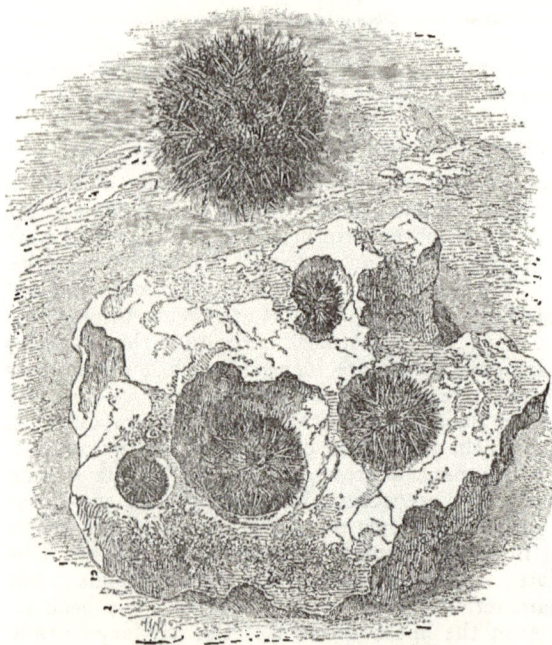

MODERN SEA HEDGE-HOGS (*Echinus delalandi*).

PRIMARY SHELL.

and that at last the highly complicated mechanism of the human body was the completion of those repeated efforts of nature towards perfection. It has been further maintained that there has been an uninterrupted succession in the animal kingdom effected by means of generation, from the earliest ages of the world to the present day; that new species and transformations have been gradually produced by the growth of new parts, originating from certain efforts of the animal to fulfil particular instincts, such as the foot of a bird becoming webbed from repeated efforts to swim; and that the ancient animals which we find in a fossil state, however different in structure they may be, were in fact the ancestors of those now living.

In the theories here advanced, I claim no originality, as the facts thus briefly stated constitute a highway over which many writers have preceded me. But they are always interesting and instructive, and I have used them in order to more clearly define the plan upon which the SAVAGE WORLD has been constructed, in which respect there is a departure from the basis of all other natural histories that I have consulted. I beg especially to refer the reader to the INTRODUCTION, in which I have endeavored to explain, generally, the objects and basis of this work, and which will be serviceable to the reader to know.

FOSSIL MOLLUSK FROM SILURIAN STRATA (*Slimonia acuminated*).

Having thus hastily glanced at the primitive condition of the globe and the probable mutations through which it has passed, my purpose now is to describe, as best I can, the animal life with which we are most familiar, using the word *animal* in a specific sense, to designate creatures that have powers of locomotion and sensibility highly developed, as contradistinguished from what is called the protozoan, or lowest order of animal creation. Nor do I esteem it as being of practical importance to include in this work descriptions of such animals of minute size as are rarely met with, but rather to confine myself to those creatures, the nobler animals, that subserve a useful purpose in the economy of nature most readily comprehensible by the mind of the average reader and thinker. But having in view the idea of acquainting my readers, by practical illustration, with the theory of the development of species, I have thought it appropriate to include here, in the early pages of this work, descriptions of some of the representative types of the lower orders of creation, by which it will be more readily discovered upon what hypothesis the theory of development is based. Passing over, therefore, microscopic organism, we reach the class coming thereafter which has been called, though not with any definiteness,

Mollusca, a term meaning *without shell*, i. e., soft, fleshy, from the Latin *mollis*, soft. The term is misleading, for if literally correct, man himself would be so classed. A better definition, if not a more suitable term is, an order of invertebrates, or creatures having no backbone, and apparently without joints, with feet and locomotive powers usually located at or near the head. This definition needs qualification in that several of the species are protected by a shell covering, such as the oyster, mussel, snail,. nautilus and others, but this protection in most cases may be cast and is therefore not a part of the animal itself.

Of the order of Mollusca, or Mollusks, there are two subdivisions, known as the *Mollusca Ordinaire*, or common mollusks, and the *Anthoid Mollusks ;* in the former being four, and in the latter five classes, which are again divisible into no less than fifty species, some of which we shall notice.

So widely different, not only in appearance, but habits as well, are the several species of mollusks, that while some are most useful, as articles of food for man, and others yielding pearls of great beauty and value, there are not a few species that by their wonderful labors become sources of extreme danger, and more than counteract the good of their more useful cousins.

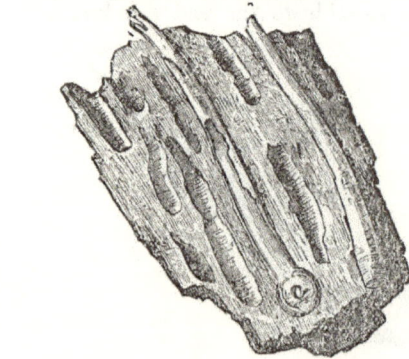

TEREDO, OR SHIP WORM (*Teredo fatalis*).

The **Teredo**, or *ship worm*, which belongs to this order, penetrates the largest ship timbers, and with constantly voracious appetite, and increase of numbers, not infrequently reduces them to a mere shell, and imperils the ship's safety.

The *teredo*, though called the *ship-worm*, from the siphons which compose its soft part, is really a bivalve. It is as destructive to wood-work as the Termite Ant, but on the other hand its method of boring is said to have suggested to Brunel the idea and the method of tunnelling the river Thames. Thus it appears that while the sluggard is admonished to go to the ant, most creatures, even those too insignificant to be of much promise, may have practical lessons to teach as well as the equally valuable lesson of religious morality.

The **Lithophytes**, or *stone borers*, not only work their way into solid rock, eroding a passage by constant application of a rough sole with which they are provided, but often concentrate in such numbers, and become so densely intertwined and strongly attached to the rocks, that they occasionally form reefs large enough to block up the entrance to harbors. They are fortunately not widely distributed, being confined to the Torrid Zone, and most common in the Pacific Ocean.

The **Coral Workers**, known to science as *polypes*, and also as *anthozoaires*, or flower animals, though so small as to be imperceptible to the unassisted eye, perform more surprising wonders and create more deadly perils to ships. Their amazing fecundity, and even more remarkable adaptation, enables them to perform the most prodigious labors, in comparison with which man, with all his boasted skill, pales into insignificance. Artisans and engineers as they are, the *polypes* build up from the greatest ocean depths, with matter either secreted from their minute bodies or gathered from the waves, mammoth structures of adamantine coral upon which the strongest ships frequently rush, unaware of

WOOD BORER.
(*Pholadi*).

danger until the shivering shock and wreck is at hand. These dangerous, though beautiful, structures rise from upheavals of more than a thousand fathoms and extend occasionally hundreds of miles, with their apexs sometimes a few feet below the ocean surface, and again scarcely level with high tide. These thus become dreadful snares for ships, but benefit sometimes results, for upon the reefs thus formed seeds scattered by strong winds, or borne thither by birds, find lodgment and, even without evidence of soil, take root. Sand also accumulates thereon from the restless sea waves, until an island is the result, when rich vegetation is germinated, and at length the land thus made becomes habitable.

Some of the more common types of *mollusks* are admirably illustrated in the accompanying engravings, to which both the common and scientific names are attached. Nearly all these creatures are habitants of the tropics or semi-tropics, where they are found in almost endless variety. The shells found so plentifully strewn along the beach, washed up by every flowing tide, are all representatives of the large mollusk family, some of which are of extinct species, whose abodes have survived the many destructive influences of vast periods of time to remind us of the marvellous changes that have occurred in the progression towards higher forms of life. Passing thus superficially over the species pretty generally known, and to which

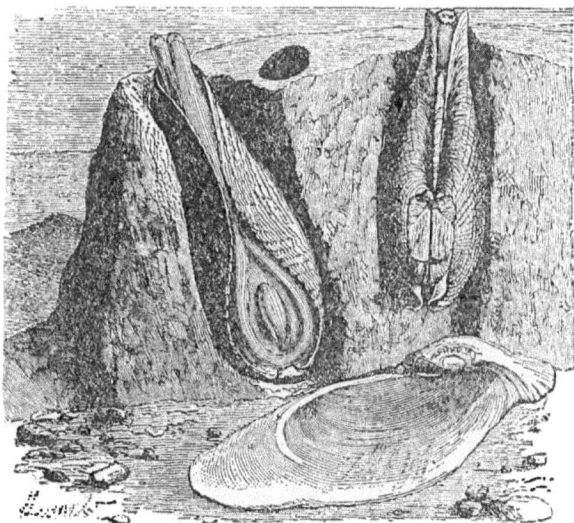

STONE BORERS (*Pholas dactylus*).

comparatively small interest can attach, we come to the larger and most important class, known as the *Cephalopoda*, a term used to denote a peculiarity of this class, which have their tentacles, or more properly feet, arranged in a circle about the head, as may be seen in the illustration. Of this class there are several species, among which are numbered some of the most beautiful of sea creatures. We have first the **Argonauta**, or *Nautilus*, the former appellation being derived from the creature's habit of lying upon the surface of the sea and being propelled by the wind, while its tentacles hang down on either side in fanciful resemblance to oars. Thus it was likened to the fabled Argo and her precious cargo. The latter name, *nautilus*, was applied because of the frailty of the shell. This little

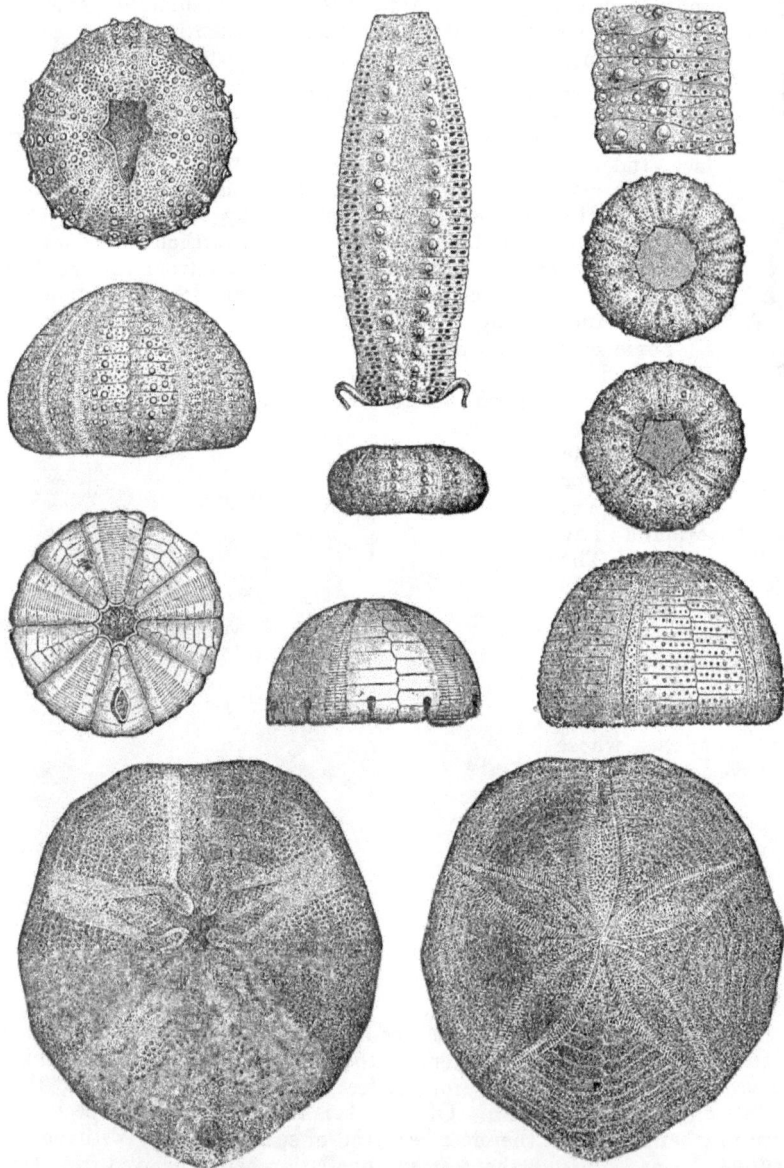

SPECIMENS OF SEA-URCHINS TAKEN FROM THE CHALK FORMATION.

animal is both dainty and exquisitely beautiful, on account of which it has been made the theme of many poets, who liken it unto an animated ocean idyl.

The argonaut does not depend wholly upon favoring winds to propel him whither he wishes to go, if, indeed, he ever relies on the wind except when he wishes to ride listlessly. His proper mode of locomotion is by withdrawing himself partially within his shell, and then driving himself swiftly backward by forcing a strong stream of water through a siphon appendage with which he is provided, and is thus propelled by the reaction. It can also creep, though slowly, along the ocean bottom, where it is most frequently found attached to stones by its tentacles, which are provided with very powerful suckers. In size it rarely exceeds three or four inches in diameter. It has the power to leave its shell at will, though this privilege is very rarely exercised, and never without some exciting cause. It subsists off the smaller *polyps*, especially the coral-worm.

The Pacific Ocean has been a veritable wonderland from the moment when, in 1513, as our school geographies tell us, Vasco Nunez de Balboa saw it from Darien, and from the time—only seven or eight years later—when Magellan gave to it a place among human habitations and a name. The very meagreness of our acquaintance with it, as compared with our knowledge of the Atlantic, has possibly added to the glamour of the sea whose Golden Gate at San Francisco is known by experience to the countless throngs which now crowd their way to the coast so celebrated, and which is each day contributing more and more to the variety and excellence of our markets. Not that, as in the tropics, the luxuriance of nature has dwarfed the useful energies of man; for while we remember the benefactions of Stanford and Lick, and others who have accepted the responsibilities as well as the opportunities of great wealth; while readers continue to enjoy the inimitable products of the pen of Bret Harte; so long as lovers of wit and humor retain an acquaintance with Phœnixiana; until we lose all recollection of the audacious courage and indomitable persistence of those who conceived and executed the magical enterprise of connecting the Atlantic and the Pacific, and making what was before an intervening desert to blossom as the rose, and to be occupied by the industries of man; until we forget these and many other achievements we shall be apt to believe that, perhaps, the most valued product of the Pacific slope is its race of men. The sunlit isles of the Pacific, with their Southern vegetation, strange native tribes, queer customs and their odd animal life,

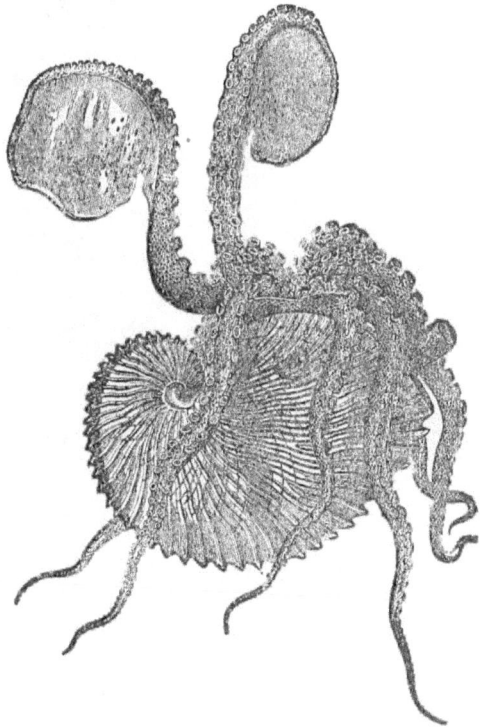

THE NAUTILUS.

are constant stimulants to the imagination, and readily suggest the wonders of life to those who find it difficult to perceive these in objects to which they are more accustomed. Many of these islands, however, are wholly the work of the coral insect, which thus offers its life that in the fulness of time it may provide a habitation for man. The fact that the coral appears unable to build at a greater depth than thirty fathoms seems to sustain the belief of scientists, who assert that the sea is only submerged land, and that now in our day this process of submerging a continent is every day going on in the Pacific. The infinite variety of shape exhibited by the coral formations, and the pellucid character of the water, suggest the most gorgeous and varied architecture of the Orient, while the myriads of jelly-fish and polyps, with their shadings of crimson and sapphire and yellow, lend the effect of coloring compared to which the gilded domes and minarets of man seem sombre. Or again, they suggest the most luxuriant and gorgeous submarine garden, whose flowers are the colored inhabitants of the sea, and whose serpentine paths are marked by the most beautiful shells. After vegetation has sprung from the seeds, brought thither by fish and bird and breeze, many an island looks at a distance as if it were merely a garland intended to add to the glories of Neptune, or to be worn by some gigantic mermaid. It has been suggested by some one with a keen instinct for turning every development to the service of man,

DIFFERENT FORMS OF CORAL.

that we shall yet see the day when we will cultivate the coral insect and make it expend its efforts in building sea-walls for our harbors. If it be remembered that man's "greatest achievements have been but the precipitation of his dreams," we shall feel less inclined to scoff at a suggestion no more preposterous than the building of the Pacific Railroad, or the spanning of the Mississippi by the great Saint Louis Bridge. It is not yet fifteen years since the use of electricity as an illuminator was unknown, and yet think for a moment of the extent to which its secrets have already been discovered and the many forms in which it is made to serve the needs and pleasures of mankind! But the coral insect serves mankind not solely by furnishing habitations for human beings. It is used not only for ornament, but its skeleton is, in its fossiliferous forms, the source of supply for the Tripoli powder so necessary in the mechanical arts for the uses of polishing metals. In Virginia the city of Richmond rests upon a stratum of coral of twenty feet in thickness.

Chili abounds in coral remains, and in addition has near Copiapo a beach now removed almost a mile from the shore and elevated nearly two hundred feet above the level of the sea; moreover, the transformation of this coral island into stone is even now going on under the very eyes of living beings. It is thought that fuller investigation will show that, in addition to the ordinary classes of aqueous, igneous, and metamorphic as divided by the physical geographer, we must add coralline rocks as composing much the greater portion of the earth's crust.

Huxley has compared the way in which corals build to that of the ancient dwellers around the Mediterranean, who erect one city upon another, as Schliemann has shown by his excavations in search of the famous city of Troy. Students of physical geography will remember that sometimes the coral builds a "fringing reef," or one which surrounds an island and is separated from it by lagoons of relatively shallow water; that again it constructs an "encircling reef" or "atoll," which almost surrounds an island; and that yet again it makes an "encircling reef," or one which must have been formed about land as it sank. There is at Mauritius a fringing reef

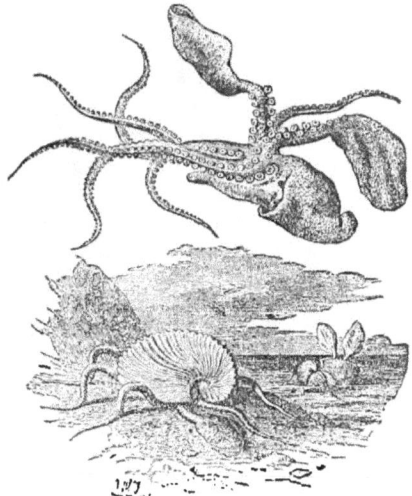

NAUTILUS IN THREE POSITIONS.

one hundred miles in circumference; and an encircling reef eleven hundred miles long.

The Nautilus, and in fact all the species which compose the order now known as *Cephalopoda*, are direct descendants from the ancient bellamites and numullites whose fossil remains are strewn so plentifully over the whole globe. Concerning this fact Rev. William Buckland, D. D., makes the following observation :

"It results from the view we have taken of the zoological affinities between living and extinct species of chambered shells, that they are all connected by one plan and organization; each forming a link in the common chain, which unites existing species with those that prevailed among the earliest conditions of life on the globe; and all attesting the identity of design that has effected so many similar ends through such a variety of instruments, the principle of whose construction is in every species fundamentally the same."

Fossil numullites are especially numerous, and though generally micro-scopic in size, their minute bodies compose the principal bulk of many moun-tains, such as the Alps, Carpathian and Pyrenees. The Pyramids and Sphinx

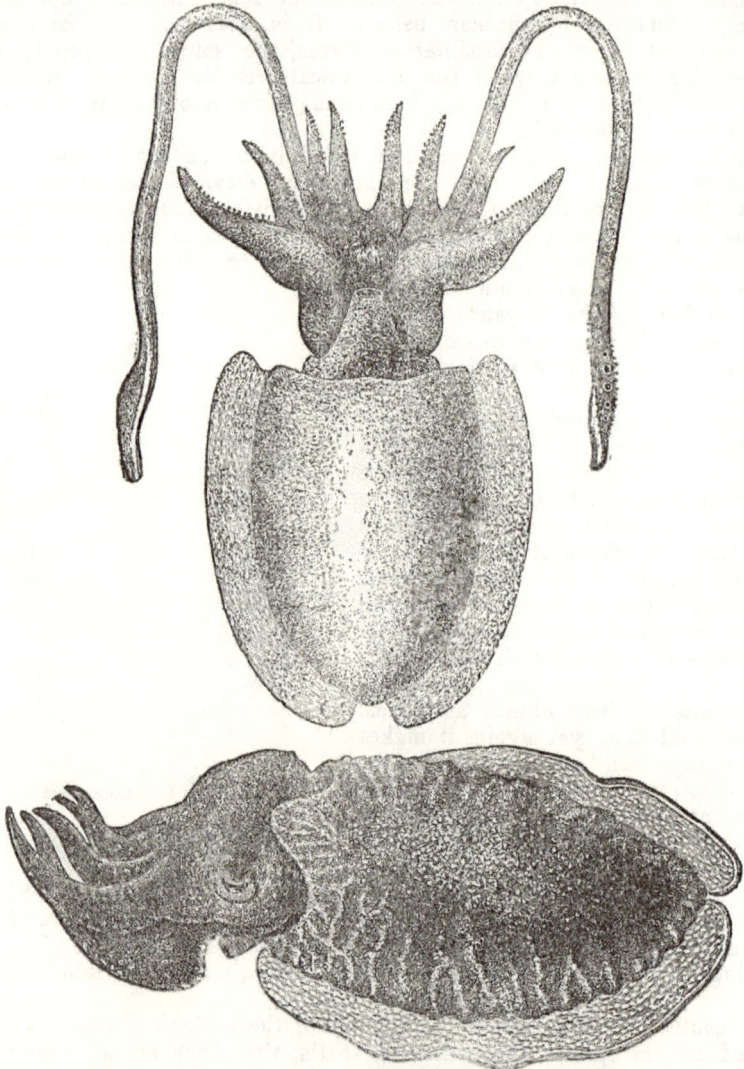

CUTTLE-FISH (*Sepia officinalis*).

of Egypt are a composite of limestone and the chalky remains of these small animals, and all chalk, wherever found, is but a composition of the fossil of these and other creatures belonging to the same genera. In the long ages of the world, however, development continued until the representative types of cephalopods now existing show greater perfection of structure, in their natural progression from the simpler towards the complex, as well as notable increase in size, with corresponding diminution of numbers. These larger existent forms will now be noticed.

The **Sepia**, or **Cuttle-fish**, also called **Squid**, claims our attention as next in the order named, and as being closely allied to the nautilus both in habits and appearance, though its power is perhaps a hundred-fold greater. They are also divisible into perhaps a dozen species, ranging in size from the smallest, pen-shaped calamary, that is often found in schools of thousands, to the giant squid measuring from tip of tentacle to tail more than a dozen feet. Indeed, it has been claimed by not a few persons, Mr. Beale among the number, that the creature grows to the extreme length of twenty feet, with power to drag an ox into the water. This animal has a soft, unprotected body, bag-shaped, and a curious fleshy appendage issuing from the sides. The head, which is about one-half the length of the body, terminates in ten tentacular arms, the under sides of which are armed with numerous saucer-shaped suckers that are used to seize and hold its prey, and by which it also attaches itself firmly to anything it may reach. The eyes are large, and near them issue two long feelers destitute of suckers except at the extreme ends, and the power that these exert is small, evidently being used by the animal to thrust into deep crevices to search out its prey. In a bag located near the heart the *sepia* carries a large supply of ink which he can eject at will, and which by discoloring the water effectually hides him from view, thus affording him an excellent means for protection against enemies.

Cephalopoda-Dibranchiata (*Two-gilled*). The name of the class is derived from the fact that its members cannot only swim, but they also rival the acrobats in their ability to walk upon their heads, along the bottom of the water. The shell is generally internal; the arms, which are provided with discs for sucking, ordinarily number from eight to ten.

The **Argonauta** represent the eight-armed genus (*Octopoda*), and are named after the mythological explorers who went forth in the Argo, and whose adventures, after stimulating the Greeks, have added to the treasure of stories for our own children, besides furnishing imagery for the poet and the orator.

The **Argonauta argo** (*Paper nautilus*) has no internal skeleton; two of its arms are outstretched and form broad sails, which seem to be unfurled to the breeze and used for propelling the animal to which they belong. Oliver Wendell Holmes' poem, called "The Chambered Nautilus," is familiar to every school-boy and may suggest the enjoyment of beauty, of useful lessons, and the inspiration which may be drawn from the animal world.

As a matter of exact science, it must be confessed that the *Argonaut* does not use his canvas-like appendages as sails, but that their function is partly that of holding animal and shell together, and partly the secretion of the substance from which the shell is formed. This shell, as doubtless is known to many of our readers, is very thin and transparent, flexible, and grooved, and so like the earlier forms of ships that it is supposed naval architects took their first lessons from the *argonaut*. The *argonaut* sometimes leaves its shell for brief periods. In repairing any damage to its dwelling, the animal wastes no effort, but if pieces of shell are obtainable, uses them so as to diminish the quantity which it must itself manufacture.

The Mediterranean is the natural home of the *argonaut*, where it may be seen sailing its mimic vessel near the shore, and frequently, when alarmed, furling its sails and sinking out of sight. Like the cuttle-fish, it ejects at pleasure an inky substance which protects it from its enemy. The eggs hatch in about three days; for two days more the shell-less young share the shell of the parent, which had previously been used as a depository for the eggs; at the end of a week the young have matured sufficiently to leave the ark and to sail life's waters for themselves.

The **Pearly Nautilus** (*Nautilus pompilius*) lives in the deep sea, though sometimes rising to the surface and, by a special arrangement for increasing or decreasing its specific gravity, can withstand a pressure and compression sufficient to force a cork into the neck of a bottle.

Of *Cephalopods* there are a great variety, and nearly all are found more or less common along the American coast in the north temperate latitudes.

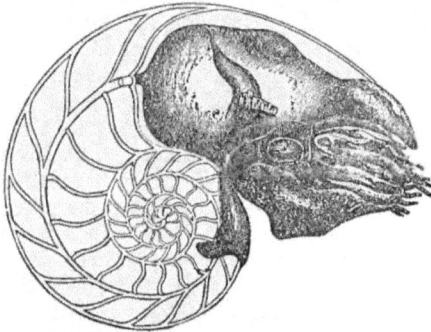

SECTION OF SHELL OF THE NAUTILUS.

The largest and most formidable of this species belongs to the order *Octopoda* (eight-footed) and *Decapoda* (ten-footed), in which classes the calamar, sepia, cuttle-fish and octopus are included. These are so very similar in habits and appearance that it is with no violence to classification that we include the three under a general description.

THE GIANT OCTOPUS.

Specimens of the *octopus* are frequently found stranded upon the beach after a storm, their soft bodies fatally injured by being dashed against the rocks. Much dispute has been indulged over the probable size attained by this animal, the ancients being very firm in the conviction that

THE OCTOPODA CUTTLE.

the creature not infrequently reached proportions that enabled it to draw a ship under water, and others maintained that in the waters about Norway they had been seen sleeping on the surface so large as to resemble an island. These stories were even circulated as facts by Aldrovanus and Pontroppidan, both of whom were regarded as learned naturalists in their day. We have also many circumstantial accounts of frightful accidents precipitated by attacks on ships by the *kraken*, which was the name formerly given to giant species of the octopus.

The **Decapoda**, or ten-armed cuttle-fish, are cylinder-shaped instead of globular, and their arms look rather like the leaves of the cactus. *Sepia officinalis*, the *Common European Cuttle-fish*, is from a foot to a foot-and-a-half in length; has a calcareous bone as an internal skeleton; a smooth white skin with brown and purplish-brown spottings. Its eyes are used as neck-pearls; its skeleton for pet birds, for the pounce used in writing and in embroidering, and for small delicate casting. It is a source of loss to the professional fisherman, as it attacks the fish in the seines or nets.

In regard to these once very popular superstitions the learned Dr. Walsh submits the following opinion:

"We cannot doubt that the depths of the sea, where vegetables flourish eight hundred feet in length, are also peopled with monstrous animals, whose organism is adapted to these unknown regions, whence they but rarely emerge. Their very real appearances have formed the basis of the mysterious traditions which, for more than two thousand years, have been transmitted from generation to generation of mariners, and which have given birth to the fantastic creation of the *kraken* and the sea serpent. While masses of small gelatinous

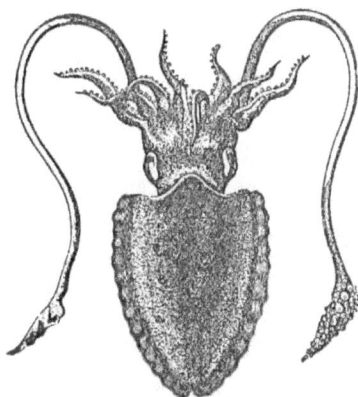

THE DECAPODA CUTTLE.

medusas floating at the surface provide food for enormous whales, there is also at the bottom of the sea an abundant prey for these prodigious animals."

THE FABLED KRAKEN.

It is an undisputed fact that there exist in the Mediterranean and other seas cuttle-fish of extraordinary size; to deny this would be to dispute the assertions of hundreds of responsible persons, as well also to deny the evidences which are contained in several museums, where specimens of this huge creature are preserved. A calamar was caught some years ago near Nice, which weighed upwards of thirty pounds. Less than forty years ago an individual of the same genus was caught in the same place that measured six feet in length, and its body is now preserved in the Museum of Natural History at Montpelier. Peron, the distinguished naturalist, asserts that he met with one off the coast of Australia that was nearly eight feet long. Two travellers, Quoy and Gaimard, picked up the skeleton of a cuttle-fish in the Atlantic Ocean, near the equator, which, when living, must have weighed at least two hundred pounds. M. Rung found in the Atlantic the body of another, which he describes as being as large as a tun cask. In this instance the tentacles were quite short, and the body of a reddish color. He secured one of its mandibles, which is still preserved in the Museum of the College of Surgeons in Paris, and is the size of a man's hand.

In 1853 a gigantic cephalopod was cast ashore on the coast of Jutland, where it perished. Some fishermen dismembered the body and bore it away in several wheelbarrow loads. The back part of the mouth of this animal is said to have been as large as the head of an infant. Another, equally great, was taken in the Atlantic in 1858, while it was engaged in a deadly combat with a whale, and parts of it may be seen in the museum at Copenhagen.

Few inhabitants of the deep with which we are now familiar, however, have furnished such opportunity for thrilling description as *Octopus vulgaris* (*Octopus* or *Cuttle-fish*). Its skeleton is confined to two dorsal, cone-shaped, horny substances. The body is round, soft and jelly-like, and has a leathery integument or covering. The arms are of extraordinary length, frequently being four or five times the length of the body. They are studded with sucking discs (frequently as many as several thousand), are reproduced if lost, and are capable of an embrace which is very difficult to resist. The mouth consists of an orifice surrounded by a circular lip, beneath which appears a beak whose longest part lies below; the mouth and jaws are supplied with powerful muscles, so that the *octopus* can easily crush the shells of crustaceans and mollusks, and carve the bodies of fishes; the tongue is adapted alike for tasting and for conducting food to the digestive apparatus; the mouth lies so as to be surrounded by the arms, which seize its food. On one side of the abdomen are two siphons used for ejecting the inky liquid which it employs to conceal itself from its enemies. They are used also for propelling the *octopus* by sucking and ejecting streams of water. It was formerly supposed that the sepia, or India ink, so much used by artists, was made from the ink of the cuttle-fish, instead, as happens to be the fact, from lamp-black and glue. The animal has large, shining eyes, placed at the base of the arms; it is keen-sighted, and, so to speak, far-sighted; finally, it is phosphorescent and one of the creators of that mysterious and interesting light on the waters which is so ordinary and so attractive a phenomenon at sea.

The *octopus* is found in every quarter of the globe, attaining its largest size, and exhibiting its greatest ferocity, however, in southern or tropical regions. The largest specimens weigh from a hundred pounds upward, and their muscular development is proportioned to their weight. They have been known to attack boats and the sailors in them, and have given rise to many an exciting story, even if the adventure is to be classed with those of the orthodox fisherman. The thrilling tale of "The Toilers of the Sea," though written by the greatest and most impassioned of Frenchmen, suffers from the author's confounding the *octopus* with the polypes.

Several specimens have been found on the Pacific coast having a radial spread of thirty feet and weighing two hundred pounds. A very large one, though in a somewhat mutilated condition, was dredged from a depth of one thousand fathoms (about one mile) by the United States Coast Survey in 1878. Another, of comparatively small size, was captured by the Fish Commission off the New England coast, and was kept alive for some time in a tank, where its habits could be observed. At all times it appeared very timid, remaining all the while at the bottom, from which it could rarely be made to stir. When aroused, however, it would dart swiftly from one side of its quarters to another, and then firmly attach itself to the bottom again by the powerful suckers along its arms. The eyes were very large but seldom opened to their fullest extent during the day; at night, however, it seemed to have greater animation and kept its eyes wide open, from which it was thought to be a nocturnal creature, as it undoubtedly was.

The Rev. Mr. Harvey described a specimen that was cast on shore in 1879, which he declares measured eighty feet, and gave a graphic account of its terrific struggles to escape

THE OCTOPUS (*Cephaloptera*).

from a pool in which a receding tide had left it; but the body was not preserved.

A great many accounts have appeared from time to time of thrilling adventures with these animals, the novelist being especially free with such descriptions, but I can call to mind only a single instance, properly authenticated, where a man has actually been attacked by any member of the *Cephalopoda* family. The account is furnished by the gentleman who had this frightful experience, and who was no other than Professor Beale, a distinguished naturalist. He relates that while engaged searching for shells on one of the Bonin Islands of the North Pacific he came upon a *rock-squid* (cuttle-fish), as it was creeping upon its eight tentacles over some rocks towards the sea. The creature's body was little larger than a man's two fists, though its arms had a spread of nearly five feet. Curious to determine the strength of the animal, Mr. Beale endeavored to arrest its progress by pressing his foot upon one of its arms, but to no avail, and his resolution becoming the stronger because of the apparent ease with which the creature successfully resisted every

attempt thus made to retain it, Prof. Beale at length seized one of the tentacles with his hands, and a veritable tug of war ensued. Small as the animal was, its hold upon the rocks was marvellously tenacious, and for a considerable time he was unable to detach it, but a supreme effort, accompanied by a jerk, finally served to separate the creature's hold, but with most disastrous results to the naturalist. The moment that its arms were free the terrible squid flew, with an amazing exhibition of passion for so small a thing, directly at the naturalist, and fixed itself with the same tenacious hold upon his bared arm. It was now Prof. Beale's time to cry for quarter, which he did most lustily, but with his cries he coupled a stout resistance, seeing that the creature was making every exertion to reach his arm with its powerful parrot-like beak, which if successful would probably take most of the flesh from that member. His cries fortunately were heard by the captain of the vessel, who was also on the beach not far distant, and who hastened to the professor's assistance. The two now tried in every way to make the creature quit its hold, but were unable to do so, and were finally compelled to proceed to the landing boat, some distance away, all the while holding the squid's horrid head, and there to secure a knife with which it was cut into many pieces before Prof. Beale was finally released from his vindictive and extraordinarily tenacious adversary. The animal had applied his powerful suckers with such force to the bared arm that blood was drawn in considerable quantity wherever the dreadful cups had been attached.

ATTACKED BY A SQUID.

If such power resides in so small a creature as is here described (and the account seems authoritative), what might one nearly twenty times larger be able to do? Imagination must pause here.

The squid and its congeners produce their young from eggs, after the manner of fishes. The eggs, however, are very much larger comparatively, and very closely resemble a large bunch of blue-black grapes. They are frequently found thrown on the beach, especially after a storm, and it requires a long exposure before their vitality is destroyed. If the eggs are taken within a few hours after deposition and placed in a tank of fresh sea-water they will germinate, after passing through many alterations in appearance, and let loose the most grotesque appearing little creatures—very harlequins—that can be conceived. Small and deformed as it appears to be, the young squid is very active immediately after birth, whisking about through the water in a most reckless manner, as if to exercise his long, pent-up limbs, until his vigor is somewhat spent, when he gradually settles down towards the bottom. Here he stops, and, drawing his siphon apparatus, blows a hole in the sandy bottom, into which he quickly sinks out of sight, and there remains for a day or more before reappearing.

There is a well-authenticated story of a sponge diver, who, upon reaching bottom in twenty-five feet of water, was suddenly alarmed by feeling something grasping him. It turned out to be an *octopus*, and it was not until after a fierce battle, during which the diver cut the body into mince-meat, that the arms, having no longer any support, ceased to embrace the intruder into submarine mysteries. But apart from the danger and the fright, the diver was confined to his bed for months from the effects of the wounds made by the *octopus*.

The fossil remains of the cuttle-fish are numerous, and indicate the need for their activity when the waters were so filled with animal life. The *Ammonites*, the fossil ancestors of the cuttle-fish, vary in size, from the minutest particle to a diameter of two or three feet. They derived their name from a fancied resemblance between their horns and those to be found on the statue of Jupiter Ammon, the supreme deity of the Libyans. Unscientific persons still cherish the belief that these fossils are genuine petrifactions, and the number of legends has been multiplied by the superstition of persons who found in the Ammonite a judgment like unto that which was executed upon Lot's wife.

The **Loligo Squids** are very common, and are most brilliant in their coloring, which they seem to be able to change at will, and with a rapidity which discredits Samuel Warren's account (in "Ten Thousand a Year") of Tittlebat Titmouse's hair-dye. Red, blues of all shades, orange and brown are quite common colors.

The **Common Squid,** or **Pen-fish** (*Loligo vulgaris*), is green, inclining to brown. Its eyes are bright emerald green; its fins are shaped like the lozenge of the geometer, and they reach from the tail to about the middle of the body. They are used in forward movements, while the mollusk, at pleasure, moves backward by contractions of its body. Its head and arms protrude from the body like the head of a turtle. The skeleton is a fac-simile of a quill pen, and hence the popular name of the creature. It lays upwards of forty thousand eggs, which are arranged upon the radii of a circle whose diameter is many inches.

The **Dotted Loligo** (or *Loligo punctata*) is quite common, and though swimming backwards (as it has no fins), is very agile. The species called the *Sea-arrow* is a favorite bait among those whose life is spent in the cod fisheries. This species jumps from the water and seems to fly. It has been known to leap as high as the deck of a large ship. The Hooked Squid is so large and aggressive as to be dreaded by pearl fishers.

The **Common Garden Snail** (or *Helix aspera*), like the other members of its order, breathes not by means of gills, but by means of lungs, not aerated water, but atmospheric air. The eyes are placed at the extremities of the four tentacles. As an embryo it is provided with a shell, so that Shakespeare's allusion to the snail, which always was provided with a house, was very happy. It manages its locomotion by means of a single disc-shaped foot attached to the central surface. It is distinguished alike for longevity and for a singular tenacity of life. It is said that after many years' enforced sojourn in the cases of a museum, the snail has been known to proceed as though there had been no interruption of his opportunities and no disuse of his services. Their eggs are produced in the necks of the snails, just as if even at this early period the

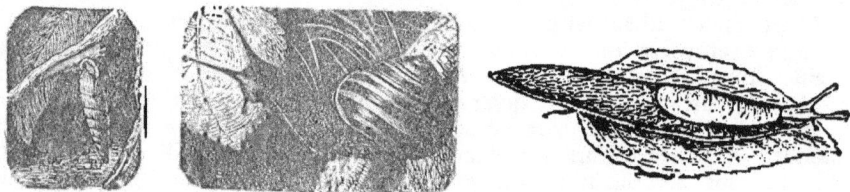

parent would gather her young to her bosom. As the snail grows it builds additional stories or whorls in its shell, and as it needs at each increase yet more commodious quarters, the front whorls are of greater size.

The edible **Heart Mussel** (*Cardium edule*) prefers brackish water to salt, and hence is so plentiful about the Thames as to give rise to an almost separate industry. It is not so very palatable, but from its great numbers and the ease with which it is caught, it forms in English life what the Americans would call an "institution."

If I were to describe conscientiously the several classes of creatures that furnish an unbroken chain—in which there would be many links—connecting the cephalopods with fishes, it would be necessary to pass over a very wide field, in which we would find not only vast numbers of animals, but many distinct *orders* as well. I would have to give space to a description of hundreds of mollusks, which are interesting because of the beautiful shells in which many of them have their houses, but tedious, at best, when it is attempted to portray them; next to these would be the slugs and linnets, members of the same family, but without charming shells to recommend them; and the gasteropods, or sea snails, that are as uninviting as our common garden species; and myriads of the *Bryozoa*, or "moss animals," which are so nearly vegetable that one may be grafted upon another, or one may be converted into several by separation, but which are interesting only to the microscopist. The coral workers, called *Polyzoa*, would next claim our attention, to pursue which would lead into many devious passages where we could hardly find our way without the light of a classical dictionary. And so, to prevent wearying the reader, and to avoid the charge of attempting the compilation of a work for the scientist rather than for the masses, I have resolved to confine myself, as stated in the introduction, to those forms of life with which we have more concern than mere idle curiosity, and thus provide a practical work for every day consultation, instead of a book suited only to those who have abundance of time for deep investigation and indulgence.

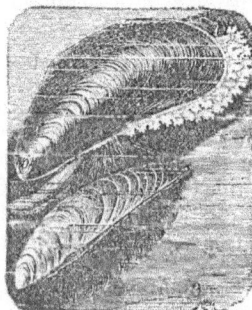

HEART MUSSEL.

With the reader's attention thus called to the missing rounds in the ladder of development up which we are gradually climbing, but with pieces of the broken rungs projecting barely enough for a footing, we will proceed to a consideration of some of the representative types of the lower orders of life in the sea, from which we will find a more perfect development in the higher existing classes, but in which there are characteristics connecting the two plainly distinguishable.

The **Echinoderms** (or *urchin-skinned animals*) take their name from the resemblance of their spines to those of the hedge-hog. They glide along like unreal beings, owing to their almost unlimited number of little tentacles, each terminating in a sucker.

The **Sea Cucumber** is cylindrical and has a leathery integument; it is an article of extended commerce among the Chinese. One species when not at ease will practise the hari-kari of the Orient; but

SEA CUCUMBER.

unlike the less skilful human being, it can reconstruct itself and renew its mundane existence.

The **Sea-Urchins** proper have their upper portions covered by a shell, which is curious alike from its appearance, and for the mechanical skill required for its construction. This shell is made up of a great number of curved plates, so that increase in size without loss of form can be attained by calcareous deposit on their edges. The spines furnish one of the wonders of the microscopic world; each one is movable at the will of the animal, and has the same joining as the upper arm of a human being. In the metamorphic period the creature is at first globular. It then puts forth a dome-like part, supported by frail legs, which might do honor to the slender supports of the furniture of a French Louis.

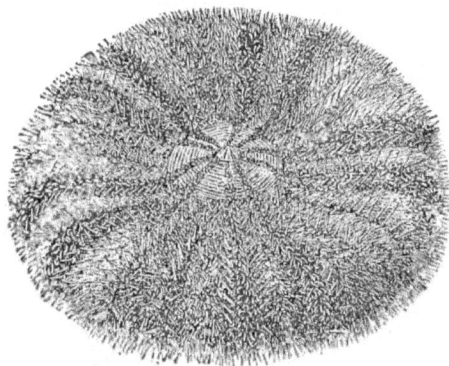

LEATHER SEA-URCHIN (*Asthenosoma hystrix*).

It then assumes the form of a clock, with the regulation position of hands, which presently expand into the typical *Echinoderm*.

The ordinary **Star-Fish**, helpless as it is in the hand of a human being, has been successfully rapacious beyond belief. It is specially predaceous among the bivalves. Its voracity is such that if it swallows a bivalve it will pay no attention to the shell, which it retains amidst the other indigestible matter. It can sustain the loss of members, which it quickly repairs. It moves with a glide, compared to which the steps taught by a dancing-master are clumsy; moreover, it can adjust itself to emergencies, and by retiring trouble-some members seems able to crawl through a "knot-hole." Its structure puts to shame the oft-vaunted cloistered cathedrals of Europe; columns arise in wondrous beauty and purity, and the vaulted aisles dwarf the skill of our most celebrated ecclesiastical architects. The *Star-fish* protect their young until these are able to provide for themselves. The species are many, and so varied as to suggest no consanguinity. Though apparently without organs of sight, scent, or sound, they are quick to perceive the vicinity of bait. They seem able to compress themselves at pleasure, so as to pass through apertures apparently smaller than themselves. The *Bird's-foot* resembles in form the foot of a duck, while its coloring of scarlet and yellow is a feast to the sight.

The **Leather Sea-Urchin** (*Asthenosoma hystrix*) can live more than a thousand fathoms, or more than a mile below the surface of the

SEA STARS (*Asteracanthion rubens*); SEA CUCUMBER (*Cucumaria doliolum*); AND CLIMBING URCHINS (*Echinus microtuberculatus*).

water, and a moment's reflection upon the fearful pressure of the sea at that depth will suggest the great strength possessed by this seemingly frail inhabitant of the deep. It is generally globular in form, and resembles the little knit caps worn by girls. Its bright coloring, relieved by the white linear markings,.

renders it a beautiful creature. It is, however, dangerous to handle, since its sharp spines both sting and benumb.

The **Marsupial Sea-Urchin** (*Hemiaster Philippii*) is remarkable for having four deep cavities, in which, like the opossum, it receives its young.

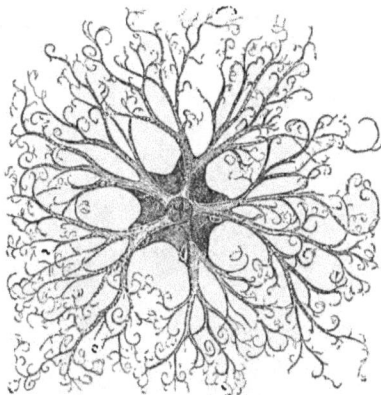

MEDUSA HEAD (*Astra caput medusæ*).

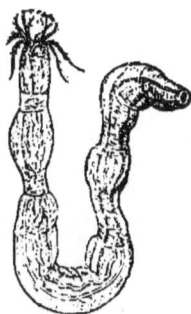

BURDOCK HOLOTHURIA.
(*Synapta inhærcus.*)

PURPLE HEART URCHIN.
(*Spatangus purpureus.*)

Snake Star, or *Pluteus*, is the name given to the embryonic sea-urchin, and is of great interest because so unlike the Echinoderm, into which it is to be metamorphosed. It has a bilateral (two-sided), symmetrical (side corresponding to side), non-radiated body, which is mostly obscured by an apparently helpless tangle of eight long arms, supported by slight calcareous rods. As these arms

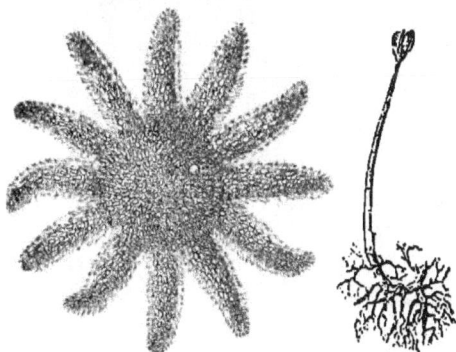

SUN STAR-FISH (*Solaster papposus*) AND YOUNG.

YOUNG HAIR STARS.
(*Comatula mediterranea.*)

vibrate they suggest some of the most grotesque and horrible drawings of Gustave Doré. The creature swims by a parachute-like opening and closing of its arms.

The **Brittle Star-fish** is constantly engaged in breaking itself into pieces, but it repairs the damage as speedily as it occasions it.

The **Shetland Argus** is remarkable for the multiplication and flexibility of its arms. In the pages of history, as these are written in the rocks of the geologist, the star fishes are called *Encrierites*.

The **Sun Star** (*Solaster papposus*) is rare, but it is caught at times by fishermen off Newfoundland. The Medusa's head (*Astrophyton caput Medusæ*), by its ever-waving arms, suggests the serpentine locks of the Medusa of Grecian mythology. It will be remembered that Medusa lost her great beauty through

SCHOOL OF JELLY FISH.

the envy of Minerva, and that her handsome tresses were turned into snaky locks and her attractiveness of feature into an ugliness which petrified the beholder. It is unnecessary to urge that the *sun star* exercises no such diabolical power, and that the ceaseless movement of its arms threatens no such danger as is to be looked for from the octopus. It seemingly is the sole survivor of numerous species to be found among fossils.

The **Phosphorescent Tunicate** (*Salpa maxima*) is specially remarkable from the fact that the single parent produces a community which exists only when joined into a chain; and, on the other hand, the individual links of this chain produce individuals which lead a solitary existence, so that the animal represents an "alternate generation."

The **Hair Star-fish**, or **Feathered Star** (*Canatula mediterranea*) is found in the Mediterranean, and frequently appears clinging to a piece of sea-weed.

The **Discophora** (*Disc-bearers*), such as **Jelly-fish, Sea-nettle**, etc., are a very numerous, attractive, and under some circumstances troublesome, community in the waters of the ocean. They seem restricted to no temperature or clime. They exhibit the most infinite variation in size, form and color. Many of them add to their personal charms the phosphorescence which is so charming a feature of evenings at the seaside. Their delicate tissues assume all imaginable forms, and rival the magic of the kaleidoscope. The arms proceed from beneath the umbrella-like disc, and resemble the four posts of some grotesque arbor. The mouth, when existing, is placed in the lower wall of the disc and is furnished with tentacles. A common species has a multitude of filamentary tentacles which resemble a fringe dropped from the seat of a chair; these it entwines about any object of contact. As each one of them is endowed with the ability to sting, it can make its presence felt, and its memory a "possession forever." One species is frequently two feet in diameter, and moves about in schools or shoals, which oftentimes are sufficient to interfere with the progress of boats; iridescent in the sunlight and phosphorescent in the twilight and darkness, their course is a path of light.

The *jelly-fish* is the glass umbrella of the sea, and in place of the handle are numerous delicate filamentary tentacles. By the contraction and expansion of the muscular,

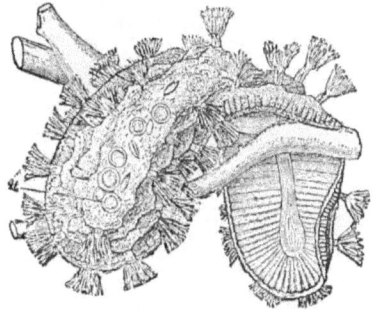

FRESH WATER MOSS ANIMAL.
(*Cristatella statoblastem.*)

umbrella-like body, the creature makes its way through the water. Its tentacles, however, contain a sting and a poison-cell, so that while admiring its beauty one must remember the maxim, "Do not touch me." All animals' bodies are partially water, but the jelly-fish contains only about thirty grains of solid matter out of a possible ten pounds of weight. Many a person who has admired the beauty of the floating jelly-fish has been surprised to find it almost vanish after it had been caught. A story is told of a thrifty farmer who collected loads of jelly-fish, thinking to fertilize his land therewith, but found that he had rather discovered a new method of salt-water irrigation.

The **Moss Animal** (*Cristatella mucede*) has the lower part of its body disc-like, so that it is capable of moving from place to place. It loves the meridian time of day, and as it basks in the sunlight has a great resemblance to the hairy caterpillar.

The **Bird's-Head Coralline** (*Bugula aviaclaria*) is curious, because until quite recent times its family was supposed to belong to the vegetable world. Externally, like the coral insect, it falls into another class when examined with reference to structure. Its activity is such as to excite attention in the localities where it abounds.

The Swimming Snail (*Pterotrachea*) is a wing-footed mollusk allied to the Cephalopods. The name is derived from portions of the feet being expanded and fitted for swimming. The two arms, corresponding to the head and tail, are

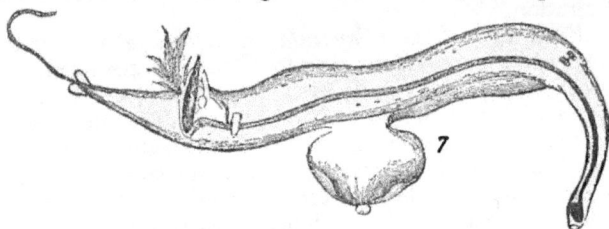

THE SWIMMING SNAIL (*Pterotrachea coronata*).

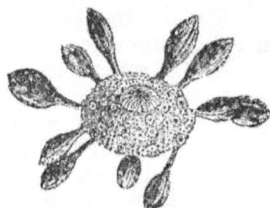

CLUB-BEARING URCHIN.
(*Cidaris clavigera.*)

supplied with suckers like those of the cuttle-fish, while the body appears dotted, each dot, however, being a protrusible arm in which is a suction disc. Its length rarely exceeds one inch.

MUSSEL SHELL.

Mouth-Closing Snail (*Clausilia biplicita*). These are found about the margins of streams, but rarely enter the water, their principal habitat being in the neighborhood of the Mediterranean. They are widely distributed over Europe, Asia and Africa and number seven hundred species. The shape is cylindrical fusiform, and the mouth appears sealed during dry seasons.

Mussel Shell (*Spondylus spinosus*). A very large variety is included under this head, the name being given to designate the thorny appearance of the shells. Among these is a very rare shell found in the East Indies under the specific name of the king mussel (*S. regius*).

The Rooted Hair Star-fish belongs to the order of *Rhizo crinadæ*, which takes a place midway between the sea-urchins and snails. It is a deep-water dweller, taken by dredge, at a thousand fathoms, in the Atlantic. It is somewhat rare now, but was very abundant in past geological ages.

The Antedons, or Climbing Sea-Urchins, belong to the same species as

ANTEDON CRINOID.

the above, both being classed under the head of *branchiata* (many-armed), because of the numerous branches into which the creature is divided, like those of a plant.

The **Shield Porcupine** (*Clypeaster rosaceus*) is a member of the very large family under which the sea-urchins are classed. The species here specially named is about the size of a silver dollar, and presents many beautiful markings which make it an interesting study, the most curious being the prickly spines with which it is covered, from whence the name is derived. It is not a very common species, and appears more frequently on the Pacific coast.

SHIELD PORCUPINE.
(*Clypeaster rosaceus.*)

The **Finny-Legged Mollusk** (*Hyalea tridentata*) is remarkable, because, like the glow-worm, it lights up the darkness of the night.

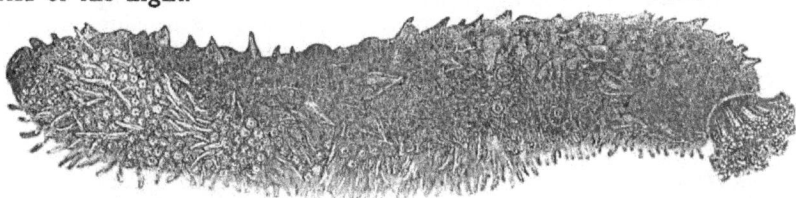

TUBULAR HOLOTHURIA (*H. tubulosa*).

The **True Kauri** (*Cypræa moneta*) is used in Africa as currency, and this fact has established its capture and exportation as a commercial interest.

The **Grape Mollusk** (*Botryllus*) is peculiar, because though an apparently single star-fish it is really a community of many individuals which have, with the exception of the organs of respiration, individual development.

The **Tree-like Snail** (*Dendronotus arborescens*) has shrub-like gills upon the back, while its tentacles take the semblance of the true gills.

The **Edible Mussel** (*Mytilus edulis*) is distributed with the greatest profusion throughout European waters, and an unexpected encounter as it clings to its support is very lacerating to the feelings of a human being. Its value as food is more than open to question, but it is always good to "feed fish withal."

TREE SNAIL.

We now hasten on to the next higher form, which, in the most natural sequence, will be

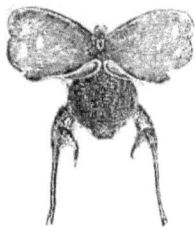

HYALEA TRIDENTATA.

THE CRUSTACEA.

Under this heading are classed those creatures that have a shell covering, or crust, in both legs and body, and which breathe by gills, and not by air tubes, as do the Cephalopods. To this order belong a large variety of species, most prominent of which are lobsters, crabs, cray-fish, shrimps, water fleas, barnacles, snails, sea-urchins, star fish, etc.

The **Coronet Barnacle** (*Bolones crenatus*) is the name given to a curious parasite that fixes itself in the skin of Arctic whales, and sometimes multiplies to such an extent as to worry the poor creature into acts of frenzy. To these parasites have the rare attacks of whales on ships been attributed. The worried mammal, being driven mad by these tormenting parasites, will rush on any object at hand, or even wildly throw itself against an iceberg. The Burrowing Barnacle is another species that infests whales in the Antarctic sea. These are also very annoying, occasionally eating their way through the epidermis and far into the blubber, but do not appear to produce such pain as the former species.

The **Goose Barnacle** (*Lepas anatifera*) is more commonly met with on account of its habit of clinging tenaciously to the bottoms of ships, measurably retarding the speed, and requiring at times the docking of ships in order to scrape off the adhering *barnacles* from their bottoms. The young of these creatures are free and very active little creatures, disporting themselves with apparently hilarious freedom, in imitation somewhat of the well-known fresh-water whirligig. As they grow older, however, their activity diminishes, until at length they attach themselves to some rock or wooden support, and there remain the rest of their lives. Their appearance in the mean time has very materially changed from that of a beetle to a mussel, so that their identity is entirely lost save to those who are familiar with the metamorphosis.

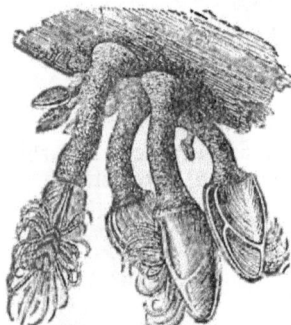

GOOSE BARNACLE.

While adhering to a support the adult *barnacle* is provided with a set of arms called *cirri*, which resemble hairs protruding some inches out of the point of the shell. These the creature uses to draw into its mouth the microscopic animals that these arms attract. Being henceforth fixed in one spot, the adult *barnacle* loses the eyes that are well developed in the young, and the same marvellous change is noticeable in losing the rudimentary limbs with which the creature is provided in its infancy. They multiply at an astonishing rate, and at maturity attach themselves, by instantaneous contact, to any substance at hand, whether animate or inanimate, so that they are frequently seen adhering to turtles, and even fishes, which are unable to divest themselves of these tormentors.

Fish Lice, or *Pœcilopoda*, is an order of crustacea that are parasites on various fishes. The Greek name given them signifies *many-footed*, on account of a provision which enables them to walk, swim, bite, and to breathe either in or out of water. Their mouths have sharp mandibles, out of which projects a horny tube, which the animal strikes into its prey, reaching to the small blood vessels, which it drains while holding on tightly by means of the mandibles. Many different species exist in both fresh and salt waters, and may frequently be seen stuck fast under the pectoral fins of a great many of our common fishes. Other species infest the whale, and even the lobster's cuirass does not protect him from a louse, which seems to have been created for his worriment.

The **Sand Flea** is a curious little creature found at times in great numbers along American beaches, looking very much like a shrimp, though hardly so large. In size it ranges from one-half to one inch in length, and is covered with a white, semi-transparent, leathery carapax. It progresses by hopping, and can attain considerable speed. When apprehended it curls up and simulates death, but is off again very soon after being released.

The **Fairy Shrimp.** This little animal is the connecting link between the water lice and shrimps, having characteristics common to both, especially in resemblance. He is so called because of his extremely attenuate and transparent body, rendering him invisible save to the critical eye. His length is about one inch, and the creature is made for progression by swimming on the back, using the tail like a crawfish, so that he can dart through the water at great speed. There are several species that are very closely allied to the *fairy shrimp*, nearly all of which are peculiar to European shores.

Mantis Shrimp (*Squilla Mantis*) is a name given to a species found only in the Indian Ocean, I believe, being most plentiful about the shores of Mauritius. It is some three inches in length, and is cov-

DREDGING FOR SEA SHRIMP.

ered with a shell like our common crawfish, though the body is very much longer comparatively. A specimen was caught by Dr. Lukis, who observed its habits in an aquarium, and reports his experience as follows: "It sported about, and after a first approach exhibited a boldness rather unexpected. When first alarmed it sprang backwards with great velocity, after which it assumed a menacing attitude, which would rather have excited the fear of exposing the hand to it. The prominent appearance of the eyes, their brilliancy and attentive watching, the feeling power of the long antennæ, evinced quick apprehension and instinct. I brought a silver teaspoon near it, which it struck out of

my hand with a suddenness and force comparable to an electric shock. The blow
was effected by the large arms, which were closed and projected with the quickness of lightning."

Phyllosome, or *Leaf-shaped Shrimp*, also known as the glass crab, is a rare
creature, found only occasionally about the Channel Islands. It is small in
size, and but for its resemblance to a spider, would be considered a

PERCH LOUSE.
(*Achtheres pescarum.*)

LEAF FISH LOUSE.
(*Argulus foliaceus.*)

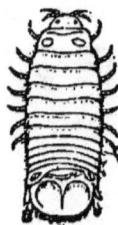

BORE WORM.
(*Limnoria terebrans.*)

LOUSE CRAB.
(*Cyniothoa astrum.*)

very beautiful animal. The legs are eight in number, radiating from the body,
and from the second joint of each issues a second limb, the anterior being rudimentary, while the posterior have even
greater development than the true legs.
The body is harp-shaped and so transparent as to be almost invisible, except
for the two brilliant blue eyes that are
set on the projecting ends of antennæ,
like the snail's, though they are not extensible.

TRILOBITE FROM
SILURIAN FORMATION.

WALL WORM, OR SOW BUG.
(*Oniscus murarius.*)

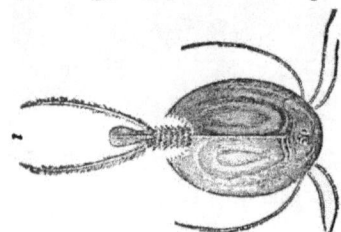

The **Nauplus** is a larval form that
was long mistaken for a species.

The **Gill-foot** (*Crenilabus tinea*) is notable
because the eggs will hatch however long the
process be delayed. It is a small green species of
the European connor.

The **Mussel Crab** (*Notodromus monachus*) is
a queer-looking creature, whose body is covered
by an egg-shaped shell, with little hairs at the
back part, and in front an upper and lower pair
of fringed antennæ.

GILL FOOT.

The **One-eyed Canthocarpoid** (*Cyclops can-
thocarpoides*) is a singular crustacean, segmented,
and as one goes backward growing beautifully less.

The **Sea Trumpet** (*Triton variegatus*) is the inhabitant of the well-known
conch shell. The **Periwinkle** (*Litorinia litona*) is found upon the rocks in
British waters, and is a well-known article in the London fish market. The
land snail still continues to contribute to the luxuries of the table, for which it
was fattened in the time of the Roman gourmands.

The **Sea Hare** (*Uplysia depilans*) is a queer specimen of the snail, which
exudes a fluid resembling violet ink.

A yellow, oriental member of the snail family takes his outings under the shade of an umbrella, which fact has given rise to its name of the *Indian Umbrella*.

The **Argulus Foliaceus** is a parasite living upon various fishes, chief among which is the stickleback.

The **Ravina Dentata**, or **Frog Crab**, resembles in body the common frog, although able to draw into its shell.

Edible Prawn (*Palæmon serratus*) is a name given to a species of shrimp abounding in the northern latitudes, and may be met with even in the ice fields of the Arctic regions. It was upon these creatures that Lieut. Greeley and his companions subsisted for a time during their desperate privations. This creature is the handsomest of his genus, and is frequently made a prisoner to enhance the beauties of the aquarium, as well as more commonly cooked to charm the appetite of those who can afford such luxuries. Its body is translucent, streaked with brown, pink and gray, while the eyes are tinted with orange, purple and blue, producing a most exquisite effect as it darts through the water, or rises to catch a morsel of

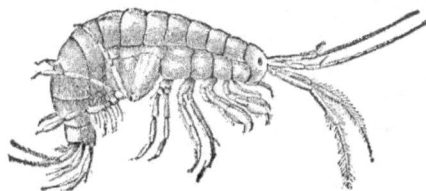

EDIBLE PRAWN.

food. Like all of the crawfish family, the female carries her eggs about within a fold at the intersection of the caudal scale with the body, and retains them until the brood appears. It is a carnivorous creature with most voracious appetite, ready to devour not only putrid flesh and offal, but to attack and eat its own species, in which respect it is identical with crabs, crawfish and lobsters.

Shrimps are used very extensively for food, but larger numbers are taken for bait, since nearly all kinds of fish bite eagerly at such morsels. There are thousands of persons engaged regularly in the work of shrimp-catching, the most general means employed being by arranging a wide-mouthed net, which is dragged over the sand at a depth of two feet, shrimps being invariably found near the shore and in shallows.

OPOSSUM SHRIMP.

The **Opossum Shrimp** (*Mysis vulgaris*) receives its common name from a pouch attached to the legs of the abdomen; this pouch, like that of the opossum, is used as a nest for the eggs, or for the young so long as they require parental care. It is specially abundant in the Arctic Sea, but species are found on the Southern coasts.

The true **Shrimp**, or **Prawn** (*Crangon vulgaris* or *Palemon vulgaris*), is esteemed a great delicacy by lovers of fish, and is a delicacy especially in Southern markets. Our marvellous facilities for transportation are rapidly obliterating distinctions of climate, and even persons of small means may now experiment upon what but a short time since were the fruits and delicacies of an almost fabulous life. Necessarily, this multiplies human industries so that plants and animals contribute to man's support in other senses than that of simply being fed upon.

The **Wall Worm** (*Oniscus murarius*) is singular from its special adaptation to a life on land. It frequents damp places and is to be looked for under moss, dead leaves, and stones. It protects itself like the hedge-hog by rolling itself into a ball.

The Sea Mouse .(*Nereis pelagica*) is a *nereid*, or centipede, and has but little resemblance to fish. Its back is covered with a substance like flax, having fastened spines and iridescent bristles, making this animal a rival of the most beautiful humming-bird. It has many bristle-like feet, which enable it to swim

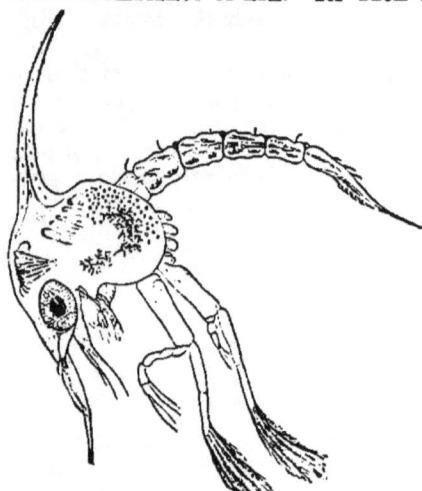

COMMON SEA MOUSE.

LARVA OF CRAB.

or crawl with the greatest facility. Its snout has jaws and fringe-like tentacles, so that altogether it has a resemblance to the body of a mouse.

The **Silurian Trilobite** is a fossil which derives its name partly from the geological formation in which it is found, and partly from its three lobes, or segments. The head is rounded in front, and often furnished at the back with long spines ; the eyes, when present, are situated on opposite sides of the head, and are single, grouped or compound. In the earliest stage the body is an oval disc, and while there is no meta-morphosis, segmentation takes place as the animal develops. Some four hundred species have been distinguished by the faithful students who have devoted themselves to the study of this animal, whose advent pre-ceded our own.

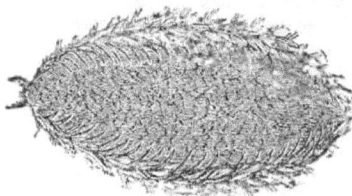

FAIRY SHRIMP.

The **Skeleton Screw** (*Caprella linearis*), a curious creature, has a skeleton-like leanness, but likewise a voracity similar to that of the greyhound. The first pair of legs terminate in a thin, razor-like joint, notched along the edges, and shutting into a groove in the enormously large joint above ; the groove has furthermore a double hedgerow of spines. It supports itself on marine plants by its several pairs of

MANTIS SHRIMP.

hind legs, straightens up its long, lean body, and continuously waves its terrible front legs, which allow no prey to escape. The antennæ, also, are spined, and probably are used in catching prey.

The **Fresh-Water Shrimp**, or **Brook Flea-Shrimp** (*Gammarus pulex*), takes its name from its active movements when upon land. It frequents the smallest brooks, the tiniest rivulets, or the most rapid and deepest streams. When frightened it conceals itself in the mud. It carries the young attached to its abdomen until they are large enough and strong enough to shift for themselves.

The **Crayfish**, or **Crawfish** (*Astacus fluviatilis*), is next, in the general order, to shrimps, which it resembles in many respects, but differs in the following: The body is longer and covered with a strong shell, in which are located all the vital organs. The extremities are flexible, but protected by a soft shell lying in flat rings, the edges of which overlap; the tail is flat, and composed of fine scales slightly overlapped so as to form a spread. In the extremity and tail reside its propelling power, the latter being capable of almost inconceivable rapidity of motion, equalling that of a fly's wing. It has eight legs springing out from the belly, where the sheath to the body is slightly separated to permit the free movement of the organs that lie along the line of separation. In front of the anterior legs and directly under the thorax are the arms, which exceed the legs in length and terminate in bony claws of great muscular power. These are its weapons of defence and the organs used to catch and destroy its prey. The eyes are set well forward and on top, and at the point are needle-like thorns, on either of which are the two antennæ, or organs of touch.

WORM CRAB.
(*Lernæa branchialis.*)

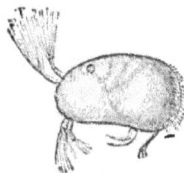

SHELL CRAB.
(*Notodromus monachus.*)

CRAWFISH.

The *crawfish*, in nearly all respects except size, and that it is a fresh-water habitant, is identical with the lobster. So nearly alike are the two that the *crawfish* is very frequently called the fresh-water lobster. Its favorite haunts are among flat rocks in the shallow places of clear streams, where it generally lies in concealment watching for any water insect that may chance to come near. But it will also feed on refuse, especially if of an animal nature, and being extremely pugnacious, will kill and devour its kind. The claws and legs are very friable, and in the fierce combats it so frequently wages these are often broken off; but the loss of any of these certainly gives it little, if any pain, and the repair is rapid, new claws growing out again in a few weeks.

Like the shrimp, its first cousin, the *crawfish* is largely used for both food and bait, and is accordingly caught by various means to supply these demands, but the more common mode is by enticing it into baited pots, into which it can easily enter but may not escape.

The **Lobster**, which is classed under the genus *Homarus*, is the most important member of this order, not only because he is largest of the genera,

but because of its importance as a factor in the commercial world. He is found all along the coast, from Labrador to Florida, but rarely south of New Jersey, and the *lobster* fisheries do not extend south of New York. Stringent laws are enforced by all the New England States against the taking of *lobsters* except at certain seasons, when their flesh is marketable. The annual total catch of these creatures in the Northeastern States is valued at about one million dollars, the Maine coast producing more than that of any State because of the laxity of the laws relating to *lobster* catching. The principal means employed by the *lobster* fishers to take their prey is similar to that used by the crawfishers, viz., the sinking of *lobster* "pots," constructed of laths made into wooden frames covered with heavy netting. A hole is left in either end, through which the *lobster* enters to the bait fixed in the centre. Several times each day these "pots" are lifted and the *lobsters* taken before they can escape, as they never atttempt to do until the pot is lifted into the boat.

A curious thing characteristic of the *lobster* is the annual shedding of his shell and the formation of a new one. The molting season is at the approach of summer, at which time he retires into a secluded spot, usually under a large stone, and there remains dormant until nature divests him of his old coat. This process is not dissimilar from that we observe in the bursting of the locust's larvæ and liberation of the fly. The *lobster's* carapax splits down the back centre, the rift gradually widening until the claw has shrunk sufficiently to permit its withdrawal through the arm opening, when the animal wriggles out of his old clothes, presenting a rather sorry and defenceless appearance. Being divested of his protective armor, he would speedily fall a prey to predaceous members of the crab family, or voracious fishes, if he exposed himself to their attacks, so, like a philosopher, he keeps indoors and awaits the formation of a new cuirass, which requires but a few weeks to complete, when he issues forth full of renewed life to continue his spoliation. Not every *lobster* is so wise as to remain concealed during the moulting season, for the weaknesses that are often perceived in mankind have their counterpart in the lower orders, and especially in the *lobster*. Not a few appear unconscious of their helpless condition, and boldly issue out of their retreats to take their chances among the many hungry maws that are watchful for such opportunities as the uncovered *lobster* affords, and many sacrifice their lives to this indiscretion.

Next to the *lobster* family, being very close in relationship, is that of the **Hermit Crab**, whose hereditary patronymic some learned classic says is *paguridea*, or shell-dweller. Of the many strange creatures that people the sea, none are more curious in their habits than the *hermit crab*. Nature never makes any mistakes without trying, at least, to atone for the omission; but when bringing forth this wonderful animal our good mother was possibly in a facetious mood, or else trying to determine how incongruous she could make a form. Her success at accomplishing the ludicrous must be conceded.

Very often, when fishing in an arm of the sea with a shrimp bait, I have drawn up two, and as many as four, small *hermit crabs* at one time. To the person who meets with them in this wise, he will think it strange that periwinkles would adhere to his bait, and unless he looks carefully he will be deceived as to what these "periwinkles" are, for the crab is sly, and will let go the bait upon being drawn into a boat, and withdrawing into the shell will not stir, as if conscious of the effect of simulating death. When they think an opportunity offers they put out their legs, lift the shell upon their backs in a most comical way and make off for the water.

The *hermit*, totally unlike his congeners, is brought forth without any protective covering. His body is quite as long, relatively, as the lobster's, but it is soft flesh, very inviting to fishes and crabs, so that to atone for making him thus helpless, nature endowed him with a cuckold's disposition, and a cunning that enables him to supply that which he ought to have had at birth. No sooner does the *hermit* issue from the egg than he makes off in search of a shell that will afford him a covering for his delicate body. Naturally he finds an abandoned periwinkle first, because these shells abound everywhere in the oceans and about the beach, and in such variety of size and shape that the most fastidious *baby hermit* might easily be suited. Having become thus entrenched, so to speak, against his foes, he goes ambling about, always carrying the periwinkle house on his back. At length, however, his body becomes too large for his first quarters, and he seeks out another that promises to accommodate him more comfortably. He is not so easily satisfied as at first, and usually continues his search a considerable time before he finds a shell which he thinks is adapted to his size and circumstances. When at length a proper abode is selected he turns it over, pokes his preternaturally large claw inside, and then drawing up within his own shell until the old and new quarters are in contact, he transfers himself with such rapidity of motion that the eye may not perceive the change.

HERMIT CRAB (*Pagurus miles.*)

As the *hermit* continues to increase in size he must make frequent removals, but the process is always as described unless it chance, as frequently happens, that two *hermits* of like size have a desire to seize upon the same shell. It is manifestly impossible that two should occupy one shell, so to determine which shall be the possessor a duel to the death is fought. In these combats the most surprising ferocity is exhibited. Their bodies being well protected, they fight with the large and lesser claw, with which each is provided, pulling, snapping and lunging, rolling about, first one having an advantage and then another, until at length the claws of one give way. When this event happens, the one thus bereft of his weapons can make no defence, but this does not protect him against further assault, since magnanimity is not a trait of crab character. The victor continues the attack until, with vindictive savageness, he drags the body of his victim from its shell, mutilates it beyond recognition, and then eats the remains.

The *hermit* grows to a length of about one foot, with body large enough to fill a conch shell. One of this size, when thus protected, presents a formidable front, his very large claw covering the mouth of the shell, with a small claw acting as a reserve, ready to give effectual reinforcement in time of need. So

4

hard is the bony sheath of his head covering, and so ponderous and powerful his claw, that no enemy, save one of his kind, can vanquish him or force him from his retreat. But cunning is generally opposed by cunning, as the wily *hermit* often falls a victim to the foes who are content to watch him until, in an unguarded moment, he projects his head too far out of the shell, when he is gobbled up in a trice by the frog fish. The octopus is also his enemy, and is

FIJI, OR ROBBER CRAB (*Birgos latro*)

most successful in luring him from his shell. To accomplish this the octopus lies flat on the bottom and stretches one of his tentacles out across the mouth of the *hermit's* shell. A slight suction by the octopus brings the *hermit* partially out of his abode to investigate the cause, probably thinking that a most appetizing morsel had suddenly dropped down within his reach. But as he ventures forth a few inches, in a trice the octopus has fastened one of his powerful suckers upon the *hermit* and drags him forth, when he is soon dispatched and eaten.

The **Diogenes Hermit Crab** is a species of the *pagurus bernhardus* just described, and differs in only a few particulars, one of which is its habit of climbing trees, and another is in the fœtid odor which it has the power to emit. When frightened or captured they give forth a noise somewhat resembling the squeak of a green frog when it is first taken. They spend considerable of their time on shore, being vegetable as well as carnivorous feeders. Their principal haunts are about the shores of Brazil and the West Indies.

The **Robber Crab**, which out of regard for its cunning in smashing and eating cocoanuts is often called the *cocoanut crab*, is found in many parts of the Indian Ocean, being most numerous about cocoanut plantations. The manner in which it breaks the shell of a cocoanut is thus described by Mr. Darwin:

"It would at first be thought impossible for a crab to open a strong cocoanut covered with the husk, but Mr. Liesk assures me he has repeatedly seen the operation effected. The crab begins by tearing the husk, fibre by fibre, and always at that end where the three eye-holes are situated. When this is accomplished, the animal commences hammering with its heavy claws on one of these holes till an opening is made; then turning round its body, by the aid of its posterior and narrow pair of pincers, it extracts the white albuminous substance of the nut."

Opposed to this statement of Mr. Darwin, which he does not confirm, having merely given the facts as related to him, is the affirmation of many travellers, who pretend to have witnessed the feats described, that the crab, after effecting an entrance through the soft eye of the cocoanut, fixes the sharp point of a claw in the aperature, with the other mandible pressed so tightly against the other end of the nut that it is held fast, while the crab dashes it violently against a stone until the shell is broken asunder. This is more probable than the account furnished by Mr. Liesk.

PALM CRAB SEIZING A GOAT.

The *robber crab* is the largest of the genera, frequently measuring three feet in length, with proportionate thickness of body. Its favorite retreat is near the seashore in a hole which it digs in the ground, usually at the root of a tree, to a depth somewhat below the sea surface, where it hides during the day and only sallies out after nightfall. In walking it has a very curious and awkward gait, going along on its tiptoes, after the manner of sand-fiddlers, in which respect it resembles a crab, though it possesses well defined characteristics of the lobster, to which it is certainly closely allied. Some persons have attributed to it the power of climbing palm trees to browse upon the young shoots, and a writer in "St. Nicholas" avers that he saw one, while suspended to a palm branch, seize a goat as it was passing beneath and lift it almost clear of the ground, a feat which we may well discredit.

The **King Crab** (*Gecarcinus ruricola*), or more commonly called the *horse-shoe crab*, is a large species found on the Atlantic coast from Maine to Florida, and also among the West Indies, where they are called Malacca crabs. The popular name, *horse-shoe*, is a very appropriate designation on account of its shape, which is very like the under surface of a horse's foot. The shell covering is composed of three pieces, viz., a hemispherical anterior, composing two-thirds of the carapax, a heart-shaped posterior, which is articulated on the frontal shell, and a long spinous tail terminating in a very sharp point, and barbed along the upper edge. The tail of this creature is frequently converted into a dangerous weapon by West Indiamen, for upon affixing a handle, or even using it without this adjunct, it becomes a veritable stiletto. A singular feature of the oval shell is the appearance of two glassy orbs, one on either side of the shell, which have such a remarkable resemblance to eyes that they are calculated to deceive any save those thoroughly familiar with the animal.

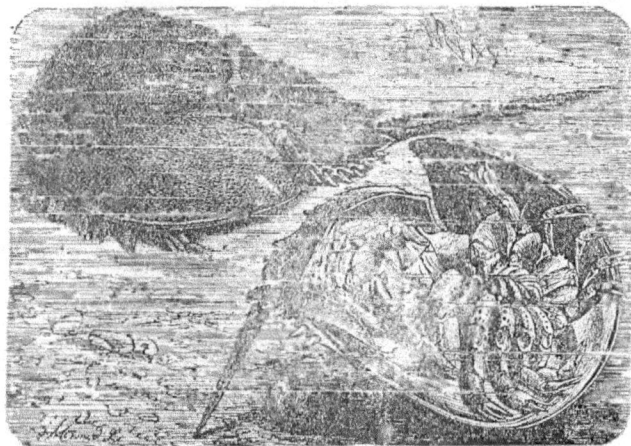

KING OR HORSE-SHOE CRAB.

The *king crab* holds a disputed ground between crab and scorpion, having been classed with both, and still continues the subject of much argument. It rarely ventures out at sea, confining its range within a short distance of the sandy beach, into which it burrows after insects. It has no large claws, but is provided with ten small nippers, sufficiently large, however, for its quiet and timid habits. While the shell sometimes attains considerable size—a foot or more in diameter— the animal itself is comparatively very small, occupying scarcely a third of the hollow, or scooped, space under the dome of its carapax. The young of the *king crab*, just before leaving the cellular membrane of the egg, and a short while after the egg-shell is ruptured, bear a rather striking appearance to the fossil trilobite.

The **Spider Crab** (*Maia squinado*) is found in many waters, but attains its greatest size in the Japan seas, where it holds the mastery as being the largest, though almost defenceless, of all the crustacea. They are so named on account of the hairy covering of the back and legs, which added to their shape gives them an appearance very spider-like. Some of the species are careful of their toilet and appear with clean bodies, but others seem very careless and become so covered with grime of the sea that they attract numerous parasites of both animal and vegetable character. So clothed at times are they with foreign bodies that scarcely any portion of the real animal can be seen as it moves about like a conglomerate mass on the ocean floor. So cumbrous do these accumulations become that the animal occasionally seems much distressed,

and but for the shedding of his shell, by which alone he may rid himself of the parasites, he would probably be rendered incapable of movement.

Another species of the *spider crab* is so unlike that just described that the classification appears arbitrary. It is peculiar in being a counterpart of the well-known insect commonly called "grand-daddy long legs," and progresses very much after the same manner.

JAPANESE SPIDER CRAB.

Specimens of this latter have been captured with bodies scarcely equalling the diameter of a dinner plate, whose spread of legs was twenty feet.

The **Porcupine Crab** is also a habitant of Japan waters, where it grows to a considerable size. It takes its common name from the sharp spines which cover its body, giving it a formidable appearance, but not sufficient to protect it from the voracity of many species of fishes.

PORCUPINE CRAB.

The **Frog Crab**, of which there are several species found in tropical waters, has a body very like that of the frog, from which fact the name has been given. The body is small, hardly exceeding the size of a man's fist, and the legs are so short as to be barely observable from a top view.

The **Death's-head Crab**, known to science as *Dromia vulgaris*, is confined to a small territory of the ocean waters, being found, so far as I have been able to determine, only about the shores of the Channel Islands. The carapax is smooth, with a knob rising from the centre having a fanciful resemblance to a skull and cross bones. The legs are hairy and generally bear about, attached to them, certain fungi until the moulting of the creature sets its limbs free.

The **Woolly Crab** is a deep-sea animal, principally confined to the Mediterranean, distinguished for the thick coating of fine hair which covers the carapax. The claws are small, and elevated on the back in a most grotesque and apparently awkward manner. The posterior legs terminate in a hooked nail, by which the creature can grasp its prey firmly, or attach itself to any object with considerable tenacity.

WOOLLY CRAB.

The **Mask Crab** is found in the India seas, but haunts the shores, where it lies concealed in the sand with only its antennæ and eyes exposed, watching for its prey like the ant-lion. The claws are well developed and possess great muscular power, which enables it to destroy creatures much larger than itself. The name is given on account of the singular markings on the carapax, thought to resemble the features of a man's face.

The **Crested Crab** is most common about Japan shores, frequently coming out upon the land in quest of food. When surprised it folds its legs and appears rigid, very much like the tumble-bug, in which position it so nearly resembles a stone as often to deceive those even who are familiar with its appearance and habits.

The **Armed Crab** belongs about the Florida reefs, where it is quite common. The name has been applied because of the armament of its shell with long and very sharp spines, making it dangerous to handle.

The **Fighting Crab** (*Gelasimus beliator*) makes his abode in marshy places near the sea coast, where he burrows to a considerable depth, but is often on the surface. It is distinguished by having one huge, abnormally developed claw, which may be either the right or left, while the other is extremely small and apparently unserviceable except to steady the animal when walking. The name is appropriately applied, for he is a pugnacious creature, and always spoiling for a fight. When running, the large claw is held aloft in a menacing manner, or occasionally waving as if signalling the approach of another of its species. As the Latin name signifies, the animal is so grotesque as to provoke laughter. Another singularity exists in the very long foot-stalks upon which the eyes are placed, being somewhat longer

WEST INDIA LAND CRAB.

in proportion than the projecting eyes of the snail. At times these crabs appear in such numbers as to fairly encumber the ground, but however thickly they may be distributed their movements are so harmonious as to always seem preconcerted, by which interference is avoided. When alarmed they move with the precision of soldiers, advancing and retreating in files of heavy columns, and never becoming disordered unless their ranks be broken by violent attack.

The **Edible Crab** (*Neptunus hostatus*) is the best known and highest prized crustacea among Americans, because it constitutes no inconsiderable part of the food product of our people, ranking next to the oyster. This species is found in great abundance from the shores of Labrador to Mexico, and though countless numbers are consumed every year, while no efforts have been made at their cultivation or protection, yet it is quite as plentiful now as at any other period. Its haunts are about the muddy shallows, and especially where grasses grow, and it may be easily taken in many ways. The common method employed, however, is by the setting of creels, or pots, into which the crab is enticed by bits of fish or tainted meat. Being a voracious feeder, it will not quit its repast until drawn out of water. In fishing about inlets it is more common to draw up a crab, holding on to the bait with its claws, than to take fishes, which makes it a very pest to angling sportsmen. Like all the crustacea, the shell is shed every spring or summer, at which time its flesh is held in the highest esteem, though many are consumed at all seasons of the year. It rarely grows to a size exceeding three inches in diameter.

The **Common European Crab** (*Cancer pagurus*) may be taken as the type of a numerous family. Its shell is round in front and narrowed as it proceeds backward. The legs are short, but the claws are disproportionately large and not symmetrical. It dwells in deep water and is caught by crab-pots, though sometimes by a net. The American *soft-shell crab* is the *lupea dicantha*, caught after shedding their old shells and before they have assumed the new. It is considered a

BLIND DEEP-SEA CRAB (*Willemœsia crucifera*).

great delicacy, although a fondness for it is probably an acquired taste. Crabbing is among the commonest amusements on the Jersey coast, and in the Southern salt-water lakes. A piece of meat tied to a cord is always a sufficient bait, and the creature, "having once put his hands upon it," will not let it go so long as it is in the water. Crab-fishing requires no skill, and involves no danger greater than a nip to the careless fisherman. The Eastern crabs are not so palatable as those found in Southern waters.

The **Worm Crab** (*Lernæa branchialis*) is found upon the gills of the

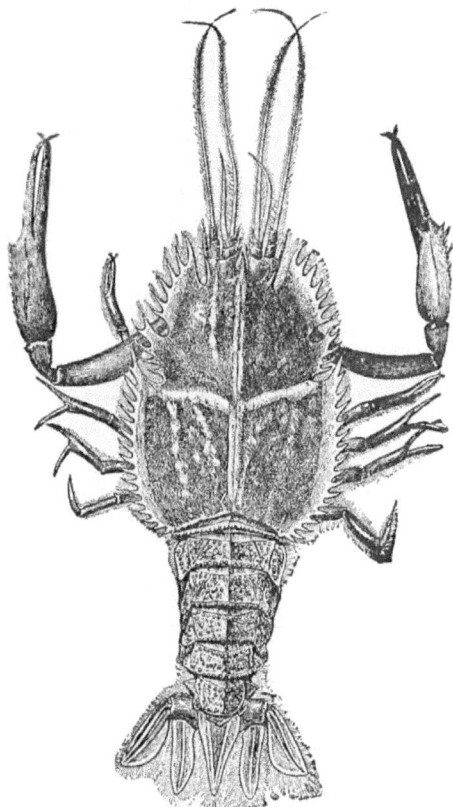

codfish. It is parasitic, and may be regarded as the type of what, in the life of fishes, is similar to the annoyances caused in higher animal life by parasitic insects. Some species devote themselves to the whale, others to the dogfish, and others yet to various representatives of the finny tribe.

The **Toulouson Black Crab**, or **Jamaica Violet Crab** (*Gecarcinus ruricola*), burrows from one to three miles from the shore, not being a true marine animal, but going seaward solely with the intention of depositing its eggs. It selects the last month of the old year and the first month of the new year, and at this time is a rare dish for the table. It hibernates in the summer, which is a more emphatic way of stating that it æstivates. During July and August it again rivals the dainty dish which was set before a king. It is an agile creature, and capable of inflicting punishment after death—its death; for some time its broken claw will continue to pinch with all the energy of a living being. It is a table delicacy, esteemed the more, possibly, because rare. It is found in the West Indies and is not aquatic.

ONE-EYED LOUSE.
(*Cyclops canthocar-poides.*)

The **One-Eyed Louse**, and **One-Eyed Sailor**, are species which may well excite the wonder of even those who profess familiarity with the many curiosities of animal life. The technical name, *Cyclops*, is appropriately applied, derivable from the popular superstition, which so generally obtained during the early ages, that on the coast of Sicily dwelt a human monster having but a single eye, which was situated in the centre of his forehead. The exposure of this myth led to the belief that though there were creatures in the insect world provided with many eyes, yet there were none whose vision was limited to a single eye. This belief was, however, like that concerning the fabled Cyclops, destroyed by the discovery of the two crabs above described, and so numerous have been the exposures of preconceived ideas during the past fifty years, revealing so many marvellous surprises in animal life, that even the wildest exaggerations of mythology and fable are shown to be exceeded by existent creatures. In this fact we observe new proofs of the well established theory of the perpetual change and progression of species.

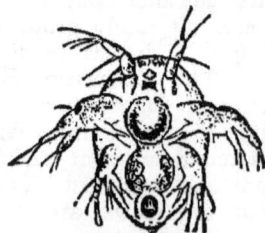

ONE-EYED SAILOR.
(*Nauplius cyclops*)

In the illustrations afforded by the Savage WORLD will be found evidences not only confirmatory of this theory, but instances in which we may perceive the successive steps of development. The succeeding chapter presents species wherein will be found the natural ascending series, manifesting a well defined and sequential progression always towards higher orders of life.

FISHES.

NOWLEDGE respecting the primary orders of life which abounded in the seas is largely speculative and made up of deductions drawn from analogy, and though the progressive steps in creation are visibly marked they are not so plainly consecutive, or perfectly adjusted in arrangement. Enough is known, however, to justify us in the classification we have adopted. The lowest orders are therefore denominated the *invertebrates*, or creatures in which the back-bone is wanting, because they are simple in form and least variable in structure. Nor is abundant evidence wanting in proof of the claim that the primitive creatures of the most ancient seas were invertebrates. Much of the proof in support of this theory has already been offered in the opening pages of this work, treating briefly the subject of the earth's development, hence we do not need to pursue it further now. From the invertebrate to the vertebrate is not so long a step as the casual reader might suppose, for in all creatures save those which belong to the very subsoil of animal life—where the vegetable fades into the animal—the back-bone exists in a rudimentary form, and must necessarily develop with the progression of species. All these links are most distinct, and the chain connecting the highest with the lowest is complete, save that single strand which is wanting to bind man with the next descending order, which still remains undiscoverable, if, indeed, it ever did or can exist. Man alone seems to unite in nature with the Deity, and remains so far exalted above all other animal kind as to be separated by immeasurable space from all affinities, save alone the spiritual, to which his mind is ever an aspirant.

If our surprise has already been excited by the wonderful forms and adaptations of species described, how much greater must be our astonishment at the wisdom of the Creator as displayed in the more complex creatures and their marvellous endowments, which must now engage our attention, as we step into the field among animals of greater sensitiveness and improved organization. Those animals which we have considered were adapted to a water habitation, but the comparative simplicity of their structure will appear when we come to contrast them with fishes, which exhibit the omniscience of God in a more surprising manner than any other creatures excepting man.

WONDERFUL ORGANISM OF THE FISH.

The term *fish* is applied to animals that breathe through the medium of water, extracting the air therefrom by means of *gills*, and which are specially adapted to an aquatic existence. It is impossible to give an exact definition without describing characteristics prominent alike in fishes and amphibians, for even the feature most peculiar to fishes, viz., gills, is possessed by creatures belonging to other and distinct orders, such as the proteus, and other animals that are supplied with gills while in the larval state. We must therefore be guided in a measure by the knowledge which every one possesses in a degree, which, with the descriptions to be given, will enable us to distinguish the several orders, so far as classification can be trusted.

The breathing organs of fishes are not in the lungs, for these are wanting, but in the gills, which are found on each side of the head, and covered by articulated arches called the *opercula*. These constitute the most sensitive parts of the fish, because they correspond in purpose to the heart, being composed of laminæ permeated by blood-vessels so near the surface as to be in contact with the water, and extract the oxygen therefrom. The process of such extraction is by receiving water into the mouth and expelling it at the gill orifices, by which action the air is absorbed and the blood accelerated, similar to the effect produced by the breathing of atmospheric air by land animals.

When a fish is deprived of water his gills become dry, and this is followed by suffocation because he can no longer absorb the air. This is proved by the fact that the constant pouring of water over the gills, even though the fish be removed from its element, will serve to sustain its life for a considerable time. That the fish absorbs air in the act of drawing water through its mouth and dispensing it over the gills, is also proved by the well-known fact that the creature cannot long exist in a small body of water unless it be frequently renewed, because the air with which the water is impregnated soon becomes exhausted, and is then to the fish what carbonic gas is to land animals. Any artificial means that will force air into the water will again revive the fish, just as fresh air will relieve suffocation arising from carbonic gas in land creatures.

The wonderful adaptation thus seen in fishes, which fits them for life in the water, is reinforced by their peculiar structure, which qualifies them for movements in their element. A fish is the very embodiment of grace, and exhibits truly marvellous design in its structure. The body is formed so as to pass through its medium with the least possible resistance, but to this advantage is added scales or skin covered with an oily secretion, which prevents friction, and facilitates its motion to such an extent that inertia may be said to be practically the only retarding influence it has to overcome. The fins are not always the same in all fishes, some having more and others fewer, but generally they are supplied with one on either side near the gills, which are called *pectoral* fins, two under the thorax, called *ventral* fins, one on the back, which projects from the spine, called the *dorsal* fin, one, and sometimes two, near the vent, called *anal* fins, and the tail, or *caudal* fin. Besides these organs of locomotion, many fishes are supplied with an air bladder, which is a rudimentary lung, by the inflation or contraction of which it is able to rise or sink at pleasure.

Scales are not always present, but when wanting their place is supplied by an excellent substitute in the form of a strong and oily epidermis very difficult to rupture. The scales are formed by concentric layers of a substance resembling neither bone, horn nor membrane, yet partaking somewhat of the nature of these three. It is by these that Agassiz established his classification of fishes, and so perfect was his knowledge in this respect that he had only to see a scale to determine the fish to which it belonged, though he may never have seen the species.

The organs of touch in fishes are either the lips or antennæ, and are very acute, as is that of sight, but the auditory nerves are rudimentary, so that their hearing is obtuse, if, indeed, they have this sense at all. They are marvellously susceptible to vibrations, as is evidenced by the excitement they exhibit at the shock produced by a person stamping upon the ground. So they have been trained to respond to the ringing of a bell, but the vibration, acting upon supersensitive nerves in the gills, may be the effect, rather than sound waves striking upon auditory nerves. The sense of smell certainly exists, but experiments indicate that it is not acute.

The sex of fishes may be determined by shape and color, but with the exception of sharks, rays, and a few others, there is no external difference between the two. The manner of propagation is not by contact, as in other vertebrates, but by a singular process which is yet imperfectly understood. Both male and female are provided with lobes lying along the intestinal canal, in which are contained what is called *roe*. At the breeding season the female voids the *roe*, which, on examination, are seen to be millions of very minute eggs, and these the male impregnates by voiding the *roe* which he carries, but which has now assumed a fluid state, somewhat resembling milk. It is not necessary that the sexes should be near each other at this time, as it has been shown by the fish culturists that eggs may be conveyed to any distance and then impregnated by squeezing the *milt*, as it is called, of the male upon them.

Some fishes are viviparous, retaining the eggs within the oviduct until the young are sufficiently developed to appear. How impregnation is accomplished in such cases is a problem naturalists have not been able to solve, though it is probably by the female absorbing the milt of the male through the vent.

Generally, fishes are predaceous in their habits, being so voracious that they turn cannibal with the most natural instinct, and feed off the young of all kinds, not even respecting their own. In this, however, we see a wise provision, since but for this check upon their multiplication the waters would soon become choked by their bodies, as no other creatures increase so rapidly, except it be a few insects.

Of all creatures, fishes are most difficult to classify, because characteristics of form, structure and organization are so blended as to appear peculiar to hardly a single species. Every attempt thus far made, even by Agassiz himself, has proven so unsatisfactory that I will not undertake a work which others much more competent than myself have given over as impossible, but will strive to increase the general interest in fishes by presenting them as they seem most naturally to occur in the order of progression.

NEST-BUILDING FISHES.

It has only been within comparatively recent years that the discovery was made that certain fishes construct nests in which to rear their broods. The species first noticed as performing this singular labor were the *Sticklebacks*, which are quite common in English aquaria. For a long while it was supposed that this was the only nidifying species, but more recent investigation discloses the fact that several fishes are peculiar in this respect in both fresh and salt water, as we shall see.

The **Sticklebacks** are a small species belonging to the family *Gasterosteidæ*, so called because of the sharp spines that project along the back. In some these spines are three in number, and again five, nine, or a dozen, each being a distinct species. In size they rank among the smallest of fishes, but are found in both fresh and salt water, and are common in Europe and America. The manner of constructing their nest, a service always performed by the male, is certainly curious. For this purpose the male is provided with an organ filled with a gelatinous secretion, which it voids through an opening in front of the vent, and which coagulates upon coming in contact with water, but which after voiding he carries about attached to his side.

At the breeding season, but before mating, the male selects a spot among aquatic plants, where a gentle current is continuous. He then begins his work by biting off small bits of plants and carrying them to the place of deposit, placing the first ones in an upright position to serve as pillars. Other bits are now carried, each piece being glued fast by the substance which he secretes, until a layer is formed. His next step is to carry pebbles to the spot, with which he weights the layer and

NESTS OF THE STICKLEBACKS.

makes it substantial. Other layers, alternating with bits of plant and pebbles, are laid until the nest is made, arched over the top, leaving a cylindrical hole through the centre, which serves as the exit and entrance. The inner side is now well plastered with the glue from his body, and worked very smooth until it resembles varnish. Having completed his nest, he goes in search of a mate, to whom he makes love in an approved manner by a display of much vanity, involving a degree of activity which he manifests at no other time. When the female accepts his advances she immediately enters the nest and deposits her spawn, which usually occupies about five minutes. She then departs by the back door, when the male enters and ejects his milt over them in a much shorter time, then reappears and goes in search of another female, for he is a confirmed polygamist. In this wise he secures, perhaps, a dozen wives, each of whom deposits her eggs in his nest, and are followed by him as in the manner described, until the nest has its compliment of fertilized eggs, which are now covered by his secretion, and then left for nature to hatch.

The Toad Fish is a habitant of the American coast, rarely exceeding a foot in length, and of most repugnant aspect. The name is given on account of the peculiar-shaped head, which is of a calloused, wrinkled appearance, somewhat resembling a toad. The pectoral fins have a large, fan-like spread, and under the lower lip are several wattles, which probably serve as sensitive organs of touch. Small as it is, the fish has very powerful jaws, and is tenacious of its hold on anything seized.

TOAD FISH AND ITS NEST (*Batrachus grunniens*).

It is generally found lurking in oyster beds, where it makes its nest, and in the breeding season remains so constantly within the nest as to give rise to the impression that the eggs are incubated. Even after the young are hatched it remains for some time within the hole, and until its brood are strong enough to provide for themselves. This species are said to be poisonous.

FROG FISH (*Lophius piscatorius*).

The Frog Fish belongs to the same species as above, but is considerably larger and its habits different. It is also called *fishing frog*, and *angler*. They occasionally grow to a length of two feet, but of this length the head constitutes one-half and is phenomenally broad. It has a very singular habit of half burrowing in the mud and projecting very long antennæ, with which it is provided, at the end of which are flashy lobes that shine like silver. This attracts small fishes, which, when within reach, the *angler* pounces upon with certain aim and crushes between his powerful jaws. It is from this habit that the name *angler* has been given. The jaws are said to be strong enough to crush an oyster shell, but the statement lacks confirmation. Some maintain that it retires into holes and there rears its brood 'like the *toad fish*, but naturalists have not been able to determine the truth of this claim.

The **Black Goby**, or **Rock Fish**, is found widely distributed, and several species are more or less satisfactorily known. It is of a singular form, being like the frog fish, large of head and tapering towards the tail. The eyes are so protuberant as to appear as if situated on a foot stalk. They are

TOAD FISH OF THE SARAGOSSA SEA, WITH ITS NEST (*Anneatarius vespertillio*).

ROCK FISH, OR BLACK GOBY.

expert climbers, and spend much of their time on shore, mounting the rocks and climbing over the roots of trees. Their progression is by jumping, and so swiftly can they proceed that it requires a fair runner to overtake them. They burrow in the sand, but bring forth their young by constructing a ball of the sea-weed, in which the eggs are deposited. For some time after the young are hatched they continue near the nest, ready to fly back again and hide in their globular house. They are found chiefly along the shores of the East Indies, though one species is quite common about the California coast. It is hunted by the Malays and Chinese, who consider its flesh a great delicacy. About the coast of Borneo it is especially plentiful.

The **Sun Fish** is everywhere common in American fresh waters, where it sometimes attains a weight of nearly one pound. As every person is familiar with its appearance, description is not necessary, but not every one is aware of the fact that it constructs nests in which to deposit its eggs. In this process both male and female engage. Having selected a suitable spot in still water near the shore, where small drift has settled in the mud, the two hollow out a circular space by swiftly circling, as if chasing each other, until a round nest, dipping in the centre, with twigs lining the periphery, is made. In this the female deposits her spawn while the male rises a few inches above her, and ejects his milt until the water about the nest is so impregnated as to hide the two entirely from view for some time.

The **White Fish**, peculiar to the fresh waters of Europe, are a species little larger than the sticklebacks, and their habits are not less interesting. Their method of nest building is by carrying small pebbles to a chosen spot and dropping them one by one, with patient industry, until a small pyramid is formed. As the work goes on, the female from time to time deposits her spawn in the growing pyramid until the laying is completed and the eggs are covered. The male now envelopes the nest with his milt, which appears to settle like a film over the pebbles, and may be seen for days after.

GOBY, OR JUMPING FISH (*Periophthalmus kœlreuteri*), OF BORNEO.

The **Piraya** is a species of bream found in tropical waters, and chiefly about the East India coast. It never goes into deep water, its range being restricted to inlets where vegetation is rank. The nest which this small creature builds is very curious as well as substantial, and rivals that of the tailor bird. When the breeding season arrives the male goes in search of a vine or branch suspended in the water, and having found such he begins carrying small pieces of decayed vegetation, such as leaves and sticks, to the hanging branch, and attaches them thereto by a secretion which has the adhesiveness of glue. The work progresses without cessation until the accumulation on the point of the pendant vine is nearly a foot in diameter. The female superintends the building, going backward and forward with the male, but no one, so far as I have been able to learn, has ever witnessed her in the act of depositing her eggs. That she performs this most important part, however, is evidenced by the fact that the young have been observed issuing from the nest in great numbers.

The **Lamprey Eel** is found in nearly all the waters of the north and south temperate zones, though so retiring are its habits that few persons are acquainted with its characteristics. Much as I have fished in American streams, from Maine to California, I never met with but a single specimen. While hunting along the foot-hills on the Missouri River, in Dakota, I stopped at a ranch where the keeper had just caught a large catfish, and discovered attached to the fish's side a lamprey about twelve inches in length. The sight was such an unusual one that the man was much frightened, and was upon the point of abandoning the fish, lest the strange animal might attack him. A knowledge gained by reading books of natural history taught me at once what the creature was, and at my solicitation the ranchman permitted me to retain the fish and its singular prey. It required considerable force to detach the lamprey, and when its hold was broken it seized upon the first thing in its way, which was a tin scoop, and this the creature held on to with great tenacity. At the point where it had fastened upon the fish the skin had been perforated, from which blood flowed some time after, showing what a drain upon the circulation had been established by the powerful suction.

NEST OF THE PIRAYA.

Lampreys seem to be a connecting link between the eel and blood-leech, the organs by which it is enabled to apply such suctorial force being almost

identical with those in the leech. Unlike fishes, the *lamprey* is not provided with gills, the breathing being accomplished through seven orifices on each side, that are distinct and on a line with the eyes. Water is drawn in through these and thrown out again by each respiration. They are also peculiar in producing their young after development in the ovary, passing out through a genital pore, and then undergoing a metamorphosis somewhat like the frog, the larvæ having at one time been regarded as a distinct genus.

Though the *lamprey* is viviparous, it constructs nests, the object of which is not clearly understood. When the breeding season arrives they proceed up towards the source

NEST OF THE WHITE-FISH.

NEST OF THE SUN-FISH.

of a stream, sometimes in vast shoals, and at selected spots begin the construction of nests, probably in which to harbor the young while in the larval state. In forming these the male and female unite their labors, and seizing stones by means of their suctorial mouths, convey them to the place desired. Often the sexes will be seen attached to a larger stone than one alone would be able to move, but by joining forces they make their success certain in trans-

5

porting stones of two or more pounds weight. By this means they build up considerable piles which, as before stated, may be to provide a refuge for their young.

APODES—EELS.

From the lamprey it is but a single step to the eel, though the former does not belong to the same order. Eels exist in considerable variety, and are found in both fresh and salt water.

The Common Fresh-Water Eel is the best example of the genus, because while simplest in organization, it is also most common, being found in nearly all the streams of America and Europe. Their form is serpentine, but in all other respects they show little differences from predaceous fishes. They are extremely voracious, preferring game fishes, of which they devour incredible numbers, and hence are a source of great vexation to the owners of ponds stocked with game fish. They also eat crawfish, shrimps and other crustaceans, being very persistent in digging under stones to get at them. Few fishes can swim through the water with such wondrous speed, and none are more daring. They burrow deep in the mud, seeming to be able to dive into the mire with the ease they pass through the water. Their voracity is so great that they are accused of seizing female fishes of considerable size and sucking the spawn from them. *Eels* of the fresh-water variety spawn at the head waters of streams and bring back with them a very large brood, though many of the young are devoured by bass and pickerel. It was formerly supposed that *eels* and lampreys spawn but once in a lifetime, but this belief is no more to be credited than a hundred other idle traditions once current about the animal.

LAMPREY EELS MAKING THEIR NEST.

The Conger Eel, though a salt-water animal, is very like the species just described both in habits and organization, though it grows to be much larger, occasionally reaching a length of six feet. Both species are scaleless, and have the dorsal, caudal and anal fins confluent. The *conger eel*, while much larger, has organs of generation of much greater development than those of the common eel. In the female *conger* there is occasionally such an abnormal supply of eggs that instances have been known of the creature actually bursting from the enlargement of the ovaries.

The **Muræna** is an eel found in the Mediterranean, the flesh of which was at one time so highly esteemed by the Romans that vast aquaria were built for their rearing, and in the markets these eels were sold at so great a price that only the wealthiest could afford such luxury. The rage for

LAMPREY (*Petromyzon americanus*).

its flesh was at one time so great that captive men, women and children were killed and thrown into these aquaria in the belief that such food made the

COMMON EEL (*Anguilla vulgaris*).

Muræna's flesh more delicious. This species is hardly so large as the conger, but its habits are essentially the same. The mouth, however, is armed with a more formidable set of teeth, and the snout sharper.

The **Electric Eel** is peculiar to marsh regions and shallows of South America, where it grows to a length of six feet. In size and feature it very much resembles the conger, though the mouth is not so large nor so well armed, but it has a much greater armament in the powerful electric battery with which nature has endowed it. These electrical organs, while easily located, are no more easily described than is the electric fluid itself. It is sufficient, therefore, to say the creature possesses the power of imparting a shock equal to that of a twelve-jar battery, and may give several discharges before the storage of electricity is exhausted. People about the region in which this animal lives are very

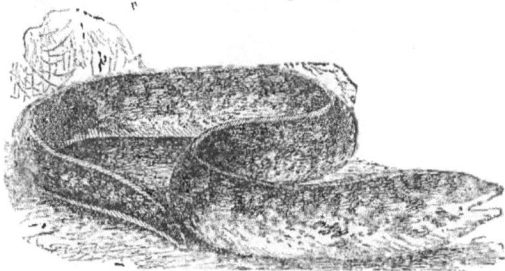

THE MURÆNA (*M. Helena*).

fond of the flesh, and make use of a singular means to effect its capture. Knowing that these eels are most plentiful in the shallow ponds in the neighborhood of salt marshes, the fishers drive into these places a herd of horses, the tramplings of which quickly arouse the eels. As the creatures are combative, they attack the horses by imparting terrific shocks, and as the horses are not permitted to come ashore the eels continue to discharge their batteries until the force is entirely expended. When this is done they are entirely helpless, and may be taken in the hand by the people without danger. This wonderful power the eel uses in capturing its prey, instead of seizing live creatures with the mouth, as do all others of the

ELECTRIC EEL (*Gymnotus electricus*).

eel species. This method of fishing is hard upon the horses, but very profitable to the fishermen. There is a non-electrical eel which is very fond of feeding upon ants, which it captures by laying its tail upon the bank and tempting them to feed upon the slime with which it is covered.

The curious organization of the eel is exceeded by that of the *Blind Fish* found in many caves in America. These fish, recently recognized as a distinct genus, are now classified under the term *amblyopsidæ* (cave fishes). They are curious in two special respects, one being in the absence of the organs of sight, for which they have no use, living, as they do, in caves where no rays of sunlight ever enter. The other curious feature is in the location of the vent, which is beneath the throat instead of toward the tail, as in all other fishes. They are small in size, though well covered with scales, and their organization is well adapted for a hard and precarious life. The young are brought forth in a condition of development fitting them for independent action, though scarcely more than a quarter of an inch in length.

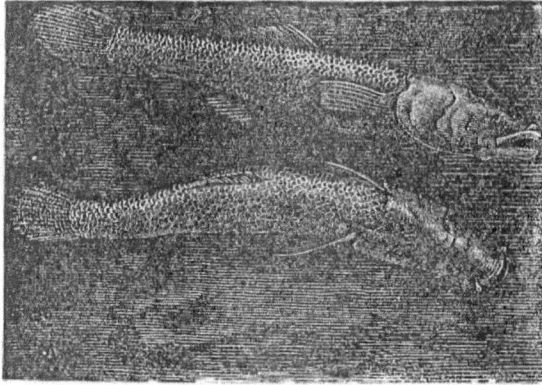

BLIND FISH (*Cyprinodons*).

Mud Whipper (*Cabitis tænia*). This properly belongs to the eel family, though not generally classed with eels. It is a small species, about one foot long, peculiar to the fresh waters of Europe and Asia. The body is decidedly eel-like, though the fins are not confluent. The head also resembles that of an eel, differing only in having eight barbels

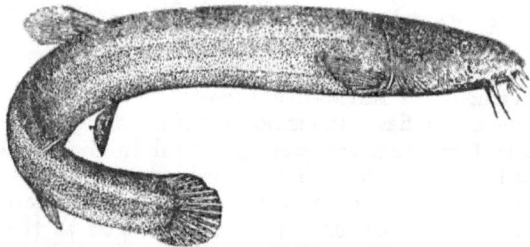

MUD WHIPPER (*Cabitis fossilis*).

depending from the upper and lower lips. The air bladder of this fish is enclosed in a bony capsule, which gives it complete protection against rupture, a provision not seen in other fishes.

FLYING FISH.

The curiosity excited by a consideration of the creatures described in the foregoing account must give way to astonishment as we proceed to a description of species which may leave their element and soar away from the crest of a wave through the air like a bird, dipping its wings in the briny crests and rising again like a wren flitting from bush to bush.

The **Flying Fish** proper form a sub-family, because many species are so nearly allied as to be admitted to the general family, though not able to fly. Those which possess this remarkable power are organized specially, and exhibit.

wonderful design. The body is slender, the bones very light and hollow, like those of birds, and all the fins abnormally developed. The pectorals, however, are of a length almost equal to that of the entire body, and have a very great spread, the web being of a very thin membrane, set on slightly-curved ribs, so that when expanded they assume the shape of an inverted dish, and thus act the part of a parachute. The caudal fin is especially of great muscular strength, the under segment being much longer than the other, by which the fish is able to rise with an immense impulse from the waves. The propulsion thus given is maintained for several minutes by a vigorous sculling movement of the tail, the parachute serving to keep the creature suspended until it drops lightly to the surface, when a beat of the tail imparts a fresh impetus, and it goes forward with increased speed again. The length of the flight may be a thousand feet, but is usually less than half that distance. There are probably a score of species that can rise from the water in this wise, all of which are small in size, the largest not exceeding eighteen inches in length. They are most unfortunate creatures in that they are remorselessly pursued in the sea by coryphenes and dolphins, while if they rise into the air they have no more merciful enemies in several species of sea birds, such as the great gull, frigate bird and sea hawk.

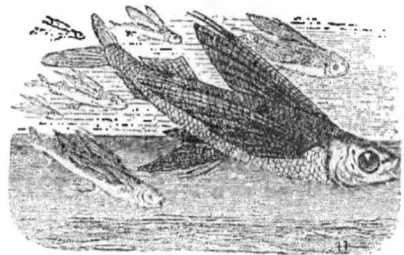

FLYING HERRINGS.

SWALLOW FISH (*Exocœtus volitans.*)

Flying Gunard. This is a species very common along the Atlantic coast, but they are so interesting that the naturalist can never tire watching their curious forms and singular habits, for this is one of the few fishes which can walk, fly or swim. The pectorals, as in other flying fishes, are abnormally developed, and preserve an identical shape. Below the ventrals are six spinous legs, or what may be so called, that are not jointed, but curved so that the points are downward. Upon these the fish creeps along the sea bottom with its head close to the mud searching for prey. The species found along the Atlantic coast is a dusky brown in color, which some persons are imaginative enough to call sapphirine. It rarely exceeds a foot in length, and though classed among the flying fish it can only leap a short distance, nor are its swimming powers specially great, though it is certainly well equipped with propulsive energies. When first taken out of the water it utters a noise somewhat like the low, suppressed croak of a frog. The true *flying gunard* is found in warmer waters, and it is particularly abundant in the Indian Ocean. In this species the head is very blunt and the pectorals spinous, but the habits of all varieties, of which there are twelve, are very similar.

The Lancet Fish, like the sponge, was for a long time unclassified as standing on the border line of the world of mollusks and the world of fishes; it was variously assigned to either. As the starting point of fish life, or the lowest fish form, it has special interest for the student of the development of animal life, and the interrelation of the

FLYING GUNARDS. ANALOGY OF MOVEMENT BETWEEN BIRDS AND FISHES.

various species which go to make up the abundant and variable life in our world. The *lancet fish* is everywhere abundant, and shows the greatest indifference to changes of temperature. Its minuteness, added to its peculiar form, tends to protect it against its enemies, and likewise necessitates the use of the microscope by one who would study its structure or habits. Its respiration is conducted by branchia, which extend through the length of its body. Resembling in appearance the worm, the skin consists of a series of minute scales or plates overlapped. The mouth and vent are both situated on the left side, each somewhat removed from the extremity. The mouth is supplied with numerous tentacles and with a little beak. The food duct is composed of the mouth, stomach and intestine (as pharynx and œsophagus can likewise be distinguished). It has the least possible skeleton, which is cartilaginous; its head is represented in the most rudimentary form by the end of the vertebral axis; its eyes, likewise, are rudimentary, seeming to be mere drops of pigment; its blood is colorless; its heart, only the enlargement of a vein; its eggs invite and reward study, as they seem to offer our nearest approach to protoplasm. The mature *lancet fish* buries itself in the sand, keeping exposed only its mouth and its tentacles. It may be seen springing from its self-selected grave and swimming toward the surface, and again diving into the sand and covering itself up. The young *lancet fish* omit the act of temporary self-interment, and, like children, seem to find satisfaction in a continuous and restless activity.

SEA CAT.

NORTHERN SEA CAT.

The **Slime Fish** (*Myxine glutinosa*) resembles an eel, and since its skeleton is cartilaginous, it is capable of the most unlimited bodily contortions. It exudes through the pores of its sides a jelly-like substance, and from this fact has taken its name. Several species are found in the Pacific, as far north as California. They are parasitic during a portion of their life A common name given them by fishermen is the *borer*, or *sea-hog*. They will attach themselves to the body of other fishes, scrape away the flesh by the use of their teeth, and, having thus boldly made an opening, will then take up their residence in the abdominal cavity of the unwilling and irritated fish. This species is quite numerous at Grand Menan, and less frequent on the Atlantic coast. It will enter the body of a fish which is dead, or which is caught by a hook, and eat out the substance of the fish, leaving only a husk, as the termite ant does with houses and furniture. For the purpose of a scoop, the *myxine* might be a useful instructor for the active and eager reporter for a sensational daily paper. Whether the creature begins its attack upon the fish by boring into it, or whether it enters through the mouth, is a matter which naturalists have not as yet been quite able to fully determine, nor is it certain whether it so burrows to eat or to secure a habitation.

The **Northern Sea-Cat** (*Chimæra monstrosa*) is gluttonously carnivorous, and has been called "The King of the Herrings," partly because of its great destruction of these food-fish, and partly because of a crown-like protuberance between its eyes. Like an unsportsman-like hunter, it is not content with killing the herring as food, but seems to take a malicious pleasure in mangling thousands which it does not attempt to feed upon. It has a long, cone-shaped snout, a long, shark-like body, green eyes, which, in the dark, resemble those of the domestic cat, and two large, wing-shaped fins. Its coloring is silvery, with brown spots, it is three or four feet in length, and its caudal fin is attenuated until it resembles the snapper of a whip-lash.

In Greek mythology the *chimæra* was a three-headed, fire-breathing monster, having the head of a lion, the middle body of a goat, and shaped like a dragon in the tail and posterior limbs. Mythology is generally regarded as a belief encrusted by superstition, and it is more than probable that the early Greek explorers and navigators met with many of the strange inhabitants of the deep, and in their ignorance of natural history, and through their tendency to polytheism, speedily converted a particle of fact into a mass of superstition. In the hands of the poets these superstitions grew beautiful, and have permeated the poetical literature of all peoples; still, none the less, from the standpoint of the naturalist these are foolish tales, exceeded even in marvellousness by the real curiosities of sea and land. In the world of our time it is believed to be more reverential, as well as more profitable, to study the wisdom of the Creator, as infinitely manifested in His works, than to replace this appreciative study by human speculations, no matter how beautiful may seem the creations of our imaginations.

FLYING PEGASUS DRAGON.

The **Angler**, or **Frog Fish** (*Lophius piscatorius*), is about three feet in length and hideous to look upon. It is not quite "all head," but, like a mis-shapen dwarf, its head is larger than its body. Its mouth stretches in width beyond the body, and is capable of taking in an animal of its own size. The mouth is lined upon jaws, tongue, fauces and palate with an armament of movable, hooked teeth, while its nose terminates in a palm-shaped excrescence, whose lustre excels that of the professional inebriate, but which, unlike the provision of man, is used to attract, not repel, the beings that it would "have more acquaintance with." It buries itself in the mud so as to expose only its long feelers, or fish lines, and is very successful in "welcoming" its prey "with hospitable hands to bloody graves."

The **White Shark** (*Carcharinus vulgaris*), frequently called the *ocean tiger*, is a man-eater, and his greed, rapacity, and quickness of movement generally reward him for his sustained pursuit of ships. Its only enemy in the sea is the sperm-whale, which attacks it apparently simply to rid itself of a rival power. The shark, were it not for his unreasonable pugnacity, might easily escape the whale, but it seems to be quite willing to be killed if it can inflict injury upon its adversary. The shark has been made to contribute to commerce its oil and its hide. His stomach seems to be even more curious than the pockets of a small boy, for he will eat even substances that are wholly without nutrition, and which are indigestible. If lacerated by hooks, he will none the less readily and greedily return

THE ANGLER (*Lophius piscatorius*).

to the bait. Shark-hooks are by preference baited with pork, and when the creature has been drawn within reach from the vessel he is harpooned, and his tail and fins withdrawn from the water by means of the hook-line and the harpoon line. He is then relatively helpless, and after a third rope has been passed around his tail, he is easily brought on deck. Having driven a handspike through his throat and cut off his tail, the sailors have succeeded in rendering their dreaded enemy powerless, and certainly are but little inclined to spare him any suffering. A singular device for killing a shark was invented by a negro boy. He heated a brick and threw it to the shark, which, though able to carry about with impunity to its stomach the head and horns of a large goat, could not contend successfully with internal combustion.

WHITE SHARK.

It is said that on one occasion thirty-three natives of Tahiti were wrecked, and undertook to save themselves on a hastily constructed raft. Their weight kept the raft somewhat below the surface of the water, and a school of sharks attacked them and killed thirty out of the thirty-three. Twenty out of twenty-two men from a wrecked vessel fell a prey to sharks, and the heroism of their commander adds to the many courageous acts of which men have proved themselves capable. Although both of his legs were bitten off by the sharks, the commander continued to give his orders for the attempted preservation of his men, and ultimately succeeded in saving two of them.

Shark fishing is a popular amusement at Nassau and along the Floridian coast. A dark night is preferred, and a steel hook having a six-inch curve is fastened to a chain, and this in turn joined to a line generally an inch in diameter. The shark generally makes several investigations before it swallows the bait, retiring to meditate after each new inspection. The moment, however, that he has swallowed the bait he starts a lightning express towards the bottom of the sea. He is now fretted, and checked, and played like a trout or a pickerel. As soon as he appears to be exhausted by his futile efforts, the line is drawn in until the head appears above water, when a noose is thrown over the pectoral fins and the tail, and when the shark is near enough to the boat he undergoes amputation, first of the tail and then of the head. We are told by the naturalist, Figuier, that there are various peoples in Africa who celebrate "The Festival of the Shark." Three or four times a year the natives row first to the middle of the river, and with odd ceremonials invoke the aid of the shark. Next they put before the sharks what might figuratively be called burnt-offerings of goats and birds. Next, an infant, consecrated at its birth and fattened during ten years for the enjoyment of the shark, is bound to a post on the sand below low-water mark.

GREAT PILGRIM SHARK.

Like Iphigenia at Aulis, the child is regarded as a votive offering, and like Andromeda, the daughter of Cepheus and Cassiopea, the child must await the attack of the sharks, though, unlike Andromeda, it is not to be rescued by any Perseus.

The **Great Pilgrim Shark** is not a man-eater; in fact, it is the most harmless and amiable of the shark family. It frequently attains the length of thirty-five feet and the weight of two tons.

The **Nurse Shark** is frequently harpooned by boys in order that it may be induced to drag them in triumph over the water. The devotion of the *pilot fish* to the shark has been the subject of many a story, and as it will not desert the shark it is sometimes netted while swimming about its master.

Whether the pilot fish really acts as a guide for the shark is disputed, but it is asserted that pilot fish have been known to find the bait, and without touching go in search of its voracious companion. Moreover, cases are mentioned where the pilot fish repeatedly steered the shark away from a baited hook. The pilot fish resembles the mackerel; its silvery gray color being similarly relieved by five dark-blue transverse bands.

The Porbeagle, or Mackerel Shark, (*Lamna cornubica*) is a great annoyance to mackerel fishermen, as it breaks their lines and carries off both bait and hooks.

The Basking Shark (*Cetorhinus maximus*) is frequently seen in schools, whose backs rise above the surface, and whose great length makes them appear formidable, although they do not voluntarily attack men, and even when irritated only occasion damage, as the whale does, by powerful blows struck by their tails.

The Thresher Shark (*Alopias vulpes*) is specially remarkable for its long tail, which frequently is nearly as long as the body. It is bluish in color and has a white belly. It is a great destroyer of herring and mackerel, and while feeding upon its prey lashes or threshes the water with its tail, apparently with the object of keeping the fish herded.

The Smooth Hound, Sea Hound, or Dog Shark (*Mustelis canis*) is sometimes called the *hammerhead*, but is only a few feet in length, and must not be confounded with the true hammerhead (*Zygæna malleus*); is harmless, feeds upon mollusks, and has received its popular name from its outward re-

BASKING SHARK.

semblance to the true dog-fish. The true hammerhead is supposed to be a man-eater, though finding parts of a human body in the stomach of one specimen does not prove absolutely that it killed the man. It reaches a length of twenty-five feet, and its head consists of a double hammer-like projection, each extremity of which is surmounted by an eye.

The Great Blue Shark (*Carcharinus glaucus*). It is found in every sea, is blue above and white below, and reaches the length of more than twenty feet.

The Swell Shark (*Scyllium ventricosum*), is so called not because of his apparel, but because he will, when caught, swallow air until swollen out of all proportion. It is found on the Pacific coast of Mexico—a land abounding in curiosities, as well as rich in products valued by mankind.

The **Cat Shark** (*Scyllium catulus*), and the **Dog Shark** (*Scyllium cani-cula*), are distinguished alike by their brown-spotted, red color, their destruc-tion of herring, their possible use as food, and because supplying the shagreen of commerce.

The **Tiger Shark**, or **Zebra Shark** (*Stegostoma tigrinum*), is an inhabitant of the Indian Ocean, and takes its name from the markings of its skin; a brownish-yellow ground is traversed by black and brown bands, or marked by dark brown spots.

The **Angel Shark**, An-gel Fish, or Monk Fish (*Rhina squatina*), is a species of ground shark found in nearly all seas. It has been known to produce twenty young at a birth, so that its voracity is a serious matter for the fish upon which it feeds. Its only claim to the name, *angel shark*, or *angel fish*, seems to arise from its wing-like pectoral fins, unless its dia-bolical aspect may have suggested some fallen angel omitted from Milton's account of the Satanic Hosts. Its smooth, round head evidently suggested to some fisherman the name of *monk fish*. Its size, rapacity and carnivorousness make it a constant terror

HAMMER-HEAD SHARK.

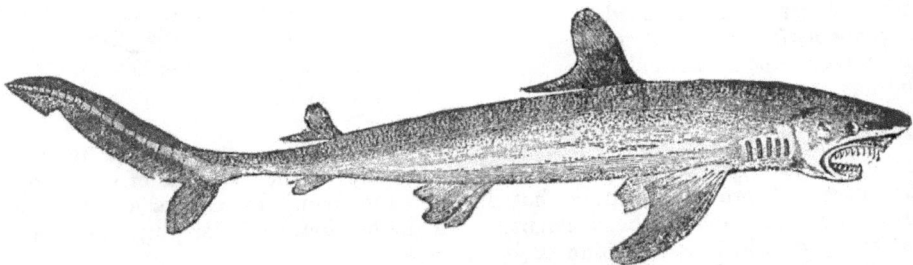

GREAT BLUE SHARK (*Carcharinus glaucus*).

to fishes, even though not dangerous to man. It is several feet in length, and frequently weighs upward of a hundred pounds. It lives on the muddy bottom, burrowing its way through the slime and weeds, and quickly seizing upon the fish which it thus disturbs. Fishermen make war upon it, not because it is useful after capture, but because it destroys their "happy hunting grounds."

ADVENTURE WITH SHARKS.

A whaling vessel having been destroyed by fire at sea, its crew was ultimately saved by a shark. After drifting about for five days, during which the rations gave out, and several of the crew died from exposure and the delirium produced by drinking sea water, a shark rose to attack the body of the latest dead man. The captain promptly harpooned the shark, and while its strength lasted, used it as a substitute for the rowing of the crew, and after its draft powers were exhausted the shark was made to furnish edibles and drinkables for the crew. But it most frequently happens that persons are *saved for the shark* instead of by a shark. Many are the incidents which may be mentioned in illustration of the shark's inhumanity to man. Within the last three years the newspapers have recited the horrible fate of Captain Mark Robinson. His vessel was capsized, and within a few moments the unfortunate officer saw his wife and child killed by sharks, and had himself lost both of his legs, bitten off one at a time. One of the crew escaped with the loss of only one arm, and another, more fortunate, lost no limb, but simply baited the shark with flesh, whose loss, while painful, was not fatal. A shark bit in half a fifteen-year-old bather at Ceylon, and carried off the lower extremities of the unfortunate youth who was so suddenly called away by so

THE MONK, OR ANGEL FISH.

horrible a death. The pearl divers, as they do not wear the armor of the ordinary divers, are constantly exposed to sudden and terrible attacks from sharks. A diver not reappearing when expected was sought out by an American, who, however, wore armor. The American found that the other diver had been rendered unconscious by a blow from the shark's tail, and that the shark was just returning to begin its feast. He plunged his blade into the body of the creature, and fortunately with fatal effect. The first diver was resuscitated, but among his recollections none ever proved so thrilling as the adventure just recited.

Shark stories are quite numerous, but possibly no more so than the frequency of the adventures in which they play a leading part. Quite frequently the sailor, tempted to take a swim, and remaining within what seems easy reach of the vessel, is bitten by a shark before his companions can pull him from the water. To render death terrible to the slaves who composed his cargo, the captain of an African slaver hung a number of corpses to ropes and merely dipped them for a moment in the sea. In an instant the sharks had bitten off all but the feet themselves. There is a well-authenticated story of a shark jumping out of water, again and again, in its efforts to capture a negro woman who had been hung to the yard-arm. The amphibious efforts of the shark not only inspired the terror and horror for which it had been intended, but was finally rewarded by securing the larger part of the culprit negro. The African slavers speedily became sources of supply to the sharks, for "The spirit of gain, in the spirit of Cain," naturally led to frequent loss of the stolen savages packed together like sardines, insufficiently fed and cared for, and like the sardine, frequently fried in their own oil. But unfortunately to the slaver, as well as to Hood's seamstress, it was pitiful "that bread should be so dear, and human life so cheap," and the sick, the maimed, the weak, were thrown into the sea with no greater reluctance than is shown by the angler in restoring to the water the fish which

BOY FATALLY BITTEN BY A SHARK.

he cares not to keep. While the shark declines no kind of food, it is evident that this want of delicacy arises from necessity, not choice, for, like the King of the Cannibal Islands, he is very fastidious, and if at liberty to pick his man, takes by preference men in the order of color, preferring the darker shades as promising greater lusciousness, and less of the leanness and muscular stringiness of the sea-faring white man. Still, although the shark draws the color line we should not recommend any visiting white person to trust altogether to the shark's being true to this preference, for like men, they sometimes forget their principles when too sorely tempted.

The Chinese, whose densely populated country has forced them to the most untiring study of domestic economy, use the fins of sharks for the manufacture of gelatine; at one fishing station some forty thousand sharks were in a single year thus converted to the use of man. Until the invention of sandpaper, the skin of the shark, under the name of *shagreen*, was extensively used.

The **Cow Shark** (*Heptanchus indicus*) is found from California to the Cape of Good Hope.

The **Eel Shark** (*Chlamydoselachus anguineus*) is found in Japanese waters, and though but a few inches in circumference, attains the length of six feet. It is sufficiently like a small specimen of the sea-serpent to have interest in connection with what has been submitted in regard to the latter creature.

The **Spiny Shark** (*Echinorhinus spinosus*) is a ground shark whose skin is covered by tubercles, whose thorn-like prickles leave a mark when handled.

The **Sleeper Shark** (*Somniosus microcephalus*) is somewhat rare, and most frequently found in the Arctic region. It is a long, heavy creature, grayish purple with white spottings. It is not a man-eater, but is a deadly enemy of the whale, upon which it preys.

Shark fishing has been described, but it may be as well to add that the negroes, in their less highly civilized state, are compelled to exhibit greater daring than the white man. Arming themselves with a long, keen knife, negroes will boldly dive into the water and swim to meet the shark, knowing that the creature, feeling secure of its prey, will await their approach. As the shark turns upon its side to devour its victim, the negro plunges his knife into the monster's belly until he who came to eat remains to be eaten.

The **Remora** is believed by credulous sailors to attend the shark as a pilot, but naturalists have ascertained that it is a more fearful enemy to the shark than man himself

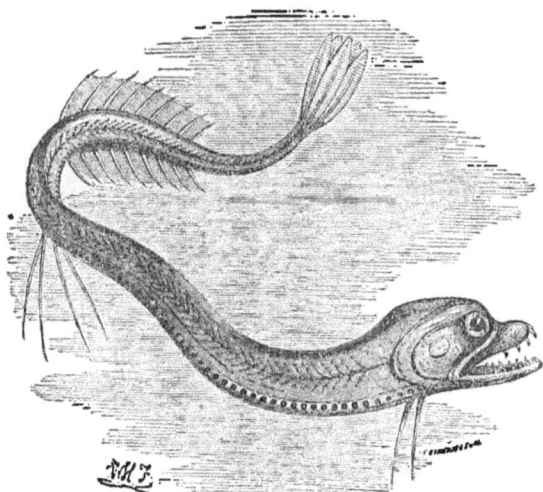

EEL SHARK, OR STOMIAS BOA.

can be. This herring-like creature is about a foot and a half in length, and suffers from constitutional weakness of the legs or fins; it is thus naturally inclined to attach itself to some other creature or object from which it requires no reciprocity of affection. On top of the head of the *remora* is an oval space, cut up into numerous partitions by many small teeth or bony filaments; these cavities, therefore, have the power of discs, or suckers, and when fastened to an object are but little likely to be removed. The shark is rarely captured without one or more of this fish being found fastened to it—a self-invited and evidently unwelcome guest. It is supposed that the *remora's* preference for the companionship of the shark arises from the fact that its presence answers as a protection, while the monster's gluttonous feasts enable the *remora* to live a life of luxurious ease upon the scraps which fall from the shark's table. This arrangement is delightful for the *remora*, but somewhat irritating to the shark, who, however, evidently submits to the inevitable. The *remora* takes its name (*echineis remora*) from the ancient superstition that it attached itself to vessels and prevented their progress through the waves. It was currently believed, for instance, that Mark Antony's vessel at the time of his defeat at sea was held fast in one position by *remoras*, which wantonly attached themselves to his keel. While these mythological stories have for the scientists only the value of stimulating to investigation, and to the discovery of more wholly natural explanations, they form so large and so beautiful a contribution to the resources of the writer and speaker that they have an office in a popular work upon natural history so long as the line is drawn between the facts of science and poetical fictions.

The **Dog Fish** is numerous alike in species and in its communities. Like the shark, it lays its eggs, attaching them to the weeds near the land by means of tendril-like projections. The *large spotted dog-fish* is common in English waters, where it is called the *bounce*; a smaller species infests the coasts of Scotland. It is abundant on the fishing banks of New York and Massachusetts, as amateur fishermen have occasion to remember, for many an anticipated catch of mackerel has been blighted by the appearance of a school of dog-fish. There are well-authenticated instances of man-eating by schools of dog-fish, where, as in one case, the victim was indulging in a salt-water bath, and in another, when a child fell from the deck of a ship. The famous Captain Paul Boynton, while swimming in the Straits of

REMORA (*Echineis neucrates*).

Messina, was pursued and finally attacked by an immense dog-fish, but by the vigorous use of his knife, the captain was able to drive the creature away. On another occasion, two boys were regularly besieged by a school of dog-fish. While amusing themselves upon a half-sunken vessel, their own skiff floated away and left them without the means of regulating the length of their stay. Presently they become aware of the presence of a school of dog-fish, which for several hours made active war upon them. The fish would jump high into the air in their attempts to reach the human prey. The poor boys grew more and more frightened at their novel and perilous situation, and during the time which elapsed before they were discovered and released by a steamer they underwent every variety of thrilling and dreadful foreboding.

The **Rays** (*Raiæ*) form a connecting link between the shark and the skate. They differ from the shark chiefly in having ventral instead of lateral gill-clefts. With its long, flat body the *ray* covers its prey, and while thus holding it down, uses its singular ventral jaws for the conversion of its prey into food. Some species are oviparous, and others bring forth their young alive; the eggs, which may frequently be found on the beach, are called *sea-purses.* During its youth the *ray* resembles the shark, but as it matures it develops disproportionately its pectoral fins and changes to a *rayed* form. It has no distinct head; the tail is long and slender; it has two dorsal fins, and sometimes a caudal one. The great size of the pectoral fins is responsible for the singular form of the fish; its powerful jaws are filled with teeth which

point back-
wards, and
they can be
protruded at
pleasure.

The Bor-
dered Ray,
the Raia Ba-
tis (notably),
the Thorn-
back, and
the Clear-
nosed Ray,
are not un-
frequently to
be found in
our markets.

The Sea
Porcupine
swells itself
up with air
until it be-
comes tem-
porarily
powerless,
and floats
around, belly
upward, un-
til it is final-
ly successful
in expelling
its incon-
venient sup-
ply.

DOG FISH (*Squalus acanthias*) AND SEA HOUND (*Mustelus canis*).

The **Balloon Fish** is frequently found on the coast of the United States. It is supplied with long, sharp spines which inflict the most poisonous wounds.

The **Puffer**, or **Swell Fish**, is likewise a source of danger to fishermen, who frequently catch them with the hook, or find them entrapped in their nets.

The **Electric Ray** (*Torpedo vulgaris*) weighs upwards of a hundred

6

pounds, and stores its batteries between its gills and the pectoral fins. It has entire control of the exercise and severity of this electric shock. It is sufficiently strong to magnetize needles, decompose certain chemical substances, and even to produce the electrical spark of the Leyden jar, requiring similarly the completion of the circuit. The positive pole of this natural battery is dorsal, and the negative pole is ventral. The *Narcacian* species is found on the Atlantic and the Pacific coasts of our country, and formerly was a tenant of European waters. Whether it has immigrated, or whether the European species has become extinct, cannot safely be stated, though the question has often been discussed.

THE STING RAY.

The **Saw Fish** (*Pristidæ*) have a shark-like body and a protruding snout, whose stout teeth make it resemble a saw. Nor does the resemblance stop here, for the *saw-fish* uses this feature as a weapon, which, reaching the length of six feet and the breadth of a foot, can do terrible execution. The species illustrated, *pristis antiquarum*, is found in the tropics, and makes war upon whales, herring and mackerel, whose flesh it saws out, since its teeth do not enable it to bite.

Its attacks upon vessels are due not so much to its envy of the work of man as to a frenzy into which it is sometimes thrown by a parasitical crustacean, which, seeking its own ends, rather than being considerate of the *saw-fish*, burrows into its flesh. Some Spanish fishermen, having unconsciously entangled a *saw-fish* in their nets, would have been dragged out to sea but for reinforcements from a man-of-war. Having been brought to the surface of the water by the united efforts of five persons, it happened to strike out towards the land, and, while thus engaged, a running bowline was thrown over its nose. Even then, with the aid of a tree as a capstan, thirty-five men failed to drag it

BALLOON FISH.

ashore, and it was not until two hundred human hands pulled on the rope that it could be landed. While in the water its lashings were awe-inspiring, and even when upon the dry land it held the whole crowd at bay until a crafty Span-iard, mounting its back, cut through the joint of the tail with the same effect as though one broke the neck of a human being. It was found to weigh 12,500 pounds (if Captain Single-ton is to be believed), to be twenty-two feet long, and eight feet in width. Its saw, or sword, was at least five or six feet in length, and but for the fact that its entanglement in the net had thrown the animal upon its back, and prevented its use of this weapon, instead of having been captured, it would have proved to be the captor. A friend of the au-thor was but a few years

SAW FISH.

ago the hero in an exciting adventure with a *saw-fish* in Floridian waters. Together with a daughter (a mere child), he was fishing for sea bass, when, immediately after get-ting a bite, he found his line playing out with the most unexam-pled velocity. He con-tinued making additions to its length un-til, after the fish had taken nearly four hun-dred feet of line, it sud-denly slackened the ten-sion, turned, and began swimming toward the boat, the gentleman meantime hauling in the slack and wonder-ing what new creature had eaten the lunch provided for the sea bass. When within six or eight feet, the crea-

ADVENTURE WITH A SAW FISH.

ture pushed its saw above the surface, and continued to strike it from side to side until it reached the row-boat and staved in the stern. Standing toward the prow, to prevent the sinking of his little boat, the doctor plunged his oar into the mouth of the infuriated creature, and presently succeeded in driving it away. The internal commotion which had been excited evidently made the fish lose its bearings for, while none the less ferocious, it failed to again hit the boat. The fisherman threw a noose over the saw and pulled most manfully for the shore, and when he reached it found his exhaustion much greater than that of the *saw-fish*. It required seven persons to drag it ashore, measured sixteen feet, and had a four-foot saw armed with twenty-two dagger-like teeth, which had not suffered from their exercise upon the boat.

The **Skate** is a broad, flat ground fish, which is caught more frequently than is desirable for the pleasure or profit of the fisherman.

The **Tobacco-Box Skate** (*Raia erinacea*) is a very common fish in the Eastern Atlantic.

The **Brier Skate** (*Raia eglanteria*) is distinguished by the possession of spines.

THORNBACK SKATE (*Raia clavata*).

The **Smooth Skate,** or **Barn-Door Skate** (*Raia levis*), is a sort of compromise between the two species last named, for when young it is spinous, but grows bald with increasing age.

The **Raiadæ** are eaten by Europeans, but our wealth of natural provisions has led Americans to prefer fish of greater delicacy.

The **Sting Rays** (*Trygonidæ*) include about fifty species, of which we illustrate *Trygon pastinaca*. The whole family bear at the base of the tail spines which can do very serious damage. These spines replace the dorsal fin, and attain a length of nine inches. As the teeth of the frontmost spine wear out it is succeeded by the spine immediately behind, and so on until the series has been completed. It is

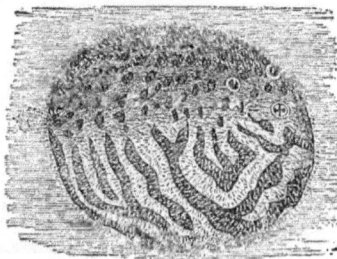

GLOBE FISH.

found in shallow water which flows over a sandy bottom, and a wound from its spine has all the effect of a bite from a venomous serpent.

The **Whip Ray** wields its tail like a horsewhip, and its sharp spines are capable of inflicting very painful wounds. The ray fish produces oblong eggs, which are provided with ribbons, by means of which they are fastened to plants, or rocks.

The **Globe Fish** is curious alike from its form and its coloring. Its egg-shaped body has three dots which represent two eyes and a mouth; a queer little tail crops out from half way up one side.

The **Sturgeon** is frequently found in European waters, is of the length of ten or twelve feet, and the weight of several hundred pounds. Its flesh resembles veal, and is

COMMON STURGEON, OR STERLET (*Acipenser sturio*).

esteemed " a dainty dish to set before a king," if one is to judge from the sumptuary legislation of King Henry, of England, who forbade all but royalty to feast upon this fish.

The species of sturgeon known as *Acipenser* has reached the weight of several thousand pounds, and is caught in the Caspian and Black Sea, and in Russia, for the sake of the isinglass to be obtained from its air bladder.

GAR PIKE (*Lepidosteus osseus*).

The Sterlet (*Acipenser ruthenus*) is useful to commerce by means of its roe, which appears upon the table in the form of what Shakespeare called the " caviare to the general." Bearing in mind the great weight attained by the sturgeon, the fact that the roe forms one-third of this, and that hundreds of thousands are annually caught, one will realize the sturgeon's contribution to the support of mankind, whether by furnishing an article of diet, or by providing employment for fishermen, merchantmen, and all engaged in transportation, commerce, or the selling of provisions and ship stores. The species found in the Hudson River is used by the poorer classes as food, and its coarse flesh is some-

SEA-BOY (*Julis vulgaris*).

times called Albany beef, but it can never be regarded as a palatable dish.

The **Gar Pikes** (*Lepidostidæ*) are, with the exception of the American species, fossil forms. They are clothed in a tough, scaly armor; have a beak-shaped mouth, and illustrate the more rudimentary forms of lungs and of gill breathing.

The common **Gar Pike**, or **Alligator Gar**, (*Lepidosteus tristæchus*), is a familiar sight in Southern waters. Against it boys wage the most incessant warfare, for it interferes alike with their bathing privileges, and with their fishing. The *gar-pike* must be looked at with the eyes of the naturalist, or else it will fail to be suspected of the mechanical beauty of which it is certainly possessed.

The **Pelican Fish** (*Eurypharynx pelecanoides*) is a species discovered only seven years ago. It is a deep-sea fish, of which only a few specimens have been caught. Its jaws are five or six times the length of its cranium, and the lower is provided with a pouch after the manner of the pelican. Its body tapers off to a tadpole-like tail, and is spined its whole length. It is a most repulsive-looking object, but holds high rank among the curiosities of fish life.

The **Catfishes** are widely distributed, and everywhere a familiar object to the fisherman. The maxillary bone forms the support for the fleshy barbel which is so distinctive a feature of the *catfish*, and which so resembles the whiskers of a cat as to have given the fish its popular name. The spines of the pectoral fins, or horns, as they are popularly named, are among the painful recollections of our boyhood's life. The effects of wounds made by the *catfish*'s spines are exceedingly painful, and not infrequently dangerous. The fishers at the

OCEAN SCORPION, OR FATHER LASHER (*Cottus bubalis*).

Southern watering places are not seldom in need of prompt attention from the physician. The fish does not poison the flesh, but makes a jagged cut which is hard to heal. The flesh of the smaller river species is found very palatable by many persons, but the large-sized *catfish* are eaten only by those who do not object to coarse food. As a study for those interested in the evolution of animal life, the *catfish* family has special interest as showing its derivation from the sturgeon.

The catfish's belly softens when the eggs have been laid, and by simply lying upon the eggs she presses them into her skin and carries them about with her. Contrary to the usual custom of fish life, the female

SLY SILURUS, OR SHEAT FISH (*S. Glanis*).

is the one which assumes the task of a nurse. There is an Indian catfish which has a sucking disc between its pectoral fins, and which it uses in resisting the strong and impetuous currents. One species, being found in the midst of matter thrown out by volcanoes, was supposed by Humboldt to live in subterranean streams, and to be unaffected by their high temperature—in short, to be like the fabled Salamander. Later naturalists, however, do not accept this explanation, but suppose that, having been poisoned by volcanic gas, they are swept along by the volcanic torrent.

The **Eel Pout** (*Clarias anguillaris*) is a singular embodiment of the catfish structure and the appearance of the eel.

The **Wels** or **Sheat Fish** (*Silurus glanis*), is one of the largest of European fishes, frequently weighing four hundred pounds. It is a great, lazy gormandizer, lying in the mud and waiting until some victim swims into contact with its immense barbels. It is not an epicure, but a glutton, eating all fish but the perch; destroying the water-dwelling birds, and not at all objecting to human flesh. Many stories of this man-eater are told, and though readily believed by the Turks and Hungarians, are considered by naturalists as not well authenticated. The dark green color of the upper portion of its body, its yellow abdomen, and the yellow and blue tints of its fins, make it seem to be clad in the gorgeousness of the mediæval knight whose armor reflected in many a color the light which glanced upon it. The North American species of catfish is smooth and naked, its thin-cut lips adorned with eight barbels; fins short, but in the case of the dorsal, as well as that of the pectoral fin, sharply spined.

The **Stone Cat** (*Noturus*) is the brook cat so common in the South and West.

The **Blind Cat** (*Gronias nigrilabris*) is a cave-dweller in Pennsylvania, and is supposed to be a recent variation.

The **Bull Head,** or **Horned Pout** (*Amiurus*), is sluggish, loves to grovel in the mud, will bite at any kind of bait, and exhibits the greatest indifference w h e n hooked.

The **Mississippi Cat** (*Amiurus ponderosus*) frequently weighs two hundred pounds; and the **Lake Cat** (*Amiurus nigricans*) is also a weighty member of the fish community.

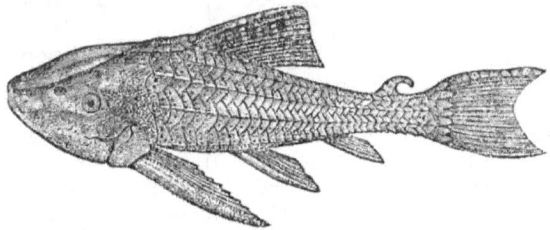

WELS, IN COAT OF MAIL (*Hypostomeus etculoculalus*).

The **Channel Cat** (*Ictalarus punctatus*) is a very excellent article of food. The male carries the eggs in his mouth until the young are born, thus always giving evidence that the father is a good provider.

The **Doradinæ** are inhabitants of South American fresh-water streams, and are very curious because they belong to the nest-building fishes, and in seasons of drouth travel in great numbers in search of water. The procession is a nocturnal one, and frequently occupies several nights. Its carpal, or wrist bones, are lengthened, and support the short, stiff, claw-like pectoral fins, and it is thus able to walk over the land.

The **Electric Catfish** (*Malapterurus electricus*) is one of the curiosities of the Nile. Its battery extends over its whole body and can magnetize an iron rod. An unsophisticated fisherman, having stuck his knife-blade into an *electric cat*, was surprised by receiving a very severe shock.

The **Trichomyteridæ** are found at great altitudes upon the Andes, and some species take up their homes in the gill cavities of larger species.

The **Hypophthalmidæ** have their eyes placed below the level of the mouth.

The **Seyphophori,** or **Cup-bearers,** are an important food fish to the dwellers in tropical Africa. They possess two rudimentary electric organs, placed one on each side of the tail.

The **Gymnarchus Niloticus** attains as much as six feet in length, and has a singular air-bladder, supposed to represent an imperfect lung. The Egyptians held in veneration the *seyphophori*, because they believed it to be one of the three fishes which, having each devoured a portion of the body of Osiris, defeated Isis in her attempt to gather together the scattered limbs of her husband. Osiris, having been slain by Typhon, was put into a chest and carried by the Nile out into the sea, where the body was mutilated by fishes. The soul of Osiris descended to the infernal regions, and continued its existence under the name of Serapis.

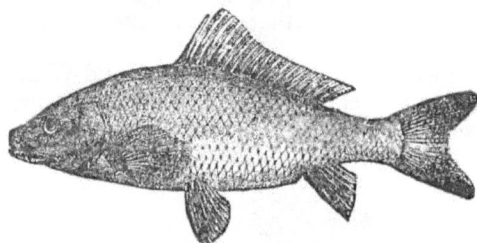

The **Carp** (*Cyprinus carpio*) has been made celebrated by writers, inasmuch as it is a common and valued fish in Europe—especially in England. It varies in size from a foot to upward of three feet, and in weight has been known to reach one hundred pounds. It is noticeable likewise for its length of life, there being a well-authenticated story of its having reached the age of a century and a half. It is cultivated in Germany, Great Britain and America, and is probably least popular in our own country. It is of an olive brown, with a white belly. It should be eaten in the fall and early spring. It loves clear water, and as it grows very tame it is a great source of amusement to children. It seems to be true that frogs are destructive of the carp, upon whose head they will ensconce themselves, and like "the old man of the mountain," ride them to death. The gold and the silver carp are familiar inhabitants of our aquariums. The orange-red color, variegated by black and brown, and the gleam of the scales make this carp quite a favorite pet. The fish was originally imported from China.

CARP (*Cyprinus carpio*).

The **Piraya** (*Serrosalmo piraya*), while small in size, is beyond compare the most voracious fish that swims in any waters. It is very numerous in the rivers of Brazil and Guiana, where vast shoals troop up and down the fresh-water courses like so many wolves ready to attack and devour every living thing. The front teeth are very sharp and set close, so that, as the jaws are powerful, it can bite out a piece of flesh with all the cleverness that might be shown by a knife. Authorities tell us that oxen are sometimes set upon and killed by these fishes before the animal can ford a stream of ten yards breadth. Travellers also claim that certain of the South American tribes place their dead in streams where the *piraya* are plentiful, in order that the flesh may be eaten from the corpse, leaving only the skeleton for sepulture. No one may venture in the waters where these voracious creatures abound, for no noise will serve to frighten them from attacking a human. The teeth are sometimes used to point and also sharpen arrows.

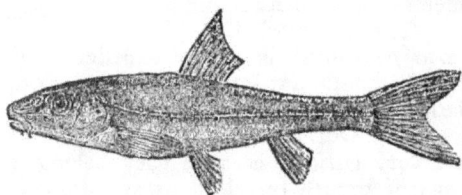

GUDGEON (*Gobio fluviatilis*).

The **Gobio**, or **Gudgeon**, is so easily caught as to have become a synonym for stupidity. It is a sightly fish, and has much to defend it against its evil reputation.

The **Sea Barbel** (so named from the wattles about its mouth), has two pair of mustache-like projections on its mouth.

The **Father Lasher**, or **Lucky Proach** (*Cottus bubalis*), will, when irritated, erect its spines like quills upon the fretful porcupine.

The **Suckers** (*Cato stomidæ*) are many in species, though possibly the buffalo-fish and the *carp sucker* are sufficiently well known to represent the family.

The **Anchovy** (*Engraulus*) may

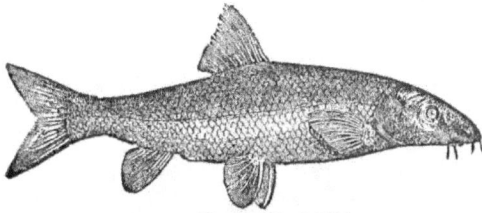

BARBEL (*Barbus fluviatilis*).

ANCHOVY.

be said to support, by itself, an industry. The Eastern coast abounds in these little fish, which are pickled and distributed throughout the country.

The **Sailor Sword-fish** can furl and unfurl its dorsal fin as though opening and shutting a fan. As those found in Ceylon are twenty feet long, it is easy to imagine the spectacle presented by these fish as they sail near the surface of the water. Its flesh is well-flavored and nourishing, and an excellent article of leather is made from its skin.

The **Sword-Fish** (*Xiphias gladius*) varies in length from ten to twenty feet, and its upper jaw is prolonged into an immense sword-like weapon. No fish is safe against its attack, and it has many a time been known to drive its sword through the massive timbers of large ships. Its flesh, when pickled, is esteemed by some persons, whose number is sufficient to convert the *sword-fish* into an article of foreign and domestic commerce, and the annual catch is valued at as much as fifty thousand dollars. The Mediterranean and New England coasts both abound in *sword-fish*, whose capture is the source of a lucrative industry. An old fisherman insisted that on one occa-

SWORD FISH.

sion, while all unthoughtful of any *sword-fish*, his dory was pierced by one of them, and the sword penetrated his trouser-leg as far up as the knee. The *sword-fish* is caught by harpooning, and furnishes all the variety of excitement to be found in the whale fishery. The *sword-fish* has an ancestry that would put to the blush the claims of a few generations. The merchantman Dreadnaught was sunk by a leak produced by a blow from the sword of a fish. In the case of a whaler it was found that the fish had pierced through the copper sheathing, four inches of planking, next through a foot of solid white-oak timber. The sword is now on exhibition at the British Museum. Three divers had a somewhat startling adventure, being attacked while under water by an immense *sword-fish*, which repeatedly knocked them down, but fortunately failed to penetrate their armor, and was finally driven away by a knife-thrust, given by one of the party.

The **Herring** (*Clupea harengus*) is abundant in American waters, which in the spawning season they irradiate with the silver sheen of a countless procession. Yarmouth, so well known to the readers of David Copperfield, is the headquarters of the herring fishery in England, and in good seasons the catch is as great as four hundred millions. The New England coast is the American centre, as those know who have been to such fishing towns as Gloucester

SAILOR SWORD FISH (*Histiophorus gladius*).

and Marblehead. The depredations of the oil factories have rendered the term "moss-bunker" a term of reproach among Eastern fishermen for the vessels, and considering "all as fish that comes into their net," have greatly injured the slender rewards of the fisherman's hard life. The yearly yield of oil is now greater in quantity and value than the whole fishery.

The **Sprat** belongs to the herring tribe, but is not so well flavored. It is, however, largely used as food, and on the coasts of Europe furnishes employment for hundreds of fishing smacks.

The **Smelt** (*Argentinidæ*) have a well recognized value as food, being rich and delicate.

The **Columbia River Smelt** is so fat that it is not uncommon to run a wick through him and use him for illuminating purposes. But in addition to this oleaginous provision, the flesh is specially eatable as pan-fish.

The **White Fish** (*Coregonus albus*) is considered a great table delicacy. It attains the weight of twenty pounds, but usually does not exceed three. It is not uncommon in the New York markets, as it abounds in Lakes Huron and Erie.

The **Grayling** (*Thymallus*) is purplish-gray, its dorsal fin banded with purple and green, and colored with roseate spots. Its beauty would seem to entitle it to the compliment paid it by one of the early Church Fathers when he termed it "the flower of fishes." Its value as an edible is quite great, so that it is useful as well as ornamental.

Every one must have heard of salmon fishing on the Pacific coast. Five distinct species have been described by the United States Fish Commission. The *Quinnat*, or *King Salmon* (Columbia River) generally weighs about twenty-two pounds, but Daddy Lamberts have been found weighing one hundred pounds. The *Blue-Black Salmon* (Frazer River) weighs only seven or eight pounds. The *Sil-*

ATTACKED BY SWORD FISH.

ver Salmon (Puget Sound) has about the same weight. The *Dog Salmon* attains a weight of twelve pounds, and the *Hump-backed Salmon* some six pounds. These species of salmon enter the rivers for the sake of spawning. Salmon-canning has developed into a giant industry, and there are but few eaters of fish who have not reason to be grateful for this addition to his Lenten fare. In a single year the Columbia River yielded nearly twenty-six million pounds of marketable salmon.

The **Rainbow Trout** (*Salmo iridens*) has numerous and large scales, is of a bluish color, silver, with red bands and red spots on the sides.

The **Salma fontinellis**, or **Brook Trout**, is suggestive to the sportsman of the most varied and exciting pleasure. Whether it be in his moments of patient waiting that he enjoys the scenery which surrounds the haunts of the hunt; whether it be the early hours and the health-giving tramp, or whether it be the excitement which this game-fish, when hooked, is able to afford in unstinted profusion, its name always awakens recollections or excites desires which give it a chief place in his esteem. Words fail to tell what its charms are for the gourmand, for no one but the gourmand himself can adequately describe the pleasures of the palate. It is not found south of Virginia, although no one has explained why our Southern brethren should be deprived of the sport and delicate flavor furnished by this famed son of the god of the waters. The *brook-trout* has many names, generally arising from his appearance, which, in turn, seems to vary in color and somewhat in form with the latitude where it is found.

SPRAT (*Clupea sprattus*) AND HERRING (*C. harengus*).

The **Common Pike** (*Esox lucius*) is a very active, powerful, rapid swimmer, as its fame would lead one to suspect. In the spring (March) it deposits its spawn in creeks, which empty into larger streams, though it is found in many lakes. It is remarkable for its voracity, ferociousness and longevity. One is said to have had an engraved ring put around it, and to have been found alive more than two hundred and fifty years afterwards. The *pike* is regarded as a well-flavored fish, and while commonly reaching fifty pounds in weight, has been known to exceed three hundred pounds.

The **American Pike**, or **Pickerel** (*Esox reticulatus*), is from a foot to three feet in length, and is common throughout North America.

EUROPEAN WHITE FISH (*Abramis brama*) AND GRAYLING (*A. vimbra*).

The **Muscalonge** (*Esox estor*) is not only the largest, but the most highly-prized member of the family. It is abundant in the great lakes. The strength of this fish may be illustrated by the story of the pike, which, though upon

PIKE SEIZING A BIRD.

land, was found securely holding a fox by the nose. They have also been known to rise from the water with a bound and seize a bird that had perched upon a limb projecting over and near the surface of a stream. The pike is so voracious and so pugnacious that it is called the fresh-water shark.

Saw Perch (*Serranus cabrilla*) is a Mediterranean species having a spinous dorsal fin, which is composed of from nine to twelve spines united.

The **Sea Devil**, (*Malthe vespertilio*), or **Bat-fish**, is a creature somewhat resembling the octopus, though its eight flat and broad arms are not provided with suckers. The illustration on page 95 gives a better idea of its appearance than any printed description could do.

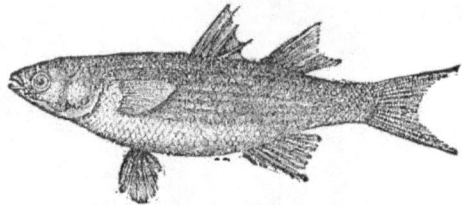

EUROPEAN SALMON (*Mugil cephalus*).

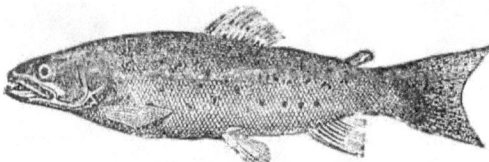

A singular adventure connected with harpooning a *sea devil* is said to have occurred. Four men, occupying two small skiffs, were fishing for *sea devils*. The boats had separated and each was pursuing its course independently of the other, when suddenly they began to bear down upon each other with a rapidity altogether beyond the oar-craft of the human arm. By skill in steering they managed to avoid a collision, but for quite a long while they would again and again approach each other as though they were a pair of goats resolved upon each other's destruction. It turned out that the crew in each boat had succeeded in fastening a harpoon into a colossal *sea devil*, and that in imitation of the whale they were dragging the boats hither and thither in a manner which rendered the story of Neptune and his dolphin steeds much less a matter of fable.

BROOK-TROUT (*Salino fario*).

The **Blue-Fish** (*Pomatomus saltatrix*) is the object sought by amateur fishermen in Atlantic waters, and has probably been responsible for greater waste of time and more frequent "drawing of the long bow" than any other fish that swims the waters.

PIKE (*Esox Lucius*).

Undoubtedly the professional guides act in perfectly good faith, and do not intentionally rob the eager sportsmen; still they have the most unhappy faculty of taking one into the midst of a school of dog-fish, or of compelling him to spend his weary hours in catching sea-robins, or find time and tide and weather unexpectedly against them. Other persons, if they are to be believed, have no such misfortunes, but the writer, though occasionally circumventing the fish (and the fisherman), is willing to confess that most frequently he has found only the proverbial fisherman's luck. It is said that *blue-*

NINE-FINNED PIKE (*Polypterus bichir*).

fish weighing fifty pounds have been taken; but four pounds is a more than average weight, and any one who has hooked a two and a half pounder has occasion to cherish the recollection both because of the grudging admiration of the professional fisherman, because of the struggle, and because of the rarity of the experience. The fish swim in schools, but the specimens caught by the writer must have been kept in while the others took their running.

The *blue-fish* is found along the whole Atlantic coast, and likewise in the

AFRICAN PIKE, FROM STANLEY'S FALLS.

Gulf of Mexico; in Europe it seems to confine itself to the Mediterranean Sea. The *blue-fish*, like birds of passage, are migratory, starting north as early as March (the proper month for inaugurating new departures), and going south again in October. The *blue-fish* is not a great favorite with those who would preserve at least specimens of the living species of fish, for it wages the most constant, relentless, and destructive warfare upon every fish that comes in its way, and seemingly is not content with gormandizing, but destroys fish with no higher motive than influences the most malicious small boy. As a food fish, the *blue-fish* is a great delicacy, but to the sportsman this is a small consideration in comparison with the healthful outdoor amusement which it provides, even when the sport is confined to the hunting and the game-bag remains empty.

THE BAT, OR DEVIL-FISH.

HAIMARA (*Erythrinus trahira*).

The *blue-fish* was unknown forty years

ago, so he may be accepted as a man of the period, thoroughly versed in watering stock, forming trusts, prospering, and absorbing the accumulations of others.

The **Pompano** (*Trachynotus carolinus*) is the choicest of Southern foodfishes, and therefore deserves at least honorable mention. It is a small fish, swimming near the surface, and in some particulars resembling the mackerel.

The **Pilot Fish** (*Naucrates ductor*) belongs here, but has already received mention in connection with

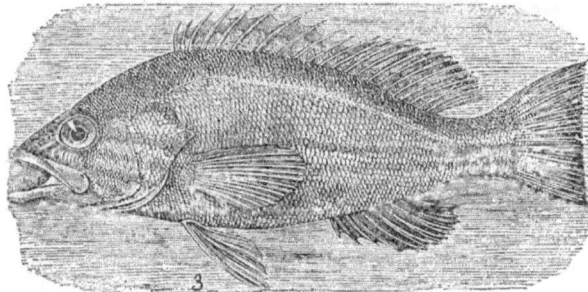

SAW PERCH (*Serranus cabrilla*).

the shark, with whom it maintains a friendly and inseparable companionship.

The **Moon Fish** (*Vomer setipinnis*) is noted for its absurd form. The body of a sun fish is supplied with the most disproportionately long face, illuminated by great, staring eyes, and terminated by a large mouth, whose under-jaw protrudes in the most melancholy fashion. Just back of the head waves a long, spine-like, single hair. The small

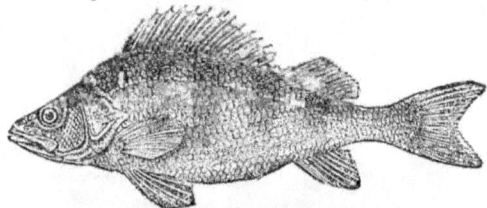

CROPPIE (*Perca fluviatilis*).

dorsal fin is set well back, and looks as though it would slide down upon the caudal fin. The belly is ornamented anteriorly by a second spine-like hair, and posteriorly by a series of small rays. Where the human being would wear his ears, the *moon fish* carries two absurdly small fins.

The **True Mackerel** of commerce (*scomber scombrus*) is the most important fish, commercially speaking, that is found about our shores. It is a very beautiful

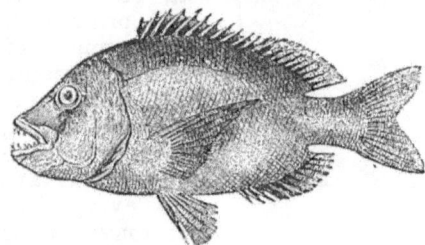

GILT HEAD (*Chrysophrys aurata*).

fish, though small in size, the color being a steel-blue, striated with undulating bands of black, while the belly is of a lustrous white, reflecting a bright silvery sheen when first taken from the water.

MACKEREL (*Scomber scombrus*).

Though frequenting our coast in vast numbers, mackerel are no less plentiful in the Baltic, North Sea, German Ocean, and all along the Scandinavian coasts, though its appearance is always erratic, since it is migratory in its habits. Their movement generally begins early in the spring, when the water temperature about harbors rises to 45 degrees. They feed on small crustacea and calamars, nor do they reject the spawn and young of other fishes, and occasionally they devour their own. When their movement begins they appear and disappear as their prey sinks or rises, the weather having no little influence on their actions.

The spawning season is in May and June, at which time they retire to rather deep water, and there deposit their eggs, which, upon fecundation, rise to the surface and are hatched by the sun. The young develop so rapidly that in two or three months their length is nearly eight inches, or within about four inches of the adult size. Though the growth is thus wonderfully rapid during the first months, it is a singular fact that thereafter the increase

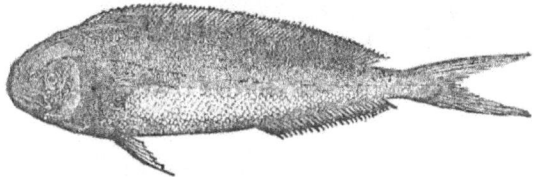

GOLDEN MACKEREL OR CORYPHENE. (*Coryphæna hippurus.*)

in size seems to be arrested, so that it requires nearly four years to attain the full size, which is from fourteen to eighteen inches.

Mackerel are generally taken by means of seines, which are spread in the spring over the American fishing grounds, extending from Cape Hatteras to Labrador. The range is at first about fifty miles off shore, but the fish move shoreward, until the best fishing grounds are from three to ten miles seaward. The extensive character of this industry may be understood when it is known that a capital of more than $2,500,000 is invested in the fishery, in which more than 5000 men are engaged. The value of the annual catch is estimated at $2,500,000, representing 111,399,855 pounds of fish taken, of which 103,142,400 pounds are pickled, 4,957,455 pounds are canned, 1,100,000 pounds used fresh, 1,100,000 pounds used for bait, and 500,000 pounds for fertilizing purposes.

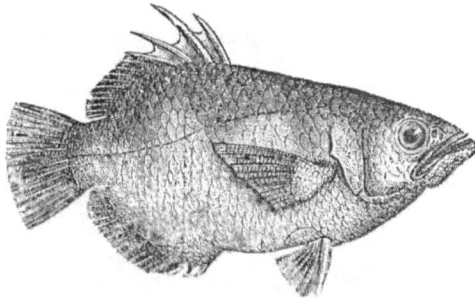

ARCHER FISH. (*Toxotes jaculator*).

The seines used in mackerel fishing are about 1200 feet long by 150 feet deep, the bottom being loaded with lead and the top buoyed with corks, which vary in size, the largest always being at the centre. When the fishing is in shallower water a smaller seine is used, some 1000 feet long and 70 feet deep; but all boats now carry both sizes for obvious reasons. Though a much greater proportion of mackerel are taken by means of seines, hand-fishing, with lines and baited hooks, is also largely followed. The bait used is either pieces of salted menhaden or small mackerel, which are caught and salted for the purpose. These are cut up into small squares that are crowded on to the hook, and then

7

scraped until only the tough skin remains. This is very bright, the belly pieces being mostly used, and, when properly put on the hook, will last for some hours' fishing. Each boat is provided with a large number of empty barrels, into which the fish are thrown when caught. During the time that a school is biting, an expert fisherman, handling a dozen lines, may take 2500 fish in the course of six or eight hours. Usually, however, the fish are so erratic that they cease biting in two or three hours.

The **Tunny** (*Orcynus thynnus*), also called *horse mackerel*, is a member of the mackerel family, and much the largest of that genera, if not indeed the largest of all coy fishes. Specimens have been taken along the American shores that weighed fifteen hundred pounds, though this size is quite uncommon. In shape the *tunny* is very like a mackerel, and the body gleams with a silvery sheen, the upper half being of a lustrous steel-blue and the lower a pearl white. Though the fins show no surprising development, it is a very rapid swimmer, and takes its prey from every species of fish found in its habitat, even gorging dog-fish of eight pounds weight, and making bold attacks on the dolphin. While its flesh is solid and said to be good for food, especially in southern Europe, the Americans seldom eat it. The *tunny* is nevertheless caught in great numbers for no other purpose than the oil it yields, as much as twenty-three gallons having been rendered out from a single fish. The means employed for its capture is generally that of harpooning, but in the Mediterranean, where the fishing for this species is carried on more extensively, they are taken in immense

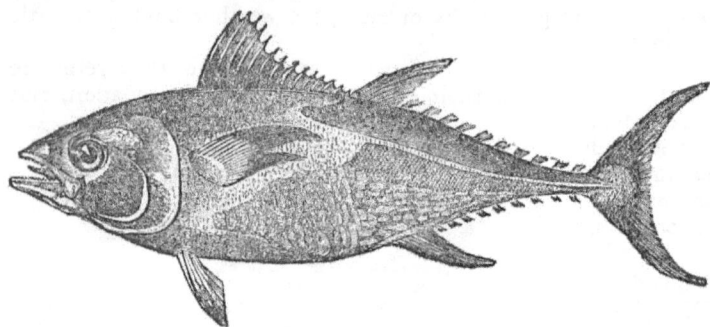

TUNNY.

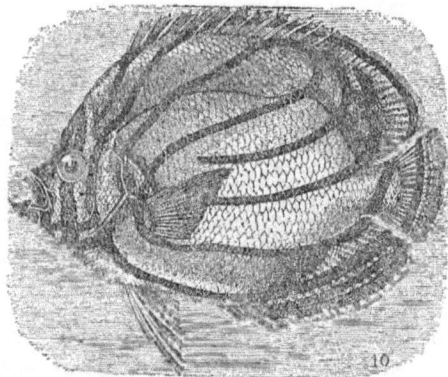

CORAL FISH. (*Chætodon meyeri*).

numbers by means of special nets strung on upright posts set in a large flooring, that after being baited and left in a proper spot for some days, is lifted by means of capstans on steamboats that are used for fishing purposes, until the fish within are brought to the surface and there killed with sharp hooks and spears.

The **Bonito** (*Sarda mediterranea*), also belongs to the mackerel family, and is found on both sides of the Atlantic. It bears a close resemblance to the tunny, except that the body is smaller and comparatively broader. The color is the same, the only distinction being several dark stripes running obliquely from the back toward the central fins, which are not observable in the tunny. Its flesh is little used.

The **Dolphin** (*Coryphæna hippurus*) is a frequent sight in tropical waters. It is about five feet in length, and has a yellowish body, with a black back, both body and back being spotted. It is often called the *golden mackerel*, and is not to be confounded with the cetacean dolphin. It is strikingly lustrous, but the iridescence of the fish when first taken from the water speedily undergoes the change which has touched the sensibilities and excited the imaginations of so many sea-going travellers.

The **Thread Fish**, or **Cutlass Fish** (*Trichiurus lepturus*) is common alike in tropical waters and in temperate. From its coloring and shape, it is sometimes called the *silver eel*. It swims on the surface and is prone to jump frequently out of the water, for what purpose is unknown. In the region of Florida it is caught for the markets.

The **King of the Herrings** has already received mention, but is introduced at this point to maintain the order of piscatorial succession.

The **Chætodipterus faber**, abundant on the coast of "Old Virginia," and called the *porgy*, is an entirely distinct species from the *porgy* of the Middle States.

THE ARCHER FISH AND CHÆTODON.

The genus **Chætodon** includes many curious fishes.

The **Fly Shooter** (*Chelmon rostratus*), found in Indian waters, is insectivorous, and as a bean-shooter rivals the most expert small boy. Upon discovering an insect it will shoot it with a drop of water ejected from its snout, and its precision of aim would put to the blush a Robin Hood of England. A species found in Java (*Toxotes jaculator*) can shoot successfully at the distance of three feet. It is about the size and resembles in appearance (if we except the proboscis), a small Sun Fish.

The well known **Porgy** (*Ephippus faber*) belongs to the same order, although its family is different and lacks the peculiarity of the fly-shooter.

The **Coral Fish**, or **Wandering Chætodon**, is a carnivorous fish; is distinguished alike by its shape, habits and color. It is circular, or disc-like; its color is a yellowish-gold, marked by purplish-brown lines. These lines are greatly varied in form and direction so as to suggest the branching of the coral.

The **Surgeon Fish** (*Tenthidae, Acanthurus chirurgus*) is a small fish, but its tail, armed as it is with a curved spine, or with stout prickles, inflicts the most painful wounds upon its incautious captor, doubtless intending to "operate upon him for cataract," that in the future his sight may be better.

SURGEON FISH.

The **Eagle Fish** (*Sciæna aquila*) abounds in European seas, especially the Mediterranean, a locality whose prominence in history is not confined to politics, state-craft, and war, but is equally pronounced in the study of animal life. It is about five feet long, swims in schools, and being held in the greatest esteem by those who live to eat, is an object of interest to the fisherman. It makes a noise as it swims, and possibly this has led to its name on account of a forced resemblance to the scream of the eagle; or, it may be that its rapacious habits are responsible for the name. The sound emitted by the *eagle-fish* is, it must be confessed, more like a porcupine grunt than like the scream of the king of birds.

Other species—the Sea Chub, the **Weak Fish**, the Louisiana **Red Fish**, the **Black Sheep's Head**, of Lake Huron, are among the most inviting articles of food, and render the observance of Lent anything but the abstinence which has given meaning to the phrase, "lenten fare."

EAGLE FISH.

The **Drum Fish**, or **Drummer Fish**, is represented by a salt-water species and by a fresh-water species.

The salt-water **Drum** (*Pogonias chromis*) wears its beard upon its neck, as one might describe the barbels of the lower jaw. The interior dorsal fin, placed about a third of the distance from head to tail, is like the wing of a butterfly, and is higher, though shorter than the posterior dorsal fin, which is even and extends very nearly to the caudal fin; the caudal fin is convex. The ventral fins are long, but narrow; the posterior one is just beneath the back dorsal fin, and the anterior one just in front of the anterior dorsal fin; it has also gill fins; all of the fins are spiny. The pharyngeal bones are joined into a triangle studded with teeth, and it is by means of these bones that the fish makes the noise which resembles the muffled beat of a drum. It has queer variations at different stages of its growth; its fins are largest when it is youngest, and the uniform silver-gray of the adult is preceded in the younger fish by several broad, dark, bar-like markings. It is to be found where the oyster makes its bed, and occasions great loss to the oysterman; it is a regular Goth, for it is not satisfied with what it requires for food, but destroys the oyster with the most wanton cruelty.

The fresh-water **Drum** (*Haplodinotus grunniens* or *Sheepshead*), is to be found in many of our rivers and lakes. It is smaller than its marine relative, and its chin is unadorned by barbels. By some its flesh is considered a delicacy, while others hold it in little esteem.

The **Scup, Scuppang,** or **New York Porgy** (*Stenotomus chrysops,* or *argyrops*), is caught in great quantities, and is marketable, although probably undervalued.

The **Sheepshead** (*Diplodus probatocephalus*) is highly esteemed at the South, and its capture is always a delight to the home angler.

The **Red Snapper** (*Lutjanus vivanus*), is another Southern fish, whose exportation has been rendered possible by our increasingly rapid transit. Those who are fond of fish regard the *red snapper* as a great delicacy, and pay in proportion for it.

The **Striped Bass** (*Roccus lineatus*) is a well-known fish of large size, averaging twenty pounds in weight and having a proportional length. Its flesh is highly prized, and though they are generally caught with a seine, lucky anglers at times enjoy the excit-

DRUMMERS (*Pogonias chromis*).

ing sport of capturing them with the hook. The fresh water *striped bass*, or *white bass* (*Roccus chrysops*) is esteemed alike by the sportsman and by the epicure. Most of the fishing stories of the Northwest have this fish for their subject.

The **Common Ocean Sun-fish** is almost circular, and its fins look as though they had accidentally been stuck into the truncated body. It sometimes weighs a quarter of a ton, and has been found five feet long by four feet thick. It is brilliantly phosphorescent, and looks like a living ball of fire. Its oil is valued by seamen as a panacea for bruises.

The **American Yellow Perch** (*Perca americana*) needs no description, but affords too much pleasure to young fishermen to be passed without mention.

The **Giant Perch**, or **Pike Perch** (*Lucioperca*), includes the blue, yellow, gray and wall-eyed perches. This fish abounds in the Saginaw fishing districts, and is highly valued alike for its flesh and for the sport with which when, spearing, it is attended. It has many an alias, *wall-eyed pike, dory, salmon, blue pike* and *Jack salmon*, being a few of them. It is a very important contributor to piscatorial commerce, or commercial fishery, and, moreover, is always of interest to the sportsman.

MILLSTONE SUN-FISH (*Mola rotundus*).

The **Black Bass** (*Micropterus*) might have been included among the nest-building fishes, since its habits are very like those of the fresh-water sun-fish. It is a voracious creature and highly esteemed as a game-fish.

The **Pumpkin Seed** (*Enneacanthus guttatus*) is a favorite fish with young fishermen because of its beauty and abundance in small streams.

The **Cichlidæ** include the **Tilapia Simonis**, so called because being found in Lake Tiberias; and the male being in the habit of carrying the eggs in his mouth, it was supposed to comply with the necessary characteristics of the fish from whose mouth Simon Peter extracted the tribute money. The **Tilapia Tiberiadis** is also noteworthy because identified with "the miraculous draught of fishes."

The **Tautog**, or **Black-fish** (*Tautoga onitis*), is alike common, abundant, and valued as a food-fish. It is the occasion of very considerable Eastern fishing industries.

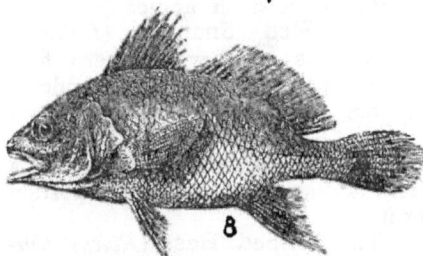

BLACK BASS (*Corvina nigra*).

The gayly-colored **Parrot Fishes** (*Scaridæ*) are a frequent and beautiful sight as they swim in and out amidst the coral. Though mentioned among the curiosities of marine life, it is equally remarkable for the antiquity of its predecessors, and for the esteem in which it was held by the ancient mariners of the Mediterranean.

GIANT PERCH (*Lucioperca sandra*).

The **Sea Hare**, or **Lump-fish** (*Cyclopterus lumpus*). This creature has peculiar characteristics not found in other fishes, and the family to which it belongs is, therefore, a small one. The body is short, and ridged with tubercles in seven longitudinal rows, some of which are warty and others bony. The average size is about five or six pounds, but specimens have been caught weighing as much as eighteen pounds. It is found along the English and American shores, where it haunts muddy bottoms, and only seeks the shallows at the spawning season. At this time the male makes a pit in the sand, invariably selecting a spot between two stones. Here the female deposits her eggs, which are vigilantly watched until the young come forth. When the young are a day old they attach themselves to the side of the male parent by means of their suctorial mouths, and do not quit their place for a week or more, until they have gained sufficient size and strength to provide for themselves. The *sea-hare*

LONGHEADED PIKE, FROM STANLEY FALLS.

is a voracious creature, preying chiefly on mollusks and the spawn of other fishes.

The **Lump Sucker** (*Cyclopterus lumpus.*) It may also be added to the "nest-builders." It has been known to lay as many as a quarter of a million eggs, so that it is certainly remarkable for its ability to multiply and occupy the (sub-marine) earth. The male is decidedly hen-pecked, for in the first place he takes entire charge of the eggs, and later, swims about with his many times twins clinging to his body.

LUMP-FISH.

The **Climbing Perch** (*Anabas scandens*) is an habitant of Indian waters, which are as full of extraordinary animal life as are its jungles. To a slight extent it is amphibious, for, being provided with internal cisterns, it can, by laying in its water supply, run about the land for days at a time. Its ability and disposition to run up tree-trunks has given it its popular name. It is edible and palatable, and therefore an object for pursuit. If as a fisherman one does not get a bite, he may become a hunter and prove to be more successful.

The Siamese have, in lieu of the Spanish bull-fights, public contests between **Fighting-fish** (*Betta pugnax*).

The **Mullet** (*Mullus barbatus*) of the ancients has long been celebrated for the enormousness of the sums expended for it by the wealthy Romans, and for the tributes which their writers have paid to its flavor.

The **Swallower** (*Chiasmodon niger*) can swallow fishes larger than itself, but it has to pay an extraordinary price for this feat of necromancy, for the distention of its stomach always kills the magician.

The **Sculpin** (*Cottus scorpio*) so frequently destroys the illusions and endangers the good temper of the fishermen that poetical justice seems to demand that the unintentional offender should receive honorable mention in a work which discusses curious, well-known or valuable fishes. Its large head and gaping mouth do not add to the beauty of the *sculpin*, but its fins and markings are both attractive to the eye when regarded from a proper distance. The *sculpin* is generally about six inches in length, and when not bothering the fishermen hides under the rocks and the sea-weed. Its vivacity leads it to fall an easy prey to those who are not fishing for the compliment, and its valuelessness being increased by the constant danger threatened by its sharp spines, leads to its harsh treatment. Perhaps its best use is as a symbol for that self-conscious irritability which is so often claimed as the " perquisite of genius."

CLIMBING PERCH.

The **Sea Raven** is only a species of the sculpin family, as is also the Great Weaver, whose spines are particularly dangerous.

The **Growling Cock** (*Trigla hirundo*) is somewhat of the color of sapphire; is abundant in European waters. It is eaten, although its flesh is somewhat dry. This species might have been described among the flying fish, although its flight is purely one of the

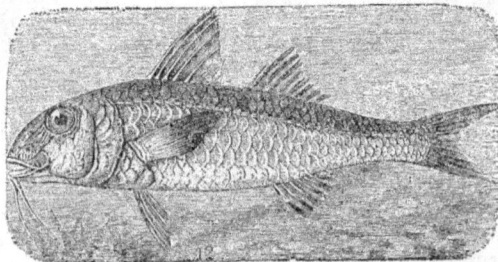

SEA BARBEL (*Mullus barbatus*).

fisherman's imagination, though another species is a flyer. The American species, the sea robin (*prionotis*), is said by some gourmands to be a table delicacy, but thus far few have been found willing to make the experiment. Any one who has visited the seashore must have become acquainted with this fish, whose only fault is that of disappointing an over-eager fisherman. The peculiar and not unmusical sound made by the sea robin is always interesting to children. It is a queer-looking object, but if one avoids uncalled-for interference with its spines it will hardly be found repulsive.

The **Swallow** Fish (*Dactyloptera volitans*) varies in length from half a foot to a foot and a quarter. The pectoral fin is sufficiently extended to admit of the fish's sustaining itself in the air for a short period, and hence appearing to fly. It is not, however, a true "flying fish." It is singular in form and appearance, and occurs in the waters of Europe and of America. It has already received sufficient notice among the flying fish.

GROWLING COCK (*Trigla hirundo*).

The **Sting Fish, Sting Bull, Great Weaver,** or **Chanticleer** (*Trachinus draco*), is a deep-water marine fish. It is spined the whole length of its back and belly, and is, therefore, a very uncanny sight to the inexperienced fisherman. It looks as though the ordinary skeleton of the fish had separated in the middle, and was escaping through the back and belly.

The **Ocean Butterfly,** or **Ocellated Blenny** (*Blennius ocellaris*), is a little fellow only about three inches in length. It lives among the weeds, and its pale-brown color makes it hard to distinguish it from them. Its dorsal fins extend from the head to the tail, and from their size and arrangement suggest the wings of the butterfly.

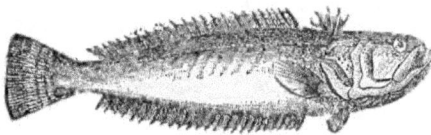

THE GREAT WEAVER (*Trachinus draco*).

The **Oar-fish** (*Regalecus banksii*) has at times been confounded with the sea-serpent. It has been found of as great length as twenty feet. Its ventral fins are simply two dagger-like prongs just back of the throat, and pointing obliquely toward the tail. The body is oar-like, and is propelled in a sinuous manner. The head is feathered with spines after the fashion of an Indian chief, and its back is adorned with spinules.

The common **Cod-fish** (*Gadus morrhua*) is doubtless the most interesting fish when regarded commercially. The largest specimen so far known weighed one hundred and sixty pounds; the average weight is from fifty to sixty pounds. The *cod* is found as far north as the Arctic Ocean, and as far south as the thirty-fifth degree of latitude. The *cod* is voracious, and by no means a delicate feeder; his perfect digestion has at times rendered him of great service to the conchologist, who procured from the stomach of the fish specimens so rare as to have been otherwise unattainable. The *cod* does not, when spawning, approach the shore, but simply entrusts its eggs to the mercies of the deep sea; when it is remembered that a single *cod* is capable of producing a family of nine millions, it will be evident that this remarkable fecundity is at least an adaptation to the use which the fish serves in human economy.

The **Haddock** (*Melanogrammus æglefinus*) is a near relation of the cod-fish, it is small by comparison, and does not swim so far northward. Under the popular name of "*finnan haddies*," the fish is doubtless known to every one who has been on the Atlantic coast, or whose traditions lead him to help maintain the fisheries from which Massachusetts has drawn so large a portion of its accumulated wealth.

BUTTERFLY FISH.

The **Pollock** (*Pollachino virens*) is another species which has not yet become so highly prized as a food fish. It is an odd fact that its young seem to be infatuated with the Robinson Crusoe habit, for they pass much of their time under the umbrellas of the jelly fish.

The **Whiting** (*Merlangus vulgaris*) is a small but delicious European cod-fish, but readers of Charles Dickens and others who describe London life will be familiar with its name and the high esteem in which it is held by the English epicures, who are exceedingly blessed by its abundance.

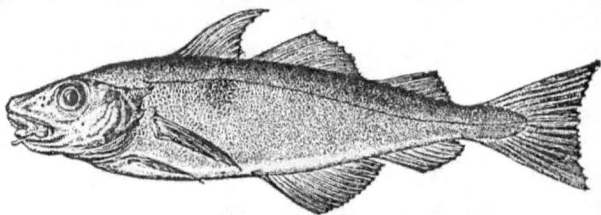

HADDOCK.

The **Hake** (*Phycis*) is abundant on the Atlantic coast and valued by fishermen.

The **Halibut** (*Hippoglossus vulgaris*) is a shallow-water fish found in the northern portions of both the Atlantic and the Pacific. It generally weighs about one hundred and forty pounds, but has been found weighing four hundred pounds. In length it varies from five to seven feet, and is usually half as broad. Though the *halibut* is not a table delicacy, it has been a source of revenue to those "who go down into the sea in ships." Till about fifty or sixty years ago Massachusetts Bay was a favorite resort of the *halibut*, until the activity of the fishermen drove them away. The fish now have to be sought in relatively deep water. The *halibut* is a ground fish, and its flatness and coloring seem an intentional provision for the life which it is to lead. It has been known to kill its prey by repeated blows of its tail.

The **Turbot** (*Psetta maxima*) has become celebrated as a table delicacy, but one must go abroad for it.

The **Plaice** (*Platessa vulgaris*) is marine, but prefers the banks or muddy bottoms. It is common in Europe, and is taken either by trawls or by spearing. It has the form of what is commonly called the sun perch, and is dotted with curious white spots.

The European **Flounder** (*Platessa flesus*) is found in the muddy bottoms at the mouths of rivers, but it will flourish in any kind of water. It is regarded as indifferent eating.

The **Platessa Plana**, found in New York, is highly prized as a delicacy; it is usually pale-green in color. The *Platessa dentata* (or summer flounder) is also a favorite with fish-fanciers in New York. Its peculiar flat form has made the fish proverbial.

The **Dab** (*Platessa limanda*) is abundant in England, and prized for the uses of the table.

Four-horned Trunk-fish (*Ostracion quadricornis*). This is a most singular fish, as a brief description will show. The body is polygonal in form, and is encased in a series of

WOLF-FISH.

hexagonal plates so rigid in their union as to permit of no laxity of the trunk. The teeth are well formed, eight above and as many below, which curve backward as in the snake. Projecting from above the eyes are four sharp horns that give at once a hideous and grotesque appearance to the creature. It is pretty generally distributed in the temperate waters, and species are common in the tropics. The flesh is sometimes used as food, people of Florida and the West Indies esteeming its flesh as a great delicacy when baked in its own shell, but not a few persons declare that the flesh is poisonous.

The **Stickleback** (*Gasterosteidœ*) has been described among the "nest-builders."

The **Tobacco-Pipe** (*Fistularia tobaccaria*) is a tropical fish, eel-like in appearance, but with a long pipe-stem-like snout, which, from its size, seems to suggest resemblance to the sword-fish, with which, however, it has no other relationship.

The **Snipe-fish** (*Macrorhamphosus scolopax*) is a common European fish, whose snout resembles the bill of the snipe, although its body is much more suggestive of a hog.

The **Sea-horse** is common in tropical waters; it has the head of a horse (or rather of the knight of the chess-board), and a prehensile tail with which it holds on to the sea-weed. The creature is a very interesting object when in an aquarium. The *sea horse* is not only interesting on account of its very singular appearance but for its habits as well, which are decidedly eccentric. The little creature, scarcely more than three inches iu length, is able to dart about with extraordinary celerity, in resemblance to the movements of the crawfish, but instead of seeking concealment it rests only upon such object as it may grasp by means of its tail. Here the grotesque creature rests until again disturbed, looking with curiosity from

9. PLAICE. (*Pleuronectes platessa.*)   10. FLOUNDER. (*Pl. flesus.*)   11. DAB. (*Pl. limanda.*)

SEA HORSE.

FOUR-HORNED TRUNK FISH.

side to side until, presto! it is gone, only to as suddenly reappear at another place, always manifesting a ludicrous eccentricity.

The **Globe Fish** (*Cirrhosomus turgidus*) is called also the swell-fish, or egg-fish.   Though when in a state of repose a somewhat elongated and symmetrical fish, it will, when caught, inflate itself with air and become an almost perfect globe, ornamented with spines.   By tickling its belly the fish can be made to repeat the process several times, after which it expires from its frog-like effort to swell itself into an ox.   Along the Atlantic coast is another species of swell-fish *tetraodon* whose habits are very similar to those of the *globe-fish*.   It is a small species, distinguished for its green eyes, rough, frog-like skin, and generally repulsive appearance.   This illustration shows the creature distended with air. The tropical representative of the *globe-fish* family (*Tetraodon fahaka*) is frequently called the *porcupine fish*.

The **African Scaly Salamander** (*Protopterus annectans*) belongs to the African tropics, but can generally be found in the collections of our museums.

THE TOAD, OR SWELL-FISH (*Tetraodon*).

It is a fresh-water fish and is insectivorous, sharing with man a fondness for fish and frogs, but adding to the rarity of its diet by including insects.   It bears a striking resemblance to the salamander, but has delicately-fringed pectoral and ventral fins, arranged in pairs towards either extremity of the body, and the gills have filament-like projections.   It is five or six feet in length, covered with cycloidal scales, and inhabits ponds, where it builds a nest in the mud.   In the absence of water it can substitute its lungs for the work done by its gills, and inspire air directly instead of indirectly through water.   The mud-fish is a species very like the salamander, with similar habits.   The engraving on the next page shows the creature in its winter burrow.

The **Eyed Pteraclis**, though not a member of the flying fish family, bears some resemblance to creatures of that class, in the remarkable size of its dorsal and ventral fins, the latter being confluent with the anal fins. The name is given

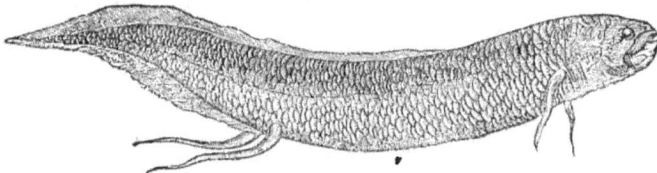

AFRICAN SCALY SALAMANDER (*Propterus annectans*).

on account of a dark-blue round spot located near the upper edge of the very expansive dorsal fin.   It is occasionally met with about our American shores, but its special habitat is along the Mozambique coast.   It is a very beautiful fish, having burnished sides and a golden gleam on the tail and pectoral fins, while the expansive fins are tinted with dark blue.   Its length is two feet.

MAMMALIA OF THE SEA.

The **Cetacea** are the water representatives of the mammals.   They breathe air directly and aerated water.   Their shape and their aquatic life is responsible for the popular notion that they are fishes.   Porpoises, dolphins, walruses, grampuses and narwals are cetaceans.

The **Straight-finned Whale** (*Orca rectipinna*) is common on the Pacific coast, but is almost valueless.   These creatures are here introduced because they make a natural transition from the *fishes*, and because they were for so long a time confounded with the *fishes*.

The **Narwhal**, or **Sea Unicorn**, is remarkable for its horn of twisted ivory. This springs from the upper jaw, is used as a weapon, or as an implement for digging, etc., and has played no unimportant part on occasions when the *narwhal* has vented its wrath upon vessels.

The *narwhal* is greatly esteemed, for, independent of its own value, it is regarded as the avant-courier of the whale. The ivory of its tusk is valued and put to many uses, among which may be mentioned the manufacture of lances used in killing *narwhals*. It descends but a little way into the sea, and after rising to the surface is so exhausted as easily to fall a victim to the fisherman. Schools of *narwhals* will surround a vessel, and remain patiently until slain for their oil and their ivory.

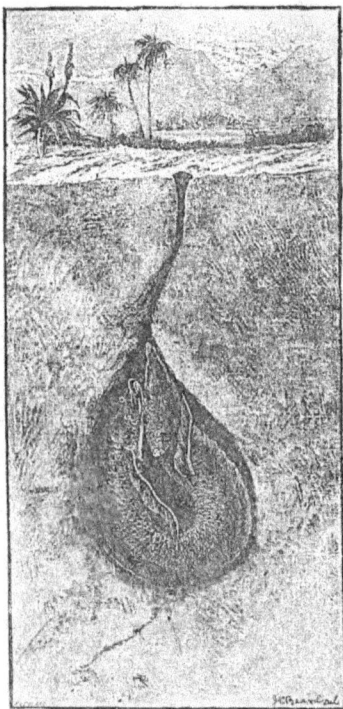

MUD-FISH (*Lepidosiren paradoxa*).

The one-tusked *narwhal*, or *sea unicorn*, is the subject of many an antique fable, and later it was supposed to possess the power of rendering poisons innocuous by the mere insertion of its tusk or horn. There was a period when there lurked within the chalice not merely the poison of wine, but poisons added by ambitious or vindictive men and women. The story of Roman and Italian politics is rendered exciting by tales of poisoning effected usually under the guise of hospitality and good cheer. Hence it is not surprising that the *unicorn's* horn commanded enormous prices, and was in great request by those who could afford its purchase. It may even be a relief to know that in spite of Tennyson's protests against "these days of the men of mind," the adulteration of this antidote, or its conversion into "a poisoned poison," was neither unknown nor uncommon. The appearance of the *unicorn* in Great Britain's coat-of-arms is undoubtedly due to a belief that the *unicorn* was useful as a friend and dangerous as an enemy. The exact office of the *narwhal's* tusk is not certainly known; it is supposed to use it as a means of effecting a landing and in its contests with the Greenland whale. It has been known to drive its tusk through the metallic sheathing and stout timbers of large vessels, but it is supposed that this was done not with any malice toward the ship, but from a frenzied irritation and possibly an unintentional collision.

The **Bay Porpoise** (*Phocæna somerina*) is a familiar sight on the Pacific coast, as it prefers bays and harbors and is quite fearless.

The **Atlantic Porpoise** (*Phocæna communis*) is a familiar sight to the sea-going travellers, and even to the thousands who daily pass up and down; they are neither strange nor without interest.

The **Dolphin** of the Pacific Ocean (*Lagenorhynchus obliquidens*) frequently appears in schools, and gambols around a vessel. It is greenish-black above, pure white below, and striped on the sides with alterante stripes of black and gray.

The **Mediterranean Dolphin** (*Coryphæna hippurus*) is bluish-green above, citron below, the pectoral fins lead and yellow, the ventra lfins black above and yellow below, the caudal fin yellow, and the iris golden; moreover, it has gold and azure reflections. Their iridescence when caught proceeds from their muscular action and is more beautiful than the most gorgeous sunset or the most vivid rainbow, but like other beautiful fishes, the luster and iridescent hues very soon fade when it is taken out of the water, especially if it be exposed to the sun.

The dolphin was regarded by the Greeks as sacred to the god Apollo,

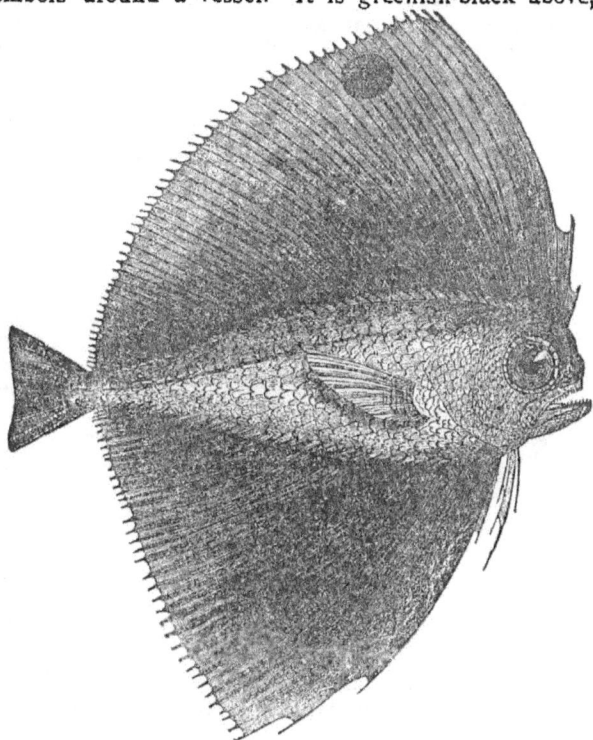

EYED PTERACLIS.

since when founding his oracle at Delphos, he appeared under the form of a dolphin. He is also represent as being borne upon the back of a dolphin and striking the harpstrings of his sweet instrument, making music to all creatures of the sea, that followed him, like the rocks and trees are fabled

SCHOOL OF NARWHALS.

to have followed Orpheus, but, unlike Orpheus, Apollo had no beautiful love doomed to early death, else the dolphin might have figured even more prominently.

Hence the dolphin became one of Apollo's symbols, and wove many a fiction of the rescue of Apollo's favorites by means of the dolphin. The poet Phalanteus having been wrecked at sea, was safely carried ashore by a dolphin, since Apollo was the protector of poets! Phalanteus, after the custom of pre-historic antiquity, then founded the Italian town of Tarentium, although except for the interposition of mythology he must have suffered from the lack of assistants. So, too, a dolphin saved Arion from the plottings of sailors. For having the bad habit of carrying with him the vast wealth of which he was possessed, Arion excited the cupidity of the sailors as he was voyaging from Tarentium to Corinth. Pleading for a last opportunity to exercise his musical skill, Arion's melodies drew together a school of fascinated dolphins (amongst whom doubtless was the rescuer of Phalanteus), and having thrown himself into the sea, Arion was, through the care of Apollo, carried triumphantly to Corinth by the obedient slaves of the beautiful god.

PORPOISE.

The experience of Arion, as it will be remembered, was completed by the dealing out of poetical justice in the execution of the sailors and the securing of his slowly amassed wealth. The *dolphin* has been honored even in modern history, having given its name to the Princess of France, to one of the fairest princes of this sunny land, and to one of the best editions of the classics. Doubtless the harmlessness of the *dolphin*, his inferior value to the fisherman, his beauty of color, and the plaintiveness of his dying moan have united to make this creature a favorite theme for the sailor, the sea-going traveller,

COMMON DOLPHIN (*Dolphinus delphis*).

and the poet. The air bladder of the *dolphin* furnishes much of the isinglass known to commerce, although the sturgeon also contributes, from which latter, however, the larger supply is obtained. A popular error has long been held that mica is a similar substance, whereas it is a mineral.

The **Sperm Whale** (*Physeter macrocephalus*), is the largest of the species, though not so long as the rorqual. It reaches a length of from seventy to eighty feet its head being at least a third of this distance. It is generally found in schools of a varying number and is a familiar sight to those who know the sea. Its remarkable endowment of mouth and teeth, and its ability to open its jaws at the most remarkable angles, together with its strength and courage, have made the *sperm whale* the frequent subject for story tellers. In addition to its supplies of oil, we owe to it the ambergris so well known in commerce as an essential ingredient of the perfumery which graces the toilet table. The "spouting" of the whale is a device for ridding itself of the water which has contained its food; the vaporized water thrown to a great height and having the form of the siphon, is one of the curiosities of life aboard ship. As is well known the harpooning is done from small boats which the whale is suffered to drag hither and thither until, exhausted by loss of blood, it yields

SPERM WHALES.

itself an easy prey. At times the whale attacks the boat and, such accidents as the smashing of whale-boats is a frequent and often fatal occurrence. On one occasion a whaleman in falling from the air into which he had been thrown by a blow of the whale's tail, landed upon the whale's back and holding to the harpoon, rode several miles before it occurred to him to let go and suffer himself to be rescued by his comrades. The oil is found in a cavity on the right side of the head, and as much as ten barrels is not an uncommon yield. **The Northern Rorqual,**

IN PURSUIT OF THE SPERM WHALE.

though slimmer than the Greenland whale, exceeds one hundred feet in length. Whales are both social and domestic in their habits. They like to

8

travel in schools, to gambol with the stimulus of emulation, and they display the greatest devotion alike towards their mates and towards their young. The herbivorous whales are gentle out of proportion to their size, but they can, in defence of their young, or when irritated by harpooning, display the most reckless and the most invincible courage. A trustworthy account is given of a whale which, having been wounded by coming into collision with a ship, deliberately attacked it again and again, until finally the ship was sunk and the crew compelled to take to their boats. It quite frequently happens in the whale fishery that boats will be destroyed and the seamen compelled to swim for their lives. Only an experienced and cautious fisherman can be trusted with the harpooning of a whale, for as it uniformly rushes off when struck, the fouling of the harpoon cable would mean utter destruction to the boat's crew. Men naturally like adventure, and readily adjust themselves to new exigencies. Hence, after one has learned the business there must be much that is exhilarating in being towed along at lightning speed by this leviathan of the deep, who, however noble his struggle, will finally discover that his immense

SOUNDING.

strength is no equivalent for the wary skill and tireless persistence of his weaker captors. After a whale is powerless it is towed to the side of the vessel and there cut up, the waste portions being left to the desiccating influence of the sea. Though whales have diminished in number, the march of progress has enabled the whalers to improve upon the antique method of

ARCTIC FIN WHALE.

harpooning by hand. They now carry a Gatling harpoon, which fires into the body of the whale not only a series of hooks or flukes, but a torpedo, which explodes with fatal effect.

The **Whalebone Whales** have, like Richard the Third, teeth when they are born, but before completing their fœtal existence replace them by plates of whalebone.

Whalebone is not really bone at all, but simply horny plates serving to retain small animals until the whale is ready to swallow them, or so that he shall not choke himself by swallowing what is not suitable for food.

The **California Gray Whale** (*Rhacianectes glaucus*) lives mostly in the shoaler water of bays and gulfs. It does not exceed forty-five feet in length and yields about twenty or thirty barrels of oil. A record of whales seen shows that as many as forty thousand of the *California gray whale* have been known to pass in a single season.

The **Hump-Backed Whale** (*Megaptera versabilis*) has a hump upon its back, a lump on its lower jaw, and a number of small eminences on the top of its head, and is certainly entitled to being regarded as " your eminence " among whales. It is an oil giver, but its whalebone is valueless. This whale is practically ubiquitous and is a very common sight at sea. As it lies on its side to suckle its cub, or as a male and female conduct their courting with love taps and other human peculiarities, the creatures are comically suggestive of an unwritten satire upon the human being.

The **Sulphur Bottom** (*Sibbaldius sulfurens*) is the largest living mammal, and is supposed to be the largest specimen that has ever existed. It is worthless to the fisherman, which is possibly quite as well, as its extraordinary strength and rapidity, and its lack of gregarious habits, would be more likely to furnish fresh illustrations of " a fisherman's luck" than to prove a profitable return for hard labor.

The **Razor Back** (*Balænoptera musculus*) is very large and powerful, colored

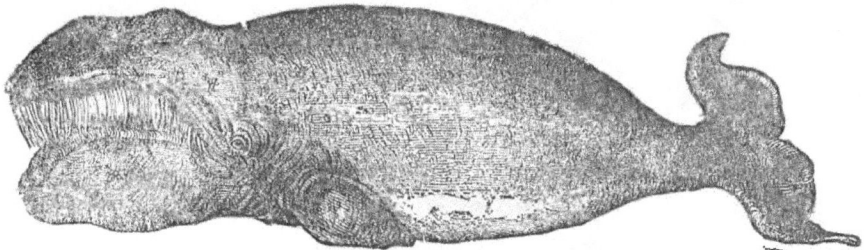

BOWHEAD GREENLAND WHALE (*Balæna mysticetus*).

black above and white beneath, and thinning out from head to tail like the edge proceeding from a thick-hefted knife. It is a producer of oil and of whalebone.

The **Bowhead, Greenland Whale, or Polar Whale** (*Balæna mysticetus*), though smaller and perhaps less curious than many other species, is the one which plays the largest part in human economy. It varies from forty to sixty feet in length, and a single whale has been known to yield eleven thousand gallons of oil and more than three thousand pounds of whalebone. It feeds upon the small crustaceans and mollusks which fall into its mouth as the whale swims along, using its baleen, or whale-bone, as a strainer. It has been known to use up a mile of harpoon rope, and has the ability to stay below the surface for more than an hour.

The **Bonnet Whale, or Right Whale** (*Balæna sieboldii*), is very much like the bowhead, but has an excrescence or bonnet on its upper jaw. It is an oil-well and artificial vertebræ manufacturer. The whale not only stands on the borderland which represents the transition of fishes to mammalia, but its history is so old and so consecutive that its name has become a household term.

As early as the time of Alfred the Great, of England, whale fisheries were known, and of modern peoples, the Spanish and Portugese, the Italians, the Dutch or Hollanders, the English and the American Yankees have succeeded to the control of the fishery. A whale-ship starts out nowadays for a three years' voyage, and carries a crew of upwards of thirty-five men. The smaller boats used for the pursuit of the whale are about thirty feet in length and six feet wide, prowed at each end. Everyone of the crew has special duties, and there is the strictest observance of that order which is "Heaven's first law."

For example, immediately upon arrival at the whaling grounds, a "lookout" is kept at masthead, and on his vigilance the crew depends for timely notice of the appearance of a whale. The whalers, having been divided into crews of from five to seven, put off at once upon word from the lookout, and each boat strives to reach the

A DEATH WOUND.

whale first. In the bow sits the harpooner ready for the first opportunity, and when this comes he seizes the harpoon in his right hand, and the coil of rope in his left, and, as with all his force he hurls the weapon, he cries, "Stem all." Then the whale dives so rapidly and so deep that men are required to see that the rope does not foul, and by constantly running water upon it preventing fire by friction. At the expiration of half an hour, or an hour, the whale reappears on the surface, and the process must be repeated.

The decreasing supply of whale oil has been compensated by the many inventions and discoveries which almost from day to day change our methods of living, and prevented an inconvenience which, before the time of gas, coal oil, and the electric light, would have been extreme.

The whale has to fear not only man, but the grampus. Commodore Wilkes gives a thrilling account of a fight between a whale and a grampus, in which unequal contest the whale could oppose nothing but his strength in resistance

WHALE ATTACKED BY A SCHOOL OF GRAMPUSES.

to the grampus, which clung with the tenacity of a bull-dog to his mouth, and gradually caused him to bleed to death. As the grampus is said to eat nothing but the whale's tongue, his gormandizing rivals the fabled dishes of Roman gourmands.

Another enemy of the whale is said to be the thresher-shark, which, like the grampus, seizes the monster, and not only bites ferociously, but thrashes the poor creature with its long tail, and thus literally rides the whale to death. The sword-fish is also said to persecute the great leviathan, as does the narwhal, by darting upward and striking its murderous weapon into the whale's belly, and continuing its unprovoked, but no less furious attack, until the great, but helpless, creature expires.

The whale fisheries towards the close of the sixteenth century had employed over two hundred and fifty Dutch ships and upwards of fourteen thousand Dutch sailors, not to speak of the smaller contributions of other nations then engaged in the whale fishing. The supply of hardy navigators required to man the many vessels of discovery, which followed in the path of Columbus, was doubtless due to the value of sea-fishing as an industry, and the desire to gain what could readily be converted into gold. No period of American history is richer than the period of discovery in interest for the lover of exciting adventures. Fascinating as it is to read about the unknown, the strange, the terrible, the search therefor requires many of man's most valued qualities, and while increasing his knowledge, adds uniformly to the accumulated wealth of mankind and to the extension of its industries. The navigator created opportunities for many and varied sorts of labor, not only directly by employing sailors and shipwrights, and by requiring an infinite number of ship-stores, but even more by the product which he brought home, and by the stimulus which his novel experiences gave to the imaginations of men. During the time of English discoveries in America, it will be remembered that Spain had grown so rich that its galleons bearing gold became a favorite object of piracy.

GREENLAND WHALE.

Sir Walter Raleigh, it will be remembered, got himself into trouble with his king, because in order to redeem his promise of returning later with gold, he selected as the location for his mining the Spanish ships. So too the Americans, during the Revolution, derived aid from the fact that the English fisheries being interrupted, those who had no other occupation were violently opposed to such Parliamentary action as destroyed their means of livelihood. Thus are the selfish interests of men overruled to beneficent ends by Divine providence. The material instinct of the whale has been less dwelt upon than similar other animals, probably because the whale is less open to observation, and because the "sailor's yarn" has come to be regarded with an unnecessary suspicion. Marvellous as are the fictions of man, the wonders of nature are even greater, especially when recited by one who restrains his imagination by well-authenticated scientific observations.

The wonders of God's universe are as astonishing as His abundant provision and constant care for His creatures. That "all is fair in war," seems to be the practical belief of animals as well as of man, and one means by which 'man takes advantage of the whale is worth mentioning—*the whale is betrayed through its maternal instinct.* The cubs are childlike in their innocence and fearlessness, and though valueless in themselves, are harpooned by whalers because they know that the mother will not desert her cub, and can thus be captured. Ordinarily the mother devotes herself to the care of her young, rising to blow whenever the less sturdy cub comes to the surface; encouraging it to swim off by itself instead of depending too entirely upon maternal care; gathering it under her fin, as a startled hen covers her brood with her wings; assisting its flight, when flight seems necessary, by supporting it by her own fin. All of this tender solicitude is delightful to the idle looker-on, and suggests many lessons which the human being would do well

HUNTING THE GREENLAND WHALE WITH HAND HARPOON.

to apply. But "in the midst of war the laws are silent," and all considerations of sentiment must be sacrificed to man's necessity for a livelihood—he does well, he believes, if he indulges in no unnecessary cruelty. The whale's affection for her cub is, as has been said, made to betray her into the loss of its life as well as of her own. Many of the most exciting adventures with whales have been occasioned by the infuriated creature's attempts to secure and protect her young—giving up her life for her young. On one occasion, a cub having been harpooned, the mother seized her young and dragged line, harpoon, boat and all several hundred fathoms. Finding her violent efforts fruitless to release her cub, she darted hither and thither after the manner of an agonized parent who sees no course to pursue. Harpoon after harpoon was plunged into her, but she seemed indifferent to pain and solicitous only for the safety of her cub. In this case the fishermen suffered no injury and were exposed to no danger. But quite frequently the enraged mother destroys the boats and causes the loss of human life in her frantic attempts to rescue her offspring. On one occasion in particular, after the mother discovered that her cub could not follow her, and that its blood which dyed the water was being shed by the enemy in the boats, she plunged many fathoms below the surface, swam some distance until she had acquired great momentum, and rising beneath one of the boats smashed it as though it had been a cockle-shell.

The crew were thrown into the water, and several of them, having become entangled, were borne to a watery grave. It not seldom happens, in spite of the military order which prevails amongst the crew of a whaling boat, and the

A WOUNDED WHALE STRIKING ITS PURSUERS.

long and varied experience of the men, that several boats will be destroyed before the frantic mother exhausts her strength and yields her life to the hunter. To be dragged by the whale many miles away from the ship, and then when thus deprived of assistance to find their boat smashed and themselves in the midst of the sea, surrounded it may be by drowning companions, has been an experience that many a whaler has undergone more than once. Such is the change which maternal solicitude or the frenzy of intolerable pain can work in a creature which, itself inoffensive, finds no enemy but man, except sword-fishes, thresher sharks, and possibly narwhals!

It has been truly said that "what is one man's meat is another man's poison," and while to our untrained or uncultivated palates the flesh of the whale would be as revolting as the bane of sick children, cod-liver oil, yet there are people who would reject with disdain what we consider delicacies and eat with avidity and relish the flesh of the whale. The inhabitants of the Arctic regions have discovered that their environment requires rich, oleaginous food, and hence use the whale as food as well as for oil and bone and ivory. The Esquimaux or Eskimos, of Kamtchatka, north-eastern Asia, the Arctic Archipelago, and the North American Arctic region take their name from being eaters of raw flesh—they call themselves not Esquimaux, but Innuit, a word which in their tongue means *the manly ones.* Through the accounts of various explorers (notably of Sir John Franklin, Elisha Kent Kane and Greeley,) we know that the Esquimaux were not vainglorious in their assumption of a name, for they have always been distinguished by many manly qualities, such as indus-

FIRING THE GUN HARPOON.

try, skill, hardship and activity. The necessities of their life have taught them to use wood and turf very sparingly, even as building material, so that their partially underground houses are, built of real bones of the whale; dress, bedding and tapestry, furnished by the seal or the reindeer; boats supplied by the whale and walrus; sleds made from skins; weapons and ornaments made from bone or ivory, and exhibiting a mechanical and artistic skill worthy of greater permanence; and finally, food and oil, if the Esquimaux would subsist at all; all these are from necessity derived from the spoils of the chase and most largely from the cetaceans which they capture. The uniform stoutness and good health of the Esquimaux go to show that in that climate, at least, one might fare royally upon flesh of the whale, which would never be "as dry as a remainder biscuit."

It is stated upon sufficient authority that from the thirteenth to the sixteenth centuries the whale's tongue was considered a delicacy by the Spaniards and Portuguese, and that the Lenten fare of the French peasant was chiefly whale meat. In passing, we should not fail to remark that the contribution of the whale to the very subsistence of the Esquimaux should be remembered when thinking of the many illustrations of the adaptation of animals not simply to their surroundings, but to render most efficient the lives of men who, however different from ourselves, are quite as much the objects of Divine care, and quite

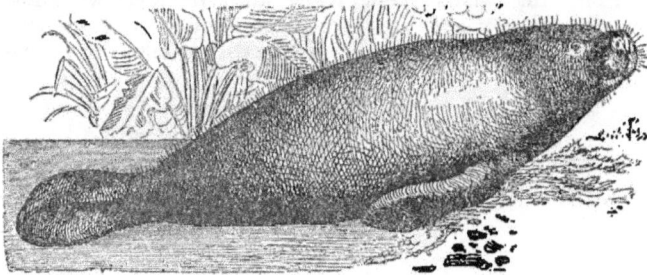

MANATUS (*Manatus Americanus*).

as conscious of their dependence. The white whale, despite the beauty of his color, the large number frequently found in a school, and its fearlessness in approaching vessels, is generally safe from attack by the whalemen. This is due, of course, to the fact that its yield of oil is not sufficient to make it a prize. Still it is quite as well perhaps, that the hard life of the sailor should be tempered at times by a perception of the beauty and playfulness of the marine monster, for though "a common sailor" may lack the most delicate sensibilities or the most susceptible imagination, he must nevertheless, in a vague way, it may be, appreciate objects of interest which do not

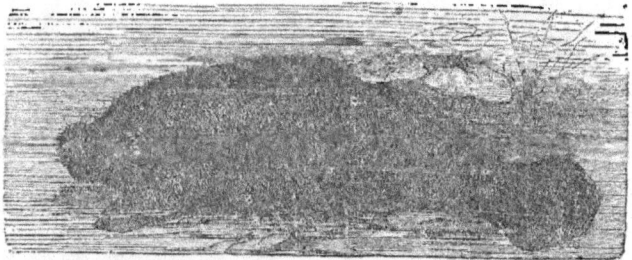

LAMANTIN (*Manatus Australis*).

directly conduce to his profit. Doubtless much of the beauty of ancient mythology is due to the etherealized telling over of the experiences of the ancient mariners, and it is easy to understand the functions and attributes with which a strange and beautiful water animal would be endorsed by the superstition of awe-struck men. There was a time when what is now mythology was religious belief, and it is not difficult to understand how the impressionable and imaginative Greek who believed honestly in Neptune's sovereignty over the sea, should assign to the sea king as servants the larger and more striking marine animals, and surround his chariot with mermaids and mermen and other strange conversions into human form and endowment with human attributes.

The Sirenia were for a long time confounded with the whale, but in the most recent systems of classification they have been assigned to an order by themselves. The *sirenia* are slow-moving, herbivorous, harmless creatures, whose ancestors belong to the fossiliferous periods. They have been commemorated in poetry and fable, being the creatures celebrated under the name of mermaids, about which so many interesting fables have clustered. Behring, for whom Behring's Strait was named, had with him on his voyages a naturalist whose study of the *sirenia* was so careful as to result in his names being given to the *Rhytina stelleri*, *Steller's Rhytina*, or *Northern Sea-Cow*. The species then abounded in the vicinity of Kamtchatka, but is now substantially extinct. The creature had a length of upwards of twenty-five feet; had a dark, hairless, thick skin; its tail resembled a pair of whalebone flukes, and its small head was ornamented by a bristled snout. It fed upon sea-weed and preferred to move about in herds (whence, possibly, its common name). It seems to be strictly monogamous, and the father, mother and two cubs of different ages generally kept together as a family. It was easily tamed, and when the sailors and natives did not kill it for its flesh they frequently converted it into an affectionate pet.

TRUNK SEAL, OR SEA ELEPHANT (*Cystophora proboscidea*).

No living specimen of this species has been found during the last hundred years, as the destruction of the animals resembled the wicked carnage which, but for the interposition of legislation, would have converted the American buffalo into an extinct species. Few governmental institutions are so unobtrusively useful as the Smithsonian, and, upon visiting Washington, one should visit its museum, which to make this remark relevant, contains various remains of the *Rhytina stelleri*.

The Sirenia, called **Manatees**, are found on the west coast of Africa, on the eastern coast of South America, and in the Floridian waters—each locality having its own species.

The **Trichechus americanus,** or **American Manatee,** has a gray hide with hair scattered over it and of a texture similar to that of the elephant. It ranges in length from ten to twenty feet, and does not confine itself to the sea, but ascends rivers with the greatest disregard of whether the water, instead

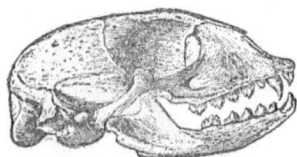

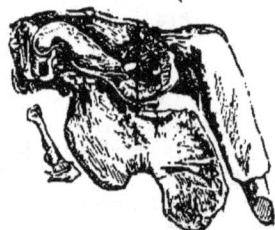

SKULL OF THE SEA HOUND.   SKULL OF THE WALRUS.   SKULL OF THE DUGONG.

of being salt, is either merely brackish or wholly pure. The few experiments upon trying to make them live in the confinement of the Zoological aquarium have been attended by only a short-lived success.

The **Dugong** (*Halicore dugong*) is the **Malayan Sirenian,** and is found as far south as Australia. It is bluish-black in color, and it is captured both for its seal-like flesh and for its oil. It is found frequenting bays, harbors and river mouths in the tropics, where it is the object of persistent pursuit.

The **American Manatee** is the species found on the eastern coast of South America. It has a grayish hide which resembles that of the elephant. In length it varies from ten

YOUNG WALRUS.

to twenty feet. It is found as far north as Florida, and is indifferent as to whether the water be salt or fresh. The animal is easily tamed, and therefore a favorite in such zoological gardens as contain specimens.

The **Walrus** (*Trichechus rosmarus*) belongs to the seal family, but is distinguished by its two immense canine teeth. It is gregarious and always keeps its sentinels posted. It is harpooned like the whale, and though yielding but little oil, is yet valuable for the ivory of its tusks.

The *walrus* is at times brave to desperation, especially when attempting to protect its young. It first endeavors to put its cubs in a place of safety, but failing in this, will clasp it to its breast and then dash themselves again and again at their persecutors. On one occasion Captain Cook, the noted navigator, had met with unusual success in hunting the *walrus*. As the boats approached the herd, the old ones seized their cubs with their fins and endeavored to escape into the sea. Several which had escaped returned for their

THE WALRUS.

young, and finding these dead, returned to the surface and bore off the lifeless bodies. One *walrus*, whose dead cub had been hauled into one of the boats, repeatedly attacked the boat, striking her teeth clear through the bottom of it. Doubtless had there been a poet among the *walrus* he would have anticipated the stirring lines of Elizabeth Barrett Browning; certain it is that the grief of the afflicted *walrus* was quite as deep and fully as sincere as that of any human parent. It may be as well to interject the remark that the cubs manifest like devotion for their parents, and often after having been assured of safety will return and join in their defence.

The Seal, though a mammal, is almost wholly aquatic, its terrene existence being limited to the brief periods during which it lies luxuriously on the shore, to which it has shambled. Passing most of its life in the water, its structure exhibits that wonderful adjustment to function and condition which converts a genuine study of natural history into the most exciting and most effective means of acquainting one with the marvellous wisdom of the Creator, and of the harmony which always exists among the laws by which He governs the universe. The elongation of the body; the legs, which are a compromise between legs and fins, with a preponderance in favor of the fin; and a skin impervious to moisture, are among the more evident of the *seal's* provision for the life which it is to lead. The *seal* is easily trained, and can be taught to

THE MORSE.

lend its services as a fisherman to its owner. The strangeness, gentleness, tractability and affectionateness of the *seal* make it a popular favorite, and in aquariums few creatures attract a more unceasing interest from the visitor. It will be remembered that even Achilles had a vulnerable spot, and the *seal's* weakness lies in its nose, upon which it is struck when captured. The *seal* is, after its sort, a vocalist, and while its moan displays even less variety than the Scottish bag-pipe, it is not unpleasant when a number exercise themselves in antiphonal choruses. More than this, the *seal* is fond of music when he is not the performer, and many a story is told of a musician finding among the *seals* an unsought but not uninterested or unappreciative audience.

The *seal* retires to the land when the tender age of its cub or cubs (for it sometimes has two), unfits it for a life on the ocean wave. When the young are about two weeks old they are taken to the sea by their solicitous mothers, whose instinct makes them much more intelligent in matters of primary education than the average human parent. The young are taught to fish and to swim—that is, they are initiated into the life which they are to lead. Fully aware of the danger of exacting too much, the mother will, at reasonable intervals, take her cub upon her back, and doubtless affording them, in addition to rest, the delight which human children find in being borne aloft by some grown person, whose strength and power seems to it gigantic. The males fight for the selection and possession of as many wives as please their fancy, but once having settled this question of relationship, the marital obligations are strictly observed, and no *seal* undertakes to invade the domestic rights of another.

The stronger males select the rocks, which they prefer for their inland residence, and the weaker must content themselves with what is left; but after any *seal* has thus taken possession of its dwelling it has no occasion to fear intrusion or expropriation on the part of the stronger or more cunning.

The American who visits Mount Desert, or whom necessity or pleasure takes to San Francisco, can find much amusement in watching the antics and in studying the habits of the *seal*. He will find the rocks covered by families of *seals*, so that he is reminded of the squatter shanties which so abound in New York city. From time to time some *seal* who has pitched his tent on

HOODED SEAL.

the very apex of some lofty rock may be seen leaping a sheer hundred feet into the sea, and then awkwardly clambering up the steep cliff apparently for the excitement of another dive. Although usually good-natured and friendly in their relations with each other, *seals* will, at times other than when struggling for the possession of their lady-loves, find some grievance not apparent to the human observer, and will then exercise that right which civilized men have disused for the courts and the ballots—those "vicarious shillalahs," as they have wittily been called. The *seal* is a great annoyance to the fisherman, for in addition to its own expertness as a fisher, it will possess itself of the fish which may be in the fishermen's seines, and leave him "to hold the bag."

The **Crested Seal, or Hooded Seal**, was doubtless the creature which the mythologists celebrated as the Triton, for the male, who, contrary to human customs, wears the bonnet or hood, has sufficient resemblance to a cowled monk

to lead the ignorant to believe it some odd species of humanity, or some supernatural being anthropomorphic in form, like the other gods of the heathen world. The Triton who gave his name to the family was, though the child of Neptune and Amphitrite, degraded to a fish-like form for some of the many misdeeds which the heathen deities were ever committing. His punishment was not simply personal, but was inflicted upon his offspring from generation to generation. The idea of heredity includes evidently the notion of the persistence of evil, and is a new presentation of the doctrine of original sin, and of the meting out to the individual a judgment extending beyond his own life, and affecting life and conditions of that of his successors. "As ye sow, so shall ye reap,"—not solely yourself, but those innocent ones whose lives are dependent upon yours.

HOME OF THE SEAL.

The *crested seal, hooded seal* or *Greenland seal,* is the most necessary of creatures to the Esquimaux; it serves as food, provides them with clothing, is used for the construction of boats, yields its air-reservoir as a buoy for floating the lance designed to destroy creatures of the sea, supplies the heads of their spears, and when needed for none of these purposes, serves as a pet. or amuses the children by its antics. The Esquimaux hunt the *seal* in two ways. The first method is to take advantage of the *seal's* known fondness for excavating a cavity in the ice, and having made therein a berth-like shelf, to pass there its times of restfulness. These *seal-caves* are always indicated to the trained eye of the hunter by an incrustation of snow, and when a slight scratching indicates the appearance of the *seal,* the hunter lying in wait

COMMON SEA HOUND (*Phoca vitulina*).

impales it with his lance. The second method is to go out in parties, and by lying down on the ice and moving so as to resemble the locomotion of the *seal,* to approach close enough to cut off the *seal's* retreat to the water. As the *seal* moves slowly and with difficulty except when in the water, it will now fall an easy prey to the hunters who attack it and seek to disable it by striking it on the nose.

The **Ringed Seal** (*Phoca fœtida*) is blackish-brown above, and yellowish-white below, and its back has markings of oval white or gray spots, somewhat resembling rings. It is abundant in the northern seas, and is a fur-yielder.

SEALS ALARMED.

The **Bearded Seal** (*Erignathus barbata*) is gray and sometimes has spottings, deriving its name from the coarse, heavy hairs that cover its nose.

The **Harp Seal** (*Phoca grœnlandica*) is so named because crescent-shaped belts of black extend from the shoulders down each side to the posterior part of the back. In the male, the head as far back as the eyes is black. The young are gray, spotted with brown, and lack the harp marking; the older seals are white or whitish-yellow, as the ground color.

The **Ribbon Seal** (*Histriophoca fasciata*) is dark-brown in coloring, but wears a yellow-white ribbon about its neck, a yellow-white girdle well back on its body, and lateral belts connecting these.

The **Sea Bear** (*Callorhinus ursinus*) is the fur-seal, so well known to luxuriously winter-clad persons. It has almost wholly disappeared from the northern coasts of our country. It is sufficiently common in zoological gardens to be a familiar sight, and, in its enforced state of domestication, it furnishes the visitor all of its attractions, unless it be the choral made by the vast schools of this animal.

FUR SEAL.

The **Sea Dog** (*Phoca vitulina*) is a seal which lives in harbors.

The **Hair Seal**, or Common Seal is frequently called the *Sea Lion*. It is polygamous, and rivals the Turks in the institution of the harem. Visitors to California have become familiar with this semi-aquatic animal, and there are many resorts, such as Mount Desert, which maintain a "summer school" of seals.

9

The **Sea Elephant** (*Crystophora proboscidea*) is the largest of the seal species. The male has a short proboscis, which has given the name to the species. Specimens thirty feet in length and eighteen feet around have been found. It is a denizen of the Southern Hemisphere, where it is hunted for the oil which it furnishes in abundance. Although a carnivorous animal, it will feed upon vegetables, and, though frequently living in salt water, is specially fond of fresh-water lakes and swamps. The young are born in June, and for two months receive the most devoted care from their mothers. After the young are brought to the sea the males fight valiantly for a selection of wives, for they are polygamous. When young the *sea elephant* is very easily domesticated. This seal sometimes goes through the experiences of Arctic explorers carried off on a floe of ice. As it cannot swim for any great length of time, it is very much perplexed when it unexpectedly finds itself floating far away from its happy hunting grounds. It is less helpless than the stranded sailor, for it can always secure food by foraging, but, unlike the sailor, the nearer it approaches civilization the greater are the dangers which threaten it. On one occasion a wandering *sea elephant* engaged, possibly, in humming to himself, "I'm afloat, I'm afloat," was picked up by a vessel and safely carried to London, which great metropolis it had not the least desire to visit, and which it failed to enjoy as a permanent residence. Such is the lack of effect of civilization upon the untutored barbarian!

MARBLED SEAL.

TRUMPET SEAL.

REPTILES.

ERE in the progression of species, which has kept pace with earth development, we find the Devonian age, or age of fishes, followed by the carboniferous, or coal-bearing period, in which reptiles first appeared. At this time vegetation was very rank, bathed as it was by dense vapors rising from the still heated earth, and the growth was generally of gigantic proportions by reason of this very great stimulation. Save where upheavals had occurred, caused by the bursting of the world's crust through the expansion and accumulation of gases generated by the fierce fires that raged within, the surface was everywhere either vast sea or morass, in which latter giant reptiles had their haunts. Of birds, mammals or insects there were none, because with their characteristics, as will be explained in appropriate divisions of this work, they could not have lived under conditions so unfavorable to their existence. But the very conditions that made other life impossible conduced to the propagation and stimulation of reptilian creatures, in which wise provision of nature we behold the special design of Providence, who in measureless wisdom adapted everything to the mutations of the developing masterpiece of His handiwork—the world.

While reptiles are widely distributed, we find them most numerous in the tropical regions, where the conditions are more nearly like those in which they had their birth, though the changes through which the earth has passed produced modifications noticeable not only in size, as compared with creatures of the carboniferous age, but their structure as well. Such changes have gradually taken place to accommodate all animal life to the subsidence of the seas, cooling of temperature and reduction of vegetation.

CHARACTERISTICS OF REPTILES.

Reptiles have been divided into no less than eleven distinct orders, but of these several classes there are certain peculiarities of organization and habits common to all. A simpler arrangement, sufficient for the purpose of the ordinary reader, is the division of reptiles into five orders, viz., **Chelonia**, or *tortoises;* **Loricata**, or *crocodiles;* **Sauria**, or *lizards;* **Ophidia**, or *serpents;* and the **Batrachia** or *frogs*. These are again divisible into several sub-classes, but to describe them separately would only serve to destroy the interest of the reader without affording any practical information.

A general characteristic of reptiles is found in that they are cold-blooded, because of a sluggish circulation, and their usual mode of locomotion is by crawling. Other singularities are noticeable in their great vitality, torpidity after eating, lethargy during cold periods, slow digestion, and muscular energy less highly developed than in the mammals. Most reptiles are oviparous, laying eggs which, however, are never incubated; but some are viviparous, like the rattlesnake, bringing forth their young so well developed that they are at once able to care for themselves. Many reptiles are also provided with shell or scales, so strong as to compose a veritable cuirass, impenetrable by common rifle ball when fired at ordinary range. Others again have minute scales so closely laid as to serve to facilitate the creature's motion through grass or water. But of the many varieties found in all countries, there are none that present a pleasing appearance, though some are clothed in a robe reflecting iridescent colors, and others are mottled with stripes and spots of splendid hues.

These, however, do not serve to lessen the natural dread in which they are held, which is felt for the harmless as well as for the venomous. But though we shrink from close familiarity with these repulsive animals, there is a curiosity, which some call morbid, which draws us almost irresistibly to gaze upon such creatures whenever opportunity presents, a desire that extends even to interest in illustrations of all reptiles, however abhorrent may be their appearance or the loathsomeness they excite. Indeed, paradoxical as it may appear, the excessively homely creatures attract us quite as much as do the beautiful, though we may not know why this is so.

FROGS AND TOADS.

Though some naturalists do not include frogs in their classification of reptiles, reserving them for a general order, called *batrachia*, I have preferred to follow the arrangement of equally reliable authorities and introduce this order as the one in most natural sequence to fishes on account of a similarity which exists between the two orders in many essentials. The two prime differences that serve to distinguish the frog and the toad is in the former being largely an inhabitant of the water, and having a smooth, flecked skin, while the latter, though not extremely averse to water, spends his life upon land, clothed in a much less inviting raiment than his aquatic first-cousin. The water frog is admirably adapted to his element by being provided with webbed feet and long hind legs, which enable him to move through the water at great speed. His toes, however, are of considerable length, by which he can cling to limb or log, and which make his footing sure on shore. Like all amphibians, the heart of a frog has a single ventricle, permitting the blood to circulate without the help of lungs, thus enabling it to remain for a long while under water.

Professor Ræsel, of Nuremberg, devoted several years of his life to a study of the frog in all its metamorphoses, and upon this subject supplies us with some very curious, as well as useful information, upon whose authority most naturalists rely for the following facts:

The common frog, in which designation both the toad and water-frog may be included, since the propagation of each is identical, chooses his mate early in

HUNTING TORTOISES ON THE COAST OF NEW GUINEA.

the season and is monogamic. The deposition of eggs begins about the middle of March, at which time the female expels nearly one thousand eggs by a single effort, which are immediately fecundated by the male, after the manner of fishes, who ejects a milky substance, beclouding the water surrounding the eggs, in which envelope they remain only a short while before impregnation is accomplished. The eggs now drop to the bottom, and after remaining there for four hours, begin to sensibly increase in size, as if expanded by gas, until they rise to the surface. Twenty-one days after being deposited a rent occurs on one side of the eggs, out of which protrudes a small tail; but not until the thirty-ninth day does the animal have motion, and two days later sheds its shell and falls to the bottom. One day later they have so increased in size and strength as to be able to rise to the surface again and feed off the jelly-like substance which had served them as a shell. A day later they assume the tadpole form, and three days thereafter fringes appear on either side of the neck, which do not develop, however, for three months. They now no longer depend upon the mucus, of which their original envelope was composed, for food, but begin a voracious consumption of pond-weed, off which they subsist until their legs are developed. It is not until the ninety-second day after hatching that two feet begin to appear at the base of the tail, and four days thereafter they refuse vegetable food, teeth having now developed, for a more substantial diet. After the legs are formed the tail remains for several days, disappearing gradually by absorption, and giving the creature the appearance of a hybrid lizard. Having at length developed, the young frog quits its former element and goes on shore in quest of insects, but rarely strays away from damp districts. If the haunt become dry by reason of a drouth, the young frogs seek shelter under rocks, logs or roots, and there continue until a shower, when they emerge from their retreat in such numbers as to have given rise to the popular, though erroneous, belief, that they have dropped down with the rain.

METAMORPHOSES OF THE FROG.

The frog lives, for the most part, out of water; but when the cold nights begin to set in, it returns to its native element, always choosing stagnant waters where it can lie without danger, concealed at the bottom. In this manner it continues torpid, or with but very little motion, all the winter; like the rest of the dormant race, it requires no food, and the circulation is slowly carried on without any assistance from the air.

Frogs live upon insects of all kinds; but they never eat any unless it be alive and have motion. They continue fixed and immovable till their prey appears; and just when it comes sufficiently near, they jump forward with great agility, dart out their tongues, and seize it with certainty. The tongue, in this animal, as in the toad, lizard and serpent kinds, is extremely long, and formed in such a manner that it swallows the point down its throat, so that a length of tongue is thus drawn, like a sword from its scabbard, to assail its prey. This tongue is furnished with a glutinous substance, and whatever

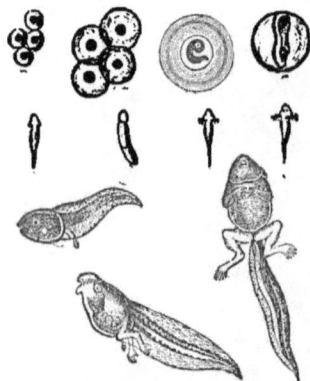

insect it touches infallibly adheres, and is thus held fast till it is drawn into the mouth.

BELLOW OF THE FROG, A TRUE PROGNOSTIC.

The croaking of frogs is a familiar sound in all tropical and temperate climates, the penetrating trill of the green-back, the guttural carping of the tree toad, and the bellowing notes of the bull frog being alike familiar to our ears, on which account the frogs of Holland are called Dutch nightingales. So loudly do these creatures bellow that they may be heard a distance of three miles. The notes are only sounded by the male, and are loudest during the coupling season, though before wet weather they are most dissonant and in full exertion. The frog is a true weather prophet, invariably croaking as a prognostic of approaching rain.

HOW FROGS EAT THE EYES OUT OF FISHES.

As frogs adhere closely to the backs of their own species, so it has been found, by repeated experience, they will also adhere to the backs of fishes. Few that have ponds but know that these animals will fasten to the backs of carp, and stick their fingers in the corner of each eye. In this manner they are often caught together, the carp blinded and wasted away. Whether this proceeds from the desires of the frog, disappointed of its proper mate, or whether it be a natural enmity between frogs and fishes, I shall not take upon me to say. A story told us by Walton might be apt to incline us to the latter opinion.

"As Dubravius, a bishop of Bohemia, was walking with a friend by a large pond in that

1. BUFO AQUA. 2. PEPA.

country, they saw a frog, while a pike lay very sleepily and quiet by the shore side, leap upon his head, and the frog having expressed malice or anger by his swollen cheeks and staring eyes, did stretch out his legs, and embraced the pike's head, and presently reached them to his eyes, tearing with them and his teeth those tender parts; the pike irritated with anguish, moves up and down the water, and rubs himself against weeds, and whatever he thought might quit him of his enemy, but all in vain, for the frog did continue to ride triumphantly, and to bite and torment the pike till his strength failed, and then the frog sunk with the pike to the bottom of the water: then presently the frog appeared again at the top and croaked, and seemed to rejoice like a conqueror; after which he presently retired to his secret hole. The bishop that had beheld the battle, called his fisherman to fetch his nets, and by all means to get the pike, that they might declare what had happened. The pike was drawn forth, and both his eyes eaten out; at which when they began to wonder, the fisherman wished them to forbear, and assured them he was certain that pikes were often so served."

GREEN BULL FROG.

The ordinary life of the frog and toad is supposed to be fifteen years, though Mr. Arscott declares he kept a toad for thirty-six years that finally lost its life by injury from a tame raven.

Many stories are told of the toad's venom and infection, but such are only idle fictions, for no creature is more harmless, or, I may add, of greater service to the gardener, since its chief subsistence is off the most noxious insects. So, also, are the stories without a grain of truth that represent living frogs as having been taken from the centre of stones and of large trees. His vitality is very great, but he perishes as quickly under the air-pump as any other reptile, though he may survive a fast of several months, as can many snakes.

Equally discreditable are the assertions made by many respectable and otherwise trustworthy persons to the effect that cancers are curable by the application of toads, which are represented as sucking out the eating virus and injecting a healing elixir.

Having thus described some of the

TREE FROG (*Hyla arborea*).

distinguishing characteristics of frogs and toads, between which there is a pronounced similarity, we may proceed to a description of the several species.

Green Bull Frog (*Rana esculenta*). At the head of the order indisputably stands the American *bull frog*, whose deeply resonant notes wake the echoes from ponds during the summer nights, and with whose succulent legs all good eaters are happily familiar. His haunt is the marsh or shallow pond, where he sits on the margin and takes such prey as comes within his reach. When insect food is difficult to procure he swims along the bottom and captures crawfish, water-beetles and occasionally minnows. The *frog*, however, is himself persistently hunted for the market, the favorite way of capturing him being with rifle bullet, net, or by fishing for him with a piece of red flannel, at which he greedily jumps and fastens his teeth, so that he may be taken before loosing his fangs.

The **Tree Frog**, also called **Spring Frog** (*Hyla arborea*), is a small but beautiful creature, considering that he is a reptile. Chameleon-like, he has the power to change his color when danger threatens, and usually adopts the hue of whatever perch he rests on. He has a pea-green back, spotted with black, and a yellow belly. His home is in the United States, east of the Rocky Mountains.

The **Flying Frog** (*Rhachophorus reinhardii*), a native of Borneo, is almost identical in appearance with the tree frog of our country, especially when in repose. Its difference is noticeable when in motion, since its toes are very long, with a web between, by which it is enabled to fly, with a slight descent, by spreading its feet and swimming through the air.

FLYING FROG.

The **Pond Frog** is very closely allied to the bull frog, its principal difference being found in a glandular sac that lies on both sides of the neck, and which are greatly distended when the animal croaks. Neither is it so large as our bull frog, but it is a much better jumper, and is extremely difficult to capture on account of its slyness.

POND FROG (*Bombinator igneus*).

The **Horned Frog** (*Ceratophrys cornutus*) is an inhabitant of South America, a country teeming with grotesque forms of animal life, but none more weirdly fantastic than this creature. There are several species, differing chiefly in size. The body is chubby, covered with a wrinkled skin that, from the tubercles thereon, seems to be eruptive. The back has a double ridge, calloused on the edges, meeting at the anus, and terminating on the head in two horns that rise up sharply above the eyes, giving the creature an impish appearance. It is very voracious, and does not hesitate to seize and gorge one of its own species, which its enormous mouth enables it to do.

The **Banded Toad** (*Alytes obstetricans*), also called the **Nurse Frog**, common to several parts of Europe, cannot rank with the horned frog for devilish features, but is infinitely more curious and interesting, rendered so by the very singular habit which the male frog has of carrying about with him the eggs laid by his mate. So soon as the female voids her spawn, her attentive companion takes immediate possession and fastens them, by means of a glutinous substance exuded from his mouth, to his legs and quarters, where they remain until the young are plainly to be seen through the transparent envelope. At this stage, which is a month after the spawning, the patient parent proceeds to the pond and there divests himself of the burden, to which he never after gives any heed. In a short while, however, the young burst their envelope and betake themselves to active life in the water, until they develop into the mature *frog*, when they repair to the land.

HORNED FROG.

The **Surinam Toad**.(*Pipa Americana*), however, is even more singular than the banded toad, in that it rears its young in a yet more curious manner. When the female lays her eggs the male is always near at hand, whose office it is to perform the very strange operation of placing them upon the female's back, to which they adhere by reason of a secretion which exudes from her skin at this time. This process is an exceedingly interesting one. The male takes the eggs in his fore feet, which he uses with all the dexterity of true hands, fixes each egg carefully in place, smooths the whole over and then apparently blesses his mate and wishes her a God-speed in her maternity. In a few days the eggs become embedded within the skin, each having a separate cell, in which it is incubated and retained until the young have all their limbs and can burst their envelope and move freely on the ground.

This very strange creature is found in Surinam, from whence it takes its name, and also in several other parts of South America. The general appearance of the *toad*, even when not enlarged and disfigured by the young

SURINAM TOAD.

in its back, is very repulsive. The head is shapeless, in that it resembles nothing else in nature, the nose terminating in a flexible snout, and the body being covered with horny projections, on which account it is sometimes called the *wart toad*.

The **Pouch Frog** (*Monotrima marsupiatum*) is a small species found in Mexico and South America. It is generally of a green color, flecked with black. Its chief peculiarity consists in the female being provided with a pouch in which she carries her eggs, very much after the manner of marsupials carrying their young, except that the pouch is located on the frog's back. When fully dilated with eggs the sack covers the entire back and gives to the creature a balloon-like appearance, quite comical to behold.

BANDED TOAD.

The **Rope**, or **Painted Frog** (*Rana temporaria*), is a pretty creature found throughout the Levant and along the Nile. It derives its name from the beautiful striping of its body and bands of rich brown on its legs. It is peculiar in being at home in either fresh or salt water. The coloring is generally olive green, spotted with white, and longitudinal streaks of gray. Sometimes the skin is smooth and velvety, and again covered by warty excrescences.

POUCH FROG.

The **Common Garden Toad** of Europe (*Bufo vulgaris*), is a repulsive creature, much less endurable than the ground toad of America. Its color is ashy, with darker splotches extending backward for an inch behind the eye, while the skin is rough and pimpled. In size it equals the American toad.

The **Horned Sand Toad** (*Phrynosoma orbiculare*) is a native of the extreme southwest and northern Mexico, where it is popularly called the *Mexican frog*. In size it is scarcely so large as our common toad, but though small its appearance is quite as horrible as anything found in nature. It very much resembles the moloch of Australia and the wart toad of Fernando Po. The body is covered with tubercles, out of which rise horny spines, a ridge of these running down the back, while around the neck and at the apex of the head are six specially prominent spines. More properly this creature belongs to the *lizard* species, as it has a short, thick tail, but in other respects it retains all the batrachian characteristics. As indicated by the name, its home is in dry, sandy districts, where it burrows during the dry season.

ROPE FROG.

The **Solitary Frog** is found in various parts of the great west, but generally preferring a sandy region. It is also covered with tubercles, though these do not terminate with spines. On the nozzle, however, is a spine which it uses to scoop out the earth, something after the manner of hogs in rooting. It also has the power of burrowing in the sand tail foremost, in doing which it works its way downward, very like a crab, and makes a hole some six inches in depth. Its color is a pale olive, spotted with brown, while down the back run stripes of pale yellow. Its eyes are large, across which run two black lines at right ángles. In size it rarely exceeds two inches long, by one inch in height. Its movements are slow and apparently laborious.

THE LIZARD SPECIES.

Under this head, the species technically known as *Lacertiliæ*, from the Latin *lacertæ*, meaning *moving quickly*, a Latin term also for lizard, are included. These creatures are the link connecting batrachians with ophidians or snakes, since in many respects they resemble both. They are usually of an elongated form, and while all have four legs, in some species these are only rudimentary or even externally absent. But even in those which have legs well developed the muscular power is so small that they rarely lift the animal's body from the ground, but are used rather to push the creature forward. In some, however, the feet are fashioned for grasping, as the chameleon's, while in yet others the toes terminate in sucker discs, which enable the animal to adhere to smooth walls. A majority lead an

GARDEN TOAD (*Bufo vulgaris*).

arboreal life, while two species, the *flying gecko* and *flying drake*, have the power of half-flying, half-leaping, from tree to tree. Nearly all are oviparous, laying their eggs in rude nests chosen in dry places and about dead timber; the eggs are never numerous, and are usually connected in a chain. Some few species, however, are oviparous, producing their young so well developed that they need little parental care. There is but one species recognized by naturalists as being venomous, viz., the *heloderma*, commonly called the *Gila monster*, an inhabitant of Arizona, New Mexico and northern Mexico, and it is greatly to be doubted whether even this creature is poisonous, as will here-

HORNED SAND TOAD.

after be explained. The water species of lizards, or newts, belong more directly to the batrachian class, since they produce their young after the manner of frogs, though the eggs are never connected, being laid singly and hatched in succession.

The **Flying Gecko** (*Ptychozoon homalocephalum*). This singular reptile, found nearly everywhere in the tropics, varies in size from six to twelve inches in length. Though repulsive in shape, their marking is exceedingly attractive, banded as they are with zig-zag streaks of deep black from tip of nose to tail. The toes are broadly spread and webbed like the feet of the common blue mud-hen, though for what purpose is not known, as the animal does not enter the water. The most singular feature observable in the *gecko* is the dermal expansion that extends on both sides, from the mouth to the tail, terminating in a foliated, or leaf-shaped tip. This expansion, when the animal is quiet, appears as if the skin was too large for the body and the surplus lay loosely along its sides. Towards the tail, however, the skin is not continuous, but appears like a succession of plates. So far from this seeming surplus of skin being a burden to the animal it serves the most useful purpose of enabling the creature to fly through the air for considerable distances, though always at a descending degree. At the moment the *gecko* leaps, these skin expansions are spread so that it sails in a manner identical with the flying squirrel.

The **Flying Drake** or **Dragon** (*Draco volens*) bears a close analogy to the gecko, though it is a much less pleasing creature. It possesses, though in a lesser degree, the power described in the gecko, being able, by means of the expansion of its surplus skin, to sail from limb to limb. The *flying drake* is also peculiar to the tropics, where it attains a considerable size, and in Africa is frequently met with more than two feet in length.

SPECIES OF CREEPING, CLIMBING AND WALL-RUNNING LIZARDS.

The **Chameleon** (*Chamæleo vulgaris*) is also a tropical reptile, but most common to Africa, where its range is from the Mediterranean to a line some distance below the equator. This creature is entirely arboreal, quitting the trees only to deposit its eggs under a small hillock of leaves gathered for the purpose. It possesses many points of singularity which make it possibly the most curious animal in nature, as well as one of the most hideous. The head is large and angular, covered with shields, and highly ridged, while the body is covered with a shagreen skin very like a shark's. The tongue is extensile and club-shaped, which the animal may dart out half the length of its body, and with such lightning-like rapidity that the eye cannot detect the

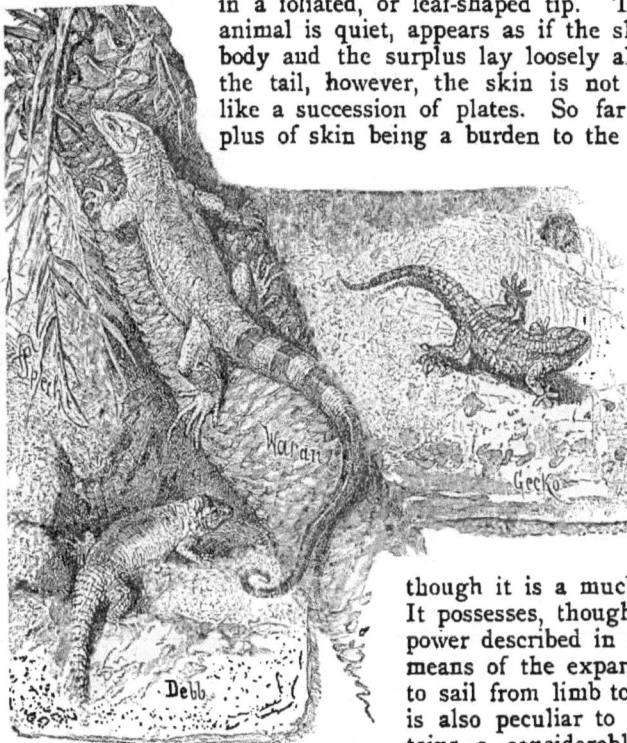

motion. Its food is flies and other insects which it takes by striking them with the tongue, to which the prey adheres on account of the glutinous excretion with which it is provided. Singular as these several features are, the eyes of this animal are yet more curious; they are of prodigious size compared with the body, yet over the balls is a growth of true skin very similar to that which covers the body, in the centre of which is a small hole opening upon the pupil. The two eye-balls are entirely independent in their action, by which the animal is enabled to look at two objects in opposite directions at the same time. Besides these singularities the *chameleon* is provided with hand-like claws and prehensile tail, with which it can cling most tenaciously to the limb of a tree,

FOOT OF THE GECKO—UNDER SIDE.

WALL GECKO.

defying all efforts to shake it off. Much has been said of the *chameleon's* power to change its color at will, but while this faculty may be exercised to some extent, the power is not nearly so pronounced as it is in many other of both the frog and lizard species. This strange ability is due to the fact, as Kingsley states, that all such creatures are provided with two or more layers of pigment cells underlying the transparent epidermis, a lighter and a darker, changeable at the will of the animal, or stimulated by surrounding objects.

The **Iguana** (*Iguana tuberculata*) comprises several species of a lizard peculiar to the West Indies and South America. Its appearance is decidedly uncouth, especially the tuberculated kind, which is distinguishable by a pyramidal head, high dorsal ridge armed with bony spines, and a dewlap, or throat, with long pendant folds, and a mouth that presents a hideous grin. The tail being very long, when

FLYING GECKO.

the animal is asleep on the bough of a tree, it hangs down in remarkable similitude of a snake. The natives, who hunt the animal extensively for food, profit by this habit, for by this exhibition of the creature's tail they are most easily discovered. The animal is very susceptible to music, a disposition which the natives take advantage of, for when approached, if the reptile is on the point of retreat, the natives begin to whistle briskly, at the sound of which the creature stops and remains · listening until a hair noose can be slipped over its head. With a jerk he is then brought to the ground, whereupon his anger is manifested by blowing himself up like the frog that tried to rival the ox.

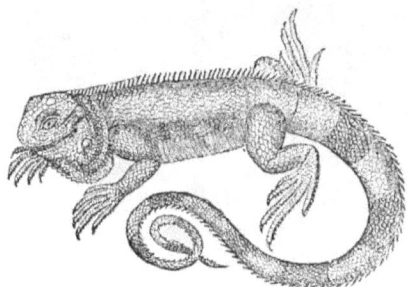

CHAMELEONS.

The **Common American Iguana** is from four to five feet long. It is very common in all the warm parts of America, where it remains in the woods, at the environs of rivers and sources of spring water. It passes

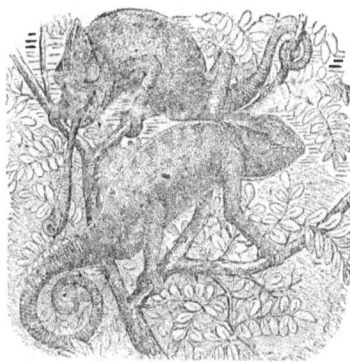

IGUANA.

most part of its time on trees, sometimes going to the water, and living on fruits, grain and leaves. Without being either venomous or dangerous, its bite is exceedingly painful; and when it is angry, the goitre, which it has under its neck, becomes distended and expanded. This reptile has great tenacity and endurance of life, and will resist the blows of a stick or cudgel very well. Accordingly, it is usually hunted with the bow or the gun. The females are smaller than the males, but their colors are much more brilliant. They lay eggs in the sand, about as large as those of pigeons, but a little longer, and of equal thickness at both ends. The shell of these eggs is white, even and soft. They are entirely filled by the yolk, and can hardly be said to have any albumen. They never harden by fire, but only become a little pasty. But their flavor is very agreeable, and they are constantly eaten in Surinam and Guiana. A single female will lay about six dozen.

The flesh of the *iguana* is considered as delicious, and is in great estimation throughout all the warm parts of America. It is white and delicate. Many persons, however, consider it as

FLYING DRAGON, OR DRAKE.

unwholesome, especially for those who are infected with syphilis, some symptoms of which, such as pains in the bones, etc., it is supposed to aggravate or cause the return of. At Paramaraibo, it is sold extremely dear, and highly thought of by epicures. Pisou, and many others of the old travellers in America, have spoken in high terms of the virtues of the bezoar of the *iguana*, a kind of stone, found, say they, in the stomach or cranium of this reptile. But, at the present day, this substance is fallen into the most absolute disrepute among all medical practitioners.

NILE MONITOR.

The **Slate-colored Iguana** is but three feet in length. It inhabits the same places as the former species, and may be merely a variety of it, in age or sex. Seba derives it from the island of Formosa.

The **Horned Iguana** of St. Domingo is about four feet long. It is frequently found in the hills of St. Domingo, between Artibonite and Gonaives. It lives on fruits, insects and small birds, which it seizes with marvellous agility, and during the day it couches on trees and rocks to watch for its prey. During the night, and the entire season of the great heats, it retires into the hollows of rocks, or into the holes of old trees, and it passes about five or six months of the year there in a state of lethargy. This reptile is considered by the negroes as a delicious meat, and is accordingly sought after by them with great avidity. According to the report of the colonists, its flesh resembles in flavor that of the roebuck, and the maroon dogs make great slaughter among these reptiles. The colors of this *iguana* are not precisely known.

The Monitor (*Monitor nilolicus*), so called from the fanciful idea that it gives timely warning of the prox-

HEDGE LIZARD.

imity of any venomous serpent, is a habitant of the Nile, and also the marsh regions of India. It grows to a length of about six feet, and has both the power and disposition to do mankind inestimable service in destroying the eggs and young of the crocodile. Its tail is equal to the length of the body, and, being flat, propels the animal swiftly through the water. The head is sharp-pointed, the neck is thick and strong, and its feet are armed with powerful talons. The mouth is provided with small but sharp teeth, and, singular enough, its tongue is round and forked like that of a snake. On account of its usefulness the Nile dwellers have great regard for its life, but many of the lowest caste of Hindoos hunt it persistently for food.

1. VARAN OF THE NILE.   2. VARAN OF THE DESERT.

The Hedge Lizard (*Lacerta sterpium*) is a harmless little creature that may be found in all parts of America, making his abode under the bark of decaying logs, and basking in the sun on rail fences. His coat is gray and rough, and his tail so friable that the slightest stroke will break it. But being bereft of his tail by accident gives him little concern, for the damage is speedily repaired by the growth of a new one. I have often found their nests under the bark of the lowermost rail of a worm fence, and tried to hatch them out, but without success. The eggs are usually six in number, and connected together like beads. I do not think the creature gives any attention to the eggs after they are laid, but yet a disturbance of them seems to destroy all vitality, judging by my invariable failure to artificially hatch them.

The **Metallic-backed House Lizard** is another familiar reptile throughout the United States, making its home under door-sills, in rotted logs and under the bark of yard trees. Its skin is bright, with variable colors, and shines with splendor when reflecting the sun's rays. It is generally dreaded, on account of its somewhat snaky appearance, yet nothing can be more harmless, or be more deserving of our protection.

SKINK.

The **Skink** (*Scincus officinalis*), an animal of Africa, possesses the singular power of burying itself in the sand with such rapidity as apparently to disappear in a previously-formed burrow. How this is accomplished does not clearly appear, though the fact cannot be gainsaid. The animal is thick of body, striped with black and white, and has four legs seemingly poorly developed, for which reason it travels slowly, and therefore seldom strays far from its abode. Formerly this reptile commanded the attention of the civilized world, by reason of the claim made that its body, reduced to a powder, possessed the most astonishing remedial virtue, and which was prescribed for nearly all imaginary ills.

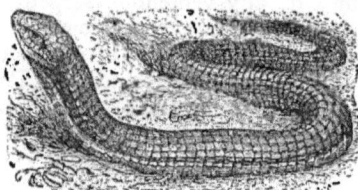

PALE-SNAKE LIZARD.

The **Pale-snake Lizard** (*Pseudopus pallasii*) is a singular creature, combining the characteristics of the *lizard* and the serpent, as its name implies. It is a native of Europe and Asia, and finds its abode in the darkest recesses of the woods, where it spends the time in quest of the eggs and young of birds. It grows to the length of two feet and has all the appearance of a snake, except that it has the head of a lizard and the rudiments of limbs. It is an extremely timid creature, and being capable of great speed, is difficult of capture.

AMPHISBŒNA.

The **Amphisbœna** (*Amphisbæna alba*), or **Two-headed Snake**, so named by Pliny because of its ability to move with equal celerity either forwards or backwards, is native to the warmer portions of Europe. For many years it was seriously believed the creature had a head at either end, and though this has long been disproved it does not yet appear quite plain how it manages to crawl so naturally backwards, especially as the head and eyes are so distinct, yet apparently not vitally important, so far as its travelling or search for food is concerned. It rarely grows to a greater length than a foot, lives among decayed brush a portion of the time, but more generally it burrows in the earth like an angle-worm. Other species are common to South America.

The **Seps Chalcidica**, a word from the Greek, signifying *corruption*, inhabits regions bordering on the eastern Mediterranean, where in early years it was looked upon with great dread from the belief that the creature inoculated cattle, during their sleep, with a poison which caused the flesh to slough off until death supervened to end the victim's sufferings. From this groundless belief the name is derived. It is in fact a perfectly harmless little animal, somewhat longer than the common *lizards* of America, but small in body and with such

SEPS, OR METALLIC BACK.

diminutive legs that it scarcely ranks above a worm, especially since these organs are of little use in the creature's locomotion. It is of a gray color, with four dark stripes running from head to tail, and subsists on slugs, worms and insects. It is now most common in Dalmatia.

The **Wallowing Worm** (*Siphonops annulatus*) is a native of the South American tropics, but even in this restricted region it is by no means common, and its habits are comparatively little known. Its color is dark, with rings of white from neck to tail. The head is neither that of snake nor lizard, but bearing a close resemblance to that of the blind-worm, except for its mouth, which is prominent. It grows to the length of one foot, and while sometimes burrowing, more commonly lies concealed under leaves in damp places.

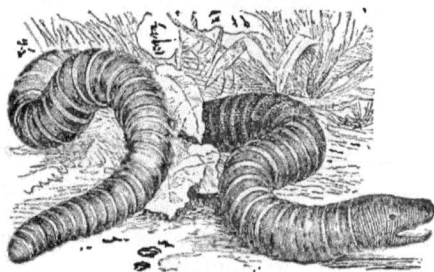

WALLOWING WORM.

The **Blind Worm**, also called **Slow Worm**, on account of its sluggish movements, is very common in England, where it haunts the hedges and heathers in quest of slugs, its favorite food. Notwithstanding that it is one of the most numerous of English reptiles, its true character is so little understood by the farmer folks that many of them regard it with intense dread, pronouncing it so venomous that no antidote can counteract it.

The name *blind worm* is also a misnomer, since it is provided with small, though very bright and sharp eyes, by which alone it could catch the insects on which it feeds. In length it rarely exceeds fifteen inches, but on account of the fierce-appearing habit of thrusting out its forked tongue, at the same time rearing its head in a bold attitude at the first intimation of danger, it certainly looks formidable. Instead of verifying its appearance, however, the worm will not stand to fight, but to facilitate its escape will frequently throw off its tail with a snap, and leave this portion to its enemy while the body finds refuge in flight. This fragility of the *blind worm* is a most curious characteristic, from which fact has no doubt arisen a belief in the existence of the glass snake, having power to break itself into pieces and afterwards to unite the parts together again. In the *blind worm* this power certainly does reside, but is restricted to the separation of its tail, which constitutes something more than one-half of the creature's entire length. Where the tail is thus thrown off, it continues active for about half an hour, a provision of nature to enable the animal to deceive its enemy and effect its escape. The tail seems to be attached to the body rather than to constitute a continuous growth, but when shed another speedily supplants the one lost.

The Grotto Proteus (*Proteus anguineus*). This curious creature, while classed with the lizard order, is more truly a newt, its home being in the water, out of which it cannot live above a few hours. It was first discovered in a grotto near A d e l s b e r g, Prussia, which extended many hundred feet underground, where profound darkness was continuous. At the bottom of this grotto was a small lake,

WORM SNAKE (*Catamaria albiventer*).

the shores of which were covered with soft mud, in which the *proteus* was found crawling. Several of these animals have been taken to England, one of which was kept by Mr. Beale in an aquarium for five years, in all of which time it is not known to have taken any food.

The *proteus* seldom attains a length of more than one foot. Its color is a very pale gray, with the faintest showing of a flesh tint. If it possesses eyes they are not discernible, though rudiments of orbits are observable in the skull; as it seeks the dark invariably, and is most uneasy in a place where any light is present, the probability is that it is sightless. But an equally singular characteristic is seen in the very curious gills with which it is provided, resembling, as they do, the delicate mosses that grow in stagnant water, and so transparent that the circulation of blood is plainly visible to the naked eye. Its mouth is armed with sharp teeth, and the jaws are so powerful that the creature has been known to bite through a fish, taking flesh, bone and fin at a single snap. In the aquarium the *proteus* has been fed on gold-fish, fresh meat and frogs, on which it thrived extremely well, by which its natural food is suggested.

The **Basilisk** (*Basilicus mitratus*) is found along and north of the equator, in the Western Hemisphere, as far as Mexico. Many singular superstitions have been prevalent concerning this creature, some of which are believed by the more ignorant even to this day. By not a few it is still spoken of as the king of reptiles, whose sovereignty is attested by the crown (so called) he wears. He was also allied to nameless things, having some occult powers conferred by reason of having eaten an egg laid by a cock, upon which a snake had set. Others as strongly maintained that the egg thus laid, out of the course of nature, was incubated by a toad. A glance of his eye was believed to be the arrow of death, while his breath infected the air with a poison so virulent as to kill not only the animal life that fell within its influence, but to destroy vegetation as well. As one naturalist in the seventeenth century wrote: "This poison of the *basilisk* so infecteth the air, and the air so infected killeth all living things, and likewise all green things, fruits and plants of

PROTEUS.

the earth; it burneth up the grass whereupon it goeth or creepeth, and the foules of the air fall down dead when they come near his den or lodging. Sometimes he biteth a man or beast, and by that wound the blood turneth into choler, and so the whole body becometh yellow or gold, presently killing all that touch it or come near it." The cock was the only creature before whom this terrible animal would retreat, hence travellers in the regions where it was common

LIZARD OF THE STEPPES, GROUND HENS AND HORNED VIPER.

rarely ventured upon a journey without carrying with them one or more roosters. All these idle stories have long since been exploded, until we now know this much-abused animal better, and close acquaintance establishes his innocence under all circumstances. The *basilisk*, full grown, is some three feet in length, his unsightly appearance being rendered more hideous by reason of a peaked crown he wears upon his head, not entirely unlike the regulation clown's hat. His back, too, is not more inviting, for from the spinal column rise long dorsal spines like the fins on the backs of certain fishes. The tail is also similarly armed while the skin is rougher than shagreen. When excited, the animal distends his crown to surprising dimensions, at the same time raising his dorsal spines, which give to him a grueful and horrible aspect, quite enough to stimulate credulous people to the creation of fables such as those just mentioned.

The **Frilled Lizard** (*Chlamydosaurus kingii*) is about the same size as the basilisk, and its skin is also similar, but the resemblance is not carried further. This lizard is a native of Australia, where it is quite common, and its habits well known, since no superstitious awe is manifested for it. The adult *lizard* would look very much like our common species of wood lizard, but for a most curious appendage which grows about his neck, and from which is derived his popular name. Concerning this animal Captain Grey thus writes:

FRILLED LIZARD (*Chlamydosaurus kingii*).

"As we were pursuing our walk in the afternoon, we fell in with a specimen of the remarkable *frilled-lizard*. It lives principally in trees, though it can run very swiftly along the ground. When not provoked or disturbed it moves quietly about, with its frill lying back in plaits upon the body; but it is very irascible, and, directly it is frightened, it elevates the frill or ruff, and makes for a tree, where, if overtaken, it throws itself on its stern, raising its head and chest as high as it can upon the fore legs, then, doubling its head underneath the body, and displaying a very formidable set of teeth from the cavity of its large frill, it boldly faces an opponent, biting furiously whatever is presented to it, and even venturing so far in its rage as to fairly make a charge at its enemy."

The **Water-dog** (*Protonopsis horrida*) is a creature pretty well known all along water-courses of the Middle States, but in latter years it has become much less common than formerly. When a boy I frequently caught them when fishing in the Ohio river, and especially in creeks, for which they have a greater liking, but I am told that it is very seldom one is now caught in places where I once found *water-dog* fishing good. A very excellent description of this creature, by Messrs. Townsend and Frear, was published in the *American Naturalist* some few years ago, as follows:

"The *protonopsis*, called *water-dog*, is an exceedingly voracious animal, feeding on fish, worms, craw-fish, etc.; some of those taken by me disgorged craw-fish shortly after being caught. May it not be a scavenger of the water? All my specimens were caught in a creek in western Pennsylvania. It is well known to those accustomed to fishing the streams of this region, from its troublesome habit of taking bait placed in the water for nobler game. When thus hooked, its vicious biting and squirming, together with the slime

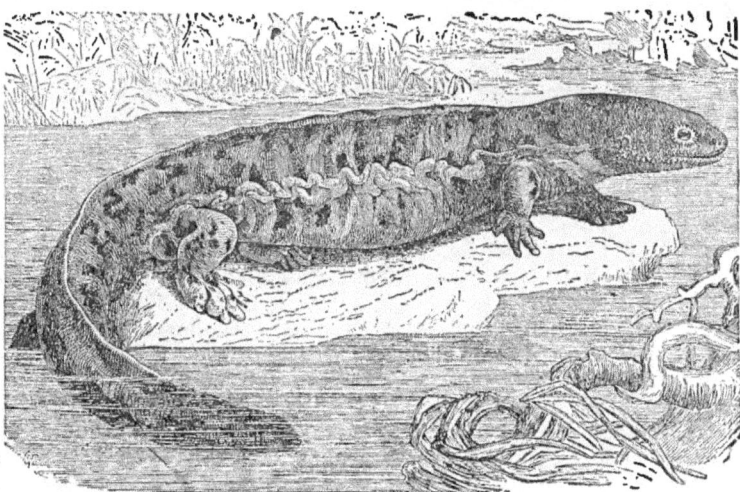

WATER-DOG

its skin secretes, render it extremely disagreeable as well as difficult to handle. It is often hooked while bottom fishing for catfish, and to avoid the trouble of handling the creature its head is cut off to facilitate the extraction of the hook.

"In the early summer, when the water is clear, *water-dogs* are often to be seen on the bottom in considerable numbers. Once, when fishing with some friends from a large rock in Loyalhanna creek, we saw quite a school of them moving sluggishly about among stones on the bottom. They would quickly take our hooks baited with meat or a piece of fish head. In one instance two large ones laid hold of the same bait and were landed on the rock. Last August I fished on the same spot for them, but without success. Acting on the advice of a 'native,' I dropped some pieces of fish near certain rocks and this brought out the retired *water-dogs*, so that I soon caught ten. Those taken measured from ten to eighteen inches in length, but fishermen say they have frequently caught them measuring two feet.

"They are remarkably tenacious of life. I carried my specimens six miles in a bag behind me on horseback, under a blazing hot sun, and kept them five weeks in a tub of water without a morsel to eat, and when I came to put them in alcohol they seemed almost as fresh as ever. During their confinement in the tub, two of the females deposited a large amount of spawn. This spawn was something similar to frog-spawn in its general appearance, but the mass had not the dark colors of the latter. The ova were exuded in strings, and were much farther apart than frog's eggs. They were of a yellow color, while the glutinous mass which connected them had a grayish appearance. The spawn seemed to expand greatly by absorption of water. It lay in the tub among the animals for a week, but was not disturbed by them."

ADULT AXOLOTL.

The **Spotted Lizard** (*Lacerta oscellata*) is a native of the Old World, found most common in the warm and barren region of Central Asia and Arabia. It is beautifully spotted with white on a dusky brown background, and is a swift swimmer, notwithstanding the unusual length of tail it is compelled to drag. Its home is upon the ground, which it seldom leaves, except to scamper over rocks or logs. In length it rarely exceeds ten inches, of which the tail is considerably more than half.

The **Green Lizard** (*Lacerta viridis*), peculiar to regions about the Mediterranean, is most beautiful in its gorgeous livery of brilliant green, as well as sprightly, inquisitive and courageous, though entirely harmless. It is a frequenter of old ruins, on the fallen glory of whose monuments it delights to bask. It is usually about six inches in length.

The **Waran** and **Debb** are both natives of Germany, the former being arboreal in its habits, while the latter, though somewhat resembling the gecko, has its toes adapted to ground dwelling, being short and stout, enabling it to travel rapidly through the brush and crevices of rock, where its haunt is chiefly made. The *waran* grows occasionally to the length of one foot, but the *debb* is a thick creature, and rarely

LARVA OF THE AXOLOTL.

exceeds six inches; its appearance is that of a poisonous reptile, but is harmless.

The **Amblystoma**, also called **Axolotl**, is found distributed pretty widely throughout the American tropics, but its most popular haunt is Mexico, where, notwithstanding its most repulsive appearance, its flesh is used rather extensively for food. There are fifteen species of this creature in Central America, Mexico and along the Pacific coast, and a single species in Japan. In size they vary from two to twelve inches in length. A most singular thing concerning this creature is the fact that the larva or tadpole is much larger than the adult animal, for which reason it was formerly believed that the creature reversed the invariable process of nature by furnishing the one single example of a regressive transition, from the adult to the larva, as wondrous indeed as was the fountain of youth, or the alchemist's elixir. The larva of this strange animal somewhat resembles the common frog-tadpole, except that it possesses gills very like those of the proteus, in which the blood circulation is as easily observable with the naked eye. These observations apply to the single species known as the *axolotl*, however, since in the others the tadpole is of ordinary size, nor have they the singular gills peculiar to the species just described.

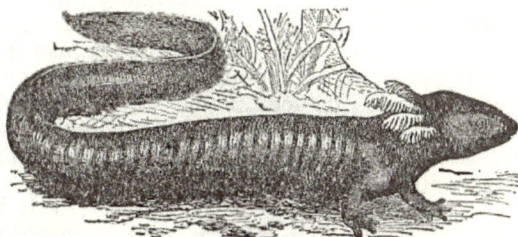

MUD EEL.

The **Mud-Eel** (*Siren lacertina*), or pseudo-lizard, is also a habitant of North America, its northern range being the latitude of St. Louis, but so very rare is the creature that it has not often fallen under scientific observation. It is usually found in ditches and swamps of the Southern States. In color it is a very dark lead, with indistinct rings of yellow about the body. It varies in size according to locality, since in the south it is occasionally as much as two feet in length, while those found in the north are scarcely half as long. It has two well-developed fore-feet and three rose-tinted gills on either side of the head, which the poor creature sometimes loses in attacks made upon it by fishes that regard the gills as delicate morsels.

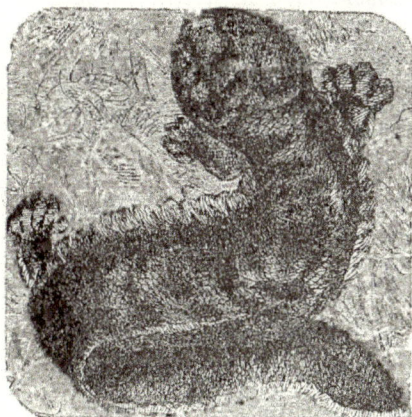

GIANT SALAMANDER.

The **Giant Salamander** (*Crypto branchus japonicus*) is the largest of the lizard family, and so dreadfully repulsive that it is fortunate the species should be confined to Japan, where it lives in lakes and pools about the basaltic mountain ranges. The first specimen taken from that country was conveyed to Europe by Dr. Von Siebold, of Leyden, who secured a pair, but on the journey the male devoured his bride for want of more desirable food. In the summer of 1887, a still larger specimen was secured for the Philadelphia Zoological Garden by some gentleman who brought it over in safety from Japan, together with another somewhat smaller species, known as the *yellow salamander*. I had the pleasure of examining these shortly after their arrival. The length of the larger specimen was said to be five feet, though I hardly think it was above four, but the size was certainly formidable, considering the creature's horrible aspect. I viewed them as they lay in a well-lighted glass tank, so that every part of the animal was plainly discernible. The larger one was a dirty black, and covered with excrescences, while the skin was so wrinkled and lay in folds at the side as if the reptile had put on a coat very much too large for him. The head was considerably greater in diameter than the body, both head and body being flat, or perhaps four times greater in width than in thickness. The eyes were so small that I was a long time examining the creature before being able to discover them, as they were not only apparently immoderately small but lustreless as well. The animal had eaten nothing since its capture up to the time that I saw it, by which I judge it may survive many months of fasting without great inconvenience. Its food in the natural state is eels and fish, which it contrives to capture, notwithstanding the very sluggish nature that it exhibits. In captivity the *salamander* will no doubt develop an appetite for fresh meat, as most carnivorous creatures do.

Large as is the *giant salamander*, it is but the prototype of a progenitor now extinct, which in life was, perhaps, many times greater. In 1726 there was discovered near Emingen the skeleton of what was for a long time supposed to be a man of giant proportions, who had perished in the deluge, and was profitably exhibited as such until critical examination by a scientist showed the bones to be those of a *salamander*. The feet of the *salamander* are thick, soft, and without terminating in claws, contrary to most of the lizard kind. It also differs from the lizard in that its young are produced alive, and from the frogs in that the young do not pass through a larval state, but are perfectly formed, and sometimes as many as fifty are brought forth at a birth.

There are many species of the *salamander* found in nearly all parts of Europe, but though differing greatly in appearance and size, their habits are very similar. They are all amphibious, and appear to live with equal comfort for an indefinite time either in or out of the water. While in the water it is said that during spring these creatures shed their skins every fourth or fifth day, and in the winter every fortnight, a statement which I have had no opportunity of verifying. Their vitality is very great, perhaps exceeding that of any other animal. The loss of a limb seems to give them small inconvenience, and they even survive the loss of the head for several hours. The creature has been known to live under dissection until its complete dismemberment was accomplished, the tail being last to cease moving. Salt seems to be much more efficacious in destroying these animals than the knife; for upon being sprinkled with it, the whole body emits a viscous liquor, and the reptile dies in three minutes, in great agonies.

The whole of the lizard kind are also tenacious of life in another respect, and the *salamander* among the number. They sustain the want of food in a surprising manner. One of them brought from the Indies lived nine months without any other food than what it received from licking a piece of earth, on which it was brought over; another was kept by Seba in an empty vial for six months, without any nourishment; and Rhedi talks of a large one, brought from Africa, that lived for eight months without taking any nourishment whatever. Indeed, as many of this kind, both *salamanders* and lizards, are torpid, or nearly so, during the winter, the loss of their appetite for so long a time is the less surprising.

Many curious fables have long been current about the *salamander*, not the least popular being its power to withstand heat and pass harmless through the hottest fire. But as nearly all the superstitions have some small base of truth to stand upon, so has that which represents the *salamander* as proof against fire. When the creature is thrown into the fire, as has been done by many persons out of curiosity to test what truth there is in so surprising an assertion, it ejects a copious flow of viscous fluid, which no doubt for the moment prevents the serious burning of the animal, and enables it to scramble from the fire little the worse for the harsh experiment. This result, however, can happen only under the most favorable circumstances. Pliny subjected a *salamander* to the ordeal and reports that it burned quickly into a powder, as it must invariably do unless a chance is offered for its immediate escape from the embers.

The **Fiery or Dotted Salamander** (*Salamandra maculata*) is the most conspicuous European species, to whose unhappy lot has fallen the *fiery* experiences above described, from whence the name is derived. In France it is regarded as a most venomous reptile, the very breath of which is supposed to be as fatal as that of the basilisk of ancient legend. Old naturalists, who invariably had special regard for the wonderful and suppositious, upon the theory that it pleased men to learn of the marvellous dangers by which all mankind are perpetually surrounded, spread such alarm of this unfortunate creature as has not yet sub-

FIERY OR DOTTED SALAMANDER.

sided in certain districts. They even represented that should a person set the heel of his shoe upon the head of a *salamander* all the hair would at once fall from his face and cranium. The mere crawling of a *salamander* on the branches of an apple tree would blast all the fruit in an orchard.

All this calumny is due to the slimy and repulsive appearance of the animal, its resemblance to the water-dog being striking. Its color is black, splotched with pale yellow, with tubercles like warts along its sides. This species is terrestrial, but deposits its young in the water, where they remain only a few days before entering upon an independent existence upon shore. The creature affects the darkest places, usually making its abode in deep crevices about damp spots, from whence it rarely ventures except on very dark, rainy days, and after nightfall, when it goes in search of slugs and other insects, which are its common food. As previously stated in the general observations concerning *salamanders*, this species whenever excited emits a viscous fluid, so copious as to possibly afford protection against the touch of fire, but it serves the creature better by protecting it against enemies, in which respect it is armed like the common toad whose secretion is so acrid as to cause a dog that has attacked it to drop the quarry promptly.

The *salamander* hibernates, spending the severe months of winter within the hollow of trees, or under stones where frost is least likely to reach him, nor appearing again until spring is considerably advanced.

Besides the species above described, there are several others common to the United States, such as the *yellow salamander* of the Alleghanies, the *red-backed*, common to the Eastern States, the *two-striped* of the Western States, and the *purple*, or *aquatic salamander* found in the Alleghany Mountain region, distinguished from its congeners by the courage with which it resists the attacks of any enemy. But none of the several species are provided with fangs, and their bite is harmless.

We come now to the most dreaded of the lizard species, viz., the so-called **Gila Monster** (*Heloderma horrida*), a native of the desert regions of Mexico, Arizona and Lower California. It has only recently come under the notice of naturalists, and we may therefore be little surprised to learn that no two writers agree as to the habits or venomous nature of the reptile. To show how far apart are they in their theories, I will quote from the best authorities. J. G. Wood says: "As the pointed teeth are set as in the deadly snakes, the natives of Mexico believe the reptile's bite to be fatal. This belief, however, is without any foundation, as the reptile really possesses no poisonous fangs.

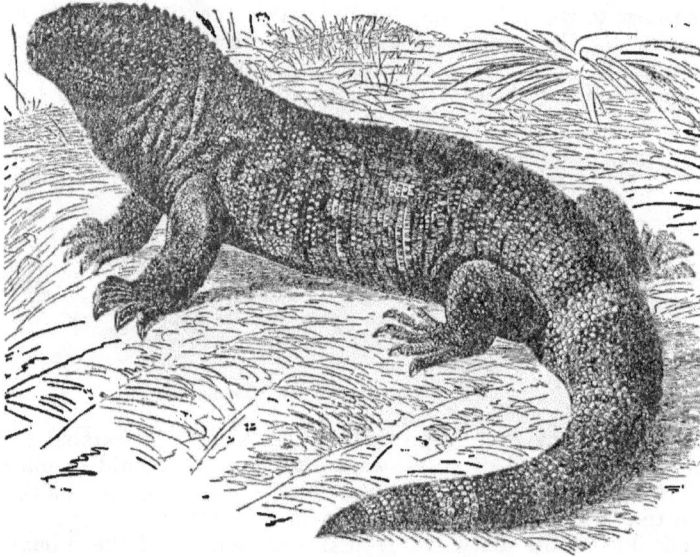

GILÁ MONSTER.

Like some frogs, the *heloderma* has a penetrating scent, and when disturbed it ejects an odorous saliva from its mouth. * * * It attains a length of nearly three feet three inches."

Opposed to this is the opinion of John S. Kingsley, who devotes considerable space to a description of the animal, quoting among other things the experiences of Dr. Shufeldt, as follows:

"Though the more incredulous scientist has questioned the character given this animal by the superstitious Indians and Mexicans, who regard it with the utmost fear, maintaining that it possesses venom of a most virulent nature, a test was recently made by Dr. Shufeldt, which is of considerable interest. He says, in giving an account of an animal at the National Museum: 'It was in capital health and at first I handled it with great care, holding it in my left hand, examining special parts with my right. At the close of this examination I was about to return the fellow to his temporary quarters, when my left hand slipped slightly, and the now highly indignant and irritated *heloderma* made a dart forward and seized my right thumb in his mouth, inflicting a severe lacerated wound, sinking the teeth in his upper maxilla to the very bone. He loosed his hold immediately and I replaced him in his cage with far greater haste, perhaps, than I removed him from it.

" 'By suction with my mouth, I drew not a little blood from the wound, but the bleeding soon ceased entirely, to be followed in a few moments by very severe shooting pains up my arm and down the corresponding side. The severity of these pains was so unexpected that, added to the nervous shock already experienced, no doubt, and a rapid swelling of the parts that now set in, caused me to become so faint as to fall, and Dr. Gill's study was reached with no little difficulty. The action of the skin was greatly increased, and the perspiration flowed profusely. A small quantity of whiskey was administered. This is about a fair statement of the immediate symptoms; the same night the pain allowed of no rest, although the hand was kept in ice and laudanum, but the swelling was confined to this member alone, not passing beyond the wrist. Next morning this was considerably reduced, and further reduction was assisted by the use of a lead-water wash.

" 'In a few days the wound healed kindly, and in all probability will leave no scar. All other symptoms subsided without treatment, beyond the wearing, for about forty-eight hours, so much of a kid glove as covered the parts involved.

" 'After the bite our specimen was dull and sluggish, simulating the torpidity of the venom-

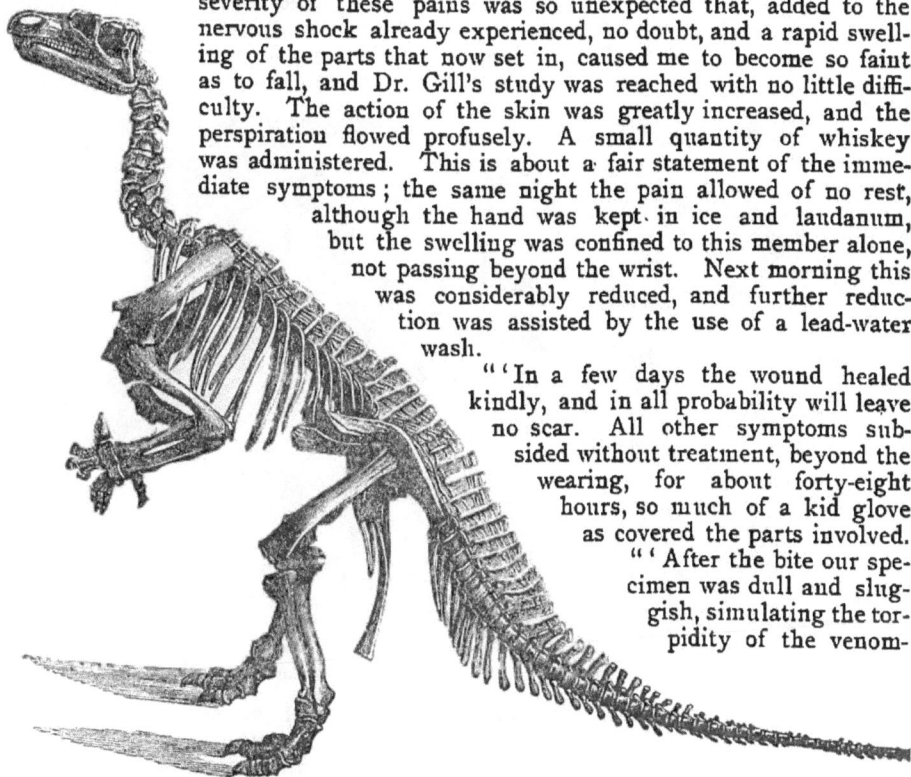

THE IGUANODON.

ous serpent after it has inflicted its deadly wound, but it soon resumed its usual action and appearance, crawling in rather an awkward manner about its cage.' "

" Dr. Shufeldt's conclusions, however, that the symptoms were no other than usually follow the bite of an irritated animal, seem to be given a little prematurely. The same reptile was afterward induced to bite the edge of a saucer, into which, during the action, a secretion dribbled. This secretion, which was of a distinctly alkaline nature, in contrast to the serpent-venoms, which are acid, was, in a small quantity, injected into the breast of a healthy pigeon, and produced death in seven minutes. On a second trial a small quantity was injected into the carotid artery of a rabbit, the animal dying in one minute and thirty-five seconds. Different from the action of serpent-poison, which affects the respiratory functions, the poison of the *heloderma* attacks the heart and the spinal cord. The power of this portion of the nervous system to respond to even powerful electric currents is abruptly annihilated."

Though the opinions of these two distinguished naturalists are in opposition, yet they are both right from the standpoint of their respective examinations. I have seen dozens of these creatures, and have regarded them with no more fear than our ordinary lizards, nor have I ever seen any fear of them exhibited by the natives of the region in which the *heloderma* is found. Still, I would not deny they are venomous, though the declaration needs some quali-

ICHTHYOSAURUS, PLESIOSAURUS AND PTERODACTYL.

fication. I have never seen the experiment made, but reasoning from observation I am of the opinion that the secretion of the common toad is poisonous, and if injected into the breast of a pigeon might produce serious symptoms, though probably not death. The same is true of the secretion of the salamander, which is undoubtedly a provision of nature to protect the creature from its enemies. But this fact is not sufficient to class them among the venomous reptiles. On two different occasions I sought to provoke a company of six *helodermas* to anger, with the view of discovering the manner in which they use, and the strength of, their jaws. For this purpose I used a stick and piece of heavy wire, but notwithstanding the teasing and tormenting not one offered the slightest resistance. I next made a careful examination of the mouth of the largest one, in which I found no evidence of teeth, the armament consisting of nothing more than a sharp jawbone and bony palate. In prying the jaws apart I found the maxillary muscles to be surprisingly strong, showing that its power to bite is very great, though indisposed to exercise this protective provision. The only excitement the animals exhibited under the provocation which I gave them, was by emitting a slight hissing noise, but though I examined carefully, no appearance of any liquid on the sticks thrust into their mouths was visible. My conclusions, founded upon my experience and such experiments as Dr. Shufeldt reports, therefore are, that the animal may secrete a poisonous fluid from its maxillary glands, such secretion being very like that secreted by other batrachians, but that the power to use this venom is so seldom exercised that the animal is hardly deserving of the reputation given to it by Dr. Shufeldt and Prof. Kingsley.

The *heloderma* is of a dirty brown color, thickly covered with small yellowish spots. The skin is rough and hard, coated with horny tubercles. Its head resembles that of the toad except the crown covered with prominent tubercles, which add much to its repulsiveness. The legs are strong and body and tail thick, around the latter being indistinct rings of pale yellow. It lives off insects, and burrows in the sand during the rainy season. I have never seen a specimen that exceeded fifteen inches in length, and am doubtful of the claim that they occasionally grow to be much larger.

THE CROCODILE, OR TORTOISE LIZARD.
SAURIANS OF THE ANCIENT WORLD.

From a consideration of the *batrachian*, the frog, and the *lacertilian*, or lizard, species, we come next and most naturally to the *crocodilia*, or *crocodile* species of reptiles, which is the next step in the progressive order of creation. As already stated, in the preliminary remarks introducing the division of reptiles, development in nature has been rather towards diminution in size in nearly all species to accommodate them to the changes which a gradual cooling of the earth produced. We find nature most prodigal in tropical regions, where vegetation is not only rankest but where animal forms of life are largest and most numerous. Should the temperature of the earth grow one degree cooler each year at the tropics we would perceive a gradual change taking place in the size and forms of life there, both vegetable and animal; for, as vegetation became less rank food would become less plentiful, when, in the reciprocal relations invariable in nature, to compensate by adaptation, animals would as gradually diminish in size, the larger becoming extinct first or dwarfing in proportion and appetite, so that their demands might not be greater than nature could supply. At a period when the earth was nearly covered with warm waters the conditions were favorable to creatures of most extraordinary proportions, as compared with animals of to-day, and to this enormity of size was added the most terrible and ferocious aspects. Herein we perceive another rule of nature, established in the beginning and prominent still even to the most casual observer, viz., that ferocity in animal kind is usually proportionate to the size of the creature. That there are exceptions to this rule we must freely admit, yet generally it is as invariable as rules most commonly are, so true, indeed, that the exceptions serve to emphasize its verity.

The age of fishes being followed by that of reptiles, we find the creatures thus adapted to life either in or out of the waters, of gigantic proportions, and being diverse as well as immense, their armaments were commensurate with their

power. Of these tremendous, savage and mighty amphibians the saurians were
most prominent, of which original species none now exist, but their bones have
been uncovered, and such remains revealed as afford to the naturalist a means
for determining their size, appearance and habits.

Of the **Ichthyosauri** (*fish lizard*) many skeletons have been found in the
tertiary deposits of Europe, measuring as much as forty feet in length. They
possessed characteristics peculiar alike to whales and to lizards, the marks of reptile being distinguishable only in the skull, eyes and backbone. It had the
teeth of a crocodile, the head and breast of a lizard, the vertebræ of a fish
and the flippers of a whale. Terrible as this creature must have been, its congener, the *plesiosaurus* (nearly perfect lizard), was yet more remarkable, for to an

PLESIOSAURUS OF THE PREHISTORIC WORLD.

equally mammoth body it joined the long neck of a serpent, terminating in the
head of a lizard, armed with powerful teeth. The remains of this creature are
so abundant in England as to prove that during the secondary period it must
have existed there in great numbers. Concerning the habits of these mighty
animals Mangin says :

"It is supposed that these monstrous amphibians discharged at the Jurassic
(middle secondary) epoch the function which nowadays devolves upon the cetaceans, viz., that of checking the excessive multiplication in the ocean of the mollusks and fish. The *ichthyosauri* were specially designed for this destruction;
their eyes were of extraordinary magnitude, while their powerful vision enabled

them to both discover their prey at great distances, and to remove it during the night to the obscurest depths of the sea. The skulls of the *ichthyosauri* have been discovered whose orbital cavities measured from ten to twelve inches in diameter. In the largest species the jaws, armed with sharp teeth, yawned for a width of nearly seven feet. The voracity of these animals exposed them to the frequent loss of their teeth; but these, as is the case with the crocodile, were quickly replaced. . . . .

"As for the *plesiosaurus*, the small dimensions of its head, and its thin, elongated neck, would seem to indicate that its appetite resembled that of our huge serpents. . . . . It is probable that this strange creature, whose extraordinary long neck would prevent it from moving rapidly through the water, swam upon the surface, or kept close to the shore in shallow water, where, concealed among the algæ (sea-weed), it might both ensnare its prey and hide itself from the piercing gaze of the *ichthyosauri*, its most formidable enemy."

Buckland, the great naturalist, who has done so much for science, while pursuing his researches in West England discovered the remains of a marine crocodile, to which

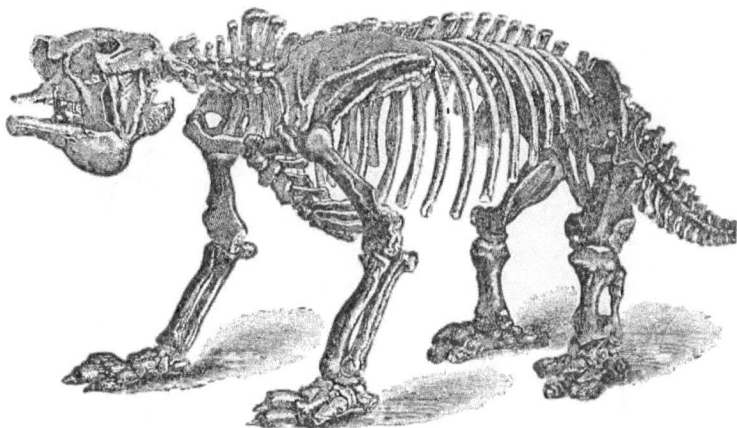

SKELETON OF THE MEGATHERIUM.

he gave the name of *megalosaurus* (great lizard). This animal is the most remarkable of any that is known to have existed. Its form was somewhat like a crocodile, but more nearly resembling the Nile monitor, its teeth being almost identical in appearance to those of the monitor, but it was so very large, in proportion to the crocodile, that it must have exceeded seventy feet in length —a lizard large as a whale. Owen, however, thinks it did not exceed thirty feet.

The celebrated gravel pits near Maestricht have disclosed the bones of creatures scarcely less wonderful, among others being those of a lizard but little inferior to the *megalosaurus*. This gigantic species was called the *mesosaurus* (monitor lizard) while to its gigantic size of twenty-five feet in length was added a dreadful armament of immense teeth, arranged in rows like those of the shark.

Dr. Mantell, another enthusiastic palæontologist, discovered the remains of an animal which evidently belongs to the same family as the *megalosaurus*, but its material differences entitle it to the distinct name which has been given to it, viz., *iguanodon* (a name given to indicate its resemblance to the iguana). It was an herbivorous lizard, and its teeth and toes particularly so closely resembled the iguana, that hence the name. Its stature was about twenty-eight feet, and it is believed to have been the tallest of all creatures on the Eastern Continent; its length was nearly thirty feet. A splendid plaster speci- men of this remarkable reptile may be seen in the Smithsonian Institute at Washington City.

We have next to describe the most grotesque and horrifying creature that inhabited the ancient world, a wild phantasm of nature, more terrible in its appearance than a nightmare conception. Its hybridity was so remarkable that it was reptile, bird and bat all at once, having the characteristics and sem- blance of each. The scientific appellation of this mongrel monstrosity is *ptero- dactylus*, which is a Greek word that implies wing-toed. The name was given it because the fifth toe of its anterior limbs was enormously elongated into a ribbed stem, intended to support a membrane which made the wing. This

PTERODACTYL RESTORED.

SKELETON OF PTERODACTYL.

wing very much resembled the bat's, except that the phalanges were much stronger in proportion, as were also the muscles, so that its flight was very much swifter. The nose was prolonged into a beak which was severely armed with teeth.

Dr. Buckland, in his "Bridgewater Treatise," expresses the opinion that the *pterodactyl* possessed the faculty of swimming, and also that it fed on fishes, which it caught by dashing down upon them after the manner of various fish-catching sea-birds. Cuvier judged it to be nocturnal, from the extraordinary size of its eyes, and this probability is increased by its other bat-like charac- teristics. Mangin says:

"The size and shape of the feet prove that these animals could stand erect with firmness, their wings folded, and that they thus possessed a mode of pro- gression analogous to that of birds; like them, also, they could perch upon trees, while at the same time they had the faculty of climbing along rocks and cliffs, assisting themselves with their feet and fingers like our modern bats and lizards."

"The most striking peculiarity of this animal," says Dr. Hoefer, "is the curious assemblage of vigorous wings, joined to a reptile's body; the imagination of poets alone has hitherto framed anything resembling it. Hence the description of those dragons which fable represents to us as having, in the early ages of the world, disputed with man the sovereignty of the earth, and whose destruction was one of the glorious attributes of the mythic heroes, gods and demigods."

There was one other monster which I must not omit to mention, since its importance among the antediluvian inhabitants of the deep can hardly

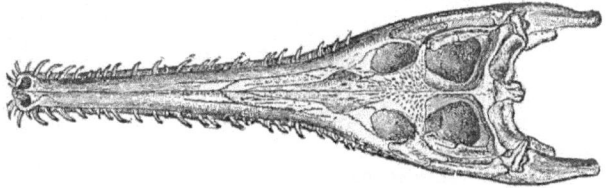

SKULL OF THE TELEOSAURUS.

be over-estimated. I refer to the *Teleosaurus*, or perfect lizard, *teleo* being the Greek word for complete, perfect. The *plesiosaurus* and *ichthyosaurus* were destitute of scaly covering, but the *teleosaurus* was clothed with an adamantine coat of mail, which would have been impervious to the heaviest rifle-ball of to-day. It was also armed with tremendous teeth, and its massive jaws, which it could open to a distance of six feet, made it capable of swallowing the largest ox. This fearful animal was thirty feet in length, and is supposed to have been the most destructive monster of the mighty deep.

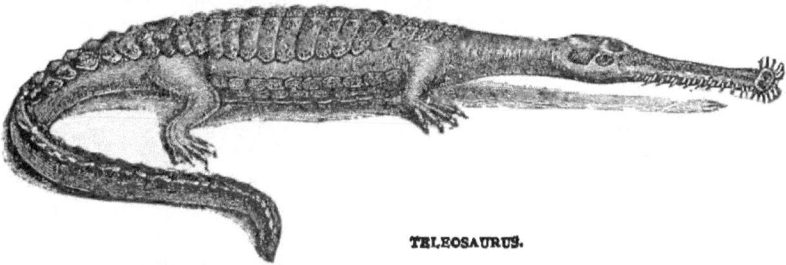

TELEOSAURUS.

Another saurian known as the *Dicynodon* (from the Greek meaning "two tusks") was formerly a habitant of African regions, the fossil remains of one having been found by Mr. Bain, in that country, in 1845. The skull of this creature presented characteristics alike common to the crocodile, tortoise and lizard, while the teeth were clearly those of a mammal. The lower jaw was remarkably tortoise-like, carrying the resemblance even to having the anterior part sheathed with horn. In size it was equal to a walrus, and like the walrus was armed with two tusks growing downward, which it probably used to dig up roots of water plants upon which it no doubt fed, as the teeth indicate that it was both a vegetable and flesh-eating animal. The *Belodon* was another creature of appalling aspect, representing the immediate link between the *Dicynodon* and crocodile of the present period, as the illustration shows. Its length was about thirty feet.

From the saurian creatures above described the crocodile and alligator descended, diminished in size in consonance with the changes which time has

BELODON RESTORED (*from the trias*).

wrought, as already explained. Though smaller than their ancestors, and modified materially in aspect, they are still the most formidable of amphibians, being quite as dangerous in the water as the tiger is on land.

The Crocodile. In the ancient species described we find them all provided with paddles instead of feet, because there was little land in

THE MYLODON.

existence during the secondary period, and even the land that was formed had not yet begun to be prolific with plant life. But the land gradually encroached upon the water, the temperature became lower, and vegetation sprang up while animal life increased rapidly along the newly formed shores. This change was followed by others equally pronounced in the animal world, for the saurians gradually exchanged their paddles for feet; or to be more definite, the saurians of the secondary period, finding their feeding grounds growing less, in their unappeased voracity fell to devouring each other, while many others were caught in pools and being unable to effect their escape across land into deeper water perished as the pools dried up. Thus from several causes the last of the mighty creatures, whose fossil remains we now study with so much interest, became extinct, but were succeeded by their milder congeners which the Creator provided with feet instead of flippers, and thus adapted them to conditions comfortable with the regions in which they were appointed to live.

The True Crocodile. This reptile is peculiar to the Nile river, where, though not nearly so numerous as formerly, it is still quite common. The animal is also found in the Niger river, and in other water-courses of mid-Africa, though some naturalists claim that the species is somewhat different from those in the Nile, a point which, however, still remains in dispute, with probabilities favoring their identity. In the times of the Jewish oppression,

SAURIANS OF THE ANCIENT WORLD.

when Egypt was in its glory and the Pharaohs ruled amid surroundings of unexampled splendor, the Nile *crocodiles* were worshipped as embodiments of supreme power. A special temple was erected at Memphis and dedicated, with

CROCODILES OF THE NILE.

many solemn ceremonies of veneration, to the adoration of this creature, and after death the body was most carefully embalmed and saturated with sweet spices.

In the crocodile proper there is an uninterrupted series of teeth round both jaws, by which feature it is distinguishable from the alligator, which has every fourth tooth of the under jaw fitting into a corresponding socket of the upper one.

In Central Africa the crocodile not unfrequently attains a length of thirty feet, and Sir Samuel Baker declares that he shot one that measured thirty-three feet, which is believed to be the maximum. The food of the crocodile is preferably putrid meat, but it will bolt whole fish and small quadrupeds. On the under side of the lower jaw is an opening from which the animal has the ability to force at will a liquid that has an overpowering odor of musk, which Mr. Bell declares the creature uses to attract fish to its haunt.

The **Double Crested Crocodile** is a most common species of India, found principally near the mouths of rivers that lead into the Indian Ocean. It is also fairly plentiful in the running streams of Java, and occasionally in those of China. It derives its name from the fact that it is furnished with two prominent ridges which extend over the jaw near the eyes. This species is extremely cowardly, and takes to precipitate flight at the sight of man. Should they be found in a shallow basin where escape is impossible, ostrich-like they will hide the head by driving it into the mud and remain unconscious of the exposure of the body. Concerning this disposition to burrow in the mud at the approach of danger, Sir Edward Tennent writes: "Some years ago, during the progress of the pearl fishery, Mr. Horton employed men to drag for crocodiles in a pond which was infested by them in the immediate vicinity of Aripo. The pool was about fifty yards in length by ten or twelve wide, shallowing gradually toward the edge, and not exceeding four or five feet in the deepest part. As the party approached the pond, from twenty to thirty reptiles, which had been basking in the sun, rose and fled to the water. A net specially weighted so as to sink its lower edge to the bottom was then stretched from bank to bank, and dragged to the further end of the pond followed by a line of men with poles to drive the crocodiles forward. So complete was the arrangement that it appeared not one of the creatures could avoid the net, yet to the astonishment of the party not one of the reptiles was entangled. Their escape could only have been effected by burrowing in the soft mud."

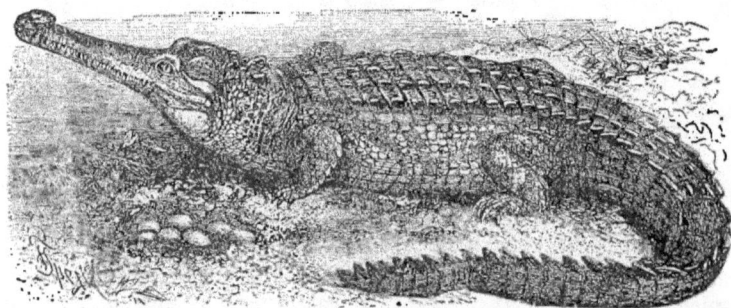

GAVIAL (*G. gangeticus*).

The **Marsh Crocodile** is found in various parts of Asia, but its principal habitat is Australia, where they have been killed measuring above thirty feet in length. This species has also a habit of burrowing in the mud, where it remains during dry seasons and until rains come to soften its bed and give a covering of water. The adult animal is shy, and seldom commits any ravages, but its young are particularly ferocious, apparently being unconscious of fear. Those not exceeding ten inches in length will seize and hold fast on to a stick thrust at them, and a finger would invite an attack no less readily.

The Gavial is the best known, because most ferocious and voracious of the several crocodile species in India. It retains a remarkable resemblance to the ancient *teleosaurus*, of which creature it is certainly a direct descendant. The Ganges river fairly swarms with these monster saurians, and to their great number and the depredations they commit is added a superstitious reverence for the animal. The practice of sacrificing infants to this monster is now uncommon, on account of British influence dominating in the native customs, but in former years the sight was frequent about Benares of a mother carrying her child towards the Ganges to offer it as a sacrifice to the *gavial* god; the fond mother, believing she was preparing a flowery way to heaven for her child, would pause upon the river's bank and cover it with passionate kisses, fondle it in a thousand ways, as if deferring the dread act about to be committed, until a fairly bursting heart was overcome by religious devotion, when she would toss the innocent offspring to the cruel monsters that were waiting for the sacrificial feast. Never more than a single cry would fall upon her ears, for in an instant the little innocent would be torn into a hundred pieces, and only a bloody dye on the surface remain to show, for a few moments, where the tragedy occurred.

The *gavial* differs in appearance from others of the crocodilian family by having a much

ALLIGATORS OF AMERICA.

greater prolongation of nose, upon the end of which is a prominent, wart-like nodule, which it uses to root in the mud, like a hog.

The Margined Crocodile is found principally in South Africa, its distinguishing feature being a compression, amounting to an indentation of the forehead, while the nozzle is somewhat shorter. But in other respects it is identical with the Nile species.

The Alligator proper is found only in waters of the Western Hemisphere, where there are several species called indiscriminately *alligator, cayman, crocodile* and *jacara*. The difference between these, if real difference does exist, is so slight that no naturalist has as yet attempted to show wherein it lies. In fact, aside from very small differences in appearance, and the disparity in size, alligators and crocodiles are practically the same animal, the larger being more ferocious, but in habits, aside from courage, the identity is pronounced. Generally, all naturalists recognize the impossibility of separating the several species by well-defined differences, on which account alligators and crocodiles are described as inhabiting alike the waters of the Old and New World.

We will now proceed to describe more particularly the appearance of the several species as a whole, and to illustrate their habits by incidents and adventures from the experiences of those who have hunted the animal, or fallen accidentally within its power. With the differences given it will still be seen how close is the family resemblance to the *teleosaurus*, the *belodon*, and to one another.

ADVENTURES WITH CROCODILES.

The most pronounced distinguishing characteristic in the crocodile species is in the shape of the head; gavials have the longest muzzle, the true crocodiles the next, while the nose of the alligator is shortest. In the latter the body is thick, teeth irregular in length and size, webs of the toes only half the length, and their haunts are chiefly in the fresh water of lagoons and river mouths. All the several species are covered with a thick, tessellated or checkered skin, that resembles scales, but the true scale is wanting. These squares are so indurated as to deflect a rifle-ball, when fired at ordinary range, though should the ball strike a crease, which lies between the squares of bony skin, it easily penetrates. It is this fact that has caused a diversity of opinion as to the vulnerableness of the alligator.

All of the several species live upon similar food, and take their prey in an identical manner. At the point of the long nose are the nasal cavities, by which the creature is able to project its muzzle scarcely above the surface of the water and yet breathe freely; by this provision of its nose it is also able to approach its prey unperceived. Being of remarkable activity in water the animal will seize the nose of a drinking quadruped, and with a spasmodic motion of almost lightning celerity incurve its body and strike the prey a powerful blow with its jagged tail. If the victim is large enough to wage a contest it is dragged under water and there held until drowned. The food, however taken, is not eaten under water, but conveyed to some spot and hidden until decomposition is somewhat advanced, when it is dragged out upon the shore and there devoured, for the crocodile cannot breathe under water, though an inflation of its lungs will serve it for some considerable time.

In the Ganges the crocodiles may be seen almost constantly lying on shore, or floating, with the nose scarcely observable, with the stream, watching for carcases in a state of putrefaction. Sometimes a carrion bird may be seen hovering above or stationed upon a human body which the Hindoos in their superstitious attachments have committed to the stream, which they look upon as the roadway to Paradise. But the vultures are not long permitted to enjoy their feast unmolested, for the watchful crocodiles make haste to dispute possession, which they more readily accomplish by dragging the body first under water and then conveying it to the shore.

BATTLE WITH THE CROCODILES.

It is a very common thing for the native princes of India, living in the neighborhood of large rivers where crocodiles abound, to have them caught for the purpose of entertaining their court and guests, by making them fight, or causing them to be attacked by other animals. Captain Basil Hall has given the following animated account of a fight of this kind, got up for the amusement of the Admiral, Sir R. Hood, and performed by a corps of Malays in the British service :

"Very early (he says) in the morning, the party were summoned from their beds, to set forth on the expedition. In other countries, the hour of getting up may be left to choice; in India, when anything active is to be done, it is a matter of necessity; for after the sun has gained even a few degrees of altitude, the heat and discomfort, as well as the danger of exposure, become so great, that all pleasure is at an end. The day, therefore, had scarcely begun to dawn, when we all cantered up to the scene of action.

"The ground lay as flat as a marsh for many leagues, and was spotted with small stagnant lakes connected by sluggish streams, scarcely moving over beds of mud, between banks fringed with a rank crop of draggled weeds. The chill atmosphere of the morning felt so thick and clammy it was impossible not to think of agues, jungle-fevers, and all the hopeful family of malaria. The hardy native soldiers, who had occupied the ground during the night, were drawn up to receive the Admiral, and a very queer guard of honor they formed. The whole regiment had stripped off their uniform, and every other stitch of clothing, save a pair of short trousers, and a kind of sandal. In place of a firelock, each man bore in his hand a slender pole, about six feet in length, to the extremity of which was attached the bayonet of his musket. His only other weapon was the formidable Malay creese, a sort of dagger, or small two-edged sword.

"Soon after the commander-in-chief came to the ground, the regiment was divided into two main parties, and a body of reserves. The principal columns facing, one to the right, the other to the left, proceeded to occupy different points in one of the sluggish

HARPOONING AN ALLIGATOR.

canals, connecting the pools scattered over the plain. These detachments being stationed about a mile from one another, enclosed an interval where, from some peculiar circumstances known only to the Malays, who are passionately fond of the sport, the crocodiles were sure to be found in great numbers. The troops formed themselves across the canals in three parallel lines, ten to twelve feet apart; but the men in each line stood side by side, merely leaving room enough to wield their pikes. The canal may have been about four or five feet deep, in the middle of the stream, if stream it can be called, which scarcely moved at all. The color of the water, when undisturbed, was a shade between ink and coffee; but no sooner had the triple line of Malays set themselves in motion, than the consistence and color became like that of pea-soup.

"On everything being reported ready, the soldiers planted their pikes before them in the mud, each man crossing his neighbor's weapon, and at the word 'March,' away they all started in full cry, sending forth a shout, or war-whoop, sufficient to curdle the blood of those on land, whatever effect it may have had on the inhabitants of the deep. As the two divisions of the invading army gradually approached each other in pretty close column, screaming, and yelling, and striking their pikes deep in the slime before them, the startled animals naturally retired towards the unoccupied centre. Generally speaking, the alligators, or crocodiles, had sense enough to turn their long tails upon their assailants, and to scuttle off, as fast as they could, towards the middle part of the canal. But every now and then, one of the terrified monsters floundered backwards, and, by retreating in the wrong direction, broke through the first, second, and even third line of pikes. This was the perfection of sport to the delighted Malays. A double circle of soldiers was speedily formed round the wretched aquatic who had presumed to pass the barrier. By means of well-directed thrusts with numberless bayonets, and the pressure of some dozens of feet, the poor brute was often fairly driven beneath his native mud. When once there, his enemies half-choked and half-spitted him, till at last they put an end to his miserable days, in regions quite out of sight, and in a manner as inglorious as can well be conceived.

"The intermediate space was now pretty well crowded with crocodiles swimming about in the utmost terror, at times diving below, and anon showing their noses above the surface of the dirty stream; or occasionally making a furious bolt, in sheer despair, right at the phalanx of Malays. On these occasions, half-a-dozen of the soldiers were often upset, and their pikes either broken or twisted out of their hands, to the infinite amusement of their companions, who speedily closed up the broken ranks. There were none killed, but many wounded; yet no man flinched in the least.

"The perfection of the sport appeared to consist in detaching a single crocodile from the rest, surrounding and attacking him separately, and spearing him until he was almost dead. The Malays then, by main strength, forked him aloft over their heads on the end of a dozen pikes, and by a sudden jerk, pitched the conquered monster far on the shore. As the crocodiles are amphibious, they kept to the water no longer than they found they had an advantage in that element; but on the two columns of their enemy closing up, the monsters lost all discipline, floundered up the weedy banks, scuttling away to the right and left, helter-skelter. 'Sauve qui peut!' seemed to be the fatal watchword for their total rout. That prudent cry, would no doubt, have saved many of them, had not the Malays judiciously placed beforehand their reserve on each side of the river, to receive the distracted fugitives, who, bathed in mud, and half dead with terror, but still in a prodigious fury, dashed off at right angles from the canal, in hopes of gaining the shelter of a swampy pool, overgrown with reeds and bulrushes, but which most of the poor beasts were doomed never to reach. The concluding battle between these retreating and desperate crocodiles and the Malays of the reserve, was formidable enough. Indeed, had not the one party been fresh, the other exhausted, one confident, the other broken in spirit, it is quite possible that the crocodiles might have worsted the Malays. It was difficult, indeed, to say which of the two looked at that moment the more savage; the triumphant natives, or the flying troop of crocodiles walloping away from the water. Many on both sides were wounded, and all covered with slime and weeds. There could not have been fewer than thirty or forty crocodiles killed, though they were generally small, the largest hardly exceeding ten feet in length."

Concerning the similarity that exists between the several species Goldsmith, following Buffon as his guide, thus writes confirmatively of what I have said:

" Of this terrible animal there are two kinds, the crocodile, properly so called, and the cayman or alligator. Travellers, however, have rather made the distinctions than nature, for in the general outline and in the nature of these two animals they are entirely the same. It would be speaking more properly to call these animals the crocodiles of the Eastern and Western world; for, in books of voyages, they are so entirely confounded together that there is no knowing whether the Asiatic animal be the crocodile of Asia or the alligator of the Western world. The distinctions usually made between the crocodile and the alligator are these: the body of the crocodile is more slender than that of the alligator, its snout runs off tapering from the forehead, like that of a greyhound, while that of the other is indented, like the nose of a lap-dog. The crocodile has a much wider swallow, and is of an ash color; the alligator is black, varied with white, and is thought not to be so mischievous. All these distinctions, however, are very slight, and can be reckoned little more than minute variations."

THE TIGER AND THE CROCODILE.

It frequently happens, in its depredations along the bank, that the crocodile seizes on a creature as formidable as itself, and meets with a most desperate

AN ALLIGATOR TRAP.

resistance. We are told of frequent combats between the crocodile and the tiger. All creatures of the tiger kind are continually oppressed by a parching thirst, which keeps them in the vicinity of great rivers, whither they descend to drink very frequently. It is upon these occasions that they are seized by the crocodile, and they die not unrevenged. The instant they are seized upon they turn with the greatest agility and force their claws into the crocodile's eyes, while he plunges with his fierce antagonist into the river. There they continue to struggle for some time, till at last the tiger is drowned.

In this manner the crocodile seizes and destroys all animals, and is equally dreaded by all. There is no animal but man alone that can combat it with success. We are assured by Labat that a negro, with no other weapons than a knife in his right hand and his left arm wrapped round with a cowhide, ventures boldly to attack this animal in his own element. As soon as he approaches the crocodile he presents his left arm, which the animal swallows most greedily, but sticking in his throat, the negro has time to give the creature several stabs in the softer parts of the neck, while the water getting in the mouth, thus held involuntarily open, the animal is soon dispatched. Though we have Labat as authority, we can no less doubt the truth of this relation.

HUNTING THE CROCODILE.

A common means of killing the crocodile practised by the natives of north-western Africa is thus described by Dr. Ruppell:

"The most favorable season is either the winter, when the animal usually sleeps on sand-banks, luxuriating in the rays of the sun, or the spring, after the pairing time, when the female regularly watches the sand-islands, where she has buried her eggs. The natives find out the place, and on the south side of it, that is to the leeward, dig a hole in the sand, throwing the earth to the side which they expect the animal to take. Then they conceal themselves, and the crocodile comes to its accustomed spot and soon falls asleep. The hunters then dart their harpoons with all their force at the animal, for in order that the strokes may be successful, the harpoon head ought to penetrate to the depth of at least four inches, that the barb may be firmly fixed in the flesh. Upon being wounded the crocodile rushes for the water, and the hunters retreat to their canoes. A piece of wood, attached to the harpoon line, swims on the water and indicates the direction in which the crocodile is moving. The huntsmen, by pulling on the line, drag the beast to the surface of the water where it is struck with other harpoons until destroyed."

A decidedly novel method of effecting its capture is often put into execution by the colored people living near the bayous of Louisiana and Florida, who affix a strong rope having a noose at the pendant end to the top of a stiff sapling, which is then bent over by the force of three or four men until the noose can be made fast around a circle of sticks placed in the ground some six inches high, and set with triggers, baited with a piece of putrid flesh. When the alligator seizes the meat the noose is loosed and catching him about the neck raises the victim half off the ground, where all his struggles can avail nothing, and he soon strangles.

The shooting of alligators has long been a favorite pastime with sport-loving tourists who visit the swamps of Florida, while thousands are annually killed for their hides, which are now used for many useful purposes. Indeed these uses have served to make the people of Florida more regardful of the life of the alligator and they now protect him from indiscriminate slaughter by severe prohibitory laws.

HOW THE CROCODILE BREEDS.

All crocodiles breed near fresh waters; and though they are sometimes found at sea, yet that may be considered rather as a place of excursion than abode.

They produce their young by eggs, and for this purpose the female, when she comes to lay, chooses a place by the side of a river, or some fresh water lake, to deposit her brood in. She always pitches upon an extensive sandy shore, where she may dig a hole without danger of detection from the ground being fresh turned up. The shore must also be gentle and shelving to the water, for the greater convenience of the animal's going and returning; and a convenient place must be found near the edge of the stream, that the young may have a shorter way to go. When all these requisites are adjusted, the animal is seen cautiously stealing upon the shore to deposit her burden. The presence of a man, a beast, or even a bird, is sufficient to deter her at that time; and if she perceives any creature looking on, she infallibly returns. If, however, nothing appears, she then goes to work, scratching up the sand with her fore-paws, and making a hole pretty deep in the shore. There she deposits from eighty to a hundred eggs, of the size of a tennis-ball, and of the same figure, covered with a tough white skin, like parchment. She takes above an hour to perform this task; and then covering up the place so artfully that it can scarcely be perceived, she goes back to return again the next day. Upon her return, with the same precaution as before, she lays about the same number of eggs; and the day following also a like number. Thus having deposited her whole quantity, and having covered them close up in the sand, they are soon vivified by the heat of the sun; at the end of thirty days the young ones begin to break open the shell. At this time the female is instinctively taught that her young ones want relief; and she goes upon land to scratch away the sand, and set them free. Her brood quickly avail themselves of their liberty; a part run unguided to the water; another part ascend the back of the female, and are carried thither in greater safety. But the moment they arrive at the water, all natural connection is at an end; when the female has introduced her young to their natural element, not only she, but the male, becomes among the number of their most formidable enemies, and devour as many of them as they can. The whole brood scatters into different parts of the bottom; by far the greater number is destroyed, and the rest find safety in their agility or minuteness.

But it is not the crocodile alone that is thus found to thin their numbers; the eggs of this animal are not only a delicious feast to the savage, but are eagerly sought after by every beast and bird of prey. The ichneumon was erected into a deity among the ancients for its success in destroying the eggs of these monsters; at present that species of the vulture called the Gallinazo is their most prevailing enemy. All along the banks of great rivers, for thousands of miles, the crocodile is seen to propagate in numbers that would soon overrun the earth, but for the vulture, that seems appointed by Providence to abridge its fecundity. These birds are ever found in greatest numbers where the crocodile is most numerous; and hiding themselves within the thick branches of the trees that shade the banks of the river, they watch the female in silence, and permit her to lay all her eggs without interruption. Then when she has retired, they encourage each other with cries to the spoil; and flocking all together upon the hidden treasure, tear up the eggs, and devour them in a much quicker time than they were deposited. Nor are they less diligent in attending the female while she is carrying her young to the water; for if any one of them happen to drop by the way, it is sure to receive no mercy.

OPHIDIA—SNAKES.

In the natural sequence of changes produced by a gradual cooling of the earth, as already explained, land animals became more numerous as the waters receded, and following amphibious saurians, other creeping and no less repulsive creatures appeared, adaptable to the modified temperature and conditions. Reptiles in this next order are classed by naturalists as *Ophidia*, a Greek word meaning *serpent*, and under this head therefore, must follow descriptions of all the numerous species, which have been divided by Lacepede into eight genera, each of which are again subdivided into several classes. The eight genera comprise the *Boa*, which contains 11 species; the *Vipers*, of which there are 196 species; *Rattlesnakes*, with 26 species; *Snakes* (under which name the harmless find classification), 24 species; the *Amphisbœna*, a double-headed, 5 species; *Langrata*, 1, *Cœcilia* 2, and *Acrocboid*, 1.

All true snakes, whether venomous or harmless, are clothed with a scaly skin which they shed at least once each year, and during this season they are irritable and sluggish, giving every indication that the change of skin is accompanied by more or less physical disturbance, if not actual pain. The old skin begins to peel first from the head, splitting on top and fleecing down over the eyes and then breaking in a line along the back. This dead integument sometimes comes off in perfect condition, and with the exception of a single rift along the back has all the appearance, save its transparency, of the snake itself, every scale, and even the delicate covering of the eyes having a most natural appearance.

Legs, or even the rudiments of limbs, are wanting except in a few of the largest species, where a posterior protuberance, as in the *Boa*, is barely perceptible. Yet while possessing no limbs these reptiles move with astonishing ease and some few with very great celerity, whether on land or in water. Their movements are accomplished by the use of the scales upon the belly, called *scutes*. These scutes perform the service of legs, by which the creature can erect and firmly hold itself to any rough surface, and by using them consecutively drag itself along. All snakes, however, are unable to move forward over a perfectly smooth surface. To the strange provisions made to enable the serpent reptiles to travel, is added a wonderful formation of structure, by which the body is capable of remarkable recurvation and flexibility, due to the ball-joints of both ribs and vertebræ.

The heart in reptiles is so constructed, that at each of its contractions only a portion of the blood which it receives is transmitted to the lungs, the remainder of this fluid is returned to circulate again, without having passed into the lungs and, consequently, without having been subject to respiration; hence it results that the action of oxygen on the blood is greatly less than in mamminiferous animals and birds, where all the blood by passing through their lungs is exposed to the action of the air. Consequently, as respiration causes the heat in the blood and gives to the muscular fibre its susceptibility for nervous irritation, the temperature of reptiles is much lower, and their muscular power greatly weaker than that of the mammalia and birds. Therefore they are said to be cold-blooded animals. Their general habits are also much less energetic, almost all their motions consisting of crawling and swimming, and although several species run or leap, at times with considerable facility, yet upon the whole their actions are sluggish.

MONSTER SERPENTS OF THE ANCIENT WORLD.

A greater variety of forms exists among reptiles than is found among warm-blooded creatures and it is in the production of these forms that nature seems to have imagined shapes of the most weirdly fantastic description, and modifying in every possible manner the general plan which she has prescribed to herself in the mammalia class of animals. To these hideous forms is added a deadly venom which has served to render all serpents detestable in the eye of man, whose hand is ever raised against them, as the Creator decreed after the fall of Adam and Eve.

Though snakes are found inhabiting nearly all regions of the earth, like others of the reptile family they are most numerous, venomous and formidable in the tropics, and in places where, from the exuberance of vegetation, man has been able to make the least progress. For this reason we may not reject as entirely improbable the stories that have come down to us from the ancients of the ravages committed by monster serpents that roamed the solitudes of the world before man had become equipped with effective weapons with which to oppose them.

The finding of fossil remains of gigantic saurians, such as have been described, gives us good reason for supposing that at one time there existed serpents of equally surprising proportions, and indeed the analogy which exists between all creatures of the reptilian order lead irresistibly to this conclusion, even though their fossilized remains have not as yet been discovered. The nature of the forests which once clothed the earth, and the other animal life that revelled therein make it extremely probable that serpents grew to lengths of one hundred feet or more, in which event it would be most difficult for man, with the rude and insufficient weapons with which he was armed, to destroy them. To such enormous and powerful creatures the lion, tiger, or even elephant itself, would be but a feeble opponent. As Goldsmith observes:

"The dreadful monster spread desolation round him; every creature that had life was devoured, or fled to a distance. That horrible *fœtor* [odor], which even the commonest and the most harmless snakes are still found to diffuse, might, in these larger ones, become too powerful for any living being to withstand; and while they preyed without distinction, they might thus also have poisoned the atmosphere around them. In this manner, having for ages lived in the hidden and unpeopled forest, and finding as their appetites were more powerful, the quantity of their prey decreasing, it is possible they might venture boldly from their retreats, into the more cultivated parts of the country, and carry consternation among mankind, as they had before desolation among the lower ranks of nature."

Indeed, we have many histories of antiquity presenting us such a picture, and exhibiting a whole nation sinking under the ravages of a single serpent. At that time, man had not learned the art of uniting the efforts of many to effect one great purpose. Opposing multitudes only added new victims to the general calamity, and increased mutual embarrassment and terror. The animal was, therefore, to be singly opposed by him who had the greatest strength, the best armor, and the most undaunted courage. In such an encounter hundreds must have fallen; till one, more lucky than the rest, by a fortunate blow, or by taking the monster in its torpid interval, and surcharged with spoil, might kill, and thus rid his country of the destroyer, as Hercules is represented to have done, and as St. George is said to have accomplished. Such was the original occupation of heroes: and those who first obtained that name, from their destroying the ravagers of the earth, gained it much more deservedly than their successors, who acquired their reputation only for their skill in destroying each other. But as we descend into more enlightened antiquity, we find these animals less formidable, as being attacked in a more successful manner. We are told, that while Regulus led his army along the banks of the river Bagrada, in Africa, an enormous serpent disputed his passage over. We are assured by Pliny, who says, that he himself saw the skin, that it was a hundred and twenty feet long, and that it had destroyed many of the army. At last, however, the battering engines were brought out against it; and these assailing it at a distance, the reptile was soon destroyed. Its spoils were carried to Rome, and the general was decreed an ovation for his success. There are, perhaps, few facts better ascertained in history than this. An ovation was the most distinguished honor conferred by the Romans, and was only given for some signal exploit, therefore it is scarcely possible that any historian could invent a story of such ovation being given without being exposed. The skin of the serpent destroyed by the army of Regulus was taken to Rome and kept on exhibition in the capitol for at least fifteen years, where Pliny affirms that he frequently saw it.

HOW SERPENTS SWALLOW CREATURES LARGER THAN THEMSELVES.

Among the most singular characteristics peculiar to serpents is the power which they alone possess of swallowing a prey the diameter of whose body exceeds many fold that of their own. This wonderful ability is due to the fact that their jaws are not fitted with sockets so as to work on a hinge as are those of other creatures; on the contrary their union is by elastic muscles capable of very great expansion, while the throat and gullet are capable of like dilatation.

All species are provided with teeth, the non-venomous having a row on either side of re-curved, hollow and immovable grasping teeth, while the venomous have two fangs in a movable palate by which, like the shark, they can erect or depress them flat upon the roof of the mouth at will. At the base of the fang is a sac in which the virus is secreted. When the creature bites, a pressure on this sac forces the virus out and it flows down through a duct, which is concealed in a groove, along the inner side of the fang and into the wound. If the fangs be extracted, which is a simple operation to perform, the reptile remains harmless forever after, though when the non-venomous lose their teeth they are very soon replaced by others.

The eyes of all snakes are small and malignant in expression, and, owing to the crystalline humor within the globe, are hard as horn. Some few possess upper lids and wink naturally, while in others there is only an under lid, and in yet others both lids are wanting.

The aural vents are scarcely distinguishable and the conduits for smelling are entirely absent, by which it is supposed that serpents do not possess the sense of smell and that their hearing is doubtful. They probably have to trust to a keen vision and their extreme sensibility to vibrations, as do the fishes.

The tongue is generally long and bifurcated (forked), most sensitive to touch, and may be projected by a provision of two extensible tendons at the root, enabling the creature to shoot out that organ with great rapidity and at considerable length. It was formerly supposed, by persons ill informed, that serpents inflicted a poisonous sting with the tongue, a fancy perpetuated by many references made in standard works to the serpent's sting. It is hardly necessary to say now that the tongue is quite as harmless as the track the creature makes in crawling.

HOW SNAKES BRING FORTH AND PROTECT THEIR YOUNG.

Many snakes bring forth their young alive, the eggs having been hatched within the body, while others, notably the Boa, deposit their eggs in a nest and incubate them with the small amount of animal heat given off by the creature's body. Father Labat gives us the following remarkable experience with regard to the manner in which the viper produces its young:

"I took a serpent of the viper kind, that was nine feet long, and ordered it to be opened in my presence. I then saw the manner in which the eggs of these animals lie in the womb. In this creature there were six eggs, each of the size of a goose egg, but longer, more pointed, and covered with a membranous skin, by which also they were united to each other. Each of these eggs contained from thirteen to fifteen young ones, about six inches long, and as thick as a goose-quill. Though the female from which they were taken was spotted, the young seemed to have a variety of colors very different from the parent; and this led me to suppose that the color was no characteristic mark among serpents. These little mischievous animals were no sooner let loose from the shell then they crept about, and put themselves into a threatening posture, coiling themselves up, and biting the stick with which I was destroying them. In this manner I killed seventy-four young ones; those that were contained in one of the eggs escaped at the place where the female was killed, by the bursting of the egg, and their getting among the bushes."

It has long been a question for debate among naturalists themselves whether or not any of the serpent species give protection to their young by providing a retreat for them through the mouth and into the stomach. Many stoutly deny, while others, backed by experience, affirm the truth, among the latter disputants being a majority of those whose opinions are considered authority. In further proof of the assertion I may add my own experience, which differs somewhat from that of others that I have read, in that while the assertion is that such protection is only given by venomous species, the story I have to tell is of a harmless striped ground snake, so beneficial to the farmer in ridding his fields of ground mice and noxious insects: One day I was walking near the river bank with a companion when our eyes fell upon a ground snake just as it was making its escape under a large board that had been deposited on the bank by a rise of the river. We naturally sought its life and shifting the board easily killed it before the reptile could reach another retreat, particularly as its movements seemed unnaturally slow. After mashing the head and giving the body a few strokes I picked the carcase up by the tail, when to our astonishment a dozen or more of its young, all marked exactly like the parent, ran out at the bruised mouth, but as I dropped the dead snake instantly about one half of them, after emerging, ran back again into their former receptacle, where I killed them. Thus the evidence furnished to my own eyes of the statement as to the young of certain snakes seeking protection within the parent's mouth and stomach is absolutely incontestable.

PROCESS OF A SNAKE SWALLOWING A FISH.

I also had the pleasure once of beholding, and thus determining, the extraordinary amount of distention a snake is capable of, confirming my assertion that they can swallow a creature whose diameter is many times greater than their own. I was one time fishing in a slough, some five miles below Warsaw, Ill., in company with my father-in-law, when, having met with poor success, at the noon hour we retired to the shelter of a corn-crib, in which we had our lunch. The day was extremely sultry, on which account we did not resume fishing for probably two hours. I had left my pole thrust into the muddy bank, with the line cast and baited with angle-worm, hoping that during my temporary absence some wandering fish might find the morsel I had thus set out and become fast, a hope which was not disappointed. Returning to the spot at length I was much delighted at seeing my pole bending under a considerable strain, and eagerly I rushed down to complete the capture. What was my surprise, upon jerking up my line, to see dangling to my hook a water-snake (of the moccasin species), less than two feet in length. Great as was my surprise at this discovery, my astonishment was very much increased upon examination to find that the snake was not hooked, but that it had seized upon and swallowed a cat-fish some six inches in length that had first become fast on the hook. The fish's body had entirely disappeared down the snake's gullet, leaving on either side of it, a few inches below the mouth, a remarkable distention of the skin, through which it seemed the sharp, pectoral fins of the fish must surely cut their way. All who have done any fishing in our Western waters are familiar with the cat-fish, and most of my readers can testify to its ability to give a severe wound with the sharp and very hard fins that set so rigidly just behind the gills. These fins may be moved very slightly backwards, but the muscles will not permit of their movement forward, hence, as the fish was swallowed tail first and the spurs (fins) from point to point were certainly four inches, the distention of the snake's mouth was most extraordinary to take in such an unyielding body; ten-fold more difficult than it would be to swallow a body twice the size, whose increase from the tail upward was gradual, thus giving the snake a chance to distend its jaws slowly, instead of by a spasmodic exertion expand them to the required extent, as must have been done.

Since witnessing this remarkable feat I have been credulous enough to believe that a snake can swallow anything if only sufficient time be given him.

Having thus briefly noticed some of the general characteristics of serpents, I will now proceed to a particular description of the more important species, such as the venomous and formidable, with casual notice of the common and harmless species, in order to enable the reader to distinguish those which are capable of inflicting deadly bites.

The general rule, but to which there are not a few exceptions, is that all venomous snakes have thick bodies and the head short, the skull rising abruptly from the neck and being angular, with slight compression below the eyes. Their movements, too, are usually, though not always, slow. The non-venomous are slimmer, very long for the diameter, with tails terminating in a sharp point, colors brighter, and head long and shapely with pointed nozzle.

Having last considered amphibious reptiles, in treating of snakes we will first describe some of the species that make their home chiefly in the water, among which we find only a single species, the moccasin, that is venomous, and another, the anaconda, that is otherwise formidable.

The **Water Moccasin** (*Ancistrodon piscivorus*) is found distributed over a large district of the United States, from the Mississippi valley to the Atlantic coast, and from the Gulf to the Northern States. But though many authorities declare it to be a most venomous reptile, without making any distinction, I must controvert the assertion and show wherein the error lies. Wood unhesitatingly pronounces all the varieties, including the *water moccasin, black moccasin,* and *water viper,* as extremely poisonous, but residents along the Mississippi and other rivers and creeks of the North will dissent from this opinion. The true *moccasin,* most commonly called the *swamp moccasin,* is confined to bayous and swamp regions of the South, and is never found in clear running water. It is considerably thicker than others of the *moccasin* species, has a broader head, and the skin is lighter in color. Its bite is poisonous, though not nearly so dangerous as the rattlesnake or copperhead; in fact, it is doubtful if a single bite from it ever produced the death of a strong man, though the pain following is very severe, and does not abate for weeks.

Of the latter two species, the black *moccasin* and water viper, nothing may be said to their discredit, so far as their ability to do any considerable harm is concerned. So far as my observations go the two are identical; distinction being made by some naturalists by reason of the fact that the snake does not always keep one color, but becomes darker towards midsummer, due probably to its more constant exposure to the sun. Having often heard that the snake was poisonous, after killing a large one once

WATER VIPER.

I made a careful examination of its mouth in which there was not discoverable either fangs or poison sacs, nor any other of the characteristics which distinguish venomous reptiles. Its teeth were well developed and eight on a side, all recurved, which serve the species most admirably in taking and retaining its prey, consisting of fish, frogs, young muskrats, slugs, or probably any small animal that might venture near its haunts. My home when a boy was in a village on the Ohio River, and curving half round the town was a creek in which there were fairly myriads of *moccasins* that showed themselves in great profusion during the hot summer months, basking on projecting rocks, floating logs, or the limbs of trees overhanging the water. They were also frequently found along the river shore, though rarely more than one at a time, whereas I have seen a dozen or more lying in confusion together on perches near the creek. So little fear was felt for these, however, that the boys went freely in bathing regardless of the many snakes that had dropped in the water at the approach of the bathers.

The *swamp moccasin* rarely reaches a length of more than two feet, while the Northern species not infrequently attain a growth of four feet.

The **Chittul Sea Snake** (*Hydrophis cyanocincta*), a venomous reptile, is found in the warm waters of the Malay Archipelago, where it is quite common. This snake is peculiar in that its body has its perpendicular flat, by which formation the creature can swim with greater rapidity than any other water species. In color it is olive green on the back, shading to an orange below, while across the back are numerous bars of deep black. In length it rarely

WATER MOCCASIN.

exceeds six feet, though specimens have been captured at least one foot longer. The creature, though living in the water and taking its prey therefrom, seems to be unable to inflict a poisonous bite while under water.

Yellow-bellied Sea Snake (*Pelamis bicolor*). This snake in shape resembles the one just noted, but differs in all other respects. Its length is about three feet and the coloring is less pleasing. The upper half of the body is dirty black, the lower half being of a light green, while the tail, spatulate in shape, is spotted with black. It is very numerous along the Australian shores and in all parts of the Indian Ocean. In this wide dispersion we see a considerable variation in the coloring, while the structure in all remains substantially identical. They produce their young so well developed as to need no care from the parent.

The Eyed Sea Snake (*Oculus natrix*) is another habitant of the Archipelago, being most common in the bays of Southwest Australia. It is very beautifully marked with layers of dull yellow and green,

CHITTUL SEA SNAKE.

spotted with black, somewhat in resemblance of the brook trout, from which coloring the name is derived. Its length is not above four feet, and its bite harmless.

YELLOW-BELLIED SEA SNAKE.

The Coral Snake (*Tortrix scytale*) is found only in tropical America, where it is regarded both favorably and with fear, according to locality. But wherever found it is entirely harmless, as is attested by the fact that it is often used by native women as a necklace, though, perhaps, in a spirit of bravado rather than of pride. In coloring it presents a pleasing contrast between black and old gold, the two colors being in transverse and alternate rings. This reptile is not only pleasing in its bright covering but is also curious, in that the body does not taper towards the head and tail as in all other species that rank above worms, but retains its cylindrical shape, being blunt at both ends. The length of this reptile is less than three feet and its principal food is insects, slugs, beetles, worms and caterpillars.

WART SNAKE.

The Wart Snake (*Achochorde javanicus*) is most populous about the shores of Java, and is hunted considerably by the natives who consider its flesh as most palatable. It differs from all other ophidia in that its diet is said to be exclusively vegetarian, an opinion supported by every investigation made by those who have dissected its body. The head is large, spreading out abruptly from the neck, after the manner of poisonous serpents, while the tail begins sharply at the vent and quickly terminates in a sharp point. The head is ridged, nostrils very close together, and the reptile has a habit of inflating its body to double the natural size, at which time the scales are separated and the creature appears to be covered with tubercles, from whence the name is given. It usually grows to a length of five feet, and is

CORAL SNAKE.

harmless; but its appearance is grotesquely horrid and even awe inspiring.

The **Anaconda**. The number of snakes whose natural habitat is the waters is few, those above described being the most interesting specimens, and as we have seen they are generally of an inoffensive nature. Following closely upon the water species comes an amphibious reptile which, though innocuous, is none the less dreaded, because of its gigantic size and crushing power. It is native to the equatorial regions of South America, and most common along the Orinoco and Amazon Rivers. Concerning this great reptile Sir Robert Porter writes: "The *anaconda* is not venomous, nor is it known to injure men; however, the natives stand in great fear of it, never bathing in waters where it is known to exist. Its common haunt is invariably near lakes, swamps and rivers; likewise close to ravines produced by inundations of periodical rains; hence, from its aquatic habits is the common

ANACONDA.

appellation, *water snake*, given. Fish, and those animals which repair there to drink, are the objects of its prey. The creature lurks watchfully under cover of the water, and while an unsuspecting animal is drinking, suddenly makes a dart at the nose, and with a grip of its back-reclining double row of teeth, never fails to secure the terrified beast beyond power of escape."

When the prey is secured the *anaconda* does not proceed immediately to swallow it, particularly if the prey be of any considerable size, but entwining its folds about the helpless victim it crushes every bone in the body, leaving the head alone intact. This the reptile performs in no undue haste, but with a kind of measured deliberation, after which it slowly uncoils its dreadful folds, crawls around the body as if to more perfectly determine the size, at the same time frequently touching the body with its tongue, but not as once supposed, to deposit a slimy saliva on the body as a coating to facilitate deglutition. It then begins the feast by swallowing the head of the prey first, much time being required in getting this within the reptile's throat. But even when once started down the œsophagus the food disappears slowly and by jerks, corresponding to a muscular exertion that is imparted by twitches.

IN THE TOILS.

The *anaconda*, like a great many other snakes, can emit at will an extremely fœtid odor, which in former times was believed to be a pestilential breath; but it has since been discovered that this evil effluvia is produced by the emission of a liquid secreted in two glands that lie near the vent, the power and provision being very similar to that of the pole-cat, the crocodile, and the muskrat.

In color the *anaconda* is beautifully marked, being of a very rich brown with a series of pale yellow rings edged with deep black along the sides, while upon the back are two rows of prominent circular black spots that present a most pleasing appearance. This reptile occasionally attains a length of eighteen feet and a circumference, about the largest part, of nearly two feet. Like others of the *boa* species, to which it properly belongs, the *anaconda* is oviparous, depositing its eggs in a dry spot near the margin of a river and incubating them, though for how long a period has not been determined.

The Boa Constrictor. As previously stated, there are no less than eleven different species of the *boa*, to each of which a local name has been applied, so that several designations are given to each, and according to the locality in which they are found. The generic name is *python*, derived from the fabled serpent Apollo is said to have slain near Delphi. The term *python*, however, is applied only to the species found in Africa and the East Indies, while *boa* is used to designate those of the species inhabiting the American tropics. The only difference between the two is that the former are provided with teeth in the intermaxillary bone, while the latter have the teeth confined to the jaws proper. This difference is, however, so very small that they may very properly be classed as members of the same species. Recognizing the local names as the best means for distinguishing the several species I will describe them under their respective appellations.

The **Rock Snake** is found in both Africa and India, but grows to a greater size in the former country, where specimens have been killed measuring twenty

RINGED BOA.

feet, while in India it rarely exceeds half that length. Though capable of exerting great muscular force, the reptile is very shy, and will submit to the mastery of a very small enemy. On one occasion a chicken was introduced into the cage of a large *rock snake* with the view of furnishing the reptile with a dinner; but when the serpent advanced to strike, the chicken showed fight, and in a few moments was pecking so vigorously at the snake's head that it had to be removed to save the reptile's life. On another occasion, however, the experiment with one of the African species was more exciting. The reptile, measuring something more than fifteen feet, had been fasting for a considerable period and was, at the time in question, shedding its skin, and half blind in consequence, as snakes invariably are at the moulting season. The keeper went to its cage with a fowl in his hand, but when offering the prey the snake struck at its intended food, but missing this seized upon the keeper's thumb. In another instant it had flung its dreadful coils about the man, and despite his struggles he would have speedily fallen a victim to the crushing power of the snake but for the arrival of timely assistance in the person of two other keepers, who had to break the serpent's teeth before forcing it to quit its hold.

The **Carpet Snake** is a member of the *boa* species, found in Australia with its congener, the diamond boa, the two being very similar. The former, however, is variable in coloring, while in the latter the markings are of a diamond shape, from whence the names are derived. They seldom exceed a dozen feet in length, and are little feared by the natives.

The **Ringed Boa** is a species somewhat common in Central America, but in former years it was extremely numerous owing to the reverence in which it was held by the Mexicans, who took every means for its protection. It is said that the Aztecs not only worshipped this snake but sacrificed human beings to propitiate its anger or to secure its supposed divine influence. This superstition no doubt grew out of the power which this truly enormous species possessed, and the insufficient weapons of the people to contend with it. Fear, therefore, prompted the Aztecs to propitiate it, in which effort priests were appointed to administer to its comforts, and in performing these duties some of the bolder succeeded in taming certain of the snakes, thereby gaining to themselves the reputation of possessing supernatural powers.

The *ringed boa* frequently attains a length of fifteen feet, and to its naturally great power of compression it multiplies its strength by seizing such prey as comes within reach and winding its coils *one over another* about it. Other species coil spirally about their victims and even in this wise are able to crush the ribs of an ox; what, therefore, must be the muscular force of a gigantic snake that re-enforces its strength by throwing coil around coil to double and treble its muscular power? To contend successfully with the *ringed boa*, it is only necessary to strike the creature a heavy blow on the tail, or better still, cut off a part of the extremity, when the reptile becomes almost powerless. Its colorings are not always the same, changing somewhat with the seasons, but are generally of a deep brown, or chocolate, with a series of circular spots of a yet darker color along the back and sides. The head is distinguished by having five dark streaks on the top and sides, extending down to the angle of the mouth. While all members of the *boa* family are provided with the rudiments of posterior legs, cropping out from beneath the skin in the form of tubercles, called spurs, the *ringed boa* has them more prominent, and in the male they are more conspicuous than in the female.

The **Dog-Headed Boa** is somewhat larger than the species just described, and makes its haunts among other giant reptiles amid the Brazilian jungles. It is beautifully marked upon a background of brilliant green, and has a fancifully striped head, while to attractive coloring is added a shape which some persons think resembles that of a dog, though it is a stretch of the imagination.

The **Boiguacu**, or **Boa Constrictor** proper, the largest and most powerful of the several species, is a native of South America and of Java, and though by no means common in any part of the world, is most frequently met with in forests that are under the equator. Like the ringed boa already described, it was once worshipped by a people who were practically defenceless against its ravages, and who sought to conciliate its ferocity by reverential attention.

The *boa* is handsome in a skin of singular markings and glistening surface. The body is covered with a chain of dark and pale white spots alternating, the white being oval and the dark compressed at the side by an overlapping of white. Im-mediately after shedding, the creature is really beautiful to be-hold, but grows lustreless towards the moulting sea-son. About the eyes is a circle of prominent scales which are evidently de-signed as a pro-tection to the or-gans of sight, enabling it to dart the head among brush or thorns without fear of injury to the eyes.

The *boa* is not often found exceeding twenty feet in length, though we have incontestable evi-dence of their occasionally reaching thirty, and even thirty-five feet. Con-

DOG-HEADED BOA (*Xiphosoma canium*).

cerning the process of swallowing prey, in which respect the *boa* is most interesting, Kingsley thus writes:

"Pythons generally prefer those localities which border on some quiet pool where they lie in wait, either suspended from an overhanging limb, or hid in the luxurious vegetation of the ground, or possibly partly submerged in the water, waiting the arrival of some small animal, which, as it is about to drink, the reptile seizes by the snout, and after wrapping several coils of its body about it, strangles it. Finally having crushed the larger bones, the process of deglutition is begun, which may last for several hours; the head invariably being the first to pass into the gullet, the body following. As the teeth all point inwards, and the jaws are successively and alternately pushed forward and drawn back, the prey, if not too large, is thus of necessity drawn into the mouth. The reptile may, however, find that its food is not suitable, or it may need to take breath, and though the prey has passed some way down the œsophagus, it is not unfrequently disgorged, making its appearance as a most frightfully contorted mass, covered with mucus from the alimentary tract; its slimy appearance having undoubtedly given rise to the false notion that the animal covers its prey, previous to deglutition, with saliva. For some time after the reptile has taken a large meal, it is, either from fatigue or from the effects of so loading its stomach, extremely lazy and inactive, being not infrequently quite indifferent to what may be going on about it. The inactivity of menagerie specimens, however, is due to the enfeebling effect of a cold climate, rather than torpor resulting from overfeeding, or gentleness from kind treatment. It is in their native forests that these forms must be studied to be admired. Not only are the caged animals inactive, but the purple bloom, so characteristic of the healthy animal, is invariably defective or lost; the rough treatment to which they are subjected, as well as a disease of the jaw—caries—rendering them indifferent and unhealthy.

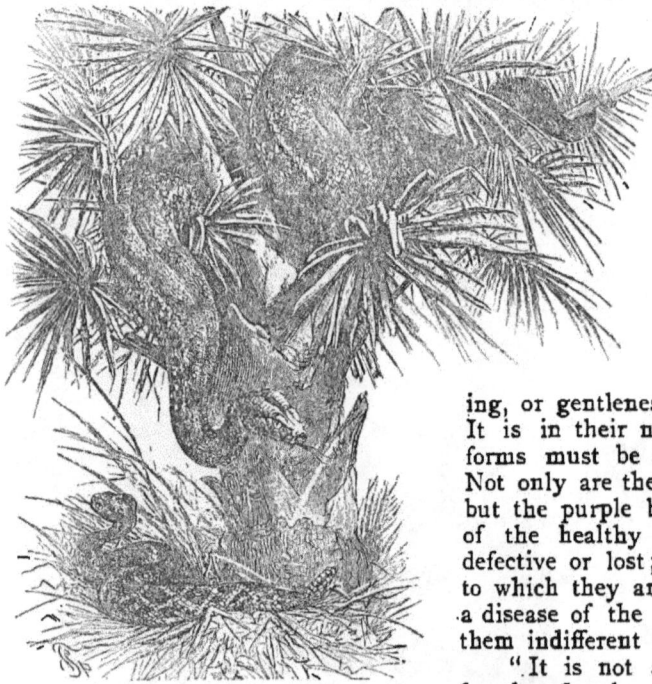

GIANT BOA AND RATTLESNAKE.

"It is not an unusual occurrence for the female python, which exceeds the male in size, to deposit her eggs while in confinement and watch over them with the most zealous care. Observations have been made which prove that the eggs are actually incubated. The mother, after arranging them in a convenient pile, coils her body, the temperature of which is considerably above the normal, around and over them, remaining in this position until the eggs, at the end of about three months, are hatched. We have here among the reptiles an undoubted instance of maternal solicitude." Mr. Kingsley might have added that the python's solicitude is imitated by a great many snakes, as already shown.

Goldsmith, to illustrate the crushing power of the *boa*, relates the following incident: "There are stories of the boa constrictor destroying even the buffalo and the tiger, by crushing them in this manner by the astonishing force of its muscles. We shall confine ourselves at present to a well-authenticated account of the voracious appetite of a serpent of this species, which was brought from Batavia, in the year 1817, on board a vessel which conveyed Lord Amherst and his suite to England. This serpent was of large dimensions, though not of the very largest. A living goat was placed in his cage. He viewed his prey for a few seconds, felt it with his tongue, and then withdrawing his head, darted at the throat. But the goat, displaying a courage worthy of a better fate, received the monster on his horns. The serpent retreated, to return to the combat with more deadly certainty. He seized the goat by the leg, pulled it violently down, and twisted himself with astonishing rapidity round the body, throwing his principal weight upon the neck. The goat was so overpowered that he could not even struggle for escape. For some minutes after his victim was dead the serpent did not change his posture. At length he gradually slackened his grasp, and having entirely disengaged himself, he prepared to swallow the lifeless body. Feeling it about with his tongue, he began to draw

A BOA CRUSHING A TIGER.

the head into his throat; but the horns, which were four inches in length, rendered the gorging of the head a difficult task. In about two hours the whole body had disappeared. During the continuance of this extraordinary exertion the appearance of the serpent was truly hideous; he seemed to be suffering strangulation; his cheeks looked as if they were bursting; and the horns appeared ready to protrude through the monster's scales. After he had accomplished his task, the *boa* measured double his ordinary diameter. He did not move from his posture for several days, and no irritation could rouse him from his torpor."

Woods also tells us that there are well-authenticated instances of men having been killed and devoured by monster boas, though he omits mention of particulars. While it is barely possible that such fatalities may have occurred, suspicion may well attach to such reports. It is a well-known fact that man has dominion over all creatures, and few there are, however ferocious, that will attack him except under provocation, and fewer still will eat man even after killing him. Snakes are even more timid than other creatures, the sight of a man being sufficient to impart dread in the largest. In addition to this I can call to mind no really authentic account of the death of anyone from the attack of a *boa*, though my reading of exploration and life in the tropics has been most extensive. Still, I cannot positively deny what the Rev. Dr. Woods declares to be true.

It is also a matter of very grave doubt whether a *boa* can swallow so large an animal as a buffalo, unless perchance it be a very small calf. The jaws of a *boa* and its power of extensibility, are not different from other snakes, and certainly not greater in proportion to size. A snake five feet long may swallow a rat, but we can hardly believe that one this length would attempt to swallow a dog, or even a cat, yet such a snake could as easily swallow a dog of ordinary size as a *boa* six times the length could swallow an ox. Such capacity is most certainly imaginary. I must also regard in like manner the stories frequently told of the *boa* killing and devouring a tiger, though this would hardly be so difficult an undertaking as the bolting of a full grown buffalo. A tiger could wage desperate resistance, but the vitality of all snakes is very great, and if the attack of a *boa* be made upon an unsuspecting animal, as is usually the case, the victim would be within its powerful folds almost upon the instant, and serious defence rendered impossible. Yet, for all this, I none the less doubt whether a *boa* ever killed and swallowed a tiger.

In strange contradiction to the above, though with that want of particulars and authority before spoken of which does very much towards discrediting the story, is the following which I take from *Goldsmith's Animated Nature*:

"In the East Indies they grow also to an enormous size; particularly in the

PYTHON, ECHIDNA, AND FENNEC (*Megalotis*).

island of Java, where, we are assured that one of them will destroy and devour a buffalo. In a letter, printed in a German Ephemerides, we have an account of a combat between an enormous serpent and a buffalo, by a person who assures us that he was himself a spectator. The serpent had, for some time, been waiting near the brink of a pool, in expectation of its prey, when a buffalo was the first that offered. Having darted upon the affrighted animal, it instantly began to wrap it round with its voluminous twistings; and at every twist, the bones of the buffalo were heard to crack almost as loud as the report of a cannon. It was in vain that the poor animal struggled and bellowed; its enormous enemy entwined it too closely to get free; till, at length, all its bones being mashed to pieces, like those of a malefactor on the wheel, and the whole body reduced to one uniform mass, the serpent untwined its folds to swallow its prey at leisure. To prepare for this, and in order to make the body slip down the throat more glibly, it was seen to lick the whole body over, and thus cover it with its mucus. It then began to swallow it at that end that offered least resistance, while its length of body was dilated to receive its prey, and thus took in at once a morsel that was three times its own thickness. We are assured by travellers, that these animals are often found with the body of a stag in their gullet, while the horns, which they are unable to swallow, keep sticking out of their mouths.

"Other creatures have a choice in their provision; but the serpent indiscriminately preys upon all; the buffalo, the tiger, and the gazelle. One would think that the porcupine's quills

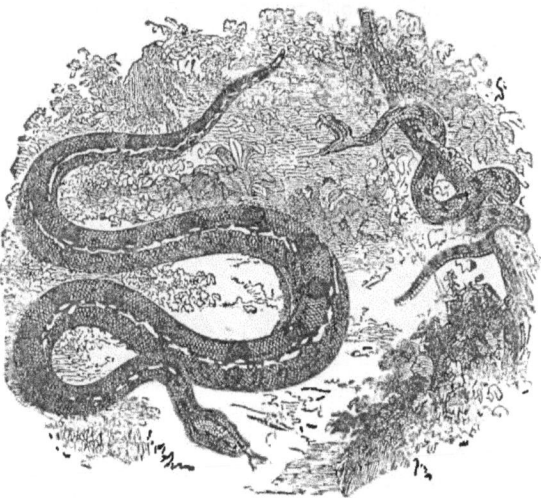

BOA AND RATTLESNAKE; COMPARATIVE SIZE.

might be sufficient to protect it, but whatever has life serves to appease the hunger of these devouring creatures; porcupines with all their quills have frequently been found in their stomachs, when killed and opened; nay, they most frequently are seen to devour each other."

POISONOUS REPTILES.

The largest species of snakes just described, while formidable and inspiring very great dread, are really much less dangerous than many comparatively small yet for this reason more insidious reptiles. Having therefore given the characteristics of the former we may now very properly turn to a consideration of the latter, which will be found distributed over an extensive area in both the old and new worlds, and consequently all the more important to be understood.

Most dangerous of all ophidians, though not the most considerable in size, is an India reptile known by different names, but commonly by that of *cobra*.

Cobra de Capello, or **Hooded Snake** (*Naja tripudians*). This dreaded snake has a large range over the East Indies, where it is everywhere regarded with the horror of a pestilence, a feeling most natural when we consider that not less than 5000 persons annually perish from its bite in the Indian peninsula alone. Notwithstanding this dangerous foe, which lurks in jungle, brake, roadway and even houses, and the spread of knowledge that has driven out the bats of superstition in nearly all parts of the earth, the Buddhists hold the venomous creature in reverential respect, and resent any attempt made at its destruction.

In coloring this reptile differs very materially, according to season and locality. More frequently it is of a deep olive on the back, white below, with hooded neck on which appear markings somewhat resembling eye-glasses, except that the bow is longer. But again it may be brown with black eye-glasses; or the body may be a deep brown with no eye-glass mark, or only two white spots on the neck; again the reptile appears with black body with a single white spot on the neck. Numerous and variable as these markings are they all belong to a single species.

The most peculiar thing at once observable about the *cobra* is the curious dilatable neck, which it spreads upon the slightest irritation. This power is afforded by means of elongated ribs which may be thrown out or depressed at pleasure. When expanded the neck is several times the width of the head, by which a curved hood is formed, on the back of which is a most singular decoration already referred to as resembling eye-glasses, from which fact it is sometimes called the *spectacled snake*.

COBRA DE CAPELLO.

A STRANGE SUPERSTITION.

The superstitious Buddhists regard this marking as the imprint of Buddha, and explain it in this wise. They declare that when Buddha became weary with the work of creation, he sought repose under the shade of a tree, but the sun moving while he thus slept soon poured its rays into his face. A *cobra*, perceiving the exposure of the god, spread its hood and remained bent over him, thus shading his face, until his sleep was finished. Upon waking, Buddha perceived what kind office the serpent had rendered him and thereupon promised to repay the *cobra* with a like kindness. This promise, however, he soon forgot, but was reminded of it under curious circumstances. Years afterwards a great bird hovered over the world and began devouring the *cobras*, until the species became well-nigh extinct. Thereupon, in a spirit of despair, the *cobra* that had shielded Buddha besought him, and referring to his promise, petitioned the god to give protection to his species, which Buddha did by placing the markings upon the *cobra's* hood, thereby frightening the great bird and preventing its further attacks. The natives therefore regard the reptile that once received the favor of their deity as being too sacred for molestation.

The *cobra's* bite is so venomous that very few ever recover, and those so fortunate as to survive are left subject to recurring pains of excruciating severity. If no immediate antidote is applied or given, the whole system becomes affected, the blood seems to congeal, and the body bloats to great proportions until death intervenes, which is usually within three hours. To this dreadfully venomous character is added a most wicked disposition, for the *cobra* will ordinarily attack on small provocation, and will almost instantly kill any other species of snake that may be confined with it.

EFFECTS OF MUSIC ON SERPENTS.

The curious structure, singular markings, deadly character, and wondrous superstitions which conspire to lend remarkable interest to the creature, are utilized by Hindoo jugglers and snake-charmers most profitably. It is a strange fact, not less difficult to understand than the accepted science of mesmerism, that music exerts a fascinating power over many creatures, and sensibly and powerfully on the *cobra*. Sometimes, it is true, the fangs of these performing reptiles are drawn, but not always, nor by any save what we may designate as the counterfeits, or impostors. Those who have studied thoroughly the habits and disposition of this snake are able to handle it with impunity, and to seemingly make it dance to their pipings. Thus may be seen in every part of India Hindoos going about from place to place with baskets filled with *cobras*. Wherever an audience seems promising the baskets are deposited, music of pipe, tambour and drum starts up, at which the snakes crawl forth of their own volition, and go writhing among the charmers.

INDIA SERPENT CHARMERS.

Concerning the influence of music on serpents, a distinguished authority thus writes:

"The incantation of serpents is one of the most curious and interesting facts in Natural History. This wonderful art, which disarms the fury and soothes the wrath of the deadliest snake, and renders it obedient to the charmer's voice, is not an invention of modern times; for we discover manifest traces of it in the remotest antiquity. It is asserted, that Orpheus, who probably flourished soon after letters were introduced into Greece, knew how to still the hissing of the approaching snake, and to extinguish the poison of the creeping serpent. The Argonauts are said to have subdued, by the power of song, the terrible dragon that guarded the golden fleece; and Ovid ascribes the same effect to the soporific influence of certain herbs and magic sentences. It was the custom of others to fascinate the serpent by touching it with the hand. Of this method Virgil takes notice in the seventh book of the Æneid. But it seems to have been the general persuasion of the ancients that the principal power of the charmer lay in the sweetness of the music. Pliny says, accordingly, that serpents were drawn from their lurking places by the power of music. Seneca held the same opinion.

EGYPTIAN SNAKE CHARMER.

"The wonderful effect which music produces on the serpent tribes is confirmed by the testimony of several respectable moderns. Adders swell at the sound of a flute, raising themselves up on the one half of their body, turning themselves around, beating proper time and following the instrument. The head, naturally round and long like an eel, becomes broad and flat like a fan. The tame serpents, many of which the Orientals keep in their houses, are known to leave their holes in hot weather, at the sound of a musical instrument, and to run upon the performer. Dr. Shaw had an opportunity of seeing a number of serpents keep exact time with the dervishes in their circulating dances, running over their heads and arms, turning when they turned, and stopping when they stopped.

"The rattlesnake acknowledges the power of music as much as any of his family, of which the following instance is a decisive proof. When Chateaubriand was in Canada, a snake of this species entered their encampment; a young Canadian, one of the party who could play on the flute, to divert his associates, advanced against the serpent with his new species of weapon. On the approach of his enemy, the haughty reptile curled himself into a spiral line, flattened his head, inflated his cheeks, contracted his lips, displayed his envenomed fangs, and his bloody throat; his double tongue glowed like two flames of fire; his eyes were burning coals; his body, swollen with rage, rose and fell like the bellows of a forge; his dilated skin assumed a dull and scaly appearance, and his tail, which sounded the denunciation of death, vibrated with so great rapidity as to resemble a light vapor. The Canadian now began to play upon his flute; the serpent started with surprise, and drew back his head. In proportion as he was struck with the magic effect his eyes lost their fierceness, the oscillations of his tail became slower,

SERPENT-EATING HOMODRYAS, OR NAJA.

and the sound which it emitted became weaker, and gradually died away. Less perpendicular upon their spiral line, the rings of the fasciated serpent were by degrees expanded, and sunk one after another upon the ground in concentric circles. The shades of azure green, white, and gold, recovered their brilliancy on his quivering skin, and slightly turning his head, he remained motionless in the attitude of attention and pleasure. At this moment the Canadian advanced a few steps, producing with his flute sweet and simple notes. The reptile inclining his variegated neck, opened a passage with his head through the high grass, and began to creep after the musician, stopping when he stopped, and beginning to follow him again as soon as he moved forward. In this manner he was led out of the camp, attended by a great number of spectators, both savages and Europeans, who could scarcely believe their eyes, when they beheld this wonderful effect of harmony. The assembly unanimously decreed that the serpent which had so highly entertained them, should be permitted to escape."—*Natural History of the Bible.*

The **Serpent-eating Homodryas** (*Homodryas elaps*) is a species nearly allied to the cobra, being almost equally deadly, and possessing a like distensible neck, though the workings are somewhat different, as will be seen by reference to the accompanying engraving. This serpent is peculiar from all others in that its sole food seems to be reptiles, of which lizards and snakes, whether venomous or otherwise, constitute the principal part. It is industrious and courageous in pursuit of its prey, being specially interesting to the observer when seeking to make a victim of another snake. At such a time the *homodryas* erects its head very high and with a dreadful hissing expands the hood, following with piercing eye and moving head every motion of its prey. At a favorable moment it launches upon and at the same time gives its victim a bite, which produces death, from poisoning, within a few moments. The swallowing next follows, after which the reptile is lethargic for twelve hours, but it does not make a second meal usually for ten days or two weeks after. Its bite is equally as fatal as that of the cobra, being sufficient to produce death in an elephant within two hours after being bitten. This snake is also variable in coloring, some being of an olive hue on the back and a dull orange below, while others are a dirty brown with cross bands of white. Like the cobra it is oviparous, and has for its enemies certain birds that destroy the eggs, and the ichneumon which will not only eat the young, but does not hesitate to attack the largest snake. A fight between the ichneumon, generally called the mongoose, and the cobra, or *homodryas*, affords a sight memorable for the cunning, wariness, celerity of action, and ferocity displayed. The serpent, apparently unmindful of the character of its enemy, feels confident of its venomous power, and promptly engages the quadruped. The ichneumon, however, better advised of the ability of its antagonist, manifests a cunning curious to see. It nimbly skips about the snake, constantly inviting an attack but always skilfully, and, it appears, luckily, escaping every stroke of what must soon be the victim. Sometimes this attack and strategy continues for half an hour, but at length the snake, worried by its futile attacks, becomes less watchful, and at a favorable moment the ichneumon now leaps high in the air and alights with certainty upon its victim, and, seizing the reptile's neck, crushes the vertebræ with a speedily fatal grip of its sharp teeth. The *homodryas* frequently attains a surprising growth, exceeding fifteen feet.

The **Spitting-Snake,** or **African Cobra,** resembles its East India congeners, showing a material distinction in only one singular respect, and from which the name has been given. Quite as venomous as its Asiatic brother, it is somewhat more to be dreaded by reason of its habit of projecting its poison to a considerable distance and with almost unerring aim. It is easily provoked, and not only attacks viciously but will pursue its enemy, and when unable to bite expels its venom in a small stream very much after the manner of the archer fish.

Gordon Cummings, the African hunter, was first to bring this wonderful snake to the attention of the civilized world, in the following brief allusion to his experience with the reptile:

"A horrid snake, which Kleinberg had tried to kill with his loading-rod, flew up at my eye and spat poison in it. I endured great pain all night; the next day the eye came all right again."

Since this report, which was much ridiculed at the time, the snake has come to be well known through specimens brought to England. Its bite is deadly, but the poison which it expels by a heavy respiration, and with a slight hissing noise, is not fatal unless it should fall upon an abraded surface, so as to reach the circulation.

The *spitting snake*, also caled *Haje*, is as variable in coloring as the Asiatic cobra, and attains about the same length. Like the homodryas it climbs trees in quest of birds and eggs, and is not averse to water, which it frequently enters, but is not known to eat fish.

THE RATTLE-SNAKE.

Next to the cobra and its species, considered

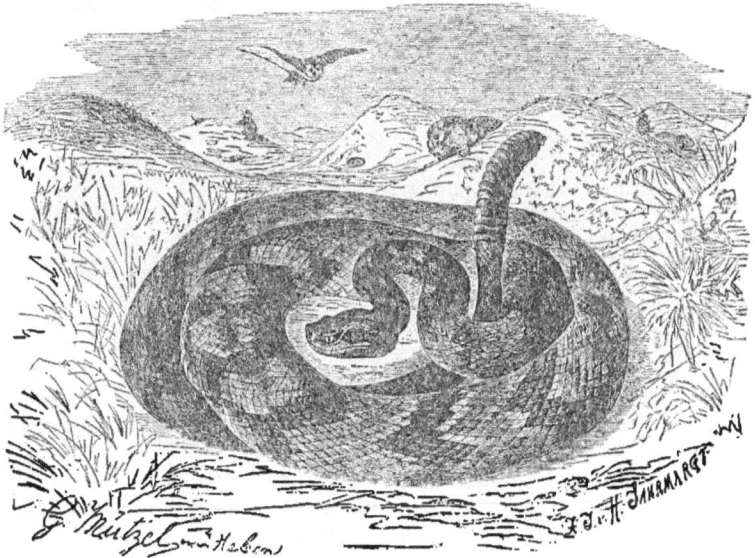

PRAIRIE RATTLESNAKE (*Crotalus durissus*).

ered for the deadly effect of its bite, must come the *rattlesnake* of North America, which is known to be one of the most venomous of God's creatures, certainly ranking next to the dangerous reptiles just described. As most of my readers know, the name *rattlesnake* is given to it on account of the curious termination of the tail, corresponding to dry, horny, hollow rings, loosely joined at the edges, which rattle at the least motion. These rings vary in number from two or three to as many as two dozen, this being the limit, so far as observation shows. It is a popular belief among the Indians that a *rattlesnake* adds a ring to his rattles every time a human being becomes a victim to its venom. Others, including a majority of naturalists, believe that these rings are an index of the reptile's life, a ring being added each year. Both of these opinions are without proof, for *rattlesnakes* in captivity have been known to add as many as four rings to their rattle in one year, while in others a new ring was added only once in three years. From this fact we must conclude that on some of the species the rings are increased much more frequently than on others, just as the beard grows much faster and thicker on some men's faces than on others.

The purpose of the rattle is difficult to determine, since it has been definitely ascertained that the reptile uses it apparently only to give warning of its presence, and not, as once supposed, to allure or fascinate its prey. I have frequently heard the warning sound, springing out of a bunch of grass on the prairies, and always associated it with the song of the grasshopper, which it much resembles. The tail trembles with such rapidity, when producing the sound, that the extremity appears blurred. At such a time the snake will always be found lying coiled up with the tail rising perpendicularly out of the centre, and the head slightly elevated, ready to attack. Unlike the spitting cobra, the *rattlesnake* is not quick to take offence, but will resist any aggression. It sometimes appears so good natured as to allow itself to be handled, as I have frequently seen, but never acquired confidence by such sights to attempt a like liberty myself. Early in the spring it is much less likely to be resentful, and at that season, too, its venom is not nearly so powerful. On the other hand, at the approach of fall its anger is easily excited and the poison which it secretes is then much more deadly.

The bite of the *rattlesnake* is not necessarily fatal if proper remedies are at hand, though very much depends on the constitution and general health of the person bitten, and on the species and season of the year when the wound is inflicted. Persons have been known to die within a few minutes after receiving the bite, while others have survived, though bitten in an equally vulnerable part, and not treated for hours afterwards. Whiskey is recommended as a sovereign antidote, and certainly possesses much antidotal virtue, but suction of the wound and bathing with ammonia may be used with equally good results, provided they are employed before the poison has ramified the system, in which event, perhaps, whiskey would fail to effect a cure.

The *rattlesnake* is said to possess the undefined power of *charming* such prey as birds, rats and even persons, while instances are reported of the *rattlesnake* lying coiled at the foot of a large tree and, by fixing its gaze intently upon a squirrel high in the branches overhead, luring the poor creature into its deadly jaws. Concerning this power of fascination, which in some respects appears identical with mesmerism, or what is more properly called *hypnotism*, Woods writes :

"Birds, especially, are more sensitive in their nature, and can be fascinated in a manner by anyone who chooses to try the experiment. Let any bird be taken, laid on its back, and the finger pointed at its eyes. The whole frame of the creature will begin to stiffen, the legs will be drawn up, and if the hand be gently removed, the bird will be motionless on its back for any length of time. I always employ this method of managing my canaries when I give them their periodical dressing of insect destroying powder. . . . There is another way of fascinating the bird, equally simple. Put it on a slate or dark board, draw a white chalk line on the board, set the bird longitudinally upon the line, put its beak on the white mark, and you may go away for hours, and when you return the bird will be found fixed in the same position, there held by some subtle and mysterious influence which is as yet unexplained."

The experiments thus reported by Prof. Woods are so singular, while he is regarded as such excellent authority, that I am more inclined to believe than ever before that snakes may exercise this power of fascinating their prey, though my experience has been to the contrary. Still, one man's experience is insufficient proof upon which to base a denial. I have frequently seen birds, such as the thrush and cat-bird battling heroically with a black snake that had attempted to despoil their nests. Yet it is claimed that the black snake can exercise the power of fascination, especially over birds, stronger than any other of the snake kind. In the cases falling under my observation the birds invariably beat off the intruder without themselves falling a prey.

When the winter season approaches the *rattlesnakes* retire into some close place affording complete protection from the biting frosts. Most commonly they seek shelter in a cave, or beneath heavy mosses, or in large hollow logs where their hibernation is least likely to be disturbed. Nor is this retirement in isolation, but on the other hand they seem to prefer communities, so that it frequently happens when their winter haunt is discovered hundreds will be found lying in a tangle, but with heads always pointing outward.

Rattlesnakes are peculiar to North America, where no less than eighteen species are found, some of which I will now describe:

The **Diamond Rattlesnake**, so called from the diamond-shaped markings of white over a dusky brown background, is usually found in marshy places, where it subsists off frogs, slugs,

DIAMOND RATTLESNAKE.

the young of muskrats, birds or any small creature that comes within its way. Concerning this reptile Holbrook says: "A more disgusting or terrific animal cannot be imagined than this; its dusky color, bloated body, and sinister eyes of a sparkling gray and yellow, with the projecting orbital plates, combine to form an expression of sullen ferocity unsurpassed in the brute creation."

The **Banded Rattlesnake** has its range from Maine to Texas, being once very populous in the East, but now rarely seen east of the Mississippi. In Texas it is still frequently met with, but the advance of civilization is rapidly diminishing their number. *Rattlesnakes* have their worst foe in the hog, which tramples it with impunity and feasts off the remains. This species prefers dry places, and its favorite food is rats, mice, frogs, young rabbits and ground squirrels.

The **Horned Rattlesnake** takes its name from the fact that its horrid head is rendered yet more terrible by being surmounted by a pair of horns rising about one-half inch above the cranium, in which respect it very much resembles the horned viper. Its habitat is the desert regions of Arizona, Mexico and Southern California. It is a very sluggish reptile and moves laterally instead of directly forward as do others of its genus. Its principal food is insects and sand lizards.

The **Prairie Rattlesnake** is still fairly plentiful along the Missouri and its tributaries, but its favorite haunt is in the Rocky Mountains. They spend the hot season amid ravines or cañons, being specially fond of willows, since among such growth they are chiefly found. When the fall approaches they seek some inviting hole, and very often in this search they take up their abode in the holes of prairie-dogs. From this well-known fact has come the belief that the *rattlesnake*, owl and prairie-dog have some natural affinity for each other, and that the three live in harmony like a happy family. The snake is an intruder into the home of the prairie-dog, but is too powerful an adversary to be expelled. But he crawls into his winter quarters at a time when the season of fasting is at hand, and when partial stupefaction from cold destroys his spirit so that he cannot be provoked into an attack unless first warmed into action. The owl, however, is a violator of the prairie-dog's house, since he does not consider the hospitality of the host, but visits the groundling with sinister motives. no less than an intent to make a meal off the young members of the rodent's family. Thus two very diverse purposes bring the *rattlesnake* and the prairie-dog together.

The **Massassauga Rattlesnake** is more distinctively a prairie dweller than the one just described, for he never leaves the prairie at any season, and this adaptation enables him to abstain from water for a very long time. Of the species mentioned, none grow to exceed six feet in length, but the *massassauga* has been frequently seen fully eight feet long. Being the largest of the *Crotalus* species, its poison is most virulent, but the extreme sluggishness of the creature renders it little likely to do injury to a human being, who is sure to receive ample warning of its presence by a loud rattling. Extremely poisonous as this reptile is, it responds to kind and careful treatment, and in a measure may become domesticated. I have a friend who, while living for a time in central Nebraska, captured a *massassauga rattler*, and as he had a great love for pets of any kind, in his lonely condition he resolved to see what effect careful attention to its wants would produce. Accordingly he secured the snake and deposited it in one corner of his rude cabin, fenced off, so to speak, by some boxing, so as to prevent it from escaping, or doing him harm. Every day he watched it attentively, introduced every kind of food, and talked to the reptile as he would to a human being. The snake at first resented all such advances, but gradually came to regard his master with more kindly concern, and at length would take such food as it had a liking for directly it was offered. By gradual approaches he finally secured an amiable recognition from his singular and deadly pet. A winter thus passed, the snake being kept during the time in comfortably warm quarters, and when

MASSASSAUGA RATTLER (*C. confluentus*).

spring was somewhat advanced he released it from prison to determine what effect its freedom might now have. Its attachment, however, was not broken by this change, but to his surprise the reptile appeared to prefer his companionship and would follow him from place to place like a dog follows his master, and would retire to its bed in the corner when night approached. For eighteen months this curious relation between man and snake continued, and until an accident put an untimely end to the thoroughly domesticated reptile.

The **Ground Rattlesnake** ranges along the Atlantic coast from North Carolina southward, and on the prairies west of the Mississippi, but on account of its diminutive size, since it seldom exceeds eighteen inches in length, has received little attention from naturalists. But what it may be lacking in size is more than compensated for by its abundance and the fact that its attacks are made without the warning that is given forth by the larger species. Being seldom provided with more than one rattle and a button, the noise it makes is scarcely audible, and being most courageous it is liable to be fairly stepped on before discovery. It is probably as venomous, proportionate to size, as the other species described, but its bite is rarely attended with fatal results.

FER DE LANCE.

The several other species not specially described are so nearly like those here mentioned that only the most critical examination by a naturalist can discover any difference, hence we do not esteem them of sufficient importance to note. Like the Indian, the *rattlesnake* is fast disappearing, the capture of one now anywhere east of the Mississippi being heralded as a great exploit. Their numbers are growing rapidly less by reason of the destruction wrought among them by hogs, as before stated, as well as by man, so that a few years hence none will be left save the few specimens preserved in museums.

The *rattlesnake* is viviparous, bringing forth from five to ten young at a time. I was once a witness to the birth of seven young rattlesnakes by a mother that had been in captivity several months, though I cannot say exactly how many, though certainly for six or more. The period of gestation has not been thoroughly fixed, but is supposed to be one year. The one to which I now refer was the property of a physician in the little village where I was born and reared, and was of course an object of great curiosity to every one in town, notwithstanding the fact that *rattlesnakes* in that vicinity were then no special rarity. This snake had taken no food during its captivity, probably because its cage was too small to permit much motion. The process of giving birth occupied the greater part of one day and a night, so that I actually witnessed the birth of only two of the seven brought forth. Directly after the young were delivered they crawled vigorously about their circumscribed quarters, and I remember that on being teased with a stick one of them opened its mouth and assumed an attitude of defence, though it did not bite the stick.

The **Fer-de-lance** (*Trigonocephalus lanceolatus*) is found in lower Mexico and thence southward to Brazil, where it is very numerous, and annually causes the death of more laborers on plantations than any other reptile. Its venom is probably no more powerful than that secreted by the rattlesnake, but being very much greater in size, it injects into a wound three times the amount of virus that is de-

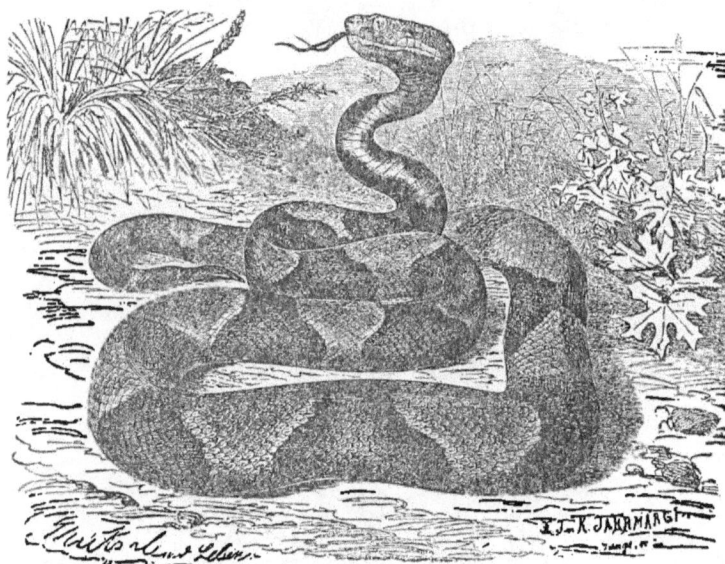

COPPERHEAD.

posited by the rattlesnake, hence its power for mischief is doubly or trebly great. Added to its intense venom is a most aggressive disposition, and making its attacks without warning, no one may escape its malignancy who comes within its path. The *fer-de-lance* also uses its venom to stupefy its prey, as it invariably kills the rats and other rodents upon which it feeds by biting before swallowing. It is a voracious feeder and somewhat compensates for its deadly ravages by the great number of rats it destroys.

The **Copperhead** (*Ancistrodon contortrix*) is another deadly habitant of North America, being found everywhere in the eastern half of the United States. The name by which it is best known has been given on account of its bronze-colored head, and dull orange hue on the body, alternating in blotches of bronze and dirty white. It is very venomous, though its bite is not so dangerous as the rattlesnake. It is frequently found under sidewalks in small villages, having a liking for the habitations of man. It is also often met with roaming through gardens in quest of frogs, mice and insects. Though cowardly in disposition, being prompt to retreat from any aggressor, it is also most subtle, and will steal upon and bite an unsuspecting person, provided it is near its hole. The *copperhead* hibernates, after the manner of rattlesnakes and European vipers, in communities of various sizes from a dozen to several hundreds, though their growing scarcity probably prevents the gathering together of so many of the latter number now. In length they never, I believe, or certainly very rarely, exceed four feet.

VIPERS.

The term *adder* and *viper* is used interchangeably between Europe and America, the former being American and the latter English, but they are both applied to the same species. In this country, however, there is but a single species, known as the *cotton-mouth adder*, and even this is more commonly called the swamp moccasin, though we often hear persons, ill-informed, speak of *spreading adders*, and applying the term to harmless snakes. The *viper*, however, is common in several parts of the Old World, many species being recognized, all of which are venomous, though their bite is seldom attended with fatal results.

COMMON VIPER.

The **Common Viper** (*Pelias veras*) is the only poisonous reptile known to inhabit England, where it is found in considerable numbers, frequently in Scotland, though it is never met with in Ireland. Since the Scotch people began the extensive raising of sheep *vipers* have become much less common, as sheep are most destructive to creeping reptiles, killing them, as the deer do, with their sharp feet. Its haunts are in heaths, dry woods, or the banks of small streams, where it feeds off frogs, shrews and birds. The color is somewhat variable, at one time appearing of a pale green, and again brown, yellow, brick-red, or

STRIPED ADDER.

even black. It is viviparous (from whence the name *viper*), usually bringing forth ten or twelve young at a birth. These are at once extremely active as well as pugnacious, and will bite viciously within a few moments after being brought into the world.

It is now a well-established fact that the mother *viper* gives protection to her young by receiving them into her mouth, and possibly into the gullet, dissection of the reptile having shown a receptacle like an elongated sac lying along and being an extensible part of the gullet, which can hardly have any other use than to provide a retreat for the young. Concerning the breeding of these reptiles Goldsmith thus writes:

"The kindness of Providence seems exerted, not only in diminishing the speed, but also the fertility of this dangerous creature. They copulate in May, and are supposed to be about three months before they bring forth, and have seldom above eleven eggs at a time. These are of the size of a blackbird's egg, and chained together in the womb like a string of beads. Each egg contains from one to four young ones; so that the whole of a brood may amount to about twenty or thirty. They continue in the womb till they come to such perfection as to be able to burst from their shell; and they are said by their own efforts to creep from their confinement into the open air, where they continue for several days without taking any food whatsoever. 'We have been assured,' says Mr. Pennant, 'by intelligent people, of the truth of a fact, that the young of the *viper*, when terrified, will run down the throat of the parent and seek shelter in its belly, in the same manner as the young of the opossum retire into the ventral pouch of the old ones. From this,' continues he, 'some have imagined that the viper is so unnatural as to devour its own young; but this deserves no credit, as these animals live upon frogs, toads, lizards, and young birds, which they often swallow whole, though the morsel is often three times as thick as their own body.'

"It does not always confine its voracity within the limits of its powers of deglutition; for I have in my possession a specimen of a small *viper*

RING-NECKED ADDER.

which was taken on Poole Heath in Dorsetshire, in a dying state, in the act of swallowing a mouse which was too large for it, the skin of the neck being so distended as to have burst in several places.

"The *viper*, like other reptilia, seeks a secret and secure place in which to hibernate during the cold months of the year. Here several are found entwined together, and in a very torpid condition; and if at this period a *viper* be made to wound an animal with its poison-fang, no injury is likely to result from it; the poison either does not exist at all, or it is inert."

The *viper* is not alone peculiar to England, but may be found distributed throughout Europe, and I saw one during my travels in Siberia a few miles west of Lake Baikal, by which, I conclude, the species described are scattered over Asia.

The **India Viper**, however, of which there are two species, does not greatly resemble its European congener. One of these species, the *Daboia russelli*, popularly called by the natives *Ticpolonga* (spotted snake), is nocturnal in its habits, and being very numerous, while exciting no reverential respect, it is very much dreaded by the people, and equally so by visiting Europeans who have given to it the name *Cobra monil*, because its venom is regarded as being next powerful to that of the cobra. Generally the color of this species is brown on the back with three rows of transverse rings edged with white, while the sides are marbled or pale yellow. Its usual length is from four to five feet. The second species, known as *Echis carinata*, differs very materially from that of the *ticpolonga*, in that it seldom exceeds two feet in length, and the bite is not dangerous, though accompanied by great pain, and its structure is by no means the same. It is quite common throughout India but never does any considerable damage, though it is regarded with much suspicion and not a little superstition, some persons claiming that its bite is most virulent, "requiring a double dose of medicine to cure," whatever that may mean.

African Puff Adder. This creature, known to scientists as the *Vipera arietans*, is found only in Africa, and usually in the southern part, especially abounding in the Kalakari Desert. Captain Drayson, traveller and naturalist, thus describes this dreadful creature:

"This formidable looking reptile is more dreaded than any other of the numerous poisonous snakes of South Africa, a fact which mainly results from its indolent nature. Whilst other and more active snakes will move rapidly away upon the approach of man, the *puff adder* will frequently lie still, either too lazy to move, or dozing beneath the warm sun of the south. This reptile attains a length of four feet six inches, and some specimens may be even longer; its circumference is as much as a man's arm. Its whole appearance is de-

AFRICAN PUFF ADDER.

cidedly indicative of venom. Its broad, ace-of-clubs shaped head, its thick body, and suddenly tapered tail and checkered back are all evidences of its poisonous nature. It derives its popular name from a practice of puffing out or swelling its body when irritated.

"An infuriated *puff adder* presents a very unprepossessing appearance. I once saw a female of this species in a most excited state. She had been disturbed in her retreat, under an old stump, by some Kaffirs, who were widening the high road through the Berea bush at Natal. She had several young ones with her, and showed fight immediately she was discovered. The Kaffirs were determined to kill the whole family, but were fearful of approaching her. Happening to pass at the time of the discovery, I organized a ring, and procuring some large stones, directed the Kaffirs to open fire. After a few moments the excited lady was killed and her body buried in a retired locality, lest some barefooted Kaffir might tread upon her head and thus meet his death."

The *puff adder* is most generally found imbedded in the sand with his cruel head projecting only a few inches, apparently watching for an unwary passer-by, be it man or beast. In such a position it presents a truly appalling appearance, more frightful indeed than are our conceptions of a demon. Not only is its bite

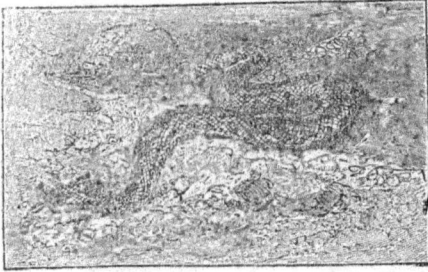

HORNED ADDER.

fatal to man, but horses or cattle die within a few hours after inoculation with its poison. Venomous as is this serpent, it speedily succumbs to nicotine poison. This fact is taken advantage of by the Kaffirs, who frequently kill the reptile by spitting tobacco juice into its opened mouth, or by irritating or inducing it to bite a stick that has first been smeared with the pasty nicotine that collects in the stem of a pipe smoked for a long time without cleaning.

The poison of the *puff adder* is used by the Kaffirs in which to steep their arrows to. make them more deadly, first covering the arrow point with a glutinous substance to make the poison adhere. The Kaffirs also drink the poison in order, as they believe, to render themselves proof against all venomous creatures.

The color of the *puff adder* is brown on the back, checkered with dirty black and white, and from the neck runs a dark red band up and over the centre of the head.

Other species of the *viper* found in Africa are the **River Jack** (*Clotho nasicornis*), and the **Berg Viper** (*clotho atropos*), both of which are very dangerous. The former possesses the curious feature of a horn on the nose, or what appears at first sight as such. This formation, however, is not horn, but is due to the development of long scales that rise up and project over

HORNED VIPER CHARMING A JERBOA.

the point of the nose, a feature found only in the male. Its color is deep black, flecked with oval spots of white along the back. The latter, or *berg viper*, is found, as its name implies, chiefly among the hills, though it often wanders into the plains, and very frequently into houses, especially on the approach of colder weather. Its length is rarely so great as two feet, and though in appearance it is most repulsive it is slow to anger and the bite less dangerous than that of the puff adder. It is commonly of an olive hue, with decorations of dark oblong spots running along the sides in four rows.

The **Horned Adder** (*clotho cornuta*), often called *cerastes*, though improperly, is another of the formidable reptiles of South Africa which, though gracefully mottled on the back presents a horrid appearance by reason of its thick, short head, from which project two sharp horns, rising directly above the eyes. The fangs are very long and when erected and mouth opened to its fullest capacity the points almost impinge upon the lower jaw. This prevents the reptile from biting anything save that it be small, or sharp-edged. This species is known to science as *Actractospis irregularis*. The *vipera cerastes* (horned viper) is very similar in size and appearance to the *actractospis*, but its habitat is Northern Africa, Arabia and Syria, and its power for evil much greater. It is supposed by many that the *cerastes* was the *asp* spoken of in history as the reptile used by Cleopatra to cure the hurt of her wounded ambition. It is natural that many superstitions should attach to this curious creature, especially by the oriental and ignorant mind that perpetually revels in the fanciful. The people of Egypt and other northern countries of Africa believe that within the two horns of this reptile re-

HORNED VIPER (*Vipera cerastes*).

side most potent virtues; in one is the store of poison to which is ascribed a marvellous potency, while the other, when pounded into a powder and laid upon the eye-lids, enables one to see spread before him all the accumulated riches of earth, although the possession of this wealth is not vouchsafed the experimenter.

The *cerastes* is found chiefly in the hottest deserts, where it has a habit of lying concealed in the sand with its wicked head protruding on the lookout for prey. But though it is vigilant and voracious it can endure an extended fast. Bruce kept two in a glass jar for two years, during which they partook of no food yet remained active throughout the whole time.

The *cerastes* is of a pale dusky white on the back, with numerous small brown spots over the body. Its length is hardly more than two feet.

The **Asp** (*Vipera aspis*) is found nearly everywhere throughout Europe, but is most plentiful in Sweden, and the whole of Scandinavia. Its length is not greater than that of the cerastes, but its bite is more dangerous, and not a few persons fall victims to its venom every year. However, the venom of this creature is not greatly to be dreaded except during the warm season, when, for reasons not clearly understood, the poison distilled by all serpents is most dangerous. But it is more noticeably so in the *asp*, and its fierceness is also much increased by thunder-storms, swift-flying clouds, or severe electrical disturbances.

The color of the creature is olive-brown, decorated with four rows of black spots, but sometimes the markings are different, showing a double chain of coalescing black spots running along the spine, very much like those on the common viper.

HARMLESS SNAKES.

Having treated a number of the most noted, because most dangerous, serpents of the Old World and the New, we will return to a consideration of a few of the species inhabiting North America that are considerable in size, often met with, but fortunately harmless, except to eggs, poultry and the small prey upon which they subsist.

The **Spreading Adder** (*Heterodon platyrhynchus*) is found in nearly all parts of the Middle and Eastern States, where it grows to a length of five feet. Its appearance is forbidding, which is very much increased by a habit it has, upon being irritated, of spreading its head and body and opening its jaws in a most defiant way. I remember when a boy, while hunting squirrels in the woods about my home, I discovered two of these snakes which, without danger, I captured, and, holding them by the tail, carried the reptiles into town and to a physician who was offering a reward of several marbles for each live snake brought to him. When I entered a drug store where the doctor was loafing every one beat a precipitate retreat, believing that the snakes I carried were of the most venomous character, an opinion which is yet generally shared, notwithstanding its fallacy.

The **Glass Snake** is also a harmless creature, about which many stories are told that belong to works of fiction rather than to a Natural History, except to expose their falsity. That such a creature does exist, capable of detaching its tail at will, like the slow, or blind worm of England, is perhaps well established, but this curious power has given rise to such fables as represent the snake being able to reunite the parts thus divested. Whatever evidence may be adduced, it is safe to deny the possibility of the reunion of the joints, since such a power would be in opposition to all laws of nature. The *glass snake*, which is of a variable color, changing with each season, is a ground creature, about two feet in length, incapable of climbing and slow in all its movements. As a means for its protection, therefore, it is endowed with the power to snap off its tail, which comprises nearly two-thirds its entire length, and leaves this portion to the mercy of its enemy while the body seeks escape in the ground or brush. The belief that the joints reunite is no doubt due to the fact that the parts thus thrown off are very soon reproduced by natural growth, just as in the lobster and other crustaceans the legs or claws, when lost, are speedily renewed.

The **Bull Snake** (*Pityophis melanoleucus*), more commonly called the *cow-snake*, is also an inhabitant of the Middle States, where it grows to a length of more than seven feet. The name is given out of belief that the reptile will spring upon a cow and while whipping her with its tail will drain her udder. I have often heard people declare that they knew of instances where cows have been ridden to death by this reptile, and their bags sucked until nothing but blood would be yielded. This is of course an idle fancy, with no more foundation than the belief that there is a reptile called *hoop-snake* that takes its tail in its mouth and goes rolling on a straight line like a hoop until it strikes the object of its wrath, when it launches forth a horn that is upon the head with direful results. A stroke of this horn, it is said, will blast a tree or kill any animal.

The *bull snake* feeds off frogs, rats, squirrels and birds. It is a good climber, and, like the blacksnake, is often found in bushes in search of birds' nests. But so timid is the nature of this reptile that it will retreat before the assaults of the birds it attempts to rob, especially if the attacking force be re-enforced as frequently happens. The markings of this snake are quite handsome, its body being covered with gray, irregular spots with brown bands between.

The **Common Blacksnake** (*Coluber constrictor*) is one of the most common snakes in the United States, and is found everywhere from the British possessions to Mexico. It is of a lustrous solid black color, and a slaty gray underneath. To this species belong the *blue racer* and *coach whip*, which are no doubt one and the same. The

COW SNAKE ATTACKED BY BIRDS.

blacksnake is courageous only when he discovers that his presence has excited fear, and has been known to pursue a fleeing person, but he is an arrant coward in the face of resistance. They grow to a length of six feet and are most graceful and active in their movements, being able to crawl at great speed. It is usually found in the woods and prefers dry places, though I have frequently met them in ravines, but generally on bushes, which it climbs with facility in quest of eggs and young birds.

The **King Snake** (*Coluber rex*) is also common to the Western and Middle States, where it is frequently confounded with the chicken snake, which it very much resembles, though its body is much slimmer. This snake derives its name from the implacable hostility it manifests to all other of the ophidian species. Nor are its attacks confined to the non-venomous, for I once saw a

14

king snake, not above four feet long, pursue a rattlesnake of equal length, which it soon caught, and a death struggle ensued. The poisonous reptile appeared to be well aware of the power possessed by its antagonist, and tried hard to escape, but when overtaken offered the best defence it was capable of giving. The *king snake* was extremely wary and moved from one side of his foe to another with wonderful celerity until an opportune moment, when it struck the rattler on the neck and there holding fast to his advantage, on the next instant entwined itself about his enemy's body, a hold that was not relinquished until the venomous reptile was squeezed to death.

The *king snake* never, I believe, exceeds five feet in length, and its food is rats, mice, frogs, and birds. Some maintain that it swallows the snakes which it kills, but this assertion is lacking of proof. On the contrary, in two instances which fell under my observation, when a *king snake* killed its adversary, one of which is described above, no attempt was made to swallow the carcase.

A SNAKE-EATING BRAZILIAN.

The **Chicken Snake** (*Coluber quadrivittatus*), as before stated, resembles the *king snake*, except that it is somewhat thicker of body. The coloring of each is a dark olive with four longitudinal bands of brown sprinkled with white. It is a frequent visitor to houses and corn cribs, and in the latter place creates great havoc among mice and rats. It is charged with being a chicken thief, but certainly does not devour anything larger than chicks, and though familiar with the reptile for many years, I know of no instance where it has molested even these.

The **Corn Snake** (*Coluber guttatus*) is not a rare reptile in the Western States, though seldom seen on account of its nocturnal habits. It is usually about four feet long and most beautifully colored, the body being of a light brown, while along the spine is a row of oblong red spots having yellow edges. The belly is of a pearly white.

The **House Snake** (*Ophibalus triangulus*), though not nearly as large as the *corn snake*, bears a striking resemblance to it in color, the markings being very similar though somewhat brighter. I have never seen a specimen that measured above two feet, though some naturalists give its length as four feet, of which I am doubtful. It prefers the dwelling places of man, and seems anxious to establish amicable relations with housewives over whose crocks of milk it exercises a zealous care, skimming off the cream whenever opportunity offers. At least this habit is charged to it though I have never seen any proof of the assertion, and have come to doubt that it has any liking for milk. This little reptile does certainly catch rats and mice, so that its presence in the kitchens and smoke-houses ought to be encouraged.

The **Ground Snake** is another reptile that should receive a better favor than is commonly meted out to it. This creature is profusely distributed throughout the Western States, especially in prairie districts. Its greatest length is perhaps three feet, and the markings are almost too indistinct for description, resembling the ground, except that there are four very faint yellowish lines traversing the body longitudinally. Its food is entirely field-mice and obnoxious insects, such as grubs, grasshoppers, and vegetable-eating beetles.

Among the other common snakes of America, familiar to nearly all my readers, are the garter snake, green snake, bead snake, ribbon snake, chain snake, thunder snake and gopher snake. The two latter are of considerable size, and both belong to the black snake species. The *thunder snake* is formidable to others of the genus, in which respect its habits are much like those of the king snake. The *gopher snake* sometimes reaches a length of eight, or even nine feet, but is extremely timid, making its home very frequently in a gopher's hole, from which fact the name is derived. In many countries, notably among the South American Indians, the harmless species of snakes are hunted for the flesh and are esteemed great delicacies.

TORTOISES AND TURTLES.

The transition from snakes to tortoises seems sudden and unnatural, especially in a work of this character, which pretends to the introduction of species in the order of their development, or supposed evolution of creation;

SKELETON OF GLYPTODON.

but to those who have made any study of zoology it will not appear so. Tortoises and turtles, under the general classification of reptiles by naturalists, are included under the term *chelonia*, and properly so because of certain identities, chief among which is the similarity that exists in the lung and heart organization which has served to distinguish them as cold-blooded creatures, and particularly because of their methods of locomotion, which is by creeping, crawling, or jumping. The distinction in these respects between reptiles and fishes is not well defined unless the general characteristics be considered, in which structure is almost as important as organization or habits. The particular line of demarcation, however, is found in the fact that fishes breathe water, so to speak, while reptiles breathe air direct. To be more explicit, a more perfect apparatus for aquatic respiration is seen in fishes; the gills are comb-like fringes supported on three or more bony or cartilaginous arches, and are composed of myriads of microscopic capillaries, by which the venous blood as it flows through them is exposed, in a state of minute sub-division, to streams of water. The gills are always covered, and the water which is taken in at the mouth escapes by the gill openings at the sides, hence the process is equivalent to *breathing water*, by which the blood is aerated, or arterialized.

While fishes have lungs, they are imperfectly developed, being no more than the air-bladder, which is wholly rudimentary. In the proteus and siren, however, which are described in this work, though under the division of *reptiles*, both gills and lungs are present, which serve to make them a connecting link between fishes and reptiles, as difficult of classification as is the bat. Lungs are relatively largest in reptiles, but the air cells are few and large, and the blood capillaries are exposed to the air on only one side. In other words, fishes inspire only, while reptiles, like birds and mammals, breathe through the mouth

EXTINCT PROTOSTEGA RESTORED.

and nose. Turtles and tortoises, whose ribs are united together so as to constitute an inflexible shield, are compelled to *swallow* the air, whereas in other vertebrates air is drawn into the lungs by expansion and contraction of the ribs.

Another distinction between fishes and reptiles is found in the sense of touch, which in the former resides in the lips, while in the latter it is most positive in the tongue.

Other points of distinction exist between the orders, but except for the student of comparative zoology they are not interesting, and require too much classical research to warrant their description here. We will, therefore, proceed with a more general consideration of the order and species which properly belong under the head of *chelonia*, a Greek word meaning *tortoise*.

TURTLES OF A PRE-ADAMITE PERIOD.

In the introductory remarks concerning the appearance and development of reptilian life, we saw how huge were the monsters that ploughed the primeval waters, and what analogy exists between species of our time and the monster saurians and ophidians whose fossil remains, or traditionary lore, enable us to form a just conception of their size and habits. It must not be supposed that the turtles, with which we are so familiar, had any less distinguished progenitors than the ravenous crocodile, however great may be the dissimilarity of disposition between the two species. The persistent search of palæontologists has brought to light many strange things during the past two centuries, by whose discoveries it has been shown that the world was one time peopled by colossal amphibians and mammals whose appearance, in life, would serve to inspire the greatest dread· in all mankind even in this day, with all the engines of destruction at our command.

The ancients affirmed that there were turtles whose shells were so large as to serve for the covering of a house, a statement ·which was regarded as highly exaggerative until remains of the gigantic *glyptodon* (carved tooth) and *protostega* were uncovered and subjected to examination, when it was found that these creatures were the equals in size of any amphibian that roamed the sea, and with a mouth large enough, as Sir John Hunter states, to admit a horse· and cart. The shell of such an animal might easily suffice to cover a house of considerable size, and but for its sluggish movements might have been a most formidable creature. So strong was its armament that a ball from any modern firearm, short of a cannon, could produce no effect upon it.

CHARACTERISTICS OF THE TORTOISE.

The most peculiar feature observable in the tortoise is the example it affords of having the skeleton on the outside, a characteristic found nowhere else save in insects and crustaceans, nor in any of these so perfectly. The tortoise, to all purposes, is enclosed within a box, with openings at the sides to permit the free exercise of its head, limbs and tail. But the creature is firmly attached to this covering, and is unable to remove any part of the back, and only very slightly the shield that protects the breast. This shell is of horny structure and divisible into two parts, the upper being called the *carapax*, and the lower, or breast-plate, the *plastron*. The difference between turtle and tortoise, in structure, is observable in that the carapax and plastron of the latter are inseparable, being a continuous growth, and the ribs are united throughout, whereas in the turtles the ends of the ribs retain their original width, and there is a difference in the composition between the carapax and plastron.

The tortoises are devoid of teeth but possess an excellent substitute therefor in having very sharp, bony-edged jaws enabling them to bite with great effect. The neck is generally long and retractile, in some species being of apparently disproportionate length and yet capable of being withdrawn entirely within the shell and concealed, as well as protected, by the anterior part of the plastron, which works upon a hinge and may be opened or closed at the pleasure of the animal.

In muscular strength the tortoise exceeds all other creatures, every bone as well as the carapax being designed with the special view of re-enforcing the muscles and giving extraordinary strength and vitality to the animal. The brain is surprisingly small and does not appear to be absolutely essential to life, if we are to believe the statement of the naturalist Redi, who has reported many strange experiments made with tortoises. Among these experiments was one to test the vitality of these creatures which he made in the month of November, on the approach of cold weather when these animals begin to prepare for the winter by digging a hole in which to hibernate. He accordingly took a land tortoise, made a large opening in its skull, and drew out all the brain, washed the cavity, so as not to leave the smallest part remaining, and then leaving the hole open, set the animal at liberty. Notwithstanding this, the tortoise marched away without seeming to have received the smallest injury; only it shut the eyes, and never opened them afterwards. Soon after the hole in the skull was seen to close; and in three days there was a complete skin covering the wound. In this manner the animal lived without a brain for six months, walking about unconcernedly, and moving its limbs as before. But the Italian philosopher, not satisfied with this experiment, carried it still farther; for he cut off the head, and the animal lived twenty-three days after its separation from the body. The head also continued to rattle the jaws, like a pair of castanets,.for above a quarter of an hour. Notwithstanding the authority, I doubt the statement.

The tortoise is oviparous, laying from eight to twelve eggs at a time, in a hole dug by her powerful claws some six inches deep, and always, I believe, in the hard earth and beneath the shade of either bush or trees. Their eggs are elliptical in shape and covered with a membrane of a leathery texture that is somewhat difficult to rupture, in which several respects they resemble the eggs of serpents, crocodiles, and lizards. The turtle, however, lays her eggs in the sand, selecting a spot where the sun's rays are strongest, depending upon the sun to hatch them. Their eggs, quite unlike those of the tortoise, are spherical and covered with a lime shell which overlies a leathery integument of the same texture as that of the tortoise egg, so that the shell may be removed without rupturing the egg. No amount of boiling will serve to harden the yolk or albumen of either.

Although there are many points of difference between the tortoise and turtle, both in structure and habit, so marked as to justify the separation of the two into distinct divisions, they are classed under one head by all naturalists, which classification I will not here depart from, especially as it will save some space to consider them together. The differences will appear as we proceed with the descriptions.

The **Lettered Tortoise** (*Emys lutaria*), also called the *morass turtle*, is a large variety, and common in many parts of the world, though most numerous in North America. It is more terrapin than tortoise, frequenting, as it does, ponds, lakes and marshy places. Its food is small snakes, lizards, frogs and worms, and in domestication, to which it is susceptible, will eat several kinds of vegetables, though its preference is for animal food. Its color is a dark-brown carapax edged with scarlet marks somewhat resembling letters, hence the name.

Chicken Tortoise (*Emys reticularia*) is very plentiful in nearly all the waters of the United States, though more common in ponds and creeks, where numbers may be seen sunning themselves on logs during the summer days. It is small in size, very awkward in movement, but extremely wary. When in the water it has a habit of stretching up its long neck and sticking the nose barely above the surface and floating about in a listless, lazy way, but quick to disappear at the sound of a foot-fall. Its flesh is said to be excellent and is occasionally sold in the markets, but it is not a popular dish. The shell is dark brown, sprinkled with yellow lines, and the bill is slightly hooked.

Salt Water Terrapin (*Malaclemys*) is a name given to the best known variety in the two Americas because of the excellence of its flesh. The head is large and covered with a spongy skin, on which account it is sometimes called the soft terrapin. Its natural habitat is salt marshes but from which it occasionally strays short distances. At the approach of winter it digs a hole at the edge of a marsh and retires therein until the warm season returns. They are very timid and active either in or out of water, being extremely difficult to take except when they begin to deposit their eggs, in the early summer months, at which time immense numbers are caught and sold in the market. Its color is variable, but most generally a dark green on top, with yellow flecks on the edge plates. The head is marked with sprinkled white and small black spots, and the lower jaw is armed with a hooked beak.

Box Tortoise (*Testudo carolina*) is a very appropriate title applied to a common variety found all over the United States east of the Western alkali regions. It is peculiar in the singular respect of being able to shut itself entirely within the shell by reason of the plastron being divided so as to work on a hinge both the anterior and posterior parts. When danger threatens, its legs and head quickly disappear

GREEK TORTOISE (*Testudo græca*).

within, the tail curled tightly behind, and no part left exposed. In this condition it may be violently used without forcing it to protrude the head or legs. Fire laid on the plastron, however, will cause it to seek escape, which cruel means is sometimes employed. It is found generally in dry places in the woods, but often comes into our gardens in search of insects, such as slugs, grubs and crickets, and is not averse to taking an egg or even a young chick. It is easily tamed and in captivity will eat apples, oranges or nearly any fruit, and when these are scarce it is content with meat and bread.

Though all the tortoise varieties are known to attain great age, the *box tortoise* is specially noted for its longevity, the duration of its life having been approximated by the cutting of a date on its shell and releasing it. This custom is quite common and has been so for many score of years, so that tortoises have been found with dates cut showing as much as seventy-five years to have elapsed since the time of the marking.

Concerning the probable age attained by this animal, Goldsmith says: "Tortoises are commonly known to exceed eighty years old; and there was one kept in the Archbishop of Canterbury's garden, at Lambeth, that was remembered above a hundred and twenty. It was at last killed by the severity of a frost, from which it had not sufficiently defended itself in its winter retreat, which was a heap of sand at the bottom of the garden."

In *Murray's Experimental Researches* we also find the following very interesting account, relating, however, to a variety of very large tortoises found in Europe:

"From a document belonging to the archives of the cathedral, called the Bishop's Barn, it is well ascertained that the tortoise at Peterborough must have been about 220 years old. Bishop Marsh's predecessor in the see of Peterborough had remembered it above sixty years, and could recognize no visible change. He was the seventh bishop who had worn the mitre during its sojourn there. If I mistake not, its sustenance and abode were provided for in this document. Its shell was perforated, in order to attach it to a tree, etc., to limit its ravages among the strawberry borders. The animal had its antipathies and predilections. It would eat endive, green peas, and even leeks; while it positively rejected asparagus, parsley and spinach.

"All animal food was discarded, nor would it take any liquid, at least neither milk nor water, and when it took a leaf that was moist, it would shake it to expel the adhering wet. This animal moved with apparent ease, though pressed by a weight of eighteen stone (252 pounds)—itself weighing 13½ pounds. In cloudy weather it would scoop out a cavity, generally in a southern exposure, where it reposed torpid and inactive, until the genial influence of the sun roused it from its slumbers. Its sense of smell was so acute that it roused from its lethargy if any person approached, even at a distance of twelve feet. About the beginning of October, or latter end of September, it began to immure itself, and had for that purpose for many years selected a particular angle of the garden; it entered in an inclined plane, excavating the earth in the manner of a mole; the depth to which it penetrated varied with the character of the approaching season, being from one to two feet, according as the winter was mild or severe. It may be added, that for nearly a month prior to this entry into its dormitory, it refused all sustenance whatever. The animal emerged about the end of April, and remained for at least a fortnight before it ventured on taking any species of food. Its skin was not perceptibly cold; its respiration, entirely effected through the nostrils, was languid. I visited the animal, for the last time, on the 9th of June, 1813, during a thunder storm; it then lay under the shelter of a cauliflower, and apparently torpid."

There are above twenty varieties of land tortoises found in both hemispheres, all of which are alike so variable in color that they are difficult to describe, each being apparently marked differently from his brother. Of these several species the following may be mentioned:

The **Wood Tortoise** (*Chelopus insculpta*), found generally in the mountainous region east of the Ohio, is not above six inches in length, and a black spot in the centre of each of its scales gives to it a very handsome appearance.

The **Elegant Tortoise** (*Pseudemys elegans*) is confined to the region lying between Illinois and the Rocky Mountains. Its most characteristic markings are a blood-red band on each side of the neck, and a yellow under-shell.

The **Speckled Tortoise** (*C. guttatus*) is common to New England, but is found as far west as Lake Michigan. The carapax is black, spotted with orange. It is easily domesticated and therefore is frequently made a pet of by young ladies who delight in being peculiar.

The **Painted Tortoise** (*Chrysemys picta*) is a very common variety in the Eastern States, where it is sometimes nicknamed *mud-turtle*. The carapax is usually black, with a greenish hue, and the marginal plates dotted with spots of bright red. •

The **Salt Marsh Terrapin** (*Malacoclenemys geographicus*) is better known as the *diamond-back*, and best appreciated for its excellent flesh. It is found everywhere along the coast from Long Island to Texas. In the cold season it hibernates in the mud, at which times its flesh is more highly prized and great numbers are taken to market, where it sells at a very high price. So profitable do tortoise catchers find them that the industry of *terrapin* farming has recently been started, which I am told yields a splendid return on the capital and labor expended. This species is of a dark olive color on the back, with dark stripes traversing the plates of the upper and lower shells.

EUROPEAN MARSH TURTLE (*Testudo lutoria*).

Yellow-bellied Terrapin (*Pseudemys troostii*) is the name of a small species found plentifully in both creeks and rivers of the West as far north as St. Louis. In color it is of a very dark green, with horn-colored lines and spots on the side plates. The plastron is yellow, splotched with black, and under the throat are several green stripes.

The **Alligator Terrapin** (*Chelydra serpentina*) is also an American species found in nearly all parts of the United States. Its preference is for stagnant ponds, where the mud is deep and of a slimy consistency, into which it burrows at the approach of danger, though it is by no means a timid animal. I have often seen them, especially the occupants of a shallow stream, with backs covered thickly with decayed vegetable matter, so as to resemble moss, as if the creature had not moved out of one spot for more than a season. They are very voracious, and apparently indifferent as to the kind of food offered, grabbing anything

SEA TORTOISE.

that is digestible. When the streams or ponds in which it has taken abode become dried up, it travels across land any distance in search of water and without inconvenience. My observations lead to the belief that its instincts for finding water are most unreliable, for I have found them in the highway and going over low-lying ground in a direction directly away from inviting ponds. While on land his motions are extremely slow and awkward, stumbling along with head held erect, unmindful of the obstacles that may lie

in his way. Walking also quickly tires him, so that at every rod of his journey he sits down, with hind legs drawn up and front ones straight, in which
position he presents a comical aspect. His appearance at best is far from pleasing; his shell is a dusty black, without markings; his tail is long and knotted,
legs long and large, terminating with powerful claws. The head is prominently
large, covered with thick wrinkled skin, neck long and studded with wart-like
tubercles, while the mouth is armed
with sharp jaws and hooked beak.
When irritated the creature gives off a
penetrating odor resembling musk,
from which fact, together with its cor-
rugated back and generally saurian
aspect, the term *Alligator terrapin* has
been given. A more appropriate name
is applied to him in the West, where
he is universally known as *snapping
turtle*, though the snapping turtle proper
is more common in the swamp regions
of the tropics. It is a most savage
reptile, with jaws powerful enough to
bite off a finger at a single snap. In
making an attack it is more tenacious of its hold than a bull-dog, for which
reason it is said not to quit its hold until it thunders. Its vitality is truly
remarkable, having been known to live for a week with the head wanting.
Notwithstanding its disgusting appearance and habits, its flesh is highly
esteemed by some people whose appetites are evidently not very delicate.

SNAPPING TURTLE.

The **Snake-Tortoise** (*Hydromedusa maximiliani*) is a species found only
in the marshes,
stagnant waters,
and occasionally
in streams of
Australia. Its
most pro-
nounced charac-
teristic is an ex-
tremely long
and snake-like
neck. The head
is also of un-
usual length,
with the eyes,
which are dis-
proportionately
large, set almost
at the nasal ter-
minus. When

THE TIGER, OR SNAPPING TURTLE OF SOUTH AMERICA.

the body is hidden from view and the long neck exposed to observation, this
tortoise has an astonishingly serpentine aspect, which is increased by a habit
which the animal has of moving the head with arched neck in wonderful simili-

tude to a snake. It lives off reptiles and fish, prey which its great speed in the water enables it to capture. The color is dark brown, and scales edged with a black line.

The **Matamata** (*Chelys matamata*) is a habitant of South American waters under and north of the equator. In appearance it is the most singular species of the tortoise tribe. The carapax has the appearance of irregular miniature hillocks, with knob projections running round the base; but while this rugged formation serves to give it a rather curious aspect, it is in the head and neck that the truly hideous characteristic is pronounced. The head is broad and most curiously surmounted by ear-like flaps of extremely tough skin, and the snout is much prolonged and sharp on the point. The neck is long and broad—almost flat—which is an odd form, but to this unusual shape is added the very grotesque feature of four rows of rather long fringed membranes, or wattles, while two prominently large wattles hang from the throat. What purpose this strange formation can subserve is not known, especially since by experiment it has been shown that they may be cut off without apparent injury to the animal.

The *matamata* attains a length of three feet, to which considerable size is added a rather ferocious and voracious disposition. It is an expert swimmer, but seldom exercises the faculty, preferring to lie in concealment under drift, or grass by the water's brink, from which it darts out upon fish, reptile or fowl, that may come near, and with the prey returns to its former place to devour it. Its flesh is considered a special delicacy.

MATAMATA

The **Soft Shell Tortoise** (*Trionyx nuticus*), commonly called *soft-shell turtle*, is found in nearly all the running streams of the United States, and is especially numerous in the Ohio and Mississippi rivers. As the name indicates, the shell is soft, or of the consistency of sole-leather when wet. The color of the carapax is that of fresh liver, and is frequently spotted with black dots, and at other times with dull yellow and pale green. The neck and head are cylindrical, terminating with a soft snout, and under the throat are pale yellow stripes and dots, though there is a great variation in the coloring. The plastron, however, is always of a light flesh color. The neck is retractile, being capable of stretching to considerable length or withdrawing until only the snout is visible. It is entirely carnivorous, feeding on fish, cray-fish, worms or any kind of meat, whether fresh or putrid. Although Woods pronounces this species "as one of the strongest and most ferocious of reptiles," confounding it with its South American congener, the *macrochelys lacertina*, its nature is directly the opposite. Though I have been familiar with them all my life, as a resident on both the Ohio and Mississippi rivers, I never knew one to attempt any revenge upon its captors. The South American species, however, which though larger and almost identical in appearance, is hardly so harmless, a captive one having been known to bite a sailor's finger off.

The true turtles are peculiar to salt water, differing in this respect from tortoises, save a few varieties described. They are also dissimilar in structure in that instead of the ribs being united throughout the length, as in the tortoise, in the turtle the ribs are flat, only a part of which are attached to the carapax, the others radiating concentrically.

The **Hawk's-bill Turtle** (*Testudo imbricata*) is the best known because the most valuable of all the turtle species, since from its shell are produced many beautiful and high-priced articles known as *tortoise shell*. It still exists in considerable numbers in the Indian Ocean, and especially on the islands of Oceanica, but a merciless pursuit and destruction which every year grows more persistent, has caused a perceptible decrease, and in a few years more, like the whale and seal, it will no doubt become very scarce. At the present time it is most numerous on the shores of New Guinea. A variety of this turtle is also found in the Caribbean Sea, but its shell is not so valuable as that taken from the East Indies species.

The *hawk's-bill turtle* gets its scientific name from the arrangement of the plates, which overlap each other like the tiles on a roof; and it gets its common English name from the partial resemblance of its mouth, seen in profile, to the bill of a hawk. Its head, neck and legs are longer in proportion to their thickness than those of the other turtles; it is more active, swimming with greater velocity, and righting itself when turned. Its eggs are eatable, but its flesh is not good, and the chief value of it to man are the plates on its back, which are the true tortoise-shell of commerce, and have been highly esteemed from the earliest ages. There are thirteen plates in the central part, surrounded by

SHIELD PLATED, OR HAWK'S BILL TURTLE.

twenty-five smaller ones. The large central plates are the finest shell, and they are often of considerable thickness; but the plates of shell do not form the entire case of the animal. The inner or supporting part is bony, and may be considered as part of the skeleton. The true skin is between the bony substance and the plates of shell. The plates are a production of that skin, and in the living state they are covered by an epidermis, or scarfskin. The common way of obtaining the plates is to heat the entire backpiece of the animal, by fire applied under the hollow on the inside. By that means the gelatine of the skin is dissolved, the skin itself swells, and the plates are easily detached entire. A turtle of about 300 pounds weight will produce about ten or twelve pounds of shell; but in the common way of obtaining the shell, the animal, which is otherwise useless in the arts, is sacrificed. In the eastern isles, where the *hawk's-bill turtle* is very abundant, the Malays, who procure large quantities of shell for the Chinese, pursue a different method. They catch the turtle alive, and retain it while they detach the central plates, so dexterously as not to lacerate the skin. The helpless creature manifests little uneasiness during the operation, and when divested it is released and makes at once for the sea where after a lapse of several months the plates are reproduced, but these are never considered so valuable.

When the shell is taken from the animal it is marketable at about five dollars per pound, and since the shell of a full grown turtle of this species will weigh about eight pounds, and the animal is easily taken, the industry must be a very profitable one. The treatment to which the shell is subjected before conversion into articles of commerce is to first boil and then steam it, by which the shell becomes soft, after which it is put into a press and under heavy pressure is compressed into a solid flat block. In this condition it is again softened and cut into thin layers, which may be again united, in any thickness desired, by compression, and moulded into any shape that is wished.

The shape of the *hawk's bill turtle* somewhat resembles a heart, and instead of being ridged as in other species, it is flat. The edges are serrated, with points directed towards the tail. In color it is a pale yellow marbled with brown on top, the plastron being of a very pale yellow hue. Like all other turtles, this species deposits its eggs in the sand, which are hatched by the sun. Usually the place of deposit is some distance from the shore, and as the young have no protection of shell, their bodies being yet soft, a great majority of those hatched out are devoured by birds while trying to make their way to the water. But

GREEN TURTLE.

for this fact their multiplication is so rapid that they would soon fill the sea. The eggs are considered a delicacy, but the flesh of the East Indies species is unpalatable, though those taken about the West Indies are highly regarded for food.

The **Green Turtle** (*Chelonia mydas*) ranks next in importance to the hawk's bill and from a gastronomic view it is more valuable. It is found in almost incredible numbers, notwithstanding the great number taken every year for the market, in all the tropical seas, but its most popular haunts are about the Antilles. The name *green turtle* has been given because of the marine hue of its, shell which, like the hawk's bill species, is heart-shaped, and though rising sharply to a ridge in the centre, the plates are smooth but not valuable. It grows to an enormous size, some specimens being as much as six feet long by four broad and weighing eight hundred pounds. The usual size, however, is from two hundred to three hundred pounds. Their flesh is so highly prized in all civilized countries, and especially in America and England, that ships with large crews are regularly engaged in the capture and transportation of the live turtles to leading markets on the coast, from which they are shipped to all the interior cities, and *green turtle* soup and steak have therefore become common dishes.

Ascension Island, in the South Atlantic, midway between Brazil and Africa, a volcanic formation but now little more than a waste of sand, is the greatest breeding place of *green turtles*, on which account it is visited by many ships and large parties of men, who spend the spring and a part of the summer season catching the animals and despoiling their nests. The industry is also carried on successfully on the Alligator Islands of the West Indies, Gallipagos of the Pacific, the Tortugas off Key West, and the northern shores of Australia, all of which places are both dreary and barren.

Of the breeding habits of these reptiles and their allies, and the methods employed for their capture, Audubon says: "The *green turtle* approaches the shores and enters the bays and inlets early in April, after having spent the winter in deep waters. It deposits its eggs in convenient places at two different times in May, and once again in June. The first deposit is the largest, and the last the least, the total quantity being at an average about two hundred and forty. The hawk's bill deposits also its eggs in two sets, once in July and again in August, though it crawls the beaches much earlier as if to look for a safe place. The average number of eggs which it lays is three hundred. The *loggerheads* visit the Tortugas in April and lay, from that period until late in June, three sets of eggs, each set averaging one hundred and seventy. The *trunk turtle*, which is sometimes of an enormous size, and which has a pouch like a pelican, reaches the shores latest. The shell and flesh are so soft that one may push his finger into them, almost as into a lump of butter. This species is the least valuable and is seldom eaten. The average number of eggs which it lays in a season, in two sets, is three hundred and fifty.

"The loggerhead and the trunk turtles are the least cautious in choosing the places in which to deposit their eggs, whereas the two other species select the wildest and most secluded spots. The green turtle resorts either to the shores of the main, between Cape Sable and Cape Florida, or enters Indian, Halifax, and other large rivers or inlets, from which it makes its retreat as speedily as possible, and betakes itself to the open sea. Great numbers, however, are killed by the turtlers and Indians, as well as by various species of carnivorous animals, as cougars, lynxes, bears, and wolves. The hawk's bill, which is still more wary, and is always the most difficult to surprise, keeps to the sea islands. All the species employ nearly the same method in depositing their eggs in the sand, and as I have several times observed them in the act, I am enabled to present you with a circumstantial account of it.

"On first nearing the shores, and mostly on fine calm moonlight nights, the turtle raises her head above the water, being still distant thirty or forty yards from the beach, looks around her, and attentively examines the objects on the shore. Should she observe nothing likely to disturb her intended operations, she emits a loud hissing sound, by which such of her many enemies as are unaccustomed to it are startled, and so are apt to remove to another place, although unseen by her. Should she hear any noise, or perceive indications of danger, she instantly sinks and goes off to a considerable distance; but should everything be quiet, she advances slowly towards the beach, crawls over it, her head raised to the full stretch of her neck; and when she has reached a place fitted for her purpose, she gazes all around in silence. Finding 'all well,' she proceeds to form a hole in the sand, which she effects by removing it from *under* her body with her *hind* flippers, scooping it out with so much dexterity that the sides seldom, if ever, fall in. The sand is raised alternately with each flipper, as with a large ladle, until it has accumulated behind her, when supporting herself with her head and fore part on the ground fronting her body, she with a spring from each flipper sends the sand around her, scattering it to the distance of several feet. In this manner the hole is dug to the depth of eighteen inches, or sometimes more than two feet. This labor I have seen performed in the short period of nine minutes. The eggs are then dropped one by one, and disposed in regular layers, to the number of a hundred and fifty, or sometimes nearly two hundred. The whole time spent in this part of the operation may be about twenty minutes. She now scrapes the loose sand back over the eggs, and so levels and smooths the surface that few persons on seeing the spot could imagine anything had been done to it. This accomplished to her mind, she retreats to the water with all possible dispatch, leaving the hatching of the eggs to the heat of the sand. When a turtle, a loggerhead for example, is in the act of dropping her eggs, she will not move although one should go up to her, or even seat himself on her back, for it seems that at this moment she finds it necessary to proceed at all events, and is unable to intermit her labor. The moment it is finished, however, off she starts; nor would it then be possible for one, unless he were as strong as a Hercules, to turn her over and secure her.

"To upset a turtle on the shore, one is obliged to fall on his knees, and, placing his shoulder behind her forearm, gradually raise her up by pushing with great force, and then with a jerk throw her over. Sometimes it requires the united strength of several men to accomplish this; and if the turtle should be of very great size, as often happens on that coast, even handspikes are employed. Some turtlers are so daring as to swim up to them while lying asleep on the surface of the water, and turn them over in their own element, when, however, a boat must be at hand to enable them to secure their prize. Few turtles can bite beyond the reach of their forelegs, and few, when once turned over, can without assistance regain their natural position; but notwithstanding this, their flippers are generally secured by ropes, so as to render their escape impossible.

"The food of the green turtle consists chiefly of marine plants, more especially the grasswrack (*Zostera marina*), which they cut near the roots to procure the most tender and succulent parts. Their feeding grounds, as I have elsewhere said, are easily discovered by floating masses of these plants on the flats, or along the shores to which they resort. The hawk-billed species feeds on seaweeds, crabs, various kinds of shell-fish, and fishes; the loggerhead mostly on the fish of conch-shells of large size, which they are enabled, by means of their powerful beak, to crush to pieces with apparently as much ease as a man cracks a walnut. One which was brought on board the Marion, and placed near the fluke of one of her anchors, made a deep indentation in that hammered piece of iron that quite surprised me. The trunk turtle feeds on mollusca, fish, crustacea, sea urchins, and various marine plants.

"All the species move through the water with surprising speed; but the green and hawk-billed in particular remind you, by their celerity and the ease of their motions, of the progress of a bird in the air. It is therefore no easy matter to strike one with a spear, and yet this is often done by an accomplished turtler.

"Turtles such as I have spoken of are caught in various ways on the coasts of the Floridas, or in estuaries and rivers. Some turtlers are in the habit of setting great nets across the entrance of streams, so as to answer the purpose either at the flow or at the ebb of the waters. These nets are formed of very large meshes, into which the turtles partially enter, when, the more they attempt to extricate themselves, the more they get entangled. Others harpoon them in the usual manner.

"When I was in the Floridas, several turtlers assured me, that any turtle taken from the depositing ground, and carried on the deck of a vessel several hundred miles, would, if then let loose, certainly be met with at the same spot, either immediately after or in the following breeding season. Should this prove true, and it certainly may, how much will be enhanced the belief of the student in the uniformity and solidity of Nature's arrangements, when he finds that the turtle, like a migratory bird, returns to the same locality, with perhaps a delight similar to that experienced by the traveller, who, after visiting distant countries, once more returns to the bosom of his cherished family!"

The **Leather Turtle** (*Dermatochelys coriacea*), often called the *Luth*, is also one of the gigantic habitants of the sea, perhaps the largest of living species,

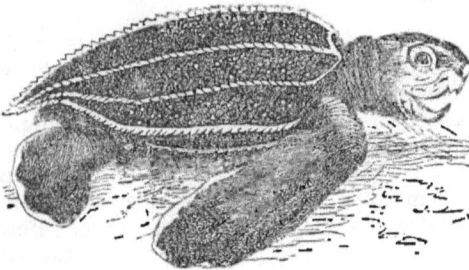

LUTH OR LEATHER TURTLE.

but it is seldom met with and therefore its habits are little known. The species spoken of by Audubon, which he calls the *trunk turtle*, may be the *luth*, as this animal has a carapax somewhat resembling the deerskin-covered trunks which were in common use some years ago, on which account it has sometimes been called the *trunk back*, but its breeding habits are not so well known as the above account would lead us to believe. Woods emphatically states that it does *not* visit the Tortugas for breeding purposes, and furthermore declares that the breeding places of the *leather turtle* are unknown.

An English officer, name not given, has furnished an account of the capture of a female *leather back* in the Ye river by some Burmese fisherman. He states that she was apprehended on a sandy beach while in the act of depositing her eggs, one hundred of which had already been laid. He represents her strength as being so great that it required the combined efforts of twelve men to arrest and carry her into the village. The eggs were spherical and 1⅝ inches in diameter, and when cooked were very agreeable to the taste, as was also the flesh, although it has heretofore been regarded as not only most disagreeable but also poisonous. Nearly one thousand eggs, in various stages of development, were taken from the body. Her length was six feet two and one-half inches.

Another large specimen was taken in a mackerel net off Cape Ann, and purchased by the Boston Society of Natural History, the flesh of which was eaten by members of the Society, and by them pronounced equal to that of the green turtle. Another specimen was taken some time before near the mouth of Boston harbor that measured eight feet in length, and weighed nearly one thousand pounds. The color of all these specimens was a deep black and glossed, somewhat resembling the back of a porpoise. It differs from others of the turtle family in having an envelope of skin, or a carapax, as before mentioned, something like stiff leather. Along the back are seven prominent ridges having slightly serrated edges. The jaws are powerful and notched, to give them the cutting and tearing power of teeth, while both upper and lower bill are hooked. The fore flippers are of great size and strength as compared with those of the posterior, in which respect the creature bears some analogy to the seal. Some few have been seen that were dark brown, flecked with spots of yellow on the back, and skin speckled with dull dots of white and black. Its extreme weight is thought

MORASS TURTLE (*Emys lutaria*).

to be eighteen hundred pounds, but only a single specimen has been found of such great weight.

The **Greek Turtle** frequents the coast of Morocco, and is pretty generally distributed throughout the Mediterranean. It is also a very large species, occasionally reaching a weight of three hundred pounds. The shell is irregular in shape, with raised scales, and bears a resemblance to the carapax of the tortoise. Its flesh is considered almost equal to that of the green turtle, and is sold not only everywhere in Morocco, but also in Spain, France, Italy, and all the countries of the Levant. The *lip turtle* is found in the streams of South America. It is rather small in size, and, being unpalatable, has nothing to recom-

LIP TURTLE (*Trionyx ferox*).

mend it beyond the curious snout with which it is provided, as shown in the accompanying engraving, and from which characteristic the name is derived.

15

INSECT LIFE.

HE subsidence of the waters prepared the way for reptilian life, as we have seen, and there afterwards succeeded other conditions preparatory to new forms of life. The growth of forests and flowers was followed by the appearance of an infinite number and variety of insect life, which exhibit the most marvellous adaptations to the conditions of the animals which preceded and were to follow in an unbroken ascending series.

Proceeding from lower to higher, from the denizens of the world in its earliest stages to their successors in more modern times (which shows in the natural times a progress no less great nor less surprising than in the world of human society), we have now reached the highest order of the invertebrates. Air, land and water are filled with these wonderful creatures which are too often neglected because of their seeming insignificance or because of the unconscious ignorance of man. To the insects the Creator has given an infinite variety of form, the most varied coloring, the most diverse offices. They offer an accessible and ever-varied field for the study of the wonders of creation, and are more closely related to human interests than the uninformed would suspect. To the student of Natural History, nature is no longer confined to speaking through landscapes, but becomes instinct with the most varied, the most marvellous, the most interesting life. Whether as a further stimulus to an intelligent appreciation of the wonders of creation, as an occupation for time which cannot be used in satisfying the direct needs of daily life, or as the means of protecting ourselves against pecuniary loss and of adding to our own and to others' wealth and comfort, the path of entomology promises to be a direct road to our goal. Insects have become articles of commerce (as the Spanish fly used medicinally, and the beetle used for personal adornment); they protect us from the ravages of pests that destroy our crops and gardens (as the potato beetle, the grasshopper, the locust); they act as scavengers, and at least decrease the impurity of the atmosphere (as the mosquito and the fly); they exhibit many of the transformations through which life passes in its progress from a lower to a higher form; they exhibit the careful provision made that the tiny being may adjust itself to its environment; they inculcate many a lesson, such as that taught Bruce by the persistent spider; they have inspired many a well-known line of poetry and many a favorite work in prose; they have held the interest of the most able, intelligent and earnest students of Natural History, such as Wallace and Lubbock and Darwin in England; Scudder and Riley in America.

With the Devonian age and the appearance of trees and vegetation, insects first appear, thus emphasizing the wonderful provision for the support of animal life and the adaptation of life to its surroundings. The mathematicians have proved the impossibility of many coincident accidents, and hence the student is forced to recognize the work of an unerring wisdom, such as belongs alone to the Creator.

The Carboniferous age, with its rank plant-life, marshes, decaying trees, furnished yet more favorable conditions for insect life, which consequently multiplied and flourished increasingly. With succeeding eras insect life continues with its changed forms of existence and its infinite adaptation to new conditions.

Entomology, or the study of insect life, offers special attractions. In the first place there is the infinite variety of species (one hundred thousand of these having already been distinguished). Then there are the many and curious transformations, beauty of coloring, diversity of form, and finally there is their wonderfully developed instinct, their adaptation to the life which they are to lead, and the service or disservice to mankind.

They furnish at once an opportunity, a provocation, and a satisfying means for the study of that process of evolution which, far from denying the agency of the First Great Cause, recognizes his method of acting rather in the initiation of nature's mysteries than by a continued and unnecessary interference when the process has once been set in motion. No student of the wonderful metamorphosis of insect life can fail to admire more intensely and more intelligently God's wisdom in providing for such adaptation to one's functions and surroundings as shall enable the smallest insect to perfectly fulfil its mission. Natural Theology is not antagonistic to doctrinal theology or inspired revelation, but rather lends the strongest support to doctrines and to the ethical teachings of Holy Writ.

Thus a moment's reflection will satisfy the reader that not merely to the naturalist and the lover of nature, but even to the speculative student of our cosmogony, the insects, as subjects of investigation, have the supremest interest and importance. But yet again, the practical value of the study can hardly be overestimated. Professor G. V. Riley, the entomologist, is admitted to have saved thousands and thousands of dollars to the agricultural, horticultural, floricultural and arboricultural interests of the United States. Truly, as has been said and sung, "knowledge comes, but wisdom lingers," and the undirected experiments of the farmer and gardener are too slow and costly to be accepted as a substitute for the exact and useful experience gathered by those whose lives and energies have been consecrated to a study of the insects. Think for a moment of the ravages of the cotton-worm, and you will realize the actual cash value of information at once exact and placed within the reach of the unscientific reader.

Many of the insects have already been made to serve economical needs, and thus have resulted in the creation of new branches of commerce with all the distribution of labor, wealth and intelligence which this implies. The silkworm supplies necessities felt by every household; the cochineal insect supplies us with dyes; the bee gathers honey not alone for itself, but even more for the pleasure and profit of human beings; so, too, the fertilization of many forms of plant-life is dependent upon the agency of the insects, while much more efficient than city contracts is the scavenger work done by others of these creatures of an hour, who yet spend their brief span of life in the most continuous and useful activity, ministering to the wants of man even while subserving their own interests.   When we shall have learned enough we shall be able to protect ourselves against the unconscious depredations of the locusts, grasshoppers, Hessian-flies, onion-flies, chinch-bugs, clothes-moths, weevils, potato-bugs, grape-vine louse and carpet-beetles.   But the first step towards so desirable a consummation must be earnest and persistent study of works such as is here attempted.

Not only is the student rewarded by the discovery of beauty so unsuspected, so varied, and so exquisite as to pale the more pretentious ostentation of human skill, but he finds the most abundant and convincing proofs of an adaptation to environment and function in the world's economy that he feels a profound significance when, in his devotions, he repeats "And all thy creatures praise Thee." Co-ordination of structure, form and relation to the other elements of the universe are brought home as a practical lesson in God's

HARLEQUIN SPIDER.

CROSS SPIDER
(*Epeira diadeinata*).

TONGUE WORM
(*Pentastomum denticulatum*).

providence which the most careless cannot but heed.   No subject has greater human interest than the processes of development, and no field is more fruitful than that of entomology.   With this brief reminder to the reader of the importance of this study, I will proceed to a description of the representative types of insect life, in which arrangement, even as to probable succession, is, however, impossible.

The **Myriapoda** are so named because of their numberless feet; they embrace the *centipedes*, *millepeds*, and the thousand-legged worms.   The *millepeds* have a round or flattened body, feet close together and inserted in pairs upon each segment, except the first three.   A comparison with fossil remains lends emphasis to the doctrines of natural selection and adaptation to changed conditions.   The existing species are found almost solely in Illinois.

The **Pamopida** mark distinctly the transition stage from the millepeds to the *chilopoda* or centipedes, and have therefore greater interest to the investigator.   They pass cleanly lives amidst the dead and fallen leaves.   They moult some nine times, and at each fresh moulting add a new segment and another pair of legs.   Their evolution seems to suggest a link between plant and animal life.   The centipedes move rapidly and are predatory in their habits. Their poison is secreted at the base of the first pair of legs. The species called *scolopendra* is the one about which so many stories have been told, and which render life a constant warfare to the visitor to the parts of the country which they infest. With the increasing luxuriance of plant life in the more southern parts of the country, animal life, and especially insect life, grows infinite in variety and numerousness. Hence the centipede increases in number and virulence as one travels toward the equator.

The **Electric Myriapod** (*Geophilus electricus*) is a phosphorescent centipede found in Europe.

The **Centipedes** have a pair of legs for each segment of the body; their antennæ are many-jointed, and at the base of the first pair of legs are poison-sacs. An Indian species is a foot in length. The bite of the insect is poisonous, although not usually fatal to mankind. The *centipedes* are a great pest in warm climates, as they take possession of one's shoes, bed-covering and other unlooked-for hiding-places, and are prepared with an ungrateful surprise for the legitimate occupant. Naturally, their ferociousness and dangerousness

SHELL MITE.

CRAB SPIDER.

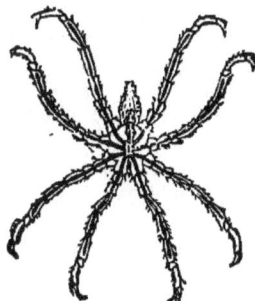

Ammothea pygnogonoides.

has been exaggerated by the tales of the romancers of the nursery and of travel; still, without being dangerous to human life, they are apt to cause very severe suffering.

The insect popularly known as the **Earwig** (although its real name is the *Earwing*, taken from a resemblance in shape to the human ear), is, contrary to popular belief, not at all dangerous to the sense of hearing. Its wings are large but exceedingly delicate, and during the day are so carefully packed away as to conceal their very existence. The *earwing*, like some other insects, is a very devoted mother. The caddis-fly, having no armor for its protection, conceals itself from its enemy, the fisher, by surrounding itself with a tube, which it makes out of earth and sticks. The saw-fly has on the under side of the abdomen a pair of saws, which it uses with great skill and effectiveness. It has a short, thick body; long, flat, cutting mandibles, and a double saw attached to the ovipositor of the female; the wings are orange. After depositing its egg, it seals the wood so that its larvæ shall not undergo the fate of the Prisoner of Chillon. The giant ichneumon-fly replaces the saws by a gimlet, and the pine trees suffer from its work as a carpenter. The common oak-galls (as well as the Dead Sea apples), are the developing eggs of the gall-fly, deposited in some part of the tree. The cochineal (*coccus cactus*) is a native of Mexico, and furnishes commerce with its most valuable, because brightest, carmines and scarlets in which the finest goods are dyed.

The **Shell Mite** (*Atax ypsilophorus*) is so named from the marking of its back, which resembles the Greek letter upsilon.

The **Scorpion.** In spite of its variety of color—gray, white, yellow, brown, green, black, ash, claret, etc.,—the *scorpion* is feared rather than admired. Though the *scorpion* flourishes in Italy and France, and is not unknown in our Southern States, yet it much more abounds in Africa, the East, and in South America. Its irascibility and combativeness are extreme; it is like many insects, a cannibal, and is perfectly willing to feast without scruple upon its own offspring. In extreme cases it will, like the stoical Roman, commit suicide by stinging itself to death.

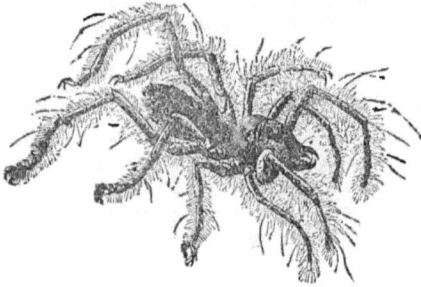

WALTZ SPIDER.

The body is elongated, the abdomen of six equal segments, forming a tail, whose tip is exactly the "tip" that the stranger does not care to receive. Respiration takes place through four stigmata, or openings, connecting with two pairs of pulmonary sacs. The young are developed within the mother, and after birth are carried about by her in the manner of Indian pappooses. Some species have upon the legs rasps, which serve as organs for producing sound. The poison gland communicates with the stinger; the sting, though not fatal to man, is exceedingly dangerous. Ammonia applied externally and internally is the most satisfactory remedy. The *scorpions* are examples of motherly affection, tenderness and devoted care; they take special pains to educate the young at home.

CYLINDRICAL SPIDER.   NATURAL SIZE.

The *scorpion* is the land octopus, and the wonderful tales told by Jules Verne and Victor Hugo may be transferred to the legendary history of this spider (for spider it is). Its appearance is repulsive, impish, devilish. Roaming abroad in the darkness, they penetrate everywhere—no pillow too soft to invite them;

no shoe too small for them to wear; no dress so coarse as to repel them. Those who live in the land of the *scorpion* have to use all precaution, even in dressing, lest they disturb this unreasonable and irascible insect, which, thorough anarchist that it is, recognizes no individual right to property. The young are frequently not content with feasting at the family hearth, but devour piecemeal their devoted parent, who thus outdoes the famed pelican. A singular species, called *Thelyphonus caudatus*, is found in Java; it is rat-tailed, and breathes through spiracles.

The **Spider**. The unwearied patience of the *spider* is the despair of the careful housewife. Like Antæus, he is thrown to earth only to rise with increased strength and determination. They are the most observant of the insects, and are always ready to turn to their own advantage any mistaken confidence of their prey. Like Argus, they have many eyes with which to guard their interests. The loss of a leg is quickly repaired by the growth of another. Their web is more ingenious than that made by the shuttle of the weaver,

GIANT CRAB SPIDER.

and is made to serve the combined uses of a trap for enemies and a castle for the *spider*. The female lays in a season about one thousand eggs, which, as soon as dry, are surrounded by a home-made bag which she fastens to her own person. Like the hen, she ceases her care over her young only when they have become able to provide for themselves.

The true *spiders* (*Araneidæ*) have the thorax and head united which contain the stomach, the centre of the nervous system, and the muscles of the legs and jaws. The abdomen is occupied by the intestines, the organs of respiration, the circulating system, the organs of reproduction, and the mechanism for spinning. The males are ordinarily smaller than the females, and have relatively longer legs and smaller abdomen; they are usually darker in color and markings. After the eggs have been laid they are easy to examine, as they develop equally well in any place, and the application of oil or of alcohol renders their shells transparent. With the bursting of the first skin the spider appears, pale, soft, hairless, and with small claws upon its feet; but when the exuviation is completed it looks like a true *spider*. The process of moulting is not confined to the earlier periods, but extends throughout the life of the *spider*. The outer spinning-organs consist of two-jointed tubes, a large number of which are to be found on each spinneret. The hind feet are used to guide the threads and to regulate the supply.

The devices used by the *spider* are so numerous and so varied as to have made it the symbol of persistent cunning. One kind makes for its nest a door of earth held together by a web, so as to simulate the firm ground; another provides an inside door, which can be used in case of house-breakers.

The *spider* commonly seen on the banks of streams is attractive from its beauty and engineering skill, as well as from the facility with which it adapts itself to all changes of locality, climate or surroundings. It is a silvery drab on the head and thorax, while in striking contrast is the abdomen, which is black, yellow, or brown, and the legs, whose color is orange ringed with black or brown. Long before the idea occurred to man did this small creature span the waters with suspension bridges. Fastening a thread to some support, it spins another series of threads, which, taken up by the wind, are finally blown

BUSH SPIDER.

across the stream and entangled in some bush or tree. Then the *spider*, using this first strand as a road-way, makes a parallel cord, and afterwards weaves with mathematical exactness, and upon the most exact engineering principles, a circular, wheel-like trap for its prey.

The **Mason Spider** (*Mygale*) builds in timber or walls. It is a native of the tropics. The *Mygale nidulans* a West Indian creature, digs an oblique hole an inch in diameter and three inches in extent. This she lines with webbing until it attains the consistency of leather, when she supplies it with a door which opens and shuts at the pleasure of the owner. The orange-white color of the web shows that this spider is less provincial than our society people, for it has anticipated them in house decorations, and has even yet an advantage over the skill of the modern house-furnisher. The door is circular and of exceptional thickness and strength, and the hinges act automatically in shutting the door. There is no need of weather strips, for this builder, unlike the clumsy human carpenter, makes his joints to fit with the greatest accuracy. Moreover, the outer side of the door is made in the pattern of a fungus, or a lichen, so that the plants themselves would be deceived, and if it lie near the surface it is colored so as to resemble the ground in which it is placed.

Another species select an elevated plot of ground so as to secure very good drainage and consequent sanitation. Digging a gallery of as great a depth as two feet, she lines this with the softest and most attractive silken tapestry.

The **Crooked-Legged Crab Spider** (*Thomisus vatens*). This is not a true spider, but a crustacean, still in form and predatory habits it resembles the spider sufficiently to deserve the popular name which it has earned.

The **Water Spider** (*Argyroneta aquatica*) is amphibious, or able to live with equal ease and

MASON SPIDER (*M. cæmentaria*).

WATER SPIDER.

comfort upon either land or water. When living in water he encases himself in a bubble of air; weaving a silken bag with an opening below, he attaches this to a plant; climbing upon the outer threads he awaits at the surface the coming of a bubble of air, which he seizes with his hind feet and bears below for the proper ventilation of his dwelling.

The **Bird Spider** (*Mygale avicularia*) is of immense size for an insect and of marked muscular power. It is found commonly in South America, being abundant in Guiana. It nests in trees, and captures the smaller birds, and also lizards or frogs, which it leaps upon after the manner of a tiger. There has been some dispute among naturalists respecting the habits of this *spider* in catching birds, but the fact is now well established. Its body is dark black and is covered with reddish brown hairs.

The **Wolf Spider** (*Lycosa iniquilina*) retains its cocoons attached to its spinnerets. Like a wolf, it inhabits a cave which it excavates for itself, and thence pounces upon its prey.

The **Mining Spider** (*Cteniza fodiens*) bores to the depth of six inches, and then constructs its dwelling at right angles to this tunnel.

The **Cylindrical Spider** (*Galeodes aranoides*) is sometimes called the waltzing spider because its walk resembles a Russian waltz.

The **Velvet Mite Spider** (*Trombidium holosericeum*) is common in gardens, hothouses, or wherever plant life is abundant. It feeds not upon the plants, however, but upon the eggs of insects. One species of the *Trombidium* is an efficient ally of man in his contest with the grasshopper, while another furnishes a valuable dye, which has long been held as among the prime and most essential articles of commerce, next, in fact, to the cochineal, the coloring matter of which it resembles.

BIRD SPIDER.

The **Marginated Mite** (*Glomeus marginata*) at times looks like a toy turbine wheel, on account of the white lines which border the segments of the body, and the recurved position in which it holds the tail.

The **Crooked-Legged Crab Spider** has a striking resemblance to a crab.

The **Crab Spider** proper (*Nygate avicularia*), also called *bird spider*, is a native of South America, where it grows to an extraordinary size, and, though not so poisonous as many species, is extremely formidable for its power to bite with serious effects. It is represented as occasionally catching birds and young chickens, but its more common diet is smaller reptiles, such as lizards, slugs, etc. In many respects this creature is more powerful for evil than the Indian scorpion, whose sting is so greatly dreaded, to which is added an offensive habit of hiding in clothing, where it is most likely to give both a sting and surpise.

INDIAN SCORPION.

The **Water Bear**, or *Macrobiotus Schultzei*, takes its English name from its resemblance to the bear. It is a microscopic form, found as a rule in the sand, but at times in the water. It has four pairs of legs, each

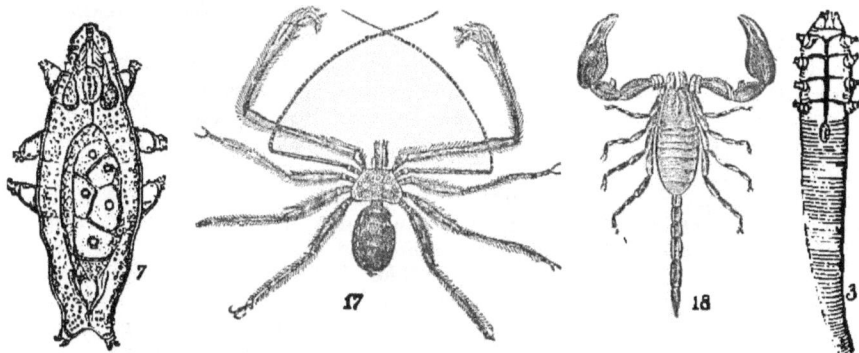

7. MACROBITUS.   17. LONG-ARMED TARANTELLA.   18. SCORPION.   3. (*Demoden folliculorum*).
(*Phrynus lunatus*).

terminated by claws; it seems to have no organs for respiration or circulation. The species is hermaphroditic; the engraving illustrates their structure.

The **Wall Spider** (*Phalangium opilio*) has, in common with the other members of the family, four principal vessels as silk repositories, which are located, not near the mouth, as in caterpillars, but near the anus. The thread is composed of the threadlets proceeding from five spinnerets, and sometimes from six. These spinnerets are composed of an infinite number of smaller tubes, for in this common insect there is recognized the truth that man has learned but slowly and painfully, that a combination of slight strands is better able to sustain a strain than is a cord whose strands are few but large. Thus, again, are we taught that the highest human wisdom consists in coming to nature in the mood of a little child—trust we of the wisdom of Providence—and applying our lessons as we learn

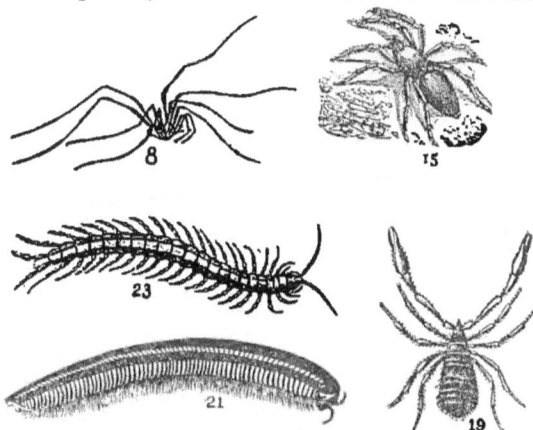

8. WALL SPIDER.   23. RIBBON-LINKED MILLEPED (*Scolopendra lucasi*).   21. MILLEPED.   15. STONE SPIDER (*Drassus lapidicola*).   19. *Obisium trombidiorides*.

them. Let the reader give a little time to observing this method of ropemakers, and then study the habits of the spider, and he will speedily be convinced of the lessons yet to be learned from the obedient creatures of the insect world, bearing in mind that it requires four millions of the common spider's threads to equal the bulk of a hair. One can appreciate the resources, the industry, and the skill of this insect, so little understood by persons, as to realize the lack of honor that is said to attend a prophet in his own country. Another advantage which the spider gains from this multiplication of threads, is the protection against total loss when a guy-rope breaks. The spider shoots out his little silken quills, whose direction he guides by the nice sense of touch, which seems to inhere in his hind-legs. After stretching lines to make the warp, the house-spider crosses these by a superposed woof, which is not interlaced; as the spider uses the ancient English method of employing her limbs as weaving rods, its web will be found to have the same lengths as its radii. Contrary to popular belief, she does not remain at the centre of her web, but quite frequently constructs on the edge of the web what the French call a *porte-cochere*, whence she comes to welcome her guests "with hospitable hands to bloody graves," though not without much affectation of style.

The **Ribbon-Linked Spider** (*Scolopendra lucasi*) has a body which looks like links of ribbons. It is about six inches in length, and is found in France and in some islands.

The **Milleped** (*Julus terrestris*) is the many-footed soldier of German entomology. It has two pairs of legs attached to each segment; its thorax and abdomen are not separate; it breathes by means of spiracles, stigmata, or air-holes; it has two pairs of paws. It is found as early as the Carboniferous age. It undergoes no true metamorphosis, but when passing from stage to stage of growth, doubles itself up, splits the integument, and bursts forth into new life.

The *Obisium trombidiorides* is a very singular-looking, top-shaped creature belonging to the scorpion family.

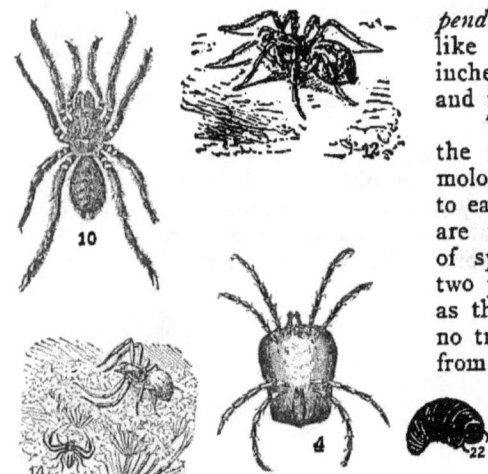

10. MINING SPIDER. 12. WOLF SPIDER. 14. CRAB SPIDER (*Thomisus ratius*). 4. VELVET MITE. 22. (*Glomeris marginata*).

The **Tarantula** (*Lycosa tarantula*) a deadly species, is named from the place of its discovery—Tarentum, in Italy—but it is uncomfortably frequent in the southern country. It is a mason-spider, living in the ground and keeping its front door closed to all intruders. Its bite is not, as supposed, fatal, but it is generally very troublesome. It is said that if unmolested the *tarantula* will cross one's body without attacking the person, though its course will be marked by small, irritating red marks. This statement, however, only applies to the centipede. The *tarantula* is as courageous as a sparrow, and when attacking man, will, it is said, aim always at the face.

The **Skin Spider** (*Demodex folliculorum*) sometimes invades the hair follicles and the sebaceous glands of the nose; it is injurious to hides and skins, but harmless to the human being. It uses skins or hides as nests for its larvæ, and frequently renders them useless for the purpose of commerce.

The **Black Spider** (*Cilax ypsilophorus*) lives in the gills of a fresh-water mussel.

The **Harlequin Spider** (*Salticus scenicus*) uses stratagem. It conceals its legs so that its body, having the appearance of a fly, deceives the insects which are to become its prey, and upon which it springs when within reach.

The **Garden Spider** is the familiar red spider; it weaves a delicate web under the surface of leaves, partly for protection against enemies, and partly for the luxurious enjoyment of their larvæ, which feed upon the leaves.

The **Beetle-like Spider** (*Scutigera coleoptera*) has a beetle-shaped body, to which are appended its long legs.

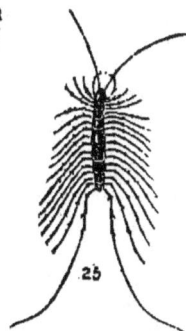

26. CLAW-FOOTED MYRIAPODA. 24. ELECTRIC MYRIAPODA. 25. (*Scutigera coleoptera*).
13. SHARP-EYED SPIDER (*Oxyopes ramosus*).

The common **Wood-tick**, which frequently feasts itself upon the blood of dogs and cattle, belongs to the spider family. There is a Persian spider which

MEMBRACEÆ, OR LITTLE DEVILS OF GEODFREY, MAGNIFIED.

attacks man, to whom its bite proves very venomous; and a family of parasites which fasten themselves to bats and beetles, all of the same species as the *wood-tick*.

The **Stone-Dwelling Spider** (*Drassus lapidicola*), lives under stones, or at times in silk tubes which it attaches to plants. It spins a little web across its nest and deposits its eggs therein.

The **Cross Myriapod** (*Epeira diadema*) is a prominent member of a brightly colored, odd-shaped family, which, while lying in wait, assumes various deceptive forms, such as that of bark, straws, etc. It is named from the resemblance of its abdominal ornamentation to the diadems of royalty.

The **Tongue Worm** (*Pentastomum denticulatum*) is a larva found in the tongue and cavities of persons and animals.

The **Dragon-Fly** (*Libellula*) is of interest from its commonness, its gay coloring, its serviceableness in the destruction of gnats, mosquitoes, and so forth, and from its transformations. It lays its eggs on the surface of water-

plants, with which it cements a friendship; in each bunch of eggs there is upwards of a hundred. The larvæ have to forage for their own food, and they are able to propel themselves through the water upon the principle of the syringe. The underlip is armed at its extremity with a pair of hooks, and is darted out at the victim. In passing to the pupa stage the larva moults and the pupa adds a pair of wings. The pupa in its turn moults, having first climbed to the surface, and the Cinderella-like grub is succeeded by the gaily-apparelled creature so well known. It is the despot of the insect world. It is erroneously suspected of annoying horses, for it is perfectly innocent of any such transgression.

The **Little Devils of Geodfrey** were named by the entomologist Geodfrey. Creatures of a day, these dragon-flies are clothed in colors so brilliant and varied as to dim the lustre of gems and jewels. The ingenuity of man in producing hues is slight in comparison with the native endowment of these insects, who multiply colors until they become indescribable by human language, though none the less appreciable by human sense. Then, again, there is an evident provision for the necessities of their lives. The *mormolycæ* insects, for example, so perfectly resemble a leaf in veining and coloring that even birds are led astray by the deception.

The **Leaf Bug** (*Mormoly phyllodes*) is three inches in length, and is to be found in Java. Color, brown, except the legs and antennæ, which are black. It is remarkable for its crab-shaped back, although it takes its name from its frequent resemblance to a leaf. It is found under the branches of trees and also haunts the tall

ANTS AT PLAY.

grasses and flowers, the juices of which constitute its principal food. Though called a bug, it is really a link between the moth and butterfly since it resembles both, and its habits are also imitative of each.

The **Ant-Lion**, when in its larval stage, digs a pit perfectly circular, using its head and jaws as a shovel, and throwing the sand out so as to form a slope about the pit. The slightest disturbance of the sand is sufficient reason for the insect to dig out more sand, and thus cause the victim to slip down within easy reach, and submit to yielding up its life's blood. Like the crab, it walks backwards. When it encounters a pebble too heavy to be thrown out by its head, it patiently struggles until it has got it upon its back, when it carries it away just as a porter would do. It builds for itself a case of sand, and lines it with a silken web, preparatory to its metamorphosis into a pupa. It finally passes into an ephemera, and closes its life similar to that of the dragon-fly. My American readers will, perhaps, more readily recognize the *ant-lion* under its common name of *doodle bug*, and will call to mind their boyhood days when among the sand of overshelving rocks they found the bug's shallow basin, and succeeded in luring it from its haunt by calling *doodle-bug, doodle-bug, doodle, doodle*, as I have very often done.

The **Orthoptera** include cockroaches (*Blattidæ*), devil-horses (*Mantidæ*), crickets (*Gryllidæ*), grasshoppers and katydids (*Locustidæ*), and the true locusts (*Acrididæ*). Though the *orthoptera* undergo various metamorphoses and frequently exuviate, there is but little difference between the first and last stages of their existence. The *orthoptera* represent the earlier forms of insect life, being found as far back as the "Devonian period," as termed by geologists; fossil remains abound in Illinois, Colorado, Wyoming and Idaho. The *blattidæ* (or cockroaches) are familiar to the sense of sight and of smell, and their wanton wastefulness does not decrease the disfavor in which they are held. Their compressible skins enable them to "crawl through a very small hole," and though naturally nocturnal in their habits, they seem to have learned how not to pass the day in slothfulness. Warmth and dampness are favorable to their growth and prosperity, and therefore the kitchens of houses are favored resorts. They have become great travellers through infesting vessels, and have been very successful in their capacity of stowaways. In this way they have become cosmopolitan, and have safely entrenched them-

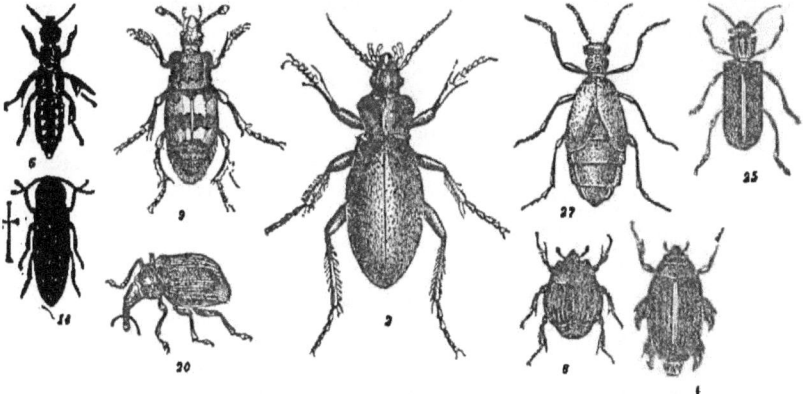

6. HUNTING BEETLE (*Staphylinus erythropterus*). 14. SNAPPING BEETLE (*Agriotis lineatus*). 2. LEATHER BUG (*Carabus coriaceus*). 9. BURYING BEETLE (*Necrophorus vespillo*). 8 DEATH SIMULATOR (*Hister unicolor*). 20. GRAPE-VINE SUCKER (*Rhynchites alni*). 27. MAY-BEETLE (*Mola proscarabeus*). 25. *Cucujus sanguinolentus*. 4. WATER-BEETLE (*Gyrinus natator*).

selves in all parts of the world. The cockroach carries about its body the egg-case until the young animals are ready to come forth.

Like the hen, she broods and is very solicitous about her young. The cockroach is nearly omnivorous, and hence the difficulty of tempting it with a pleasure of taste which shall be fatal. The German cockroach (*Ectobia germanica*) is the one which frequents our houses, infests our bakeries, and destroys our cloth-bound libraries. The oriental cockroach especially frequents our sinks. The American cockroach, though frequenting dwellings, prefers a life in the sewers and slums.

The **Mantidæ**, vulgarly called *devil-horses* and *praying insects*, bear a resemblance to leaves and twigs, and have received their name from a fancied resemblance to soothsayers or devotees; they are carnivorous, and, like the cannibals, feast themselves even upon their own kind when slain. They lie in wait for their prey, and will kill butterflies, potato-beetles, caterpillars and grasshoppers.

The **Phasmidæ**, (called walking leaves, or walking sticks) is specially interesting as illustrating the adaptation of animals to the conditions of their life, and have a close relation with the *mantidæ*.

The **Ephemeridæ**, or **May-Flies**, are so-called because of the momentary duration of their existence. Their delicacy and grace are very striking, and to some extent atone for their annoyance to one who is not engaged in the study of Natural History. They furnish a means of subsistence especially for fishes; in some parts of the world, where their number is very great, they are converted by the farmer into fertilizers. A few species are carnivorous, but most of the *ephemeridæ* surpass the vegetarian in abstinence, as they add plants to the articles of food interdicted by their laws. At first sight it would seem a needless cruelty for an animal to live so short a life, and to find his mission in furnishing food for fishes. But this life is but the flashing up of a flame before it expires, the fire having been long in burning, for in their preparatory state the *ephemeridæ* have lived from ten months to three years, so that they have had time to enjoy the pleasures

MIGRATING LOCUSTS (NAT. SIZE) (*Pachytylus migratorius*).

and fulfil the mission of a reasonably extended existence. These insects moult an unusual number of times, eight being no uncommon number and twenty-one occurring in one species.

The **Locustidæ** include winged and wingless species. The wings are supplied with a drum-like attachment, which is the means of the reveille which we hear. Of the wingless species, the so-called stone-crickets are a type. One species of the winged kind is migratory in its habits, and sometimes descends upon the valleys in numbers like the famed army of Xerxes.

The true **Grasshopper** is herbivorous, and sufficiently common to make the means of study easy. However delightful to children and interesting to entomologists, the *grasshopper* is at times incredibly destructive. Prof. C. V. Riley devoted much time to the study of these successors to the border ruffians, and a summary of his results will be found of interest. Fifteen million dollars is the loss inflicted in a single year.

MAY-BEETLE—NYMPH, LARVÆ AND FLY.

The *grasshopper* is a genuine troubadour, and his musical call to courtship is produced by a transparent membrane covering a hole at the base of the wings. Like many a singer of mediæval times, he devotes the hours not given to love to plundering, and in wanton wastefulness puts to the blush the famous Goths and Vandals.

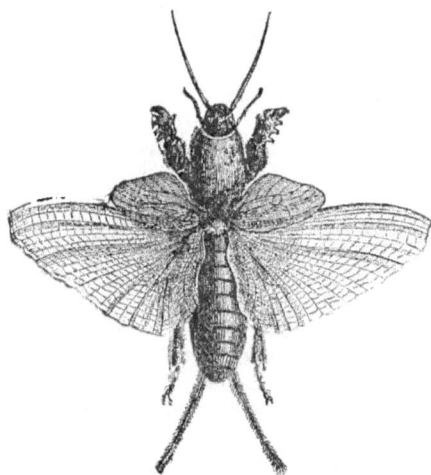

MOLE-CRICKET (*Gryllotalpa vulgaris*).

The **Acrididæ**, or **True Locusts**, is the largest family and the most destructive among insects. They have short antennæ, composed of numerous joints. The migratory species is frequently referred to in the Bible. The Rocky Mountain *locust* wrought such devastation in the seventies that it has been credited with causing the financial crisis. The fact that this *locust* is unusually developed in its digestive and reproductive system; that it is strong of wing and fitted for continued flights; that it is gregarious, and that a swarm is always perfectly organized, renders it an intruder to be feared. In a few hours they have reduced wide stretches of the most promising corn to mere stubble; they come apparently as avengers of unconscious wrong, for the resistless torrent of their approach most frequently occurs when the crop is largest and ripest; and when settled good weather has led the farmer to escape for a moment from his ever-recurring cares, and to dream those dreams which steal over him as he stands about to realize the returns for severe and long-continued toil and self-denial.

Migratory Locusts. The swarming of the armies of the world would

16

be small in comparison with that of the mighty army of *migratory locusts;* the descent of the Goths and Vandals was insignificant in the extent and rapidity of devastation when compared to the speedy and effective work of the *locust.* Towards the close of the seventeenth century, for example, three swarms of *locusts* simultaneously invaded Poland, and almost, as if by magic, they covered the ground to a depth of several feet, loaded the trees until these bent beneath the weight, and exterminated every vestige of the crops and vegetation.

In Tartary the swarming of *locusts* is frequent and their number illimitable. In Barbary the people have learned how to tear off the disguise of such a blessing and convert the intruders into delicacies for the table.

The **Gryllidæ,** or **Crickets,** have a vertical head in which are set elliptical eyes; the body is cylindrical, the antennæ (or feelers) long and their wings veined, the front pair ovate, the hind pair triangular; the legs are short and armed with spurs. In prehistoric times the cricket's chirp was heard in the land, and the beetle droned his evening song. They are the Quakers of the insect world in respect to their disbelief in war and contention. They house in burrows, and generally live, hermit-like, alone; their never-wearisome vesper song is produced by rubbing together the wings, so some naturalists say, but I doubt it. As a rule they are herbivorous, but have been known to eat animal matter.

The **Mole-Cricket** is in appearance somewhat like the mole, is of large size, and lives underground, being well fitted out for burrowing. They make endless galleries, ever changing in direction, and destroy all roots which come in their way. Laying as many as four hundred eggs, it is easy to see the rapid rate of increase. The common *house cricket* has been the frequent theme of poet's song and household story. What story or play more familiar or touching than "The Cricket on the Hearth?" What sound more home-like or more welcome than the chirping of the *cricket* by the winter's fire? Its love for warmth and moisture frequently leads it to take early possession of new houses, and to be prepared to welcome the owner amidst the discomforts of taking possession. It is a naturalized denizen of the United States, its progenitors having come from the far East, which contributes so much to the luxuries of life. Then there is the other little *field-cricket,* and yet again the more brightly colored *tree-cricket,* whose coloring, however, hardly compensates for the injury it does to vines and fruit trees.

The **Cicadæ** have two parchment-like sacs gathered into plaits and located in the cavity at the bottom of the abdomen; the ribbing has all the effect of the reeds in an organ. The species most common in the United States is black above and brown beneath; others are black or brown with yellow or red markings. The *cicada septemdecem* lives a subterranean life of seventeen years, rioting upon the sap of forest and fruit trees. After so extended a life of apparently uninterrupted enjoyment, the *cicada* digs its way to the surface, and casting off its skin is transformed into a winged creature of the air. The largest of the species is the lantern-fly of Brazil; it is greenish-yellow, spotted on the humps and on the side of the head, with rose-colored borders to its wings, which are also veined with black, and have an olive-colored spot bordered by dark brown.

The **Candle-fly,** (*Fulgoria candelaria*) is common in China and in the East Indies. They are in color orange or greenish, and have highly decorated wings. The children use them as pets. The *lystra lavata* has a peculiar method of escaping destruction. It secretes and projects from the abdomen silvery tape-like threads, which, being the most striking feature, are seized by pursuing birds. The insect bites off this, to him non-essential member, and thus enjoys an April fool, while the hunter secures but a cotton muffin.

Plant Lice are oval, green in color, and have long, slender antennæ. In many cases they are used by the ant as milch cows.

The Bed-Bug pumps instead of sucking blood. It is too familiar an insect to require a description, though it may be said that as a subject for the microscope it has an interest which it altogether lacks when regarded as an unwelcome intruder into our houses. The *bed-bug* deposits its eggs four times a year —March, May, July and September—and fifty is the usual number of a brood, which mature in eleven weeks. It appeared in England after the "Great Fire" and is supposed to have been imported with the timber used in repairing the losses. The *bed-bug* is called a "Norfolk Howard" in Great Britain, owing to a story to the following effect: A person being named Bug, and Bug being restricted to the *bed-bug*, grew sufficiently weary of constant but poor jokes and witticisms. Having secured from Parliament a change of his name to Norfolk Howard, the wicked punsters transferred the new name to the insect, and thus again proved that "the best-laid plans of men and mice gang aft aglee."

The Coleoptera (sheath-winged) Beetles are six-legged, have chewing mouth organs, horny fore-wings, and their metamorphosis is essentially radical. The wings are, when not in use, protected by sheaths (elytra); in the species whose wings are rudimentary the elytra form a protection for the abdomen. It is said that

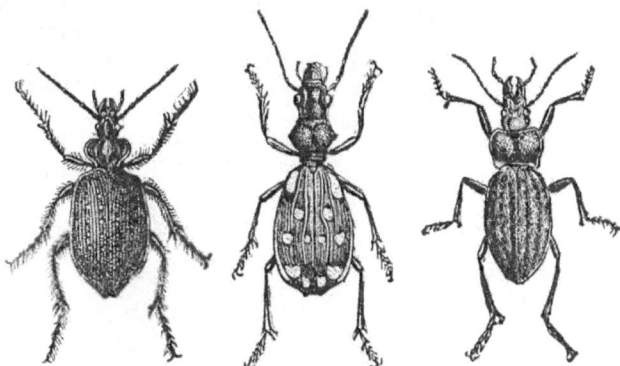

FLESH-EATING COLEOPTERA OF THE FAMILY CARABIDÆ.
Calosoma sycophanta. Anthia duodecimpunctata. Carabus gryphæus.

upwards of seventy-five thousand species have already been distinguished and described by persistent entomologists, whose delight is in the pursuit.

The *Colorado potato-bug, or beetle*, has been spread as an incident of commerce and transportation. The *carpet beetle* is a European immigrant. The *meal*, or *flour beetle* (*Tenebrio molitor*), and the *grain weevil* (*Calandra granaria*), are likewise illustrations of the importations of pauper labor from Europe. The *beetle* furnishes an inviting subject for the student of entomology because it is numerous, easily obtainable, capable of being reared, and because there remains ample opportunity for original discovery. The antennæ serve the double function of feelers and of a nose. The front wings seem to serve the use of a ship's rudder, and the hinder wings that of its propeller. The organs of hearing have not yet been located, but naturalists unite in asserting that the *beetle* has this sense. The creation of sound is effected by friction of the legs, wings or body; usually by rubbing the hind legs against a segment of the abdomen.

The *fire-fly* and some other species are luminous, but naturalists have not yet succeeded in determining the method of the production of light. *Beetles* live separately and in communities; they seem to be indifferent to locality; temperature appears to have no effect, and they are practically omnivorous. Their irreconcilable enemies are fish, toads, frogs, birds and skunks, parasitic flies and worms, wasps, mites and parasitic fungi. As a protection, many *beetles* take on the semblance to plants and to different insects. The *potato beetle* deceives the birds by its likeness to the potato leaves. Some *beetles* take on the semblance of fire-flies; some resemble seeds, twigs, and parts of plants. So, too, some species feign death as a means of escape; some protect themselves by unpleasant odors or by disagreeable secretions. The *coleoptera* minister to the pleasures or necessities of mankind, being used for blisters (*cantharides*); for medicinal remedies; for viands; for ornament (*chrysochlas*); and as gladiators. The *coleoptera* destroy myriads of noxious insects, act as scavengers, assist in the fertilization of plants, furnish dainty repasts for birds and fishes. Their indiscriminate zeal for destruction makes them unfriendly to crops, lumber, trees, books, furniture and carpets. Chief among the scav-

COLORADO POTATO-BUG, EGGS AND LARVÆ, NAT. SIZE.

engers is the *scarabæus*, though it is not known to devour putrid flesh. It is so classed, however, because of its habit of depositing its eggs in the dung of cattle, which it gathers, and by rolling converts into a round ball, which is afterwards buried. It is commonly known in this country as the *tumble-bug*, from its habit of tumbling about with its ball. In the early age of Egypt it was held sacred.

SACRED SCARABÆUS, OR TUMBLE-BUG.

The **Lady-Birds,** or **Lady-Bugs** (*Coccinellidæ*), are brilliant in coloring, which may be black, red, yellow, white or spotted; they are too common to require description.

The **Five-jointed Beetle** (*Cryptipentamera*) is a leaf eater. A highly colored species is not infrequently used as jewelry.

The **Potato Beetle** (*Coptocyda clavata*) is a frequent pest. The *crepidera* ruin the tobacco leaf; there is a *turnip beetle*, a *cucumber beetle*, a *grape-vine beetle*, as well as many others which manifest a preference for some vegetable or fruit valued by man. The Colorado *beetle*, or the potato-bug, was quick to exchange its previous diet for the potato, and has spread from the Rocky Mountains to the Atlantic coast. A natural traveller when opportunity offers, it has availed itself of the advantages of rapid transit as furnished by our railways. It lays from five hundred eggs upwards, and these require but a week to hatch. The larvæ mature within eighteen days, and then go under ground and are transformed into pupæ; at the end of ten days the pupa develops into the perfect *beetle*. These *beetles* have as many as four broods yearly, and as they hibernate, and as the larvæ as well as the *beetles* feed upon plants, their destructiveness can easily be imagined; at times their ravages have been such as to induce foreign nations to prohibit the importation of American potatoes. It has acquired new tastes in the absence of potatoes, will eat cabbage, thistles, and even oats. Within twenty-five years it has readjusted itself to every variety of climate; has changed its

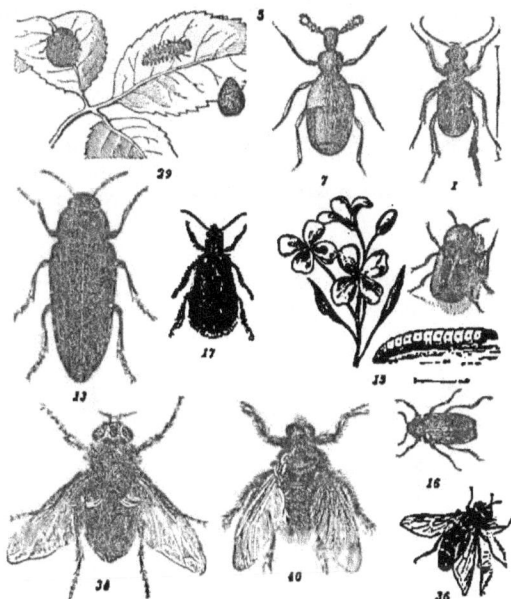

5. PESTLE WATER-BUG AND LARVÆ (*Hydrophilus piceus*). 29. LADY-BUG AND LARVÆ (*Coccinella septempunctata*). 13. PINE BEETLE (*Buprestis Marianna*). 17. WOOL BEETLE (*Lagria hirta*). 15. BRILLIANT BEETLE AND LARVÆ (*Meligethes brassicæ*). 34. HORSE-FLY (*Musca vomitoria*). 40. HORSE LOUSE-FLY (*Hippobosca equina*). 16. BACON BEETLE (*Dermestes lardarius*). 36. RING-FLY (*Syrphus festivus*).

diet, and has assumed great variety of form. Thus it strikingly illustrates the doctrine of adaptation to conditions, and Divine provision for the otherwise helpless, while the brevity of the time occupied in its development renders it a specially good subject for the modern naturalist. The common barn-yard chicken and the duck, as well as crows and frogs, render good service in taking the *potato-bug* where it shall no longer eat, but be eaten.

CORN WEEVIL AND ITS RAVAGES.

The **Corn Weevil** (*Calandria granaria*) has no preference, but feeds with equal satisfaction upon corn, oats, wheat, barley, or rice; furthermore, it prefers the grain after the farmer has prepared it for storing in his barns. The *weevil* lays its eggs inside of a grain, so that the larvæ may eat the substance and leave to the farmer the form only. In about six weeks the *weevil* attains maturity, and is ready and anxious to continue the work of his progenitors. As a single *weevil* can in four or five months raise a healthy family of six thousand young *weevils*, his troublesomeness to our agricultural interests is easy to realize. This insect, as is estimated, destroys three hundred thousand bushels each year of European wheat.

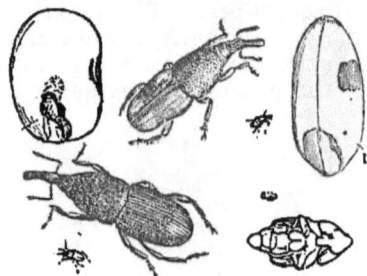

The premature falling of fruit from trees is generally due to the larvæ of the *plum-weevil*, which destroys apples, pears, plums, nectarines, peaches, apricots, cherries and quinces. As the fallen fruit contains these larvæ, it should at once be fed to hogs, or otherwise destroyed, instead of being left to rot on the ground. Nearly all the several species are more or less harmful, though

CORN WEEVIL, NATURAL SIZE AND MAG-NIFIED.

they all have their uses in the wonderful economy of nature.

The Sacred Scarabæus (*Ateuchus sacer*), vulgarly called *tumble-bug*, figures in the carved monuments of the Egyptians, by whom it was esteemed sacred. It is about an inch long, flat in form, dull black in coloring. The Egyptians used the *scarabæus* as a symbol of the earth, because of its activity from sunrise to sunset in rolling its ball of booty ; as a symbol of the sun, because of the raylike projections from the head ; as a type of the warrior; as a symbol of fecundity and as emblematic of Isis and Osiris.

The Common May-Bug (*Melolontha vulgaris*) has a pair of sheathed wings, reddish-brown (though frequently sprinkled with a white dust) ; the neck covered with a black or red plate ; forelegs adapted to burrowing, since it makes its nest half a foot under ground.

MALE, LARVÆ AND NYMPH OF MAY-BUG.

Three months are required before the egg releases a small grub, which continues a predatory existence for about three years, gradually increasing in size until it becomes the red-headed white maggot usually found in newly-dug earth. It provides itself with new clothing every year, and at the end of

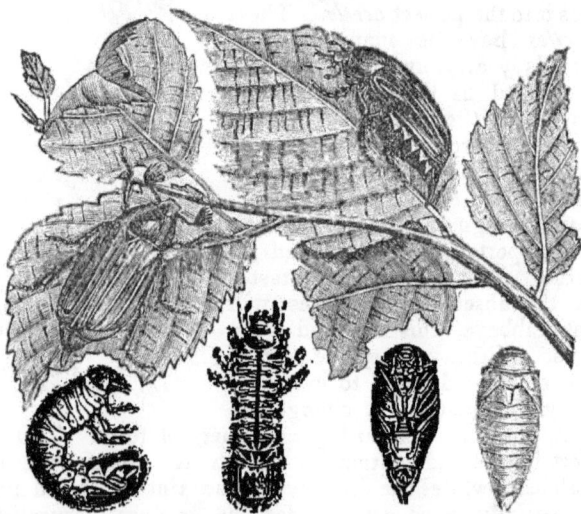

the fourth year prepares for its transformation. Returning yet further underground, it builds itself a roomy dwelling, and passes three months in accomplishing its change to a winged insect. In May they burst upon the vegetable world, and frequently blunder into houses, and annoy, rather than injure, the in-dwellers. The bug should be killed upon its first appearance, before it has had the opportunity to lay its eggs, for the great damage done by this insect takes place during its underground, larval existence.

The common *may-bug* sometimes swarms so as to obscure the sunlight and obstruct the road-way. But however uncomfortable this inconvenience may be, it is transient, whereas the devastation committed in the larval stage is beyond estimate. So serious has been their depredations in France that legislative action has been required; in a single department there were collected, in the space of a fortnight, thirty-two carloads of these insects, which had already destroyed one-fourth of the crop.

The Spanish Fly (*Lytta vesicatoria*) is three-quarters of an inch in length, bronze-green in color, common in Europe and Asia, and feeds mainly upon the ash. Under the name of cantharides, it is an article of commerce and a medicinal remedy. The earliest larval form (*triungulin*) changes to a small, six-footed white grub, whose sharp mandibles are shortened and blunted. At the end of five days this grub moults, and again a second time at the end of another five days, when its eyes disappear. It now descends into the ground, and after five days changes to a yellowish-white, after which it hibernates, until with the open-

WEAVER BEETLE (*Lamia textor*) AND FIRE-FLY, MALE, FEMALE AND NYMPH (*Lampyris noctiluca*).

ing of spring a third larval form is produced. In about two weeks it assumes the form of a beetle, and in twenty days reaches maturity.

Garden Hair Beetle (*Bibio hortulanus*) is black when male, and rusty-red when female.

The Flour-Bug, also called the **Meal Beetle** (*Tenebrio molitor*), is found about flour-mills, granaries, and bakeries.

The Mourning-Bug (*Blaps mortisaga*) lives in cellars and dark places. It is the darkest colored of the beetles, and its sombre appearance has given it its popular name.

St. John's-Bug (*Lampyris splendidula*) is a glow-worm—gray-brown in the male, and golden white in the female. Its lighting apparatus is in the body, but the source of supply or the real nature of the phosphorescence is still unknown. The pleasures of childhood would be greatly curtailed by the absence of these little star-twinklers, and their beauty and attractiveness are not neutralized by any harmful propensities.

Cucuja, or **Phosphorescent Beetle** (*Cucuja sanguinolentus*), is so luminous that a single one furnishes light enough for the reading of print, and eight of them imprisoned in a vial will enable one to read script. The illuminating

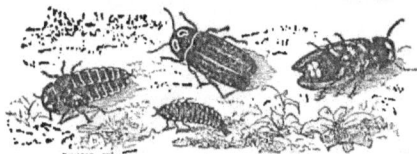

apparatus consists of two yellow patches on the thorax and two more beneath the elytra. In the Indian Archipelago these insects are so numerous as to convey the impression that the trees are illuminated by Chinese lanterns, or gemmed with rubies of the richest dye. It has a short, bright head, large eyes, and illuminates the darkness with a greenish, golden fire. The Cuban ladies frequently imprison this beetle in the tissue of their evening dresses, and thus multiply their attractions.

The **Weaver-Bug** (*Lamia textor*), a European borer into willow trees, from one to one and a quarter inches in length, and about one-half as broad; is dark brown with yellow body beneath the cased wings.

Alpine Goat-Bug, found in the high Alps, is named from a goat-like odor which it exudes. It feeds upon the willow, oak, poplar and other trees, and constructs its caves within them. It does not waste the wood which it throws out while burrowing, but uniting this with silk, which it spins, makes a thick hanging for its apartments. It hibernates, and before reaching its final metamorphosis has enjoyed a life of three years.

32. HAWK-FLY (*Asilus œlaudicus*). 31. HAIR-FLY (*Bibio hortulanus*). 19. ALPINE GOAT BEETLE (*Rosalia alpina*). 39. FLEA (*Pulex irritans*). 28. *Lina populi.*

The **Grape-vine Sucker** (*Pelidnota punctata*) is brownish yellow above, having the back and sides of the head black, a single dot on each side of the prothorax, three dots near the margin of each elytrum, greenish black beneath the body and on the legs. It is nocturnal in its habits, and hence more difficult to detect.

The **Parti-colored Ant-Bug** (*Clerus formicarius*) is ant-like in appearance, and is one of seven hundred species. The larvæ are predaceous and take up their homes in bee-hives, where they live upon the honey comb. They sometimes make their burrow among the coleoptera.

GRAPE-VINE LOUSE (*Phylloxera vastatrix*). 1. BACK OF LOUSE, 2. BELLY OF LOUSE 6. PIECE OF ROOT BORED BY LOUSE. 3. LOUSE BORING. 4. PERFECT FLY. 5. ROOT SHOWING SWELLING CAUSED BY SUCKING OF THE INSECT.

The **Seed Runner** (*Agriotus lineatus*) is brown in color, and may easily be mistaken for a part of the plant upon which it rests.

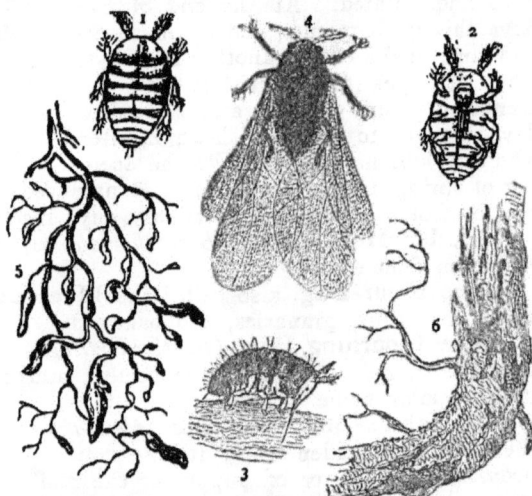

The **Brilliant Beetle** (*Meligethes brassicæ*) revels in the brightest and hottest sunshine, and distracts one's attention from his personal discomfort.

The **Bacon Beetle** (*Dermestes lardarius*) is three-quarters of an inch in length, burnished black, oval, elongated in shape, will feign death, and can be depended upon to eat anything that comes in his way. This beetle is destructive to the museum, the warehouse, the store or the pantry; drugs are grateful to them, tobacco is not distasteful, woolen or silk, feather or fur, is to them palatable and nutritious. It takes its name from its fondness for the Southern dish of bacon.

The **Wool Beetle** (*Hagria hirto*) is black with yellow wings. It lives upon vegetables and blood of insects.

The **Sand Beetle** (*Cicindella campestris*) lives on sandy banks, while their larvæ pass their lives half a foot under ground, though they come to the surface for food. Their white coloring affords protection, by rendering them difficult to be distinguished from the sand.

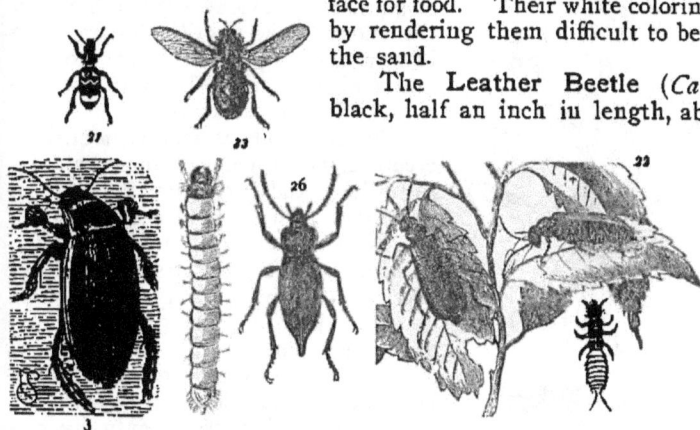

The **Leather Beetle** (*Carabus coriaceus*) is black, half an inch in length, abundant in Europe, where it feeds upon worms, caterpillars, and other insects.

The **Swimming Beetle** (*Dytiscus marginalis*) lives in, not merely upon, the water. It is guided by sight. Some species are musical. They are rarely found in salt water, but do

3. SWIMMING BEETLE (*Dyticus marginalis*). 22. MOURNING BEETLE (*Blaps mortisaga*). 26. SPANISH-FLY (*Lytta vesicatoria*). 21. PARTI-COLORED ANT BEETLE (*Clerus formicarius*). 33. *Ogcodes gibbosus*.

not object to hot springs; like the camel, they are able to carry supplies, though they select air and not food. He has the power of making flaccid his shelly covering, so as to escape from the enemy wishing to feed upon him.

The **Hunting Beetle** (*Staphylinus erythropterus*) is brown and gold in coloring, sometimes black and gray, again brown, black and yellow.

The **Pestle Beetle** (*Claviger foreolatus*) is used by the ants as a milch cow; its reproduction is by depositing from twenty to one hundred eggs in a cocoon sometimes attached to a plant, and sometimes to the parent. The larvæ prey upon one another; they seize flies and suck their blood.

The **Death Simulator** (*Hister unicolor*) is round, hard and seed-like. It lives upon decaying vegetable and animal matter. In color it is generally black. By drawing in its legs and assuming the form of a ball it successfully simulates death. This sometimes protects it against attack, and sometimes renders its prey unwary.

The **Burying-Bug** (*Necrophorus vespillio*) is a European species, which lays its eggs in animals which it has buried. It is most frequently black (or black relieved by stripes of yellow); its wings are marked with two orange-colored bands. It flies or runs with rapidity and ease, and is offensive mainly from its office as a scavenger. It does its damage by cutting the roots which may intercept its path. Its front feet are used like those of the mole, and are equally skilful and efficient. It ingeniously burrows beneath its prey, and thus causes it to bury itself, in which the eggs are then deposited. The buried body is thus made to serve the double purpose of furnishing a receptacle which promotes development of the eggs, and also provides sustenance to the larvæ as soon as they issue from the envelope.

The **Stag Beetle** (*Lucanus cervus*), or **Corn-Bug**, is a European species; it is there commonly called the cockchafer. It flies heavily, with a whirr like that of a pheasant. It has a familiar representative in the American tumble-bug, especially the larger species, which is similarly provided with a horn on the snout. The *stag beetle's* antennæ bear a striking resemblance to the horns of the deer. It burrows into trees to deposit its larvæ.

The **Rhinoceros Beetle**, a curious creature, is so named from its having a long and back-curving horn on the snout, which bears a striking resemblance to the nasal weapon of the rhinoceros. To this class also belongs the *hercules bug*, a very large creature, and probably the greatest of all the insect family. To its immense size, sometimes exceeding three inches, is added a formidable appearance, though it is dangerous only by reason of its power to inspire fright, since it has no ability to do any considerable harm beyond a slight pinch, equal to that of a small crab.

30. RING-TAILED GALLINIPPER (*Culex annulatus*). 10. RHINOCEROS-BUG (*Oryctes nasicornes*). 35. CATER-PILLAR-FLY (*Tachina fera*). 37. CATTLE-FLY (*Oestrus bovis*). 22 FLOUR BEETLE AND LARVÆ (*Tenebrio molitor*).

The **Magnificent Pine Beetle** (*Buprestis marianna*) cannot move when laid upon its back. It girdles trees, for which the beautiful coloring of the many species (green, copper, bronze, for example), is not a sufficient compensation. This insect is called magnificent, not only because of its brilliant wing sheaths, but also on account of its graceful appearance, whether at rest or in flight. It is found in many parts of the United States, and its presence is easily discovered by the marks of its ravages, or rather the ravages of its larvæ, for the adult beetle is not known to commit such depredations itself. The eggs are deposited in a rift, or beneath the bark of a tree, and there left to nature to hatch. Directly after the young appear they begin eating the inner bark, and continue their feasting in tortuous courses, and often circular, apparently taking care not to show themselves, being a wood-worm, until they emerge into a perfect beetle. A tree thus girdled rarely shows any outward effects of the damage so done until the following year, when the buds put out slowly and the tree gradually dies.

The **Tiger Beetle** is quite common and is strikingly beautiful. Its covering is an armor studded with gems and embossed with gold, and no chainmail ever adapted itself more readily and more perfectly to the graceful and rapidly changing attitudes of its wearer. To the unassisted eye it may appear to be but a dull green, but subjected to the magnifying power of a microscope it displays the most marvellous beauty of coloring and of mechanism. It is to be found throughout Asia, Europe and America.

The *Brachinus crepitans* is noteworthy because upon being touched it discharges, with a sensible explosion, a pungent vapor, which has all the effect of a mordant acid. The *beetle*, which makes its home in the Mammoth Cave of Kentucky, is blind—an evident adaptation to the conditions of its existence. The flattening of the hind legs into propelling oars of the *water-beetle*, and its ability to store away air in receptacles under the elytra, are yet other illustrations of the same intelligent Providence. The common *whirligigs* of our ponds and streams worship the sun with a constancy and intensity, so far surpassing the fabled statue of Memnon or the mythology of the sunflower, as to bring them into comparison with the human sun-worshippers of the Orient. The *devil's coach-horse*, or *rove-beetle*, is fierce out of proportion to its size, and can, like the pole-cat, render itself unexpectedly and effectively offensive. The "*shard-borne beetle*," of Gray's "Elegy in a Country Churchyard," is so described because of the prominence of its elytra. The

STAG BEETLE, MALE AND FEMALE (*Lucanus cerous*).

bronze beetle is a strikingly handsome species. There is an Indian beetle whose thorax is a burnished blue, relieved on the sides by copper-colored indentations; its elytra are cream-colored, and have patches of purple at the tip and on both sides. The *cardinal beetle* is flame-colored, and illustrates the inimitable richness of natural dyes.

Lina Populi, or **Linden Beetle**, is black, green or blue; its wings are red with black borders.

The **Seven-dotted Lady-Bug** is hemispherical, hairless, and the largest and most common European species. It is called in Europe, the lady-bird, and in America, the lady-bug. It has a black and white head and reddish elytra. The species of this country is very small, with red elytra dotted with black.

Hercules Bug (*Dynasta Herculi*) is found in tropical America; it is six inches long, black, the elytra blue-green, with black spots.

Flesh-eating Coleoptera embrace many of the families named *cicellidæ*, *carabidæ*, *dytiscidæ*, and *coccinellidæ*.

The **Diptera**, or **Two-winged Flies**, are the most numerous species of insects; their small size, and often their quiet colors, prevent their attracting such attention as they deserve, alike as a study for the entomologist and because of their relation to man's interests. Their wings, though tenuous, are capable of the most rapid vibration, as in the case of the house-fly, which reaches four hundred movements per second. Of the one hundred thousand species known, most are the benefactors or useful servants of mankind, destroying, or at least reducing, matter which would else prove the source of sickness and death; diminishing the number of insects which rob the farmer of the rewards of his toil; and illustrating many a lesson which mankind seems to require to have repeatedly urged upon his notice. In the colder climates, where vegetation is less rank, water less stagnant, and animal life less troublesome, the *diptera* are fewer in species; but as conditions unfavorable to animal or human life multiply, these insects likewise increase in number, variety and useful activity. Thus do they illustrate the wonderful way in which the Creator renders the world fit for the habitancy of his children, and increases our sense of responsibility for discharging our duties as persistently and as cheerfully as do these little beings. Gnats, mosquitoes, Hessian-flies, the wheat-fly, the horse-fly, the asilus-fly, and the bott-fly are the most interesting members of the *diptera*.

The **Gnat** (*culex pipitus*) has a long and cylindrical body; its wings are covered with scales; its antennæ are feather-like; its eyes seem to occupy the whole of its head, and are protected by a net work. The trunk, or sucker, is cylindrical, and

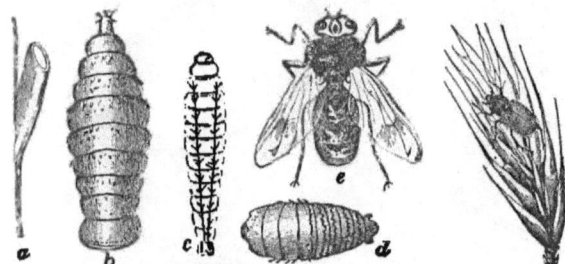

a. b. c. d. e. THE BOTT, OR GAD-FLY AND LARVA (*Gastus equi*).
11. WHEAT-BUG (*Anisoplia segetum*).

contains a bundle of stings, each of which is composed of six parts; its sting is followed up by the injection of an irritating liquid. During the larval period it appears as a worm in stagnant waters, and comes to the surface to inhale air for respiration, which it does through organs located in the last segment of the body. It sheds its skin several times. As a pupa it has no digestive organs, and replaces the posterior air tube of the larva by two ear-like appendages on the head. With its last transformation it ceases to be a water-insect; indeed, water is so fatal to it, that if touched by a liquid immediately after parting from the pupal stage, it dies at once. During this period, brief as measured by the life of man, but doubtless of painful length to the *gnat*, it sails about in its sheath, using its new body as mast and sail. A basin of water, a few gnats and a microscope, will enable those who are "inland bred" to enjoy the spectacle of a miniature sea studded with the sails of many a craft; or to dream many a legend of spectral boatmen; or to study those changes which have the profoundest significance for the biologist. When, finally, the *gnat* liberates itself from its boat-like sheath, it leaves the water forever, and joins the insects whose habitation is the dry land. As *gnats* breed throughout the whole year, and at intervals of three or four weeks, it can readily be seen that great is the need for their services, and that equally great, at times, is the need for checking their undue multiplication.

The mosquito is one of the best known species of *gnats*. Disagreeable as they are to one whose attention is absorbed by his own personal discomfort, they are among the most useful laborers in the field of animal economy. As one travels southward, the need for their services grows more pressing, and their presence and, as it seems to the ignorant, pernicious activity becomes more and more noticeable. In cities their number diminishes with improvements in drainage, but experience would seem to suggest that their absence is bought at the cost of increase in frequency and fatality of several forms of disease; the disagreeable and the helpful, as well as the pleasant and the dangerous, are not always recognized. Mosquito-bars have doubtless become too well known to excite surprise in America, but the reader may not be aware that at no greater distance than that of the Louisiana plantations, persons find it necessary to surround their rooms with a mosquito bar that they may attend to the cares of household life. It is probably better known that in Minnesota, and in forest tracks north as well as south, the lumbermen have at times to protect their heads and hands. In Africa the mosquito has been a more dangerous enemy than the savage tribes, dense jungles and dangerous streams with which Livingston and Stanley had to contend. The eggs of the mosquito are laid in the water and float in little skilfully constructed rafts upon its surface. Within a few days the larvæ are hatched, and appear in the form of what children call *wriggles*. During the pupal stage the mosquito changes its skin two or three times, and exchanges its breathing tube

COMMON GNAT, (*Culex pipiens*), AND ITS METAMORPHOSES MAGNIFIED. LARVÆ OPENING THEIR RESPIRATORY DOORS ON THE SURFACE OF THE WATER. NYMPHS AND PERFECT INSECTS.

at the hips for two others, which proceed from the thorax. In time the pupa exuviates, and the young mosquito begins his career of labor attended by song far surpassing in its constancy the music-crowned labors of the wine-press, and calculated to render a real service to mankind, instead of betraying him by delights which inebriate but do not cheer.

The larger sized species are frequently called *gallinippers*. The well-known *daddy-long-legs* belongs to this family, and the thinness of his dispropor-

tionate members would seem to be a provision for escape from the web of the hostile spider; certain it is, that the loss of one or more legs never seems to embarrass him. The *asilidæ* or robber-flies are the most predaceous of the fly family; they have scattered themselves throughout the world, and are so numerous that twenty-five hundred species have already been classified. The *Hessian-fly* takes his name from the fact that his appearance in America was simultaneous with the coming of the Hessians of the Revolutionary period. These unfortunate soldiers were not, as our forefathers supposed, willing hirelings, but having been sold by a spendthrift king to the British monarch, were impressed, though in their homes, or in the street, and left to fight or die in America; it is, however, through

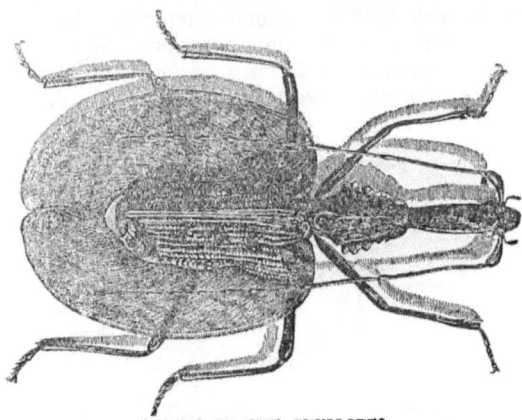

THE MORMOLYCE PHYLLODES.

no crime of theirs that our crops have been ravaged by this insect pest. It hatches two broods a year, and lays as many as a hundred eggs at a time; these eggs are deposited in the grain stalks, so that in addition to the loss from the appetite of the larvæ, the plant suffers from an obstructed or suspended circulation. The *horse-fly* is remarkable for its long-sustained flights, and though irritating to horses and cattle, is less dangerous than mosquitoes and gnats. The *oleander* hawk-fly (*asilus oleandicus*) passes its larval period underground, or buried in rotten wood. It is very predaceous, and though fond of the oleander, is not disinclined to prey upon its own species. The *bott-fly* or *gad-fly* lays its eggs in the skin of cattle, or on the hair on the knees and shoulders of horses, which

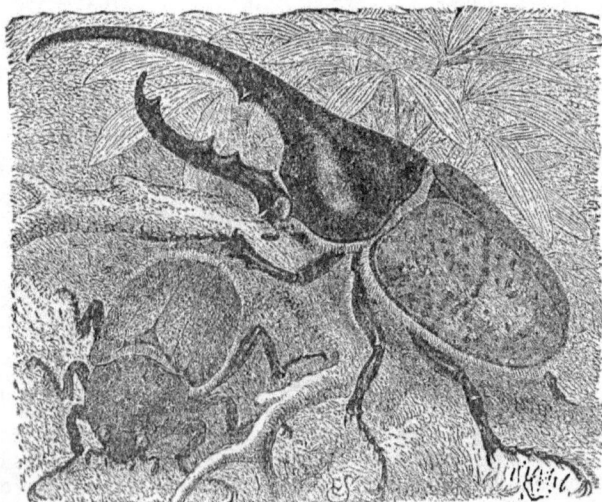

HERCULES BUG, MALE AND FEMALE (*Dynastes Hercules*).

in licking themselves swallow the eggs. The *vestrus bovis* (cattle-fly) sometimes occupies the front and nasal cavities of sheep and oxen; another species burrows in the hides of oxen. The horse-fly proper (*gastrophilus equi*) attaches a single egg to a single hair of the horse. In its larval period it feeds upon the stomach of the horse, and exhibits a strange power of apparent suspension of respiration when the air is unfit for its use. At the end of ten months the larvæ, being ready for metamorphosis into pupæ, are ejected with the excrement of the horse, and return like a frightened child to the bosom of Mother Earth.

The **House-Fly** is ash-colored and has a black face, whose sides are yellow. The forehead has yellowish black stripes. The thorax is blackened; the abdomen pale beneath and yellow on the sides. The feet are black, and the wings transparent. The proboscis is a separate organ, thick, membranous, and ending in two lobes, which form from the lower lip and the mandibles a blade-like instrument. Beginning life as a white maggot, upon approaching the chrysalis, or pupal stage, it shortens its body and hardens its skin, and for the first time develops a head and legs, and replaces its rudimentary mouth by

METAMORPHOSIS OF THE DRAGON FLY. (*Libella depressa.*)

a true mouth. Filth is to them the staff of life, and they seem to be indifferent to the wide differences between sugar and tobacco spittle. This healthful but repulsive appetite is thus made to subserve the sanitary interests of the higher animals. The larval stage is about a fortnight in duration; the egg period but twenty-four hours; the pupal age a week or two. Examined under the microscope, the wonders of their mechanism become apparent, and one cannot but be struck by the wonderful provision for the life which they are to live.

The Sheep-Tick and the Bat-Tick have aborted wings—an illustration of the economy of natural life, which, except in man, prevents one's powers from wasting because unused, by removing the organs which would serve no useful purpose. The *fungus gnat*, found under the bark of trees, when about to be transferred into pupa, will form into long processions five inches in breadth, and travel in a solid column. This fly is known as the army-worm, and its ravages are well known. The *blow-fly* (*Musca vomitoria*) is well known to housekeepers by its jet-black color, and varied by the steel-blue of the abdomen, by its ready perception of the odors of food which is being cooked, and by its persistent intrusiveness in depositing in meat or vegetables eggs which will hatch in a single day. The *gall-fly* is a very minute, very delicately constructed, and very beautiful species. The galls so frequently seen upon the oak are swellings caused by the insect, which uses them as a nest for its eggs, of which as many as sixty have been found in a

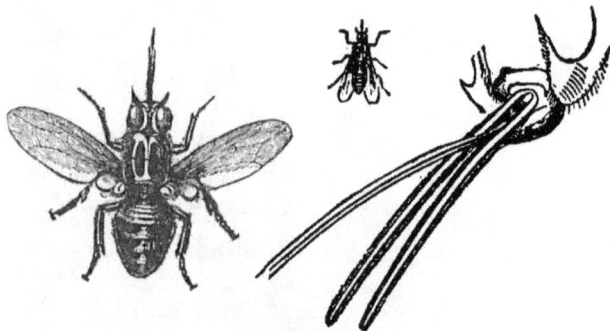

TSETSE FLY, NATURAL SIZE AND MAGNIFIED.

single gall. Though single species confine themselves to some favorite kind of plant, yet among the various species numbers of trees and plants are thus selected, and as the insect attacks every part of the plant, from the smallest to the thickest root, and from the wood of the root to the most delicate leaf, its devastations are to be feared. The *buffalo gnat* is found in incredible numbers in the Southern States, and causes the most serious loss of cattle. Another species is found in the Eastern and Middle States, and is called the *black-fly*. Sportsmen in the Adirondacks, for example, suffer great annoyance unless they respect the season which this irritating insect appropriates to itself, with such annoying results to its victims.

The **Triodite** devours the eggs of the locust. The *syrphus vestius* is half an inch in length, and is marked with yellow bands upon a green metallic ground. Its larvæ feed upon plant-lice, and thus keep their number within some sort of bounds. The *stable-fly* does not frequent dwellings, and as it lives by sucking the blood of vertebrates, its proboscis is slenderer, stronger and better adapted for piercing.

The **Tsetse Fly**, though deadly to domestic animals, is harmless to men and to wild animals.

African travellers tell of the courage of the *tsetse-fly* which disputes with man the possession of the land; it has been known even to prevent either agriculture or exploration. In form not unlike our horse-fly, it resembles the snake in the secretion of the most deadly poison. To be sure this poison is fatal only to domesticated animals (except the goat and ass), but no industries can be carried on without the aid of our beasts of burden. The *tsetse* never goes in swarms, being fortunately few in number, and therefore its effect upon domestic animals is the more surprising. The buffalo gnats of Arkansas and other southern districts will occasionally worry cattle to death, but these attack in such great swarms as to fairly drain their victims of blood. The *tsetse*, on the other hand, while a blood-sucker, does not do any great damage by the mere act of securing its supply of nourishment, but by injecting a virulent poison into the wound, from which its victim languishes, first losing all appetite and gradually growing weaker as the poison continues through the system, until complete exhaustion is soon followed by death. None of the wild animals of Africa, however, suffer from its attacks, a fact which naturalists have not been able to account for. It is never met with outside of a small district near the centre of Africa, and even in this restricted region it is not always to be seen, even though there be domestic animals to attract it.

There are yet other species of the fly which mine into leaves, burrow into fruits, and dig into the stems of plants. The *cheese-mite* is a pronounced black in color, and has the four hinder-legs yellow. From their power of leaping they are popularly called "skippers." Some mites live in wine, alcohol, or salt water. Some of them are roasted by certain tribes of Indians and are said to make a not unpalatable dish.

The **Flea**, as a guerilla, is unrivalled; agility, caution, persistency—these amiable qualities does he employ to the discomfort of those whom he persecutes with his attentions. He is the foe of the household, for not content with coveting his neighbor and his neighbor's wife, his son and his daughter, he insists also upon the pets of the household and some of the animals of the field. Troublesome throughout the United States, and specially numerous in sandy districts, it is yet far from being the pest which it is in France, Spain and Italy. Laying from eight to twelve eggs, these are hatched in the brief space of five days, and in fourteen days the small grubs pass into the stage of chrysalis; after yet sixteen days more they spring forth, like Minerva, fully armed, and at once enter upon the duties of active life. Fleas have been trained to work in harness, but though able to draw one hundred times their own weight, their avoirdupois is too slight to have converted them into useful beasts of burden. The scientific name for the flea is *pulex irritans*. It is wingless, body compressed, eyes small and round, legs stout and strong. Its eggs are laid in the hair. The larval and pupal stages are each about fourteen days in duration. In the West Indies and in South America a species called the *chigre* (or jigger) causes great trouble by burying itself under the nails of the foot. There is also a species of *chigre*, common to this country, which burrows into the flesh, and though scarcely larger than a pin-point, is so poisonous as to produce intense irritation and great swelling of the parts bitten. If this insect were so large as a common tick, with an equal increase of venom, its bite would produce death, since nothing in nature is supposed to be more poisonous, and small as it is its burrowing has been known to produce a dangerous inflammation.

17

BUTTERFLIES AND MOTHS.

Butterflies, alike from their ever-varied form and coloring, are among the most commonly observed of the insect world. In their modes of passing the winter there is also great variety. Some exist solely as eggs, which early spring is to quicken into life; others have already become *butterflies*, but without reaching maturity; yet others hibernate like the bears. The caterpillar wears his skeleton upon the outside, where, with its series of rings, it serves as a coat of mail. In this are nine pairs of openings connecting with tubes, and with

ADMIRAL BUTTERFLY (*Vanessa atalanta*). SILVER WING (*Argynnis paphia*). CHECKERED BUTTERFLY.

them composing the organs of respiration. In addition to the metamorphosis which proceeds to the grub, thence to the chrysalis, and finally to the *butterfly*, the insect sheds its skin at least eight or ten times. The evolution of *butterflies* as an order may save repetition, for peculiarities of the different species will then alone remain to be mentioned as these species pass in review. Beginning with the egg, we find this to be about as large as the head of an ordinary pin, with a shell thin and elastic. Even the forms of these eggs are said

WHITE FLECKED. NIGHT MOTH (*Noctua trapezina*). AURORA. METALLIC WINGED.

by naturalists to illustrate the most varied principles of architecture, and to suggest designs not yet realized by man. A little circle of cells on the summit have the effect of stained glass windows of definite but unusual patterns. Then follow what might be termed vestibule cells, since by their agency entrance is made for the fertilizing principle. Their shapes are varied, presenting the sphere, the hemisphere, the spindle, the sugar-loaf and many others. The process within the egg is that already pointed out—the yolk separates into many distinct cells, and the germinal cells rise to the top. The future insect now begins to appear as a band rolled up, so that the ends (head and tail) meet. Finally, in a period sometimes as short as a few days, less frequently in a month, and rarely after hibernation, the egg is developed into the caterpillar. Worm-like in appearance, it apparently consists of a head and of a body composed of successive rings—these rings being the skeleton of the insect. It has a pair of stout jaws which work sidewise. The mouth lies between the jaws and has secondary jaws; it is further supplied with a slender tube, which serves as a spinning-wheel for the silk which the caterpillar is constantly making to render yet more secure its locomotion. The antennæ are at the base of the mandibles and are furnished with a long bristle. The eyes are arranged in a curve, and the sixth one is placed some distance back of the others. The rings composing the body are thirteen in number, the three front ones furnishing support for the legs; the third, fourth, fifth, sixth and thirteenth rings bear false or temporary legs. It respires by means of *stigmata*, which

PAPILIO BROOKEANA.

occur on each side of a segment, except the second, third, twelfth and thirteenth. The segments have a covering of hairs, the arrangement of which in the first three

APOLLO BUTTERFLY.

and in the last ten segments is different. On the under side of the body, included by the lower legs and the under lip, are the organs of scent and hearing. Moths are distinguished from *butterflies* by structure; the hind legs and wings in the moth are hooked together; the antennæ are fern-like, instead of being turnip-shaped, as with the *butterfly*.

The **Admiral Butterfly** has black, velvety front wings marked with scarlet and white; the back wings are black. The eggs are ribbed so that their summits look like a succession of little hills. The thorax is very slight, and the middle segment carries the front wings and the relatively powerful muscles which must control them. The legs proceeding from the thorax, thin but jointed and strong, support the centre of gravity of the body. It builds its nest by fastening together the edges of a leaf, and sometimes uses what it does not need for support as a temporary means of subsistence. It hibernates until May or June; the caterpillar takes about ten days for his transformation, and soon provides for a new family.

The **Many-Eyed Butterfly** (*Polyommatus*) is copper-colored with eye-like spots.

The **Theckla** (*Thecla betulæ*), or **Hair-Streaked**, is found throughout the globe. Five hundred species have been classified. In coloring it represents every shade—orange, brown, blue, black, etc., and has a metallic lustre.

The **Aurora** (*Anthocharis cardaminis*) has black-spiked, orange-red front wings; the under side of the hind wings is a moss-green.

POLYOMMATUS PHLÆAS.   THECLA.

The **White Flecked** (*Vanessa C. album*) has tawny wings, black-spotted above, and brown or blue below, with a white spot resembling a capital C, a comma, or the Greek letter gamma. The showers of blood which sometimes fall in Europe are attributable to a red liquid which the butterfly exudes when migrating—not, as the superstitious peasants believe, to a special and portentous miracle.

WINDOW BLOT
(*Thyris fenestrella*).

The **Great Ice Bird** (*Limenitis populi*) is a dweller in forests, and is notable for the height of its flight.

The **Metallic Wing** (*Melitæa cinxia*) has brown-yellow and white checkered spots; on the under side, bands of white and yellow.

The **Checkered Butterfly** (*Melanargia galatea*) has a soft, hair-like covering, and the wings are beautifully checkered with colors of black and orange.

The **Swallow-Tailed** (*Papilio machaon*) has wings of yellow and black, a black body, yellow beneath and on the sides, and the hind wings notched and prolonged into tails.

The **Apollo** (*Parrassius apollo*) is found in Switzerland and in the Pyrenees. Its wings are generally white, tinted with yellow; lower wings have also crimson spots bounded by black rings; fore wings are spotted with black.

The **Leaf Formed** (*Phyllium siccifolium*) has a striking resemblance to a leaf in its wings, legs and thighs.

SWALLOW-TAIL BUTTERFLY (*Papilio machaon*).

The **Bee-Moths** (*Sesia apifumis*) take possession of cells in a bee-hive and feed upon the accumulations of others. The *mordellidæ* similarly enscouce themselves in the nests of wasps, which do not attempt to expel the intruders. The *bee-moth* is sometimes called the poplar-moth. As a larva, it passes two years, during which it is even more destructive than when fully matured. It illustrates adaptation to environment, since its jaw is suited for work upon the hard substance of the tree, and not simply upon its foliage.

The **Cabbage-Moth** (*Pæniocampa gothica*) lays its eggs upon cabbages and turnips; it is the soft, white butterfly that wears the livery of innocence "to serve the devil in;" it represents another unwelcome European immigrant. The spring broods are darker in color, and thus attract less attention. The *cabbage-moth* destroyed in Canada during a single season a quarter of a million dollars' worth of produce. The *cabbage-moth* spins a cocoon, and hence is sometimes called the spinning butterfly.

The **Gamma Eule** (*Pluria gamma*) is a nocturnal, or night-flier. It takes its

BROWN BEAR (*Arctia caja*).

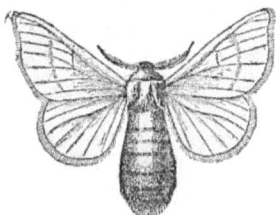

SILK-WORM MOTH (*Bombyx mori*).

name from a silvery marking which resembles the Greek letter gamma, which is found upon the front wings. The caterpillar is green, with two yellow and six white lines. It hides its cocoon in a leaf which it rolls together for this purpose.

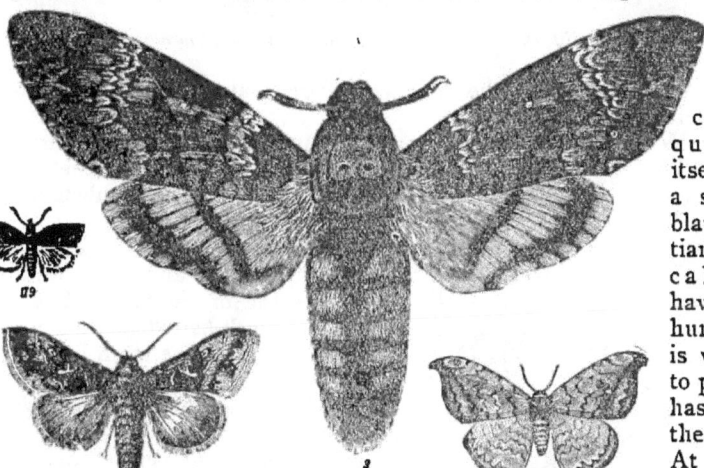

The **Pine-Moth** (*Sphinx pinastri*) was so named because it frequently erects itself, when it has a striking resemblance to the Egyptian sphinx. It is called also the hawk-moth, or the humming-moth. It is very destructive to pine forests, and has no respect for the age of the trees. At times in France a whole village will turn out by night and devote itself to

3. DEATH'S HEAD MOTH. 19. APPLE MOTH. 16. GAMMA OWL. 13. *Drepana falcataria.*

the destruction of this ravager of the finest and most stately forest trees.

The **Hop Spinner** (*Hepialus humuli*) is not an attractive insect, as it smirches itself in feeding. The male is white above and grayish-brown below; the female is light gold above and reddish-gray beneath. It flies in the early evening.

The **Twelve-Feathered Moth** (*Alucita hexadactyla*) is found both in Europe and in America. It is ash color with brown markings. It feeds upon the honeysuckles and hibernates. It takes its name from the fact that each wing is divided into six lobes or feathers.

The **Blood Drop** (*Tygena loniceræ*) belongs to tropical climates. The *tygenæ* resemble the bee-moths. They are day-fliers, and as their cocoons are put under ground, they cover them with a varnish as a protection against dampness.

Green Leaf (*Geometra papilionaria*) is of great size and of equal destruc-

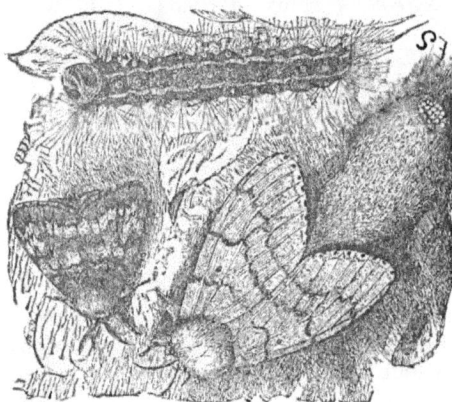

SPINNING MOTH (*Liparis dispar*).

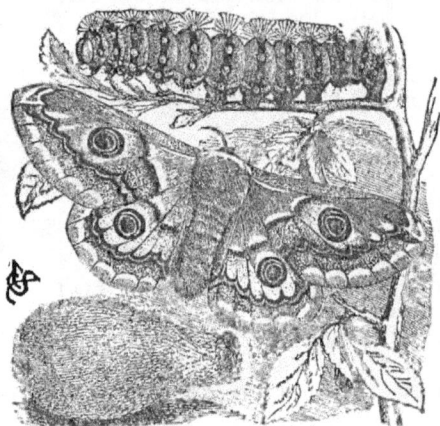

NIGHT HAWK (*Saturnia pavonia*).

tiveness. Its common name is the *measuring-worm*, and hence it has been named the geometer, or surveyor.

The **Apple Winder** (*Carpo-capsa pomonana*) occurs wherever there are apple orchards. The egg is laid in the undeveloped blossom, and the young caterpillars eat out the substance of the fruit before it matures.

GREAT ICE BIRD.

The **Brown Bear** (*Antia capa*) is beautiful in form, and in appearance resembles the brown bear. The caterpillar is black-haired, and looks like a hearth broom. The moth has brown fore wings, irregularly marked with broad white lines, and the hind wings are a lighter brown, having a map of Africa about the centre, and three large black eyes in the margin.

The **Linden** (*Smerinthas teliæ*) is common to elms and horse-chestnuts. Its upper wings are a grayish-green, banded; its banded thorax is gray.

The **Death's-Head** (*Acherontia atropos*) is found in Europe, Asia and Africa. It is from four to five inches in length, brown fore wings; hind wings yellow, banded with black. It feeds upon the tomato, potato, and similar vegetables. It has a skull-like mark on the thorax, and in flying emits a sound, whence it is regarded with superstitious veneration.

Oak Leaf (*Tortrix viridana*) is green in color, and lays its eggs upon the leaves of the oak, which it pulls downward and backward. Their design is thus to provide concealment against the enemies which prey upon them.

The **Oak Eggar** (*Lasiocampis quercus*) as a larva, hibernates. The male is chocolate-colored, with a yellow band in the centre, and a white spot on the wings; the female is a pale yellow.

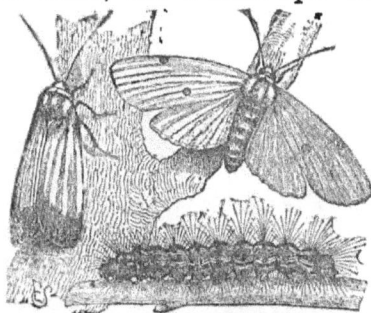

BEECH SPINNER (*Stauropus fagi*).

SQUARE, VELVET-MOTH AND CATERPILLAR (*Lithosia quadra*).

Beech Spinner (*Stauropus fagi*) frequents herbage, beeches and oaks.

The **Silk-Worm** (*Bombyx mori*) was brought by monks to Constantinople about the middle of the sixth century. Before that time silk was imported in small quantities and at great cost from China, where the silk worm had been domesticated as early as 2700 B. C. For a long period the Greeks were the special European cultivators of the *silkworm*, until Italy wrested away its sceptre. From Italy, silk culture spread to Spain and France; in the latter country the production of the eggs is a large and well-recognized industry. The English have found their Indian possessions the most satisfactory

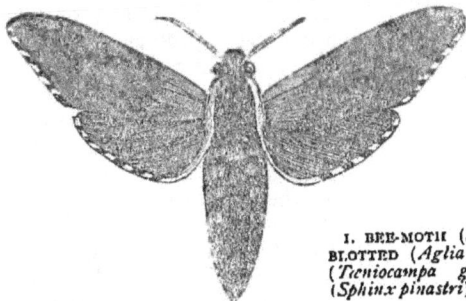

1. BEE-MOTH (*Sesia apiformis*). 9. NAIL-BLOTTED (*Aglia tau*). 15. CLIMBING-MOTH (*Tæniocampa gothica*). 2. PINE-MOTH (*Sphinx pinastri*).

place for their silk manufacture, and the Americans, while far behind their foreign competitors, yet have attained a reasonable success.

Beginning with the egg, which at first is yellow, and later blue, the worm, when hatched, is black. At the end of eleven or twelve days the worm sheds its skin, or rather peels itself, and appears as a white caterpillar.

During exuviation, the larva changes its skin five times. It soon prepares to make its cocoon. On each side of the body are two slight vessels which secrete a yellow gum, which is drawn through the mouth in the form of threads. First making a foundation of irregular threads, it constructs upon

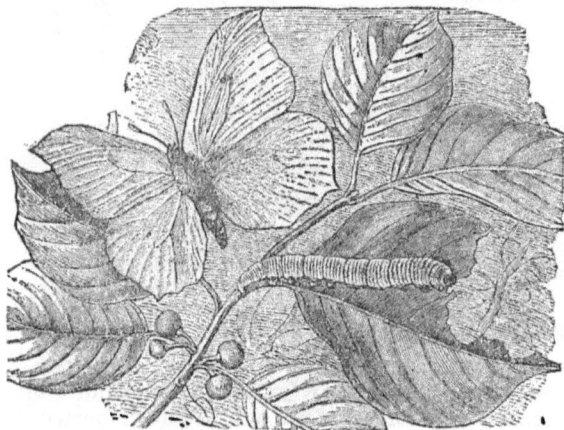

LEMON BUTTERFLY (*Rhodocera rhamni*).

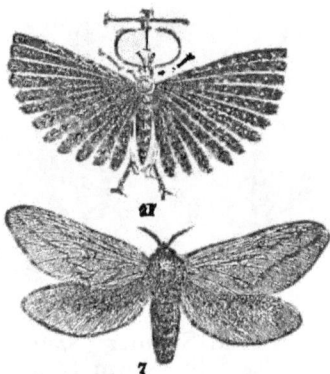

21. TWELVE-FEATHERED MOTH (*Alucita hexadactyla*). 7. HOP SPINNER (*Hepialus humuli*).

these a loose network of floss silk; finally it makes a firm yellow ball, the outside of which is smeared as a protection against the weather. After completing its task, it throws off its last skin and is metamorphosed into a chrysalis, which, at the end of a month, bursts forth as a moth, whose span of life does not exceed a few days.

More fertile than rabbits, their families vary from two hundred to five hundred. They have stout, hairy bodies; broad wings, which frequently are brilliant in coloring; lay upwards of three hundred eggs, and will feed upon the Osage orange equally as well as upon the mulberry.

OAK LEAF (*Lasiocampa quercifolia*).

The Emperor Julius Cæsar was a versatile and able man; in oratory, inferior to Cicero alone; in military affairs, still supreme; in elegance and trustworthiness as an historian, unsurpassed; in political skill and forecast, without rival. Still, to the modern inheritor of the conquests of the past, Cæsar appeals rather by his serviceableness, and among his benefits must be included not merely the reformation of the calendar, but the introduction of silk to the attention of the Romans, and consequently to the modern world. Fond of display, not so much for its own sake as for its political effect upon a wonder-loving, amazement-seeking people, he, at a cost which would have been incurred by no one but a Roman or an American, had constructed a silken tent, when the expensiveness of silk was proportionate to its rarity. This fabric was to become a necessity of human life, and to furnish employment for the inventor, the mechanic, the busy brain of the designer, the unresting hand of the weaver, the various employes of commerce, transportation, the retail dry goods, millinery, and haberdashery callings, while it opened new avenues to those living and laboring upon the farm.

As long ago as 1848, our importations of silk amounted to over five millions of dollars, with all the romance and life-sustaining power which a dollar, circulated from hand to hand, and from clime to clime, implies. Surely, the small must not be confounded with the insignificant, nor must one mistake his want of acquaintance with the populous and varied world of insects, for any want of claim on its part to his most serious attention. An oak-feeding *silk-worm* has been imported and promises well. The *cecropia* will feed upon the more common fruit and shade trees. The *ailanthus* has also been imported, and has lessened the disfavor of the tree

LEAF BUTTERFLY (*Kallima paralecta*).

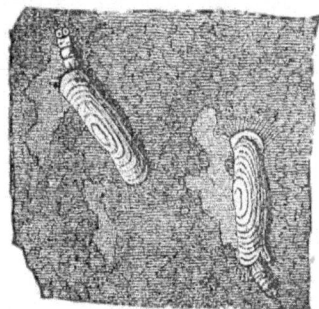

upon which it feeds. It may be remarked, in passing, that the American *silk-worm* proves a greater source of profit than the naturalized moths. Sorghum is invaded by the *Nola sorghiella*.

LARVÆ OF CLOTHES MOTH. CLOTHES MOTH (*Tinea sarcitella*).

The **Humming-Bird Hawk-Moth** is variously colored; its front wings are ashy-brown with three black wave-like lines; the hind wings are of a rusty yellow.

The **Grain Butterfly** (or *Tinea granella*) is troublesome alike while the grain is standing in the field and after it has been stored in the barn.

The **Melon Caterpillar** has iridescent white wings, with a broad band of purple or black on the outer margin.

The **Meal-Moth**, or **Flower-Moth** (*Asopia farinalis*), has front wings of yellowish-brown, bordered by chocolate-colored edges; the other wings are smoky-brown, marked by two irregular white lines and a row of spots along the edges of the wings. It is a frequent visitor to our supplies of meal, flour, etc.

The **Canker-Worm** is greenish-yellow or brown, and has pale stripes on its sides; their coloring enables them to imitate the twig so closely as often to escape detection. It destroys for food the leaves of fruit and shade trees.

The **Noctuidæ** turn night into day. The *erebus odora* is the largest American species. It is of a dark brown, speckled with gray, and its outspread wings measure six inches.

The **Cotton-Worm Moth** (*Aletia argillacea*), though found throughout the United States, is destructive especially in the cotton-belt country. In spite of its gorgeous coloring (light brown, variedly olive and claret), it is a constant enemy, and does at least thirty million dollars' worth of damage in the years when its ravages are greatest.

The **Single-dotted Sunlover** (*Heliophila unipuncta*) is another army-worm whose devastations are formidable.

The **Io Moth** is a large, showy insect; the males are distinguished by two purplish lines crossing, wave-like, a deep yellow; the hind wings are purple next to the body and at their extremities. The females have wings of purplish brown less distinctly marked by lines. The tobacco-worm moth has gray hind wings, black-spotted at the base, and with two black bands in the middle and a grayish-black margin; the front wings are white spotted and gray, lined in black. The *vanessa io* is a European moth admirable for its rich and varied coloring. Its reddish wings have eye-spots of mingled yellow,

SILK WORM (NAT. SIZE), COCOON, BUTTERFLY AND PERFECT CATERPILLAR.

white, black, lilac and rose—combinations which produce a very beautiful effect, scarcely imitable by the painter's brush.

ANTS AND BEES.

The highest species of insect life is found in the order which embraces ants, wasps and bees, and with them we find those marvellous instincts which seem to be only unconscious reason. They have been used to point the advice of the moralist, and to inspire the song of the poet; they have multiplied the pleasures and occupations of mankind; they attract by their commonness, interest by their intelligence, and illustrate with exceptional power the wisdom and providence of the Creator. To the students of evolution they offer unusual opportunities; to the lover of color, form, and to ingenious mechanism they furnish ever-fresh pleasure; to the superstitious they have been the source of many a legend, and to the husbandman and man of commerce they are an object of concern and of interest.

The *Aculeatæ* include ants, wasps and bees. The ants are called *Formicidæ ;* their eggs are elongated and yellowish white, and require about a month for being hatched. The larvæ are little grubs, conical-shaped and without legs. They are classified by the nurses and changed from room to room as their necessities demand. In a few months the larvæ develop into pupæ, and after three or four weeks into ants. The classes among ants have been thus described: "The workers are without wings and ocelli, and the thorax is narrow. The males and females have wings and ocelli—the female shedding her wings after pairing. The female has but six segments to the abdomen, the male, seven. It is believed that at the pleasure of the ants a given egg may be developed into a queen or into a worker. Ants build in different styles and of various material; they live in a single dwelling, or form settlements; they dig tunnels and construct roads. In some parts of Europe they are protected by law because of their destruction of other insects."

Ants make their toilets either unaided or with the assistance of others, which use their legs as scrapers; they teach not only the lesson of industry, but that of cleanliness in the person and in the dwelling. They never suppose that "out of sight is out of mind," and instead of allowing refuse to accumulate in their streets and alleys, at once remove it to a distance from which it cannot possibly poison their lives.

The illustration on page 271 shows the ants closing the entrance to their dwelling, immuring their queen.

Military affairs display no special resort to strategy, but the formation of lines of battle, and the hand-to-

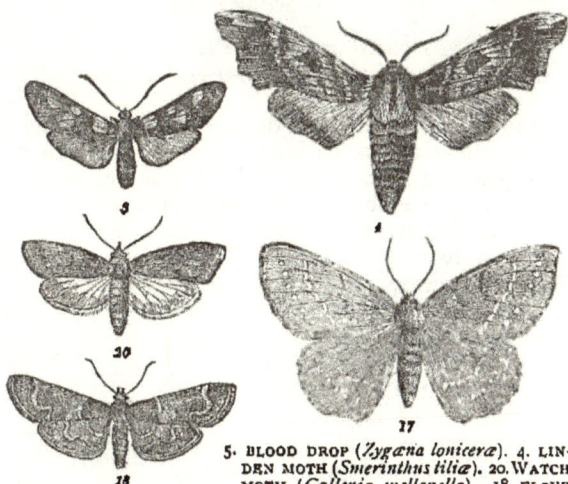

5. BLOOD DROP (*Zygæna loniceræ*). 4. LINDEN MOTH (*Smerinthus tiliæ*). 20. WATCH MOTH (*Gallenia mellonella*). 18. FLOUR MOTH (*Asopea farnialis*). 17. GREEN LEAF (*Geometra papilionaria*).

hand contests in which attacking ants engage, show a martial ardor, courage, and a knowledge of the fact that where discipline is equal, the victory will fall to the most numerous. Their movements in battle are wonderful to see, manifesting precision, strategy, and great courage, from which great generals might learn much contributive to the science of war. When the clash of battle occurs, however, ants are no longer regardful of the importance of concentrating their forces for simultaneous action, but follow up the fight in a now apparently disorganized body, each one acting independently. We therefore see them fighting in couples, while others are running off with the captives they have made. These battles are always caused by the invasion of an ant hill, being the result of piratical enterprises, in which the invaders despoil their enemies for plunder and slaves, the latter being taken away in the larval state and raised up for servitude under the direction of females appointed for their training.

The ant, from its commonness, and the industry, intelligence and persistence which it exhibits, has long been the theme of writers. Still, as some of its supposed characteristics are mythical, it may be well to refer to peculiarities which it undoubtedly possesses. In the first place, Sir John Lubbock has established the fact that the ant dislikes many colors of the spectrum, avoiding violet, green and red. There would seem to be no evidence of the ant's possessing the sense of hearing. It hunts by scent, not sight; it has remarkable powers of memory, especially in regard to direction. It would seem to be able to indulge in at least rudimentary reflection, as may be seen from the following anecdote: An ant having found an insect pinned to the ground, returned to its hill and again came forth attended by others. The insect having been removed, and therefore not to be found, it was again placed in the way of the ant, and the ant solicited the aid of its fellows; these would go no farther than to the entrance of their hill, evidently meaning to profit by their disappointing experience. Furthermore, ants have been proved to be able to recognize a member of their own family even when the stranger had been taken away in its larval period, and returned only after reaching maturity.

PHRYGANEA, LARVA AND ADULT (*Phryganea striata*).

The ant, however, displays curiosity, rather than sympathy, at any misfortune which may befall his fellows. Ants educate their young; they understand the economic value of the division of labor, so that they separate themselves into workers, soldiers, kings and queens; they go to war, make slaves, keep cows (in the shape of certain aphides); attend to their toilets; build cemeteries and fortify their dwellings. The *formica refaseas* enslaves the black ant, the *formica sanguinea*, for household purposes; in Brazil they use the leaf-bugs as beasts of burden; they make pets especially of some kinds of beetles. They cleanse themselves and each other by the use of their fore legs; they believe that life is not intended to be all work and no play, so they engage in wrestling, in games of hide and seek, and in other forms of active amusement. They build their cemeteries at a distance from their hills, though at times, if a stream pass through their dwelling, they give their dead the burial of a sailor. Among the builders is the *formica cæspitum*, which is a small black ant.

CEMETERY OF THE ANTS.

Selecting by preference a stone or a tuft of grass, these ants hollow out a series of rooms with connecting galleries. They pile up earth, using the stems of the plant for supports, and being dependent upon water as a cement; in the absence of this they construct their buildings upon the principle of the arch. Preferring a circular form, these ants yield to necessity and use other shapes. The main rooms are given up to the pupæ; others, dug below, are devoted to the service of eggs and cocoons. The yellow ants occupy numberless cells communicating with each other, and seem to pass much of their time in attending to their larvæ; they do not at all object to the presence of the wood-louse,

LARVÆ OF THE ANT.

which seems to be a welcome guest. The brown ant (*formica brunnea*) is the most fastidious of builders, executing its work with the minute attention of the maker of mosaics, rather than with the larger care of the worker in brick and stone. They adapt their buildings to surrounding conditions, and thus teach

ANTS FIGHTING.

man the folly of erecting pagodas under a winter sky. They support their roofs by columns which in perfect mechanism and mechanical beauty rival the Ionic, Doric and Corinthian columns of human art. Our modern thirteen-storied buildings are outdone by theirs of forty stories. They build in equal proportions toward the sky above and the water under the earth, for their object seems to be to secure the nicest gradations of temperature, so that their young shall lack nothing for their most perfect development. Not secreting wax, as does the bee or the wasp, the ant builds during rainy or humid seasons, so that the earth is thus united in the crevices and subsequently dried by the sun.

Furthermore, the ant seemingly never begins its work without reflection, for it always avails itself of accidental aids, such as fragments of plants placed so as to assist in forming a roof, or materials making a good beginning for vertical walls. The common ant, or pismire (*formica rufa*), is generally of a rusty brown, relieved on the head by black. The building materials it uses is infinitely varied. Beginning with a cavity in the earth, they first cover the entrance; they next mine out halls and galleries which inter-communicate, and on

POLICE ANT MAKING AN ARREST.

ceasing the work of each day close the passages, so that love's labor shall not be lost.

The *termite ant* is not a true ant, but has derived its name from its apparent resemblance to the ant family, and from their living in mounds or hills.

It makes its appearance as early in time as what the geologists call the coal-measures, and the *termite* of to-day has a well-authenticated genealogy, which puts to the blush those who date their family from so recent a period as the Norman Conquest. The *termite* is mostly an inhabitant of the tropics, but some species, commonly called white ants, are to be found everywhere, and inspire well-founded terror in the housekeeper and gardener. It mines its way through the world, and like the human miner, leaves nothing of the furniture or plants attacked except the excavation, or at best a thin covering which conceals the dangerous tunnel within. In India the ant has been known to thus mine the legs of tables, chairs and beds, and not to be discovered until the strength of materials was insufficient to support the customary weight. One species of the *termite ant* builds in trees or in the roofs of houses. The fixtures of a house become one of

WARRIOR TERMITES (*Formes bellicosus*), SOLDIER, WORKMAN, MALE, AND FEMALE SWOLLEN WITH EGGS.

its most transient features, for the ant, by working from within outward, cunningly conceals his depredations. Though not generally addicted to literature, they will verify the statement of England's Lord High Chancellor, that some books are to be tasted, others to be digested.

The nests of the *termite ant* are so many in number, and so great in size, as to produce the effect of villages of barbarous peoples. In form like a sugar-loaf (conical), ten to twenty feet in height, covered by a triple-roofed dome, which is strong enough to support an ox. Within there are numerous apartments; first, the royal chambers for the king and queen, then countless apartments used as nurseries; yet again others which serve as magazines for storing food; and finally quarters for the soldiers and quarters for the laborers. At the beginning of the rainy season the winged ants appear, and from their number each colony selects its king and queen; the rejected suitors, having no further mission in life, soon perish of neglect. The new king and queen are now immured in a cell, the entrance to which is too small to permit of their egress. Everything is theirs but freedom, for the wily ants do not trust their fortunes to the caprices of their monarchs, nor even to the accidents of parliamentary action, but act upon the belief that " fast bind, fast find." The queen lays at the rate of eighty thousand eggs a day, and the eggs, as fast as laid, are removed by the attendants to the incubator; after the eggs are hatched the young are brought up on the English nursery plan, though as soon as able they are compelled to join the workers.

One species rolls itself into a ball as a protection against its enemy; another emits an odor; still others display the tenacity of the bull-dog. Some build underground, some partly underground, and others in trees; the last tile their roofs. They lay out roads, sometimes make an arched covered way, excavate tunnels, and display great skill as architects and builders. They display the loftier and more ignoble passions of man, for some species do not dishonor a conquered foe, and others maltreat their foes after their death; some species enslave others; some are brave and pugnacious, others cowardly and thievish; some go forth singly, others move only in troops.

IMMURING THE QUEEN.

The investigation of ant-life has furnished a fresh illustration of the fact, that even though for a time the conclusions of the naturalist are different from the statements of Holy Writ, a larger knowledge will show the naturalist to be in error. Some species of the *ant* do not store up food for winter, but there are other species, such as the *Texas ant*, which has shown that the Bible is no less inspired in matters of entomology, than in those of revealed religion.

WARRIOR ANTS MAKING A COVERED WAY.

The bees, wasps, ichneumon-flies, gall-flies, saw-flies, like the *ant*, have mouths adapted to the double office of biting and sucking. They have two pairs of wings, which are nearly unveined membranes. The female is crowned with a sting, which sometimes is saw-like. These insects undergo complete metamorphosis. The different species recognize the needs of the prospective larvæ and provide them with food, or special defence, if their nature requires such provision. They never mistake the conditions necessary for the development of their larvæ; they select with the most unerring certainty the plant, or tree, or egg, or larva suited to the fertilization of their eggs and the prosperity of the yet unborn young; they provide, when necessary, for the proper nursery care and education of their offspring; they recognize the convenience of a division of labor, and waste no time in abortive attempts to change the lot assigned to them; they display statecraft, a knowledge of social economy, the qualities of thrift, and useful activity; are also acquainted with the arts of the architect, the builder, the mining engineer, the upholsterer, the house-decorator, the paper-maker, the embalmer, the soldier. They possess the senses of sight, hearing, smell; they are able to communicate with each other; they distinguish colors; they are able after the most lengthened separation to recognize members of their own community, even when this embraces a million of individuals and the returning prodigal left while still in the pupal stage.

The **Humble-Bee** or **Bumble-Bee** (*Bombas*) is found everywhere. It was imported into Australia as a fertilizer. The queens hibernate, and in the spring found new colonies. She surrounds the eggs with a mixture of pollen and honey, upon which the larvæ are to feed. At maturity the larvæ spin a silken cell, which the mother encases in wax, and are then metamorphosed into pupæ. The workers first reach full development, and at once relieve the parent of the care of providing for the family. In building, the *bee* uses such material as is most easily obtained, and shows great ingenuity in adapting the form of her dwelling to the necessities of the case.

The **Honey-Bee** is the most important member of the bee family, and is deserving of the popular interest in which it is held, since its product yields fifteen millions of dollars annually, and is rapidly increasing.

TERMITE HILLS.

The *bees* are easily distinguished as males, females, drones and workers. The male bee has at least one-fifth of its head occupied by large eyes, and the hind legs are slender at the base, and widen as they approach the top. The female and worker bees have elongated and lateral eyes, not joined together; the wings have a marginal cell; the hind legs are not spined. The queen is immured in a separate cell and waited upon by the workers, who omit no attention calculated to add to her comfort and pleasure. Her eggs, as fast as laid, are carried away by watchful attendants and placed in other cells. As there is need for change in the conditions of incubation, the eggs are shifted to chambers selected with reference to the temperature required, and no hot-house gardener is more intelligent or more watchful. As soon as a new queen is born the old queen, with a few devoted attendants, takes her flight and colonizes elsewhere. The bees seem to have the power of developing a worker, or sterile female, into a queen. The drones are produced not by the queen, but by the unfertilized queen bees, or by the workers. The wax-workers and the nurses are the two divisions of the working bee. The former is the builder and architect, as well as the one who provides honey for the honeycomb; honey thus stored up is not touched by the bees themselves, other receivers being provided for the supplies required for daily consumption., The wax-cells lie in a second stomach, which is never used for digestion. In the same manner as the various elements of human food are directed to the tissues, bones, muscles, etc., materials of which the bee can manufacture wax are separated and carried to the second stomach. It would be superfluous to dwell upon the sound mathematical and architectural princi-ples recognized in the construction of the comb and cells; this wonder of the natural world has already been described at length by naturalists, moralists, and essay-writers. It sup-plies likewise a viscous, resinous, thread-like cement, which it employs mainly as the mason employs mor-tar; but he also converts it into ropes, by which it may remove in-truding insects or troublesome sub-stances, and, as in the case of snails, uses it to safely tie an animal which it cannot remove; in the last case they embalm the corpse, not for the reason of the ancient Egyptians, but that its putrefaction may not embar-rass themselves.

The form, color, scent and suc-culence of flowers is due entirely to the agency of insects; without these workers, among whom the bee is chief, we should lose the delights furnished by plant life—the coloring, varied in hue and infinite in tone— the graceful and odd-shaped forms— the grateful perfume—the iridescence which represents the imprisoned sun-shine. The gum, out of which the bee makes its cement, is collected and carried in basket-like hollows in the middle of the hind legs. When a hive is about to be built, the bees

DWELLINGS OF THE GRASS ANT.

divide into sets of workers, whose labor is not shared by the other sets. One body confines itself to the production of materials; another works out the build-ing plan in the rough; another examines, improves and perfects the structure. All the while another band is occupied in bringing provisions to the laborers, though they exclude such as, having their field of labor abroad, can forage for themselves. The wax-workers suspend themselves in clusters, the lower ones fastening themselves to the legs of those above. Finally the " founder bee " chews and fastens his wax to a support and roughly marks out a cell; he then drops out of the crowd and is succeeded by one after another of the wax-workers. The form of the cell insures the greatest strength, together with the least waste of space or materials. A celebrated mathematician worked out the size of the angles made, by a six-sided figure whose base is a concave triangular pyramid, and it was found to be that of the angles instinctively made by the bee. The males, being larger, are provided with enlarged cells specially constructed for their use. The out-door workers carry a honey-bag, and are faithful in delivering at the hive the honey collected; some cells are reserved wholly for storage. The eggs, having been enclosed in a cell, are, as has been said, looked after by the nurse-bees until the pupa appears; this is fed for six days and then closed up in a cell, whose walls it at once begins to cover with tapestry, after which it is transformed into a bee, opens its cell and joins the older bees, who cleanse it, and put its room in order for another occupant. The young bee

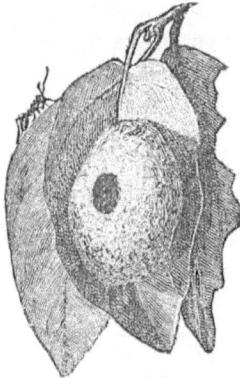

NEST OF THE POLYSTRACTIS.

now begins the labor of life, and seems to regard work, not as a curse, but as a blessing perverted by man. The old bees now prepare to move them out and the young to leave the ancestral home. After the drones have fulfilled their special mission the working-bees fall upon them and slay them.

The *mason-bee* builds against a wall and uses an earthen mortar; the *ground-bee* builds in the earth; the *leaf-cutter* uses leaves, which it varnishes red; the *wall-bee* builds a silken habitation in crevices; the *carding-bee* arranges itself in columns and passes from hand to hand (or from foot to foot) the moss which it uses in building.

The **Wasp** is too pronounced a friend of our youth to be forgotten; whatever may have been the stings of our outrageous fortune, the pain has passed and we can afford to consider the peculiarities and aptitudes of our former enemy, who always obeyed the injunction " when found, stick a pin in it." The *wasp*

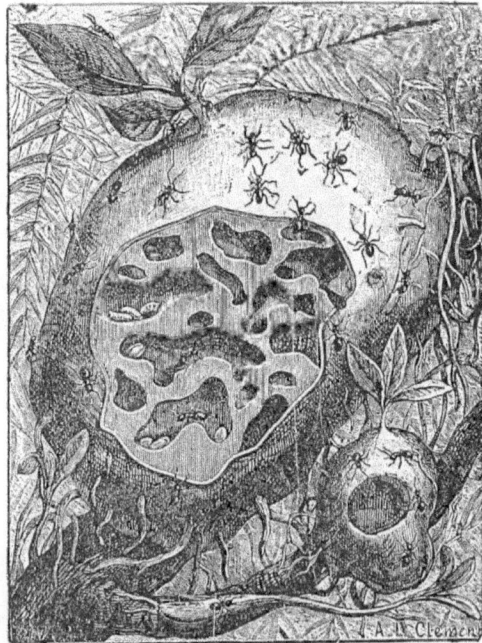

TREE NEST OF THE MYOMECODIA.

masticates bits of bark until transformed into veritable paper pulp. Making a pillar in the highest part of the vault, they next provide a roof. As soon as the cells are ready the *wasp* begins to lay, and when the larvæ appear they are fed by the parent. About the close of the larval period, the young line their cells with silk, and then undergo their metamorphoses into *wasps*. Unlike the *bees*, they frequently support one or two hundred queens. The *wasp*, like the bear, is fond of honey, which, however, he can obtain only by the plunder of bees. As supplies diminish the *wasp* puts to death the old, the infirm, the useless, and destroys larvæ which have not arrived in time but been born out of season. The *mason-wasp* resembles the mason-bee in habits. It uses its feet as trowels and fastens and seals its door. The *carpenter-wasp* constructs his home in timber; the *tree-wasp* deserts the ground; there are species

DWELLINGS OF THE TAPINOMA.

which are parasitic; some that are solitary. The *wasp* feeds itself upon the choicest fruit, and thus robs the nurseryman of his legitimate reward; it seizes with avidity upon meat, and thus annoys the butcher and the house-keeper; it does not hesitate to use its sting, and thus threatens the comfort of persons.

We have now passed in review many of the more striking insects; we have seen their intelligence, industry, adaptation to the demands of their life, metamorphoses, and must have been convinced of the constant exercise of a wisdom not finite, and of

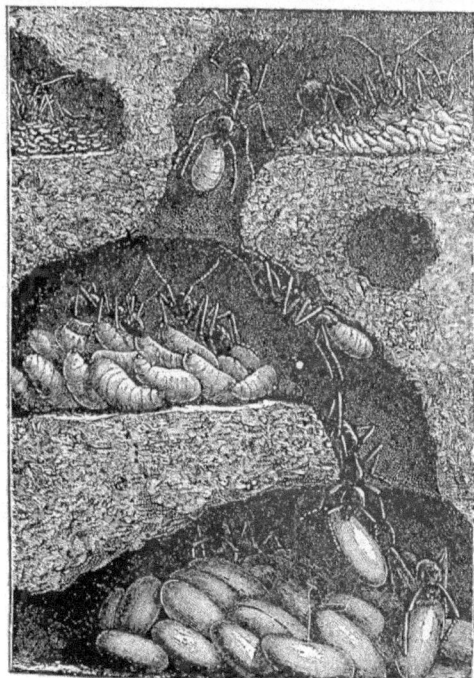

INTERIOR OF ANT NEST.

a charity larger than that which belongs to man. For such insects as are dependent thereupon, the maternal instinct is strongly developed in the parent; for those which in the larval condition require provision of food, food is supplied; for others it is not.

AVES–BIRDS.

CONNECTION BETWEEN BIRDS AND REPTILES.

FTER a consideration of reptilian life, which comprises the most dangerous and repulsive of creatures, doomed largely to crawl in the dust at the feet of their implacable foe since the gates of Eden were shut against mankind, we approach the next order of animate creation only to find a transition most remarkable, a true antithesis, yet connected by a chain in which none of the links are missing. How seemingly great is the chasm lying between the sinuous creeping, poisonous reptiles, and the bright plumaged warblers, that go whirling through space, with every dip of wing a symphony of grace, while filling the air with a harmony that makes every wood musical, and every hedge an orchestra. How great the difference between the scale of the saurian and the gossamer, sun-tinted feather of the bird; or the threatening hiss, croak or bellow of one, and the lute-like notes, the piping trill of woodland eloquence flooding the very world with melody, that is flung with generous praise from the throat of the other! Though these may be ever so widely divergent, the lines of separation are like an acute angle, the ends of which are far apart but converge again to a common centre.

The characteristic features by which the three great divisions of animals, viz., reptiles, birds and mammals, are distinguished, are scales, feathers and hairs. Other points of difference exist in that feathered creatures possess a double circulation of the blood, which is single in reptiles, while the absence of a diaphragm and mammary glands, as well as the possession of a second stomach, the crop, a grinding organ and the gizzard, serve to separate birds from mammals. This separation from mammals is very much more distinct than are the characteristics which divide birds from reptiles. While the differences are strongly marked, there still remain several homologies that serve to unite birds with reptiles. In speaking of these connecting links, the English naturalist, Kingsley, thus writes:

"* * * In these and other particulars, the birds show a near relationship to reptiles, so close, indeed, that they have been included with them in a separate group, called *sauropsida;* at any rate, birds are more nearly related to reptiles than they are to mammals, notwithstanding the beak of the duck-mole and the recent re-discovery of the fact that the echidna (*porcupine ant-eater*) lays eggs, and whatever was the origin of mammals, so much is certain, that they sprang from an ancestral stock with which birds are only remotely connected. Their position between reptiles and mammals in our linear system does not indicate any intermediate position in nature, but is simply due to our inability of expressing exact relationships.

"There are other features which frequently are attributed to the bird class as diagnostic, but which really are but of little account; for instance, the modification of the jaws into a beak sheathed with horn and destitute of teeth, for not only have the turtles and duck-moles similar beaks, but we know now that teeth were common in certain groups of extinct birds as they are in reptiles or mammals nowadays. Nor is the laying of eggs and their hatching an exclusive characteristic of the feathered tribe, for we have birds which leave the hatching to be done by the heat of decaying vegetable matter heaped upon them, while the latest indications are that the old report of the monotremes (of the duck-bill species) laying eggs, hitherto regarded as a fable, is substantially true. The so-called pneumaticity (hollowness) of the bird skeleton, or the peculiarity of the bones being hollow and filled with air through the canals, in connection with the respiratory organs, has also been regarded as belonging to birds only, but the bones of the extinct *pterosaurians* (winged lizards) and some other forms were also filled with air, air canals being present in nearly all the bones of skeletons of the larger species, while several recent birds, for instance the kiwis and penguins, are entirely destitute of pneumaticity in any part of the skeleton."

THE LINK THAT CONNECTS BIRD WITH REPTILE.

Other species of extinct reptiles, such as the *pterodactyls* (wing-fingered, or wing-toed), and *ramphorynchus* (branch-beaked), might also have been cited by Mr. Kingsley as combining characteristics common to birds. The former, which was about the size of a swan, had the powers of flight highly developed, though its structure was very similar to that of the bat. The latter, while possessing membranous wings, was only able to use them as a parachute, by the extension of which it could leap from an elevation and sail a considerable distance on a descending plane. Its size was equal to that of a crow, and in

THE RAMPHORYNCHUS.

many respects its structure was decidedly bird-like. The head and beak were those of a bird, except that the bill was long and armed with teeth, a provision no longer observable in any of the feathered tribe; but the breast and back were those of a bird, save that the extension of the back was vertebrated and terminated in a tail of considerable length. The combination of bird and reptile was strikingly illustrated in this curious creature, and well entitles it to being considered as the connecting link between the two.

Though we have the strange anomaly before us of reptiles producing their young alive, and of at least two species of quadrupeds that lay and incubate

their eggs, it is an invariable law among every species of bird that their reproduction shall be by the deposition of eggs, yet, however, as seen above, all
species do not incubate.

HOW BIRDS SAIL THROUGH THE AIR.

The most striking feature of bird physiology, as showing a marvellous
creative design that impels our reverence for a wisdom so infinite, is found in
the skeletal structure and wonderful form, substance and adaptation of the
feathery covering, by which flight is rendered at once easy and natural. While
the bones of all other animals now existent are solid, the centre being a cone
of marrow, in birds there is extreme lightness without the sacrifice of strength,
by the bones being hollow and braced, as it were, by numerous osseous capillary ducts, through which a circulation of heated air is in constant passage.
We look with mingled surprise and wonder at the hawk, buzzard, and some
other species, as they go sliding through the air with only an occasional flap
of the wing, seemingly sustained by some invisible force, and propelled by an
agency that appears difficult to understand. If one of these sailing birds be
taken and submitted to certain experiments, including dissection, it will be
found that the bones are not only hollow and extremely light, but that they
are very warm to the touch.
If the end of a wing be
lopped off, and the bird's
windpipe closed, however
tightly, it may still be seen
to breathe with the same
regularity as before being
injured. By this experi-
ment we are given to know
that both heart and lungs
are directly connected with
all the bones, that the heart
action is not confined to
pumping blood, but that it
also pumps air through all
the bony channels and feath-

DIFFERENT POSITIONS ASSUMED BY BIRDS WHEN MIGRATING. a. DUCKS.
b. EUROPEAN IBIS. c. GEESE. d. CRANES. e. BUZZARDS. f. DIVERS.
g. OYSTER-FISHERS. h. SAND-PIPERS.

ers; this air is heated by the heart and then distributed through every canal,
however small, in the entire skeleton. We find also that the normal temperature of birds is ten degrees greater than in mammals, and this normal temperature may be increased by at least ten degrees more, at the will of the bird.
This produces a rarefaction of the air within the bones sufficient to give considerable buoyancy, to which sustaining effect is added the lifting power which
resides within the feathers. This force is ill understood, but that it exists is
demonstrated by the fact that to divest the body of a bird of its feathers, though
the wings may be left intact, destroys its power of flight.

In addition to the knowledge gained by experiments which are occasionally
cruel, having to be made while the bird is alive, is the striking contrast
between birds of flight and those destitute of this ability. We know that the
bones of domestic fowls are thicker and heavier than those of the wild species,
and investigation has also shown that in such birds as the penguin, ostrich, kiwi,

and others which are unable to take wing, not only is the wing imperfectly developed, but the bones are almost solid, a fact which operates to confirm the theory here advanced, or at least lends great plausibility to it.

Next to the feathers, in point of wonderful adaptation and structure, is the eye of a bird, and in this, too, we observe another analogy, or correspondence, with reptiles. In all mammals the eye is protected by a double lid, an upper and a lower, but in birds there is a third lid, technically called, the *nictitating membrane.* This so-called lid is controlled by a muscle united to it at the outer corner of the cornea, by which the membrane may be drawn quickly across the globe of the eye and back again, by which motion the eye is kept moistened and clear of foreign substances, as is done by the winking of the lids in mammals. This nictitating membrane is noticeable in the eye-structure of a large number of reptiles, but is never present in fishes, insects or mammals. By this provision the vision of a bird and of the reptiles that possess it is rendered much more acute and superior to that of man.

As classification is much a matter of individual or arbitrary arrangement —beyond the confinement of species under appropriate heads—we will preserve a consistency in the general plan of this work by introducing, not what can be called the lowest orders, for all are equally perfect, but by first describing the smaller species, and maintaining a gradual ascent from the least to the greatest, in which respect it will bear the semblance at least of progression, if not development. Under this arrangement, therefore, we will first introduce the

TROCHILIDÆ, OR HUMMING-BIRDS.

It is a matter for some wonder that *humming-birds* are confined exclusively to the Western Continent, particularly since their range of latitude is almost equal to that of North and South America. The species are very numerous, some of which may be found widely distributed, while others are restricted to a single locality less than a dozen miles in diameter. This variable distribution is due to the fact that those of widest range are swift of wing, and can sustain themselves in a long flight, while those whose habitat is circumscribed are feeble, have short wings, and are otherwise incapable of wandering any considerable distance.

Wherever found, the *humming-bird*, so called from the whirring sound produced by its wings, is a creature that excites our admiration as being comparable to an animated flower. Not all have gorgeous plumage, but the delicacy of structure, and the gracefulness with which they poise above a flower, when extracting the nectar from its chambers; the manner in which they flit from bloom to bloom, or dart off like a bee in arrowy flight, bearing away in their crops a drink upon which the gods were once believed to regale themselves, is a habit peculiar to the several species.

In the British Museum is a collection of every known species of *humming-bird*, more than four hundred in number, but the total number of specimens there shown is perhaps half a million, a very large wing of the building being entirely donated to these beautiful creatures.

Though we generally observe these birds hovering amid the flowers, honey is not their only food, for experiments have shown that none of the genus can subsist without at least an occasional diet of insects—spiders and flies being their preferable animal food. Some of the least attractive species seem to live chiefly off insects, for I have many times watched them sitting upon a perch, usually the limb of a tree, from whence they would dart off to seize passing flies, mosquitoes, or even small case-winged insects, and these species not only exhibit a carnivorous appetite for a day, but may be seen about the same spot watching for prey in this manner day after day. Some few species emit several notes, and one South American habitant is said to sing sweetly, but generally they give forth only a single low *twit*, except when fighting, when they produce a sharp twittering noise. Small as the *humming-bird* is, he is a fearless and pugnacious creature, not only quick to attack one of his kind, but equally prompt to avenge himself upon the largest birds of prey.

The male hummer is much more gaudily clothed than his more modest mate, in which respect the male of all birds excels the female. The nests which they construct are at once marvels of structure and comfort, though some are scarcely larger than a walnut shell. Their shape is usually cup-like, and lined with some very soft material, while not infrequently there is some effort at exterior ornamentation, by the tasty addition of lichens and mosses attached by cobwebs. The place selected is usually a pendant bough or the side of a large leaf, though some even select the shelf of a rock. Professor Jamison, of Quito, mentions having seen a nest attached to a piece of suspended rope, and that as the weight of the nest being upon one side threw it out of perpendicular, the bird overcame this defect by carrying stones with which it weighted the other side until the equilibrium was restored.

NEST OF THE SWORD-BEAK.

The number of species is so large that a volume would be required to describe them all, hence we can at best only introduce some of the more striking specimens of the genera, though these will be sufficient to excite anew our interest and admiration in this, at once the daintiest and prettiest of God's creatures.

The **Ruby Throat** (*Trochilus colubris*) is the most common species found within the United States, for nearly all these living sunbeams have their homes in Mexico, Central and South America. The *ruby* is therefore hardly to be compared for beauty with many of the much more charming species of the equatorial regions, but he is a most fascinating elf withal. The *ruby* rarely arrives from his southern home before the middle of April, and his coming may be looked upon as a sure sign that the chilly days are over. The name *ruby* has been given as a distinguishing mark, because of the ruby red feathers that cover his throat. The back presents a scintillating sheen of gold and iris hues, while the tail feathers are of a purplish, steel color. But the colors of all the gayest plumaged *hummers* are variable, changing with every change of position or angle of the sunlight that falls upon them.

This refulgent property of the feathers is due, as Woods says, to certain minute furrows which are traced upon the surface, and are analogous in their mode of action to the delicate lines which give to nacre (the interior of a shell) its peculiar iridescent splendor.

The *ruby throat*, though scarcely more than two inches in length, exhibits upon occasions the most reckless courage and bravado. He is up in arms against every living thing that dares to invade his sanctuary, having been frequently seen attacking a hawk or eagle with fiery fury, and while perched upon the great bird's head driving its keen little bill with savage force into its enemy's eyes and neck. Though of a peculiarly vengeful disposition towards all other birds, the *ruby* is more easily tamed than perhaps any other wild creature. When first taken it simulates death, but will soon respond to gentle treat-

RING-NECK HUMMING-BIRD (*Typhœna Duponti*).

ment, and the moment it shows any activity will begin feeding from the hand of its captor, and has been known to return for several seasons to the congenial home where it was once held a captive. The *ruby* builds its nest about the first of May and lays but two eggs, which are as many weeks in hatching. It is said that nearly all species of humming-birds raise two broods each year, but in this respect I do not think the genera differs from other birds, as all depends upon the length of the season.

Flag-tailed Sylph (*Steganurus underwoodii*) is a species whose special habitat is near the equator, but its range is as far north as Jamaica, considerable numbers being found about a district on that island known as Bluefield Ridge, where every tree is covered with creeping flowers, thus providing a great attraction for these dainty creatures. The name *flag-tail* has been given to this species because of two very long feathers that form the tail, which, however, are not so handsomely colored as the plumage on its body. The back of this bird is of an emerald sheen, the wings a dark purple, and the throat and breast a brilliant green. Its length is some ten inches, of which the tail comprises at least seven inches.

The **Sword Bill** (*Docimastes ensifer*) is found principally at extreme elevations along the equator, the region about Quito being its favorite haunt. It is provided with a bill of extraordinary length, in which we again observe a wonderful design and adaptation, as the bird draws its sustenance almost entirely from the corollas of the brugmansiæ and other

EMERALD HUMMING-BIRD (*Chlorostilbon prasinus*).

species of the trumpet flower. A bill of extreme length is therefore absolutely essential, since by no other means could it reach the nectar which lies at the bottom of the deep cups. The *sword bill* is very handsomely clothed in plumage extremely iridescent with emerald, purple and bronze that change in hues with every movement of the bird.

Crimson Topaz (*Topaza pyra*) is confined to the extreme north of South America, and is a very beautiful creature, as the name implies. The body is of a deep crimson, the main tail feathers bronze, from which issue two very long feathers of a purplish green. It is semi-nocturnal, and being also extremely shy, is not often met with.

The Topaz (*Topaza pella*) is perhaps the most gorgeously beautiful of all the genera, probably exceeding in this respect the species just described. It, too, is a habitant of northern South America, but ranges somewhat nearer the equator than the *topaza pyra*. The color is a flaming scarlet on the back, while the head and neck feathers are of a deep velvet black, presenting a most charming contrast. The throat is an emerald green, which fades to a light opaline hue towards the tail coverts, reappearing again in a bright green with an orange gloss, while two long tail feathers of a greenish-purple complete the royal wardrobe of this fairy among the flowers. Its size, being about eight inches in length, entitles it to rank among birds belonging to larger classes, except for the characteristics which bind it to the trochilus species.

The Trailing Sylph (*Sparganura*) is sometimes called the *Comet of Sappho*, and also *bartailed*. Its home is Bolivia, where it haunts the gardens and orchards through the summer, and retires to eastern Peru when the winter season appears. It is specially distinguished for the prolongation of its tail feathers, which are of a fiery red tipped with a velvet black band. The general color of the body feathers is a sheeny green, with bronze hues on the neck, and a crimson red on the lower back. The term *comet* has been applied to it on account of the extraordinary velocity of its flight and the darting, eccentric motions it frequently indulges in.

Brilliant Elf (*Lophornis ornatus*), or *tufted coquette*, is a rather rare species

GROUP OF HUMMING-BIRDS. FLAG-TAILED SYLPH, TOPAZ HUMMING-BIRD, TRAILING SYLPH, SWORD-BEAK, BRILLIANT ELF.

occasionally met with in northern Brazil, and is of most diminutive proportions, scarcely exceeding two inches in length. The crest with which its head is ornamented is a rich chestnut-brown color, the back a bronze green, wings a purple black; while a broad band of white crosses over the lower part of the back. From the neck project plumes of snowy white, the ends of which are tipped with bright green.

Spangled Coquette (*Lophornis reginæ*) is the name of another species very similar to the *lophornis ornatus*, differing little in color, but has the power of elevating or depressing the crest upon its head, which the former does not possess, a faculty which greatly increases its value as a curiosity among naturalists.

Horned Hummer, or **Double-crested** (*Trochilus cornutus*), is also a South American species, ranging between the Amazon and Parana rivers. It is small in size, but of exquisite plumage, and distinguished for having a fan-shaped tuft of parti-colored feathers growing out from either side of the head. The back feathers are iridescent with hues of burnished gold and copper, while the breast is changeable with tints of emerald and ruby, and the long tail feathers are a pale yellow.

The **Sparkling Tail** (*Trochilus dupontii*) inhabits Mexico and some parts of Central America, where it has an affection for flowers in cultivated gardens. It is extremely small, and the two eggs which it lays are scarcely larger than peas, and of a pearly white. The nest is always made upon a slight twig, or the underside of a large leaf, to which it is made to adhere by the use of a glutinous substance and spiders' webs. This tiny bird is gorgeously bedecked with bronze-green feathers on the back, and a crescent of white crossing at the base of the tail. The tail is long, swallow-shaped, and the feathers are of velvet-black, tipped with white.

Conver's Thorntail (*Goldia conversi*) is peculiar to the region about Bogota. It is also a small species remarkable for great speed on the wing. The color is green with a bar of white running across the lower back, which presents a striking contrast. The tail, however, is its most distinguishing feature, from the shape of which (resembling sharp-pointed, radiating spears) the name *thorntail* has been given. These feathers are shining black, while the shafts are white, as are also the tips.

The **Flame-Bearer** (*Selasphorus scintilla*) has several features resembling those observable in the last species described. The tail feathers are rich in color, but short and spike-like. On the throat is a tuft of feathers of a fiery red color, from whence the name has been given. But more singular than its appearance is the fact that this species inhabits the inner side of the extinct volcano Chiriqui, New Grenada, where a luxurious vegetation is now found.

Black Warrior (*Oxypogon lindenii*). This is quite a large species, and its haunts are at elevations where snow and ice may nearly always be found. They are somewhat remarkable for the long, pointed crest which gracefully swells upward from the base of the beak, and a flowing beard of delicate feathers. Concerning this interesting creature, Mr. Linden, the discoverer, says: "I met with this species for the first time in August, 1842, while ascending the Sierra Nevada de Merida, the crests of which are the most elevated of the eastern part of the Cordilleras of Columbia. It inhabits the regions immediately below the line of perpetual congelation, at an elevation of from 12,000 to 13,000 feet above sea-level. It occasionally feeds upon the thinly scattered shrubs in this icy region, but most frequently upon the projecting ledges of rocks near the snow. Its food appears principally to consist of minute insects." Another species (*O. guerenii*), very like that described by Linden, is found at the high altitudes of the Columbian Andes. The coloring of neither is very brilliant.

Avocet Hummer (*Trochilus recurvirostris*) is the name of a species inhabiting Guiana, remarkable for the shape of its bill, which turns upward at the point, like that of the Avocet, after which it is called. On account of its comparative rarity, or residence within the deepest forests, very little is known of its habits. Swainson suggests, as a cause for this singular recurvation of bill, that the bird's principal sustenance may be drawn from the *bignoniæ* and similar plants, whose corollas are long and generally bent in their tubes; the nectar, being at the bottom, could not be reached readily either by a straight or incurved bill, though very easily by one corresponding to the shape of the flower. The species is uncommon.

The Sickle Bill (*Trochilus aquila*) presents a feature

TROCHILUS ANNA, AND CURVED-BEAK OR AVOCET.

HORNED HUMMER AND SPARKLING TAIL.

equally as curious as that observed in the foregoing species. Instead of the bill turning upward, however, it makes a sharp turn downward, which gives to it a sickle shape, hence the name. Like the previous species, too, it is very rare and little understood; whenever found it has always been at an elevation of at least ten thousand feet above sea-level.

The Trouser, or Puff-Leg (*T. cupreoventris*), is also an altitudinous dweller, being rarely found below nine thousand feet, and is confined to a very narrow strip of ground in the Andean regions. It is small in size, but of a most exquisite appearance, with head, neck and back washed with shining green, and toward the tail coverts a metallic lustre; the wings are a blending of brown and purple, and the tail black, with purple gloss. The abdomen is of a coppery lustre, from which fact it is sometimes called the *copper-bellied puff-leg;* but the most curious feature about this bird, and from which the more popular name is derived, are the puffs of white down, very much resembling the toilet puff-ball, which envelop the legs and thighs, imparting a most singular appearance.

The White-Booted Racket Tail (*Spathura underwoodii*) very much

resembles the previous species, the coloring being nearly identical, and the puff-legs present, but it differs in possessing a bifurcated tail, with two greatly elongated feathers that terminate in spatulate tips. It is comparatively plentiful about Bogota. Its flight is swift and arrow-like, and on the wing it is the spirit of grace and loveliness.

Star Throat, or Angel Hummer (*T. angelus*), is found in nearly every part of Brazil, though at no place does it seem to be common. The bill of this species is very long, to enable it to reach the bottoms of the llianas flowers to extract the nectar, but it also catches insects. The plumage is very beautiful, the head being a metallic green, which changes under reflection to ultramarine, blue and gold. The back is of a golden sheen, and the wings and tail a variable purple-black and seal-brown. The gorget, or centre of the throat, is a brilliant crimson, edged with feathers tipped with blue, and on each flank is a tuft of white feathers.

The White Cap (*T. albocoronatus*) is one of the smallest of humming-birds, and being quite rare, few naturalists have had an opportunity to study its habits. Mr. Gould, who has devoted much time to an investigation of the habits of all the genera, writes thus concerning the *white cap*: "It was in the autumn of 1852, in New Granada, that I obtained several specimens of this diminutive variety of the humming-

ANGEL HUMMER, OR STAR THROAT.

WHITE CAP AND MAGNIFICENT SUN ANGEL.

bird family. The first one I saw was perched on a twig, preening his feathers. I was doubtful for a few moments whether so small an object could be a bird, but on close examination I convinced myself of the fact and secured it. Another I encountered while bathing, and for a time I watched its movements before shooting it. This little creature would poise itself about three feet or so above the surface of the water, and then as quick as thought dart downwards so as to dip its diminutive head in the placid pool, and this action it continued so rapidly that the water was kept constantly stirred."

The Magnificent, or Sun Angel (*T. magnificus*), is one of several species that are found in the Andean range, and, with all its genus, is distinguished for the royal ornamentation of its raiment. Of this particular species, Mr. Gould thus writes: "I regard this as the most beautiful and charming of the genus *heliangelos.* It has all the charms of novelty to recommend it, and it stands alone, too, among its congeners, no other member of the genus similarly colored having been discovered up to the present time. The throat vies with the radiant topaz, while the band on the forehead rivals in brilliancy the frontlet of every other species." On account of the remarkable beauty of this little feathered gem, thousands are killed every year by means of blow-guns, and their skins sent to all parts of the world and used for ornamental purposes.

BIRDS OF PARADISE.

As the humming-bird is found only on the Western. Continent, nature has compensated for this seeming partiality by giving to the eastern world a species which have received the very appropriate designation of *Birds of Paradise,* no other expression being discoverable that would convey so excellent an idea of their transcendent, yes, marvellous beauty. It may at first glance appear somewhat singular that these charming creatures should be limited to not only a small

TROUSER OR PUFF LEG, AND SCIMETAR BILL.

district, but that this circumscribed region should be at once the wildest and most inaccessible, and therefore least liable to come under the dominion of civilized man. But this seems a wise provision, when we consider with what cruelty and reckless disregard men destroy the most beautiful birds to satisfy the condemnable pride of the age. Every lady's hat must be decorated with the head or wing of some forest warbler or animated sunbeam, while the pseudo-naturalist is not less remorseless in establishing his collection; so with net, trap, blow-gun, arrow, bird-lime and shot, the destruction goes on, and every year the woods are less resonant, the flashing luminaries of lambent wings are fewer, and melody that once woke the echoes over upland, meadow and deep tangled forest is dying away, until it is now like the strains from an instrument with nearly all the strings broken.

BOLD HONEY-SUCKER.

Fortunately, therefore, the most beautiful birds have their haunts in the wildest districts of the tropical world, and it is in the unexplored interior of New Guinea that the most exquisite of sun-birds are to be found. The species, however, are not numerous, and this fact, added to the deep seclusion of their unfamiliar haunts, has prevented such study of their habits as would afford the knowledge all naturalists are anxious to possess, considering, as they do, this genera to be the most interesting of bird creation.

The **Bold Honey-Sucker** (*Nectarinia metallica*) has been classed among the sun-birds on account of its resplendent plumage, but its general characteristics are at variance with those of the genera, who appear to be confined entirely to the Malayan Peninsula and the Oceanic Islands. The *honey-sucker* is a habitant of north-eastern Africa, small of size, but gorgeously bedecked, and distinguished also for the two long feathers of the tail that trail after the flying bird like parallel rays of iridescence. The head, neck and mantle are of a refulgent green, the rump and throat a purplish-blue and the belly a reddish-yellow.

BEAUTIFUL, OR LEGLESS BIRD OF PARADISE.

Beautiful, or Legless Bird of Paradise (*Paradisea apoda*), is found only in the Malay Archipelago, and no intelligent observer has reported it outside of New Guinea. The name *legless* was applied by the earliest voyagers to these still comparatively unknown shores out of respect for the belief among the neighboring islanders that the sunbirds spent all their time in the air. This idea was further perpetuated by Malay traders, and also by the Portuguese, who frequently bartered with the natives for the skins of these birds, to which the legs were never attached. They therefore called them *birds of the sun.* John von Linschoten, a learned Dutch naturalist, gave to them the name *Paradise birds* in 1598, and in writing of them says: "No one has seen these birds alive, for they live in the air, always turning toward the sun, and never lighting on the earth until they die; for they have neither feet nor wings, as may be seen by the birds carried to India, and sometimes to Holland."

About the year 1700, the historiographer of the Dampier Expedition, William Funnel, mentions having seen specimens of these birds at Amboyna, and was told by the natives that they came to Banda among the nutmeg groves, where they, feasting off the fruit until becoming intoxicated, would fall senseless to the ground and expire.

PARADISE WIDOW AND BEARDED FINCH (*Amadina fascinata*).

The species *P. apoda* was named by Linnæus in 1758, and he seems to have cherished the tradition that the bird was without legs, as the name implies. More recent and reliable knowledge concerning the species has been obtained by Wallace, Rosenberg, D'Albertis and Beccari. We now know that the *P. apoda* is the largest of the sun-birds, measuring in length of body about eighteen inches. The color of the back is a rich seal-brown, running into a velvet violet on the breast. The head and neck are of a bright straw-color, while the throat is of an emerald green extending to the eyes. A marvellously beautiful velvety band of deep green extends across the forehead, with ends resting over the eye tipped with yellow. Beautiful as are these blending colors, the chief glory of this magnificent bird is in its enormous tail, which is fully thirty-six inches long. The feathers with which it is so royally endowed spring from both sides of the body in a graceful sweep and fall in delicate gossamer of golden orange, with tips of pale brown, presenting the appearance of a shower of gold flakes.

Paradise Widow (*Vidua paradisea*) is a habitant of Western Africa, and not nearly so large as the preceding species. On account of its splendid plumage the name *incomparable bird* has also been given it. The body decoration is most lustrous, with sheens of deep brown, black, and buff and white on the abdomen. Like the former species, the tail constitutes the most magnificent feature about it, though unlike the former the tail consists of only two scimetar-shaped feathers that are nearly two feet long, while above these shafts are two other shorter,

WIDDAH FINCH, OR WIDOW-BIRD.

paddle-shaped feathers, the ends of which are denuded, the points of the quills projecting. The *widow-bird* builds a curious nest of vegetable fibres of a downy softness, which is divided into two compartments, in one of which the male sits on guard while the female occupies the other with her eggs or brood.

The **Shaft-tailed Widow,** or **Widdah Bird** (*Vidua reglia*), is another African species similar in appearance to the former, though not so large, and the tail, instead of having two broad feathers, is ornamented with four very slender shafts springing as central feathers from the true tail. Both species are frequently seen in European aviaries.

The **Lyre Bird** (*Menura superba*) is commonly classed among the wrens, but a glance at the illustration will serve to convince the reader that it belongs to the

19

paradise species, since appearance, rather than habits, is considered in the classification of the *paradisea* genera. This magnificent bird is native to New South Wales, and its range is restricted to the region lying between Port Phillip and Moreton Bay, where it is said to be fairly plentiful. Of this elegant bird Mr. Bennett writes: "I first saw these birds in the mountain range of the Tumat county; lately they have been very abundant among the Blue Mountain ranges thirty-five miles from Sidney. They are remarkably shy, very difficult to approach, frequenting the most inaccessible rocks and gullies, and on the slightest disturbance they dart off with surprising swiftness through the brakes, carrying their tails horizontally; but this appears to be for facilitating their passage through the brush, for when they leap or spring from branch to branch as they ascend or descend a tree the tail approaches the perpendicular. On watching them from an elevated position, playing in a gully below, they are seen to form little hillocks or mounds, by scratching up the ground around them, trampling and running flightily about, uttering their loud, shrill call, and imitating the notes of various birds."

The nest of the *lyrebird* is a large, loosely built, domed structure, composed of sticks, roots and leaves interwoven, and is oven-shaped, with the entrance in front. The lining is of down or other soft substance. A peculiarity of this bird is found in its reluctance to take flight, for though perhaps the most timid of birds, it trusts almost entirely to its legs to convey it beyond danger. As a

LYRE BIRD.

runner it probably has no equal, and as its haunts are in the densest coverts, into which it can plunge out of sight almost upon the instant, a sight of the bird is not easily gained. The color of this bird is dull and by no means inviting, but the shape of its tail is magnificently beautiful, resembling nothing so much as a lyre, from whence the name is taken. Though not a songster, it has the power of imitation very greatly developed, being able not only to mimic birds, but certain animals also.

Bird of the Gods (*Epimachus magnus*) is the name given to one of the most superbly beautiful paradise species inhabiting New Guinea. The body is small, but the tail feathers are so long that from point of beak to tip of tail the length is four feet. To quote from Mr. Lesson: "To add to the singularity of this bird, nature has placed above and below its wings feathers of an extraordinary form, and such as one does not see in other birds; she seems, moreover, to have pleased herself in painting this being, already so singular, with her most brilliant colors. The head, neck and belly are glittering green; the feathers which cover these parts possess the lustre and softness of velvet to the eye and touch; the back is changeable violet; the wings are of the same color, and appear, according to the lights in which they are held, blue, violet, or deep black, always, however, imitating velvet. The tail is composed of twelve feathers, the two middle ones the longest, and the lateral feathers gradually diminish; it is violet or changeable blue above and black beneath. These feathers shine with the brilliancy of polished metal. "The feathers above the wings are of the color of polished steel, changing into blue, terminated by a large spot of brilliant green, and forming a species of tuft or appendage at the margin of the wings. Below the wings spring long curved feathers, directed upward; these are black on the inside and brilliant green on the outside. The bill and feet are black."

Twelve-thread Plume Bird (*E. albus*) bears some resemblance to the previous species, especially from the shoulder of the wings forward, and inhabits the same region. The body color is rich violet, and the collar feathers, forming a ruff, are emerald green. In this species the tail is short and fluffy, from the downy feathers of which spring twelve thread-like shafts, from which peculiarity the name has been given. Occasional specimens of this bird have been found that were a snowy white.

BIRDS OF PARADISE. BIRD OF THE GODS (*Paradisea apoda*). KINGLY BIRD OF PARADISE (*Cicinnurus regius*). RADIANT CROWNED PARADISE (*Parotia sefilata*).

Superb Bird of Paradise (*Paradisea superba*). This species carries its most distinguishing feature on the shoulder, instead of about the tail, as in those previously described. The scapulary feathers are so developed as to form a bifurcated arch, which may be raised at the will of the bird above the head, imparting a strange umbrella-like appearance, the grotesqueness of which is much increased by a corresponding forked tuft of feathers springing out from the throat. The general color of this bird is a very deep violet shot with green, which is very reflective of the sun's rays, being almost prismatic in its bright refulgence.

Red Bird of Paradise (*P. rubra*). This bird is never found on the mainland of New Guinea, though pretty common on the smaller islands in close proximity. Its plume is hardly so rich as that of the preceding species, but it possesses, nevertheless, much beauty, the chief point of ornamentation

ARBOR OF THE SPOTTED BOWER BIRD.

being the tail. It is a plump bird, corresponding in size to a dove, and has a bold expression like that of a bantam cock. On the crown is a double crest of deep green and of a velvety softness. The head, neck and back feathers are of a beautiful golden yellow, tinged with carmine on the edges. The wings and chest are of a warm chocolate brown, in most pleasing contrast with the other ornamentations. The tail is composed of filamentous feathers of a beautiful carmine color, while from either side, at the root of the tail proper, are two filose, or thread-like processes, that curve most gracefully their length of quite two feet.

The **Spotted Bower Bird** (*Chlamydera maculata*) has also been classed among the birds of Paradise, and properly so, because of its beautiful plumage; but I have preferred to describe it among the singular nest-building species on page 359.

King Bird of Paradise (*Cicinnurus regius*) is so called on account of its imperious habits, and its apparent assumption of ·superiority over other species. But though exercising a sovereignty, this bird is very small, ranking as the least of its congeners, and scarcely rivalling a sparrow in size. It differs from others of the *paradisea* species in that it is gregarious, going usually in flocks of twenty or thirty, over which one seems to preside as king. The natives of New Guinea declare that this ruler of the flock may be distinguished by two spots, resembling eyes, on the tail feathers; they also affirm that in case the king is killed all those by which he is surrounded flock down upon the body and refuse to leave it, so that they fall easy victims to the native hunters. In color the upper parts are a rich chestnut brown, washed with delicate hues of purple, and the belly a pure white. Across the breast is a band of golden green, extending to the shoulders, from which springs a plume of feathers of dusky-brown tipped with green. From the tail coverts project two filose shafts, which expand at the tips into spiral tufts of an emerald-green color.

Radiant Crowned (*Parotia sefilata*) is the name given to a royal species of sun-birds, distinguished for their brilliant plumage and the six filamentous shafts that spring from the crown, three on a side. D'Albertis is the only traveller that I remember who has seen and described this beautiful bird, and from his account I therefore quote: "After standing still for some moments in the middle of the little glade, the beautiful bird peered about to see if all was safe, and then he began to move the long feathers of his head, and to raise and lower a small tuft of white feathers above his beak, which shone in the rays of the sun like burnished silver; he also raised and lowered the crest of stiff feathers, almost like scales, and glittering like bits of bright metal, with which his neck was adorned. He spread and contracted the long feathers on his sides in a way that made him appear now larger and again smaller than his real size, and jumping first on one end and then on the other, he placed himself proudly in an attitude of combat, as though he imagined himself fighting with an invisible foe. All this time he was uttering a curious note, as though calling on some one to admire his beauty, or perhaps challenging an enemy. The deep silence of the forest was stirred by the echoes of his voice."

Other species of sun-birds are found in the Arroo Islands, Java, Celebes and Philippine Islands, among which may be mentioned **Wallace's Bright Wing** (*Semioptera wallacii*), distinguished for having two thin but long feathers standing erect over each wing; **Schlegel's Bird** (*Schlegelia wilsonii*), which is covered with magnificent feathers of a cobalt blue, especially about the head, and velvety black on breast and tail.

The above includes all the species that have up to this time been described by naturalists, a list which appears altogether too brief, and found in a range too circumscribed for those whose love for the beautiful make them wish that the woods of every country were animate with such magnificent and splendid forms as only the barbarians of Oceanica are permitted to enjoy.

PARROTS AND COCKATOOS.

The brilliant plumage of several species of *parrots* entitles them to a position next to the birds of Paradise, notwithstanding the fact of total dissimilarity of general characteristic. The line of connection between parrot and cockatoo is, however, most distinct, and the two are therefore inseparable in the proper arrangement of species.

The most marked distinguishing features of both *parrot* and *cockatoo* are the large and strong beak, the upper mandible being sharply curved and hanging over the lower, but the lower jaw is extensible so as to permit the sharp points of the two bills to be brought together. The tongue is short and thick, free at the anterior end and capable of articulation. The claws have four toes, usually short and very muscular, which serve the double purpose of grasping a perch and conveying food to the mouth, in which latter service the claws perform the services of a hand. The power of flight is rarely great, and some species spend most of their time on the ground. The smallest of the *parrot* species are called *parrakeets*, a diminutive expression for *parrot*.

Warbling Parrakeet (*Melopsittacus undulatus*) is the name of the most beautiful of the species, and betrays such affection for its mate that it is often called *love-bird*. It is a native of Australia, but such large numbers are imported to America that the species is quite common among us. In the regions of New South Wales the bird goes in large flocks, and feeds off the grasses which cover the inland plains. Its nest is made in the hollows of dead trees, where it deposits four eggs, laying only on alternate days, and hatching its brood in about three weeks. The sounds it emits are variable, so much so that it may be said to have considerable powers of imitation. The more usual sound it produces is a low warble, which the male may often be seen producing in the very ear of his mate, as if she should have the full benefit of his vocal accomplishment. The color is green striped with black.

The **Blue-banded Parrakeet** (*Euphema chrysostoma*) is also a habitant of Australia, but in the fall migrates to Van Diemen's Land to pass the winter. It spends much of its time on the ground, where it is a swift runner. It nests in a hole of some dead tree and lays six or seven eggs. The body is green washed with brown, and the head is a beautiful azure, with yellow around the eyes.

WARBLING PARRAKEET.

The **Scaly-breasted Parrakeet** (*Trichoglossus chlorolepidotus*), or *hairy-tongued lori*, is peculiar to New South Wales, where it is very plentiful. It is the largest of the species, but feeds chiefly off the nectar of flowers, a dainty diet for such a large and vigorous bird. They seem to possess the power to distil honey, for when killed in their native haunts their crops are sure to be filled with this sweet, so that the people take the birds for the honey they yield. This species is easily tamed, and will thrive on a diet of sugar and seeds. They go together in immense flocks, sometimes settling on trees in such vast numbers as to break down the branches. The color is a rich green on the back, and the breast a light yellow with green edges, giving an appearance of scales.

The **Ground Parrakeet** (*Pezophorus formosus*) is a beautiful little creature with green and yellow coat trimmed with black flecks. It loves the ground, rarely rising higher than the low branches of a shrub. It has a pheasant shape and can run equally as swift, being able to baffle an ordinary dog. It builds no nest, but lays its eggs upon the bare ground.

Ringed Parrakeet (*Palæornis torquatus*) inhabits Africa and Asia, and is a species of a large genus that is frequently mentioned by ancient writers. It is chiefly distinguished for having a tail of extraordinary length, double that of the body, and for the confiding and affectionate disposition it displays. The color is a grass-green on the back, changing to a light blue toward the neck, while below the neck is a narrow, rose-colored band joined to one of black that swings around the breast till the points reach the eyes. The upper mandible is a beautiful coral-red, while the lower is black, being a remarkable variation from all other birds. This species has been taught to utter a few words, but is not worthy to rank among the talking parrots.

Yellow-Bellied (*Platycercus caledonicus*) and the **Rose Hill** (*P. eximius*) are both found in Van Diemen's Land, and are so nearly alike in size and habits that they may be included in a single paragraph. The former is very beautiful in its livery of rich crimson crest and mottled-green back, while the breast and throat are yellow. The latter is no less charming in a plumage of scarlet head, neck and sides, with throat of pure white. The cock is a dark green, with lighter shades on the tail coverts and central tail feathers. Both species are gregarious, appearing sometimes in such vast numbers as almost to darken the sky. Their flight, however, is short, and when aroused they quickly settle on the ground. They are as much despised by the Australians as the English sparrow is by us, being inveterate thieves, and an enemy to other more useful birds. But they are excellent food, and this fact will soon lead to a rapid diminution of their number.

RINGED PARRAKEET.

Gray Parrot (*Psittacus erythacus*). This is the common name given to the most interesting of the parrot species; not because of its plumage, but because of the marvellous powers of imitation it possesses, in which faculty it has no rivals. This species is a native of West Africa, its range being from the Gold Coast to the Gaboon river, where considerable numbers are found. Its color is an ashen gray over both back and breast, and a carmine tail, in which alone resides any beauty. Great numbers of these birds are imported to Europe and America, where they are in popular request for their linguistic accomplishments.

As they build their nests in the hollow of dead trees, the natives have little trouble discovering them, and effect a capture of the young by felling the tree, in which, however, many are killed. The natives also take the old ones by shooting them with blunt arrows, that stun but seldom kill. The young are very much preferable, as they are more easily taught if taken before they acquire the harsh notes of the older ones.

SOME AMUSING STORIES.

Many amusing stories are told illustrative of the comical situations frequently provoked by talking parrots, two of which are felicitously reported by Woods as follows:

"There was a parrot belonging to a friend of our family, a Portuguese gentleman who had married an English wife and resided in England. This parrot was a great favorite in the house, and being accustomed equally to the company of its owner and the rest of the household, was familiar with Portuguese as well as English words and phrases. The bird evidently had the power of appreciating the distinction between the two languages, for if it were addressed its reply would always be in the language employed.

"The bird learned a Portuguese song about itself and its manifold perfections, the words of which I cannot remember. But it would not sing this song if asked to do so in the English language. Saluted in Portuguese, it would answer in the same language, but was never known to confuse the two tongues together. Toward dinner-time it always became very excited, and used to call to the servant whenever she was late, 'Sarah, lay the cloth—want my dinner!' which sentence it would repeat with great volubility, and at the top of its voice.

"But as soon as its master's step was heard outside the house its tone changed, for the loud voice was disagreeable to its owner, who used to punish it for screaming by flipping its beak. So Polly would get off the perch, very humbly sit on the bottom of the cage, put its head to the floor, and instead of shouting for its dinner in the former imperious tone, would whisper in the lowest of voices, 'Want my dinner! Sarah, make haste, want my dinner!'

"In the well-known autobiography of Lord Dundonald, there is an amusing anecdote of a parrot which had picked up some nautical phrases, and had learned to use them to good effect.

SOUTH AMERICAN GRAY PARROT.

"Some ladies were paying a visit to the vessel, and were hoisted on deck as usual by means of a 'whip,' i. e., a rope passing through a block on the yard-arm, and attached to the chair on which the lady sits. Two or three had been safely brought on deck, and the chair had just been hoisted out of the boat with its fair freight, when an unlucky parrot on board suddenly shouted out, 'Let go!' The sailors who were hauling up the rope instantly obeyed the supposed order of the boatswain, and away went the poor lady, chair and all, into the sea.

"Its power of imitating all kinds of sounds is really astonishing. I have heard the same parrot imitate, or rather reproduce in rapid succession, the most dissimilar of sounds, without the least effort and with the most astonishing truthfulness. He could whistle lazily like a street idler, cry prawns and shrimps as well as any costermonger, creak like an ungreased 'sheave' in the pulley that is set in the blocks through which ropes run for sundry nautical

purposes, or keep up a quiet and gentle monologue about his own accomplishments with a simplicity of attitude that was most absurd.

"Even in the imitation of louder noises he was equally expert, and could sound the danger whistle or blow off steam with astonishing accuracy. Until I came to understand the bird, I used to wonder why some invisible person was always turning an imperceptible capstan in my close vicinity, for the parrot had also learned to imitate the grinding of the capstan bars and the metallic clink of the catch as it falls rapidly upon the cogs.

"As for the ordinary accomplishments of parrots, he possessed them in perfection, but in my mind his most perfect performance was the imitation of a dog having his foot run over by a cart-wheel. First there came the sudden, half-frightened bark, as the beast found itself in unexpected danger, and then the loud shriek of pain, followed by a series of howls that is popularly termed 'pen and ink.' Lastly, the howls grew fainter, as the dog was supposed to be limping away, and you really seemed to hear him turn the corner and retreat into the distance. The memory of the bird must have been most tenacious, and its powers of observation far beyond the common order; for he could not have been witness to such canine accidents more than once."

The *Gray Parrot* is noted also for its singular attachments, in which respect, however, there is a correspondence among all the genera. Several instances are well authenticated in which the parent has become foster-mother to fledglings of other birds, in which capacity she has given her charge the tenderest possible care, and has exhibited the greatest grief when the brood, being raised up, have deserted her. They have also been seen to converse with other birds, first using their acquired vocabulary, and manifesting great impatience at receiving no response, but afterwards communicating by what appears to be a universal bird language.

AFRICAN GRAY PARROT.

That parrots live to a very great age is a well ascertained fact, the limit falling little short of one hundred years. M. Le Vaillant gives an account of one that was kept in captivity through a period of ninety-three years. At the time that eminent naturalist saw it, it was in a state of entire decrepitude, and in a kind of lethargic condition, its sight and memory being both gone, and was fed at intervals with biscuit soaked in Madeira wine. In the time of its youth and vigor it had been distinguished for its colloquial powers and distinct enunciation, and was of so docile and obedient a disposition as to fetch its master's slippers when required, as well as to call the servants, etc. At the age of sixty its memory began to fail, and, instead of acquiring any new phrase, it began to lose those it had before attained, and to intermix, in a discordant manner, the words of its former language. It moulted regularly every year till the age of sixty-five, when this process grew irregular, and the tail became yellow, after which no further change of plumage took place; its death was easy, as though the physical forces had become gradually exhausted.

The **Hyacinth Parrot, or Arara,** is a very large species, resembling the *macaws.* It is found in the deepest forests of South America, but never associating in flocks, like others of the species. Indeed, its numbers are so few that the bird is rarely seen, and it is therefore omitted from nearly all natural histories. The peculiarity of this bird, admirably illustrated in the accompanying engraving, is found in the bill, which, instead of curving and pointing downward, as in all other members of the parrot family, the curve is continued until the point of the beak is directly under the throat, making nearly a half circle. This shape of the bill is convenient, since being largely a ground bird, subsisting off worms and insects that burrow, it is the better enabled to dig, and also to bite through decayed and fallen limbs in search for its food. The sides of the face are not bare as in *macaws,* though in other respects it bears a striking resemblance, the tail being equally long. Around the base of the lower mandible runs a sharply defined circle of flesh; bare of feathers, which forms rather a striking feature.

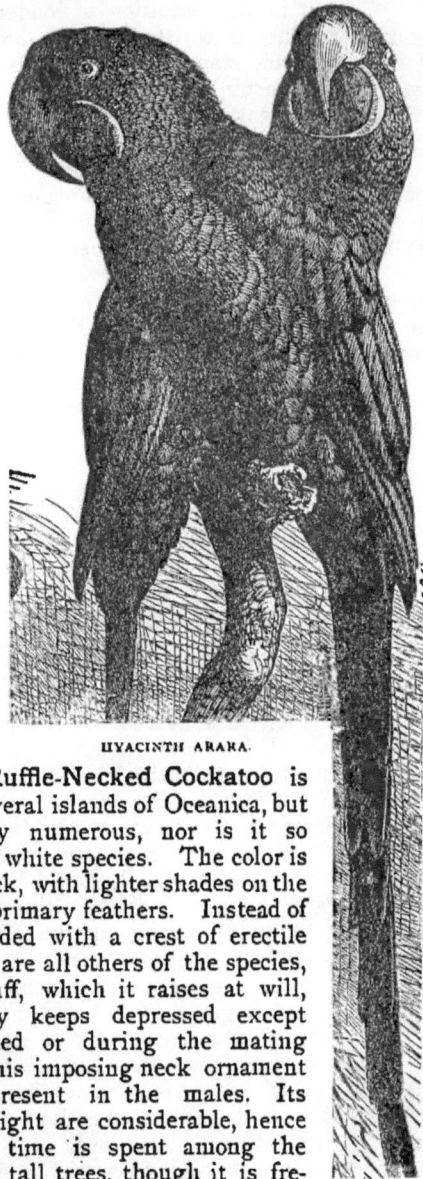
HYACINTH ARARA.

RUFFLED NECKED COCKATOO.

The **Ruffle-Necked Cockatoo** is found in several islands of Oceanica, but is not very numerous, nor is it so large as the white species. The color is a dusky black, with lighter shades on the tips of the primary feathers. Instead of being provided with a crest of erectile feathers, as are all others of the species, it has a ruff, which it raises at will, but usually keeps depressed except when excited or during the mating season. This imposing neck ornament is only present in the males. Its powers of flight are considerable, hence most of its time is spent among the branches of tall trees, though it is frequently seen upon the ground feeding upon roots and insects. Few species have been introduced into European or American aviaries.

The **Amazon Green Parrot** (*Chrysotis amazonica*) is about the size of its African congener, and its powers of linguistic expression equally great, though it is hardly so quick to learn. The coloring is somewhat more florid, the head being adorned with yellow, red and violet, ends of the wings red, and the long tail feathers yellow. Mr. Clausius relates the following anecdote:

" A certain Brazilian woman, that lived in a village two miles distant from the island on which we resided, had a parrot of this kind which was the wonder of the place. It seemed endued with such understanding as to discern and comprehend whatever she said to it. As we sometimes used to pass by that woman's house, she would call upon us to stop, promising, if we gave her a comb, or a looking-glass, that she would make her parrot sing and dance to entertain us. If we agreed to her request, as soon as she had pronounced some words to the bird it began not only to leap and skip on the perch on which it stood, but also to talk and to whistle, and to imitate the shoutings and exclamations of the Brazilians when they prepared for battle. In brief, when it came into the woman's head to bid it sing, it sang; to dance, it danced. But if, contrary to our promise, we refused to give the woman the little present agreed on, the parrot seemed to sympathize in her resentment, and was silent and immovable; neither could we, by any means, provoke it to move either foot or tongue."

Rose Parrot (*Psittacella rosicalles*) belongs to a very small species found in New Guinea, and which might with propriety be classed among the parrakeets. It is beautifully marked with scarlet plumage on head and back part of the neck, with breast of white and yellow; the back and tail are of two shades of green. It is not a very well known species. To this species also belongs

ROSE PARROT.

GREEN PARROT (*Platycercus eximius*).

the **Pigmy Parrot** (*Nasitema pygmæa*), which is of possibly more diminutive size and less gaudy plumage.

Carolina Parrot (*Conurus carolinensis*) is the only one of the genera found within the United States, and though it was at one time very numerous along the southeast coast of Florida, and was once found as far north as New York, and west to Ohio, it is not now very commonly met with, owing to its persistent slaughter by cruel huntsmen. It is a small bird, scarcely ranking above the parrakeet, of a generally green color on the back, forehead dark red, and head, neck and belly yellow.

The **Ring Parrot** (*Palæornis torquatus*) is a native of India and Africa, a species that was first brought to Greece by Alexander, and the first bird known by Europeans to possess the power of speech. It is specially numerous in India in both a free and captive state, and in the former at times becomes a great pest to agriculturists, whole fields of grain being sometimes settled upon by enormous flocks and entirely destroyed. The color is green and black, with a ring of black extending from the lower jaw back and over the neck. The tail is very long and of a yellowish hue.

Blue-crowned Parrot (*Coryllis galgulus*) is a small species found in considerable numbers in the Malay peninsula. They have imperfectly developed wings, and therefore spend much of their time on the ground, or hopping from limb to limb. The most singular characteristic is its bat-like propensity for hanging head downward when sleeping, and often when eating, this position appearing a most natural one to them. The prevailing color is green, with a crown of light blue.

The **Owl Parrot** (*Stringops habroptilus*), also called **Kakapo**, is a singular creature, combining, as it does, the features of both owl and parrot. It is found principally in New Zealand and Australia, but was not brought to the attention of naturalists until 1845. The head, especially about the eyes, is peculiarly owl-like, and its habits largely nocturnal. The wings are small, so that its

OWL PARROT.

BLUE-HEADED LORIS (*Loriculus galgulus*).

flight is restricted to short distances, and most of its time is spent upon the ground. The color is generally green, with longitudinal dashes of yellow and cross-bars of black. About the eyes are discs of radiating feathers of yellowish brown. It burrows in the ground, but builds its nest on the surface or under shelving rocks, and deposits, like all the parrot family, but two eggs. In captivity it shows a kindly disposition and all the playfulness of a young kitten. But four species are known, and these will no doubt soon become extinct, their numbers now being few, and constantly diminishing through the depredations upon them of dogs, cats and rats.

The **Loris** constitute a family called *trichoglossidæ* (hair-tongued), from the papillæ on the tongue, which resemble thin hairs. They are in all other respects like the parrot, except that they do not climb by the use of feet and bill as do all others of the genera. Usually the tail is broad, but covered by the projecting wing, but in some species, like the **Papuan Lory** (*Charmosyna papua*), the tail feathers are extremely long. The colors of this bird are remarkably rich, being a deep scarlet flecked with azure, golden yellow and pea-green, while the bill is orange-red. The length is about eighteen inches, of which the tail constitutes two-thirds.

The **Purple-capped Lory** (*Lorius domicettus*) is, like the former species, a native of New Holland (Papua), considerably larger in size of body and even more brilliantly clothed. The crown is a deep purple, and face, neck, and breast a rich scarlet, with a bright collar of yellow. The wings, too, are scarlet, tipped at the ends with black and yellow.

The **Macaws** are grouped under a separate head because of their greater size, gaudy plumage and other characteristics at variance with the true parrot, noticeable among which is their naked cheeks. They are almost entirely confined to South and Central America. Its haunts are usually in dense woods, and about swampy districts where the palm flourishes, on the fruit of which it feeds. Eighteen species are known, of which the three most familiar are the **Blue** and **Yellow Macaw** (*Ara ararauna*), the **Great Green Macaw** (*Sittace militaris*), and the **Scarlet Macaw** (*S. coccinea*), all of which range from Brazil to middle Mexico. They are extremely noisy birds, and can rarely be taught to speak even a few words, but being very courageous they are easily captured, and despite their clamorous cawings, are popular pets.

Cockatoos belong to the family *Plictolohidæ*, a name derived

GREAT GREEN MACAW.   AMAZON PARROT (*Chrysotis amazonica*).

from the Latin and Greek, to designate a characteristic peculiar to all the species, viz., the provision of a crest of feathers, which they have the power to erect or depress at will. They are confined to Australia and Oceanica. The prevailing color is white, black or brown, the former predominating.

Their habits are very sociable, having some resemblance in this respect to parrakeets. Being strong of wing, they perch on very tall trees, but feed on grain chiefly, though some species descend to the ground and dig up roots and bulbs, while yet others imitate the habits of woodpeckers, attacking decaying trees, tearing off the bark, and even biting out pieces of wood in their search for insects. There are thirty-five species classified and named by naturalists, nearly all of which produce similar notes, resembling a phonetic pronunciation of the word *cock-a-too*, from whence the name is derived.

The Pink Cockatoo (*Plictolophus leadbeateri*) was the first specimen of the genera brought to Europe, which was purchased by Mr. Leadbeater, an English naturalist, after whom the species was named. The color is a pure white, suffused with pink, while the crest is barred with crimson, yellow and white.

The Kea (*Nestor notabilis*) is confined entirely to Philipp Island, which is only five miles in extent. It is remarkable for the astonishing length of the upper bill, projecting, as it does, some two inches beyond the lower mandible. On this account it is sometimes called the Long-billed Parrot. It subsists largely from the honey it extracts from blossoms of the hibiscus, to obtain which its tongue is furnished with a long, narrow, horny scoop at the under side of the extremity, up which the sweet juices are sucked. But it is also known to dig in the earth for tender bulbs, for which purpose the very long upper mandible is well adapted. The color is brown and gray, with occasional flecks of red and yellow. This bird is supposed to be the connecting link between parrot and cockatoo.

THE PINK COCKATOO.

The Banksian Cockatoo (*Calyptorhynchus*) is an Australian bird of considerable size and brown color, sometimes dyed with richest hues of red and yellow. The name *Banksian* has been given to it on account of the seeds of the banksia being its chief subsistence, though it is also fond of insects, which it digs out of decaying trees.

The Sulphur-crested Cockatoo (*Cacatua galerita*) is a very beautiful and kindly disposed bird, and a species most commonly seen in aviaries. It is a pure white with tints of pale red, and the head is surmounted by a long crest that when erected looks like young onion stalks. It yields readily to kind treatment, and may be easily taught to perform many amusing tricks.

The Great White Cockatoo (*Cacatua cristatus*) is also a very handsome bird, almost equal in size to a guinea-hen, and is strikingly intelligent. It learns quickly and talks almost as well as the green parrot. But it becomes excited on small provocation and displays the greatest violence, both by a shocking noise and fierce attacks. The plumage of this species is white, with very slight rose tinge, and the crest is also white.

The Great Black Cockatoo (*Microglossus aterimum*) is a native of New Guinea, and the largest of the genera. It differs from others of the family in having the bill toothed and the tongue long, tubular and extensible. This formation of tongue would seem to place it among the honey-suckers, but its habits do not accord with this apparent provision, since it is a grain-feeder. The use of the tongue has therefore not yet been fully determined. The general color is black, with a raven gloss, and sometimes of a very light color, due to the accumulation of dandruff, or quill powder upon the feathers.

TOUCANS AND HORNBILLS.

Of the many singular forms and characteristics displayed by nature, perhaps none are more curious than the features observable in the birds classed under the general head of *ramphastidæ* (large billed). They are dis-

tinguished by what appear to be preternaturally large beaks, equal in length to one-half the body, and of immense circumferential proportions. Some have not only enormous beaks, but to these colossal frontispieces are added a helmet of horn overweighting the head to such an extent as would appear to constantly hold the head downward. But nature does not make her creations with amateur hands, for though she produces many queer appearing and apparently redundant things, it is always with marvellous design and wondrous adaptation. Disproportionately large as are the bills of the *toucan* family, they give no inconvenience, for they are marvels of lightness, rivalling the pearly nautilus in delicacy of structure, and serve no less admirable uses.

The **Red-billed Toucan** (*Ramphastus tucanus*), like all its congeners, is a habitant of South America, and grows to a considerable size, equal perhaps to a crow. There are some fifty species, but there is great resemblance between the varieties in size, appearance and habits, so that a general description will answer for them all. The *red-billed toucan*, and

SULPHUR-CRESTED COCKATOO.

the *great toucan* (*R. tuco*) are the largest and differ only in the color of the bill, one being red and the other orange. They build their nests in the hollow of some dead limb, scooped out by a laborious process, and, like the parrots, only lay two eggs. The young have the bill well developed as soon as they appear, but are nourished by the parents a considerable while longer than are the young of most birds.

The flight of the *toucan* is very slow and ungraceful, and on the ground its movements may be described as "straddle-legged," moving by hops, with the legs spread far apart, giving it a most awkward appearance. The cry is quite as rasping and distressing as is that of the parrot, which it somewhat resembles, though when calling the notes are *toucano, toucano*, from the sound of which the name has been given. They possess an omnivorous appetite, feeding without special preference off fruits, grain, white ants or any kind of meat. It commits great havoc among fruit gardens, but partially compensates for such depredations by the great destruction it works among ants, the largest hills of which it breaks down and devours the greater part of the inhabitants. In captivity it will kill and eat mice, rats and birds, and an instance is on record where one seized upon a soldier's cartouche box and devoured all the cartridges, but with fatal result.

When sleeping, the *toucan* shifts his head so as to lay the monstrous bill squarely upon the back, and then covers it entirely by tilting the tail up over it, the vertebræ of the tail being articulated specially so as to permit of this singular motion. When excited, or even hopping from branch to branch, the tail is kept tipping as though worked on hinges, which adds a weird effect to the otherwise curious creature.

The **Hornbills** differ from the toucans in having a double bill, the purpose of which has not yet been determined. D. G. Elliott, in an excellent monograph on these strange birds, says: "As they exist at the present day, they exhibit to us probably but a remnant of the great family which once dwelt amid the forests of that mighty Eastern Continent, of which a large portion is now beneath the waters." Mr. Elliott has succeeded in collecting sixty species, which he found distributed over Africa, India and the Austria-Malayan region, where it is sometimes called rhinoceros bird.

GREAT BLACK COCKATOO.

The *hornbill* varies in size, according to species, from that of a robin to a crow, the larger species being somewhat more plentiful. They are frugivorous, except that during the nesting period the female feeds principally on insects. Remarkable as is the appearance of this bird, its habits are no less curious, and particularly in the manner of rearing the young, in which it differs very much from all other birds. The nest is built in the hollow of a tree, with no more care than the mere scooping out of a convenient receptacle some twelve inches deep, no lining of any kind being used. The female deposits but a single egg and then begins with great earnestness the process of incubation, but with this beginning a strange thing occurs. Instead of following the customary usages of birds, leaving the nest at infrequent intervals to keep up old acquaintances, through occasional visits, she becomes an astonishing example of exclusiveness, for she goes in upon the nest and then seals up the exit, carefully plastering up the hole with her own ordure, and leaving only a small slit through which to receive the food brought to her by the male. She is therefore unable to leave the nest until the young is hatched, her liberation being then accomplished by the joint efforts of herself and mate. Should the plaster which confines her be broken before the hatching is completed, she seems to be almost helpless, and if put upon the ground is so stiff and nearly featherless that she is unable to rise even to the lowest branches.

The larger species of *hornbills* have powerfully developed wings, but their bodies are so heavy that flight is awkward and seemingly very laborious. When several take wing together they produce a noise not unlike a locomotive when getting under way with a heavy load.

THE CUCULIDÆ, OR CUCKOOS.

The order of birds under which cuckoos are classed is a very large one, numbering a hundred species, and found in nearly all countries. The popular name proceeds from the familiar note it utters most frequently at early dawn. The great variety of species presents a remarkable dissimilarity of appearance, ranging between that of a timid sparrow to a striking resemblance which at

RED BILLED TOUCAN.

least one species (*cuculus canorus*) presents to a small hawk, while several show all the markings of falcons, and also of the shrike. Only in one respect do the several varieties exhibit a common characteristic, unless we are critical enough to look below the exterior. They are all parasites (save the few American species) in respect to laying the ireggs in the nests of other birds and imposing the care of their young to these enforced foster-parents. Concerning this singular habit Mr. Seebohm, an excellent authority, thus writes:

" The cause of this curious habit is difficult to discover. It has been suggested that the hereditary impulse to leave its breeding grounds so early originally obliged it to abandon the education of its young to strangers; but the same habit is found in many species in India and Africa, which are resident and do not migrate. Others have attributed it to the polygamous habits of the *cuckoo*, but the *cuckoo* is not polygamous, it is polyandrous; the males are much more numerous than the females, and the sexes do not pair even for a season. It is said that each male has its own feeding grounds, and that each female visits in succession the half dozen males who happen to reside in the neighborhood. A more plausible explanation of the peculiar habits of the *cuckoo* is to be found in the fact that its eggs are laid at intervals of several days, and not, as is usual, on successive days. Very satisfactory evidence has been collected that the *cuckoo* lays five eggs in a season, and that they are laid at intervals of seven or eight days. The American *cuckoo* and many of the owls very often do the same.

GREAT TOUCAN.

This power has probably been gradually acquired by the *cuckoo*, so as to give the female time to find a suitable nest in which to deposit each egg. It is possible that this singular habit of the *cuckoo* has arisen from its extraordinary voracity. The sexual instincts of the male *cuckoo* appear to be entirely subordinate to his greed for food. He jealously guards his feeding grounds, and is prepared to do battle with any other male that invades them, but he seems to be a stranger to sexual jealousy. He is said to be so absorbed in his gluttony that he neglects the females, who are obliged to wander in search of birds of the opposite sex, and appear to have some difficulty in obtaining the fertilization of their ovaries. The extreme voracity of the young bird is an additional reason why the care of the five nestlings should be entrusted to as many pairs of birds.

"In its choice of a foster-parent for its offspring, it exercises more discrimination than might be supposed .from the long lists which have been published of birds in whose nests its eggs have been found. An insectivorous bird is generally chosen, and preference is given to such as build open nests. Sometimes the *cuckoo* is unable to find the nest of a suitable bird, and is obliged to deposit its egg in the nest of a granivorous

WRINKLED HORNED TOUCAN (*Buceros plicatus*).

bird, such as the various species of finches, buntings, etc., and occasionally in the nests of jays, or even owls."

Different varieties lay eggs of different color, but they do not possess the power of determining the color in order to imitate those of other birds, as some suppose. The natural food of the *cuckoo* is caterpillars, especially those of the largest and most repugnant aspect, and also bumble-bees, of which it consumes great numbers.

BEARDED CUCKOO (*Bucco flavigula*).

The mode by which the *cuckoo* contrives to deposit her eggs in the nest of sundry birds was extremely dubious, until a key was found to the problem by a chance discovery made by Le Vaillant. He had shot a female *cuckoo*, and on opening its mouth in order to stuff it with tow, he found an egg lodged very snugly within the throat.

When hatched, the proceedings of the young *cuckoo* are very strange. As in process of time it would be a comparatively large bird, the nest would soon be far too small to contain the whole family; so the young bird, almost as soon as it can scramble about the nest, sets deliberately to work to turn out all the other eggs or nestlings. This it accomplishes by getting its tail under each egg or young bird in succession, wriggling them on to its back, and then cleverly pitching them over the side of the nest. It is rather curious that in its earlier days it only throws the eggs over, its more murderous propensities not being developed until a more advanced age.

ALMOND CUCKOO (*Coracias garrula*).

There seems to be some peculiarity in the nature of the *cuckoo* which forces other birds to cater for its benefit, as even in the case of a tame and wing-clipped *cuckoo*, which was allowed to wander about a lawn, the little birds used to assemble about it with food in their mouths, and feed it as long as it chose to demand their aid.

Generally, the color of the *cuckoo* is bluish-gray above and along the back, with wings of black barred on the tips with white. The largest species is the Giant Cuckoo (*Scythrops præsagus*) of New Guinea, the characteristics of which are very much like the toucan. Other species very commonly known are admirably pictured in the accompanying engravings. The **Ant-eating Cuckoo** (*Crotophaga ani*) is confined to North America, but the large bill is very suggestive of the toucan. In the West Indies, where this bird is also found, a dozen or more have been known to build a single nest in which they deposited their eggs, sometimes thirty or more, and also do their hatching in communistic fashion, each female in turn doing her proportion of the incubating, and afterwards all uniting in the care of the young.

GIANT CUCKOO.

The **Chaparral Cock** (*Geococcyx californianus*) is an American species included among the cuckoos for want of a more appropriate classification. It is chiefly found in the chaparral regions of the southwest, from whence the name is taken. It is a good flyer but seldom quits the ground, being so swift of foot that a dog does not easily overtake it. The food of this species is insects, and the smaller lizards and mice, though some declare that it kills and eats rattlesnakes, which is most improbable.

The **Hoopoe** (*Upupa epops*) is also a bird that naturalists have long suffered to wander in lonely isolation for want of a family relationship. Some have included it among the hornbills, because of the horny substance which compose the bills, others placed it next to the cuckoo, and Woods classed it with the birds of Paradise, which seems to fall a little short of ridiculous. The *hoopoe* is distinguished for having a tuft of erectile feathers, in which respect it resembles the cockatoos, but it has also a long, slim and slightly curved bill, which characteristic belongs to the snipe family. It has no metallic colors, but is beautifully pied with white and a rusty buff, and a slight pinkish tinge on the breast. Some six species are known, all confined to the old world, with considerable range, the *upupa* being found as far north as the semi-Arctic regions, and the *wood hoopoe* (*irrisor*) as far south as lower Africa. Many superstitions were formerly connected with this bird, on account of its singular voice, but they are not now generally current. Mr. Robert Swinhoe, who has critically observed the habits of the *hoopoe*, says its strange notes are produced by the

ANT-EATING CUCKOO.　　　　　　HOOPOE BIRD.

bird puffing out the sides of the neck, and hammering on the ground as it violently exhausts the accumulated air, emitting a sound something like "*hoo-hoo-hoo*." The bird has the power of giving off a dreadfully offensive odor, but why this provision has been vouchsafed to it and denied to all other birds, is not known.

GOAT-SUCKERS AND SWALLOWS (Passeres).

A very large order, numbering some hundred or more species, includes those birds of passage which are distinguished for taking their food while on the wing, of which *goat-suckers* and *swallows* constitute the larger number. They are also called *fissirostres*, because of the wonderful structure of their mouths, by which they are enabled to capture winged insects with an ease and celerity that at once exhibit marvellous adaptation. The name *goat-sucker* was originally applied by ignorant persons, who entertained the very silly belief that these birds drained the udders of not only wild goats, but of cows and sheep also.

These silly superstitions were perpetuated by numerous stories, related by persons who pretended to be eye-witnesses of the birds' milk-loving propensities. But these accusations were not the most serious that were urged against the night swallows, since they were thought to be winged messengers of an evil power, or possibly wandering souls going about under the cover of darkness to discharge some envious commission. The whip-poor-will, which belongs to this species, is even to this day regarded as a bird of evil omen, the wandering shade of some soul that is permitted to make fitful visits from its Plutonian abode. This belief is more general than it would, perhaps, otherwise be on account of the rarity with which the bird is seen. Its voice is commonly heard during the still summer nights, but though its notes are so well known as to make the bird familiar to people of all North America, yet in fact not one person in a thousand, probably, ever caught so much as a glimpse of the creature. For this reason many superstitious persons regard the bird as a spirit.

I have myself sought this creature with great diligence, creeping with all

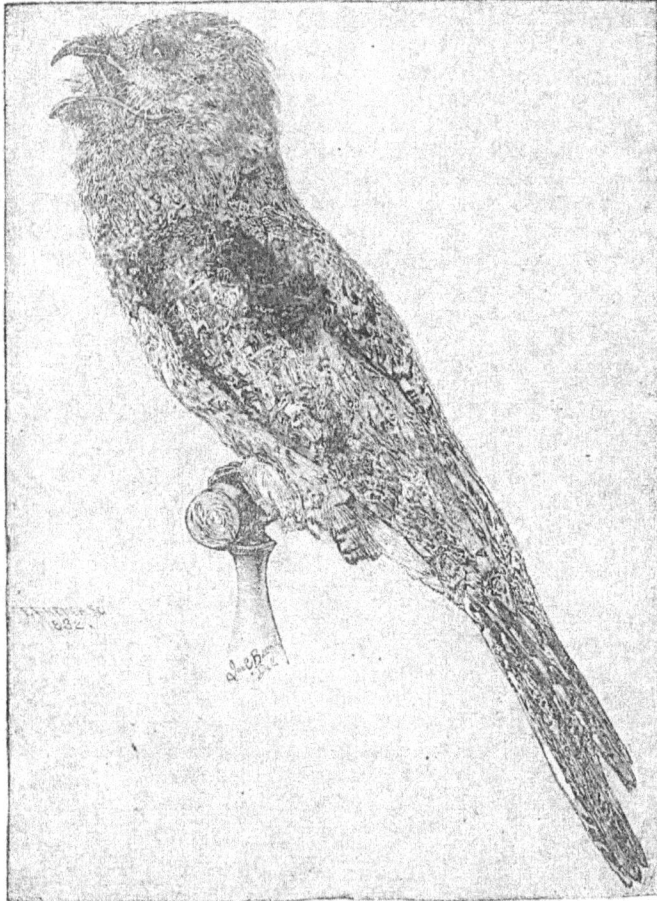

GREAT GOAT SUCKER, OR NIGHT SWALLOW.

possible care towards the spot from whence its voice seemed to proceed, but except upon two occasions I was always unable to gain a view of it. The bird is both cunning and wary, leaving its perch so quietly, and usually haunting such dense coverts, that the most acute vision is required to perceive it. The bull bat, or night swallow, is also rarely seen except when in pursuit of insects. When at rest it sits so closely upon a limb as to be almost invisible.

Though there is considerable disparity in the size of the several species, this is about the only well-defined variation, for in habits they are all very similar. Being among the swiftest of birds we may expect to find them in nearly all parts of the world, coming and going with the season's changes. In color they are uniformly sombre—black, brown or ashen-gray; the bill is short but they have an enormous gape, which is kept spread when the birds are on wing. The tail is square or forked, wings long and pointed, and legs so short that movement on the ground is both slow and awkward. The *goat-suckers* have a wonderful provision of cilia, or hairs, radiating from the jaws, that spread out to act as a funnel, somewhat like the baleen in the Greenland whale, though in the latter the purpose is that of a strainer. They usually lay their eggs, two in number, in a rudely constructed nest either upon the ground or on a flat rock, and the eggs so nearly assimilate in color to the surroundings that they are rarely found, the rudeness with which the nest is made aiding very much to prevent their discovery.

Swallows, however, build their nests either in sandy banks, as our sand-swallows; or of clay cemented to a wall or rock, with a round hole for entrance, as our mud-marten; or of sticks glued together with the birds' glutinous saliva, like our chimney-swallow; or in the hollow of some decayed tree or box, like our house-marten. The sand-marten, or bank-swallow, is very abundant about

FALLOW SWALLOW (*Glareola pratincola*).

the steep banks of American rivers, where it excavates a round hole several feet deep, in the rear end of which it builds a nest of down and lays four white eggs. The mud-marten, or barn-swallow, is about the same size as the sand-marten, but much more graceful in movement, as it is handsomer in color. The breast is white, back a glossy black, with a circlet of dull orange about the breast, and tail forked. It usually builds its nest of mud on the rafters of old barns, rarely venturing into new buildings. The clay thus used is taken from the moist shore of some pond or stream, and first worked into a round ball before being conveyed to the place of deposition. When the nest is completed the bird lines it with coarser feathers, over which it arranges a layer of down plucked from its own breast. Its eggs are also four in number, and of a pure white. The cliff-swallow constructs its nest very similar to that of the barn-swallow, except that the entrance is like the long neck of a bottle. Otherwise the habits of the two are identical.

Several species of swallows, while possessing little brilliancy of plumage, are rendered very attractive by an ostentatious display of feathers, such as the wire-tail (*hirundo filifera*), which trails behind it two very long hair-like feathers, and the crested-swallow (*dendrochelidon longipennis*), that is remarkable for the large helmet-like crest with which its head is adorned. Some species of goat-suckers are similarly arrayed, such as the leona (*macrodiptex longipennis*), and long-winged (*caprimulgus vexillarius*), both of which have a single feather in each wing equal to double the length of the body, which imparts a most graceful appearance to these swift-moving creatures. The long-tailed (*C. lyra*), rivals the long-tailed bird of Paradise in the extraordinary length of two feathers of its tail which curve inward to give a resemblance to the shape of a lyre.

CRESTED SWALLOWS (*Dendrochelidon longipennis*).

The most interesting species, because most important from a commercial point of view, is the **Esculent Swallow**, of which there are four species, viz.: the linchi (*collocalia fuciphaga*), the white-backed (*C. troglodytes*), and the gray-backed (*C. francica*), all of which, however, are natives of the Malay Peninsula, Corea and some of the neighboring islands. The nests which these birds construct are most singular, in that they appear somewhat like lichens or some other fungous growth, and not at all like nests. Their shape is irregular, and the hollow hardly great enough to retain the eggs. They are invariably built on the face of precipitous rocks, in places least accessible to man, as if the birds were conscious of the estimation in which their nests are held. The only means of reaching them, is by attaching a strong rope to some support above the rocks, and by this the nest-gatherer must be lowered over the precipice, which is always a most laborious and dangerous undertaking.

The nests when first gathered are most uninviting in appearance, but when washed thoroughly exhibit the shining, glutinous substance of which they are composed. Nests which have served a brood are of little value, but those gathered before the eggs are hatched, are so highly regarded that they bring a price equal to their weight in silver; or, to be more exact, this is the price which wealthy Chinese are prompt to pay, but other nationalities

BARN SWALLOW (*Hirundo rustica*).

consider them very differently, declaring the taste insipid if not nauseous. The method of preparing the nests for table is to first soak them thoroughly in tepid water, in which they swell and dissolve, until after a time the water and nests form a pasty consistency, which very much resembles gum arabic glue. This substance is heated, also like glue, by suspending the vessel containing it in another filled with boiling water, and when thoroughly heated, is seasoned to taste and served in small cups.

EDIBLE SWALLOW'S NEST.

It is a somewhat singular fact that though the secretion of our common chimney swallow is almost identical with that of the *esculent swallow*, yet, notwithstanding the high price paid for the latter, no attempt has ever been made to market the former.

The **Swifts,** of which there are only two species, are found native to Africa, which country they leave in summer for a sojourn of a few months in Europe. They are so nearly allied to the swallows in appearance and habits, that separate description is unnecessary. They take their name from the extraordinary velocity of their flight, and the agility which they display on the wing. The two species are the *Cypselus apus* and *C. melba*. Next to the frigate bird the *swift* is the fastest flyer of feathered creation, and is also capable of sustaining its flight for many hours; indeed, it is rarely seen at rest during the daytime.

DOVES AND PIGEONS (Columbæ).

Unlike the preceding order, the family enumerated under the general classification *Columba* is a large one, with representatives in all parts of the world,

PARROT DOVE.

whose progenitors were no less important birds than the great *dodo*, which perished forever nearly two hundred years ago. The *dodo* was once very numerous on the island of Mauritius, but was never widely distributed over any mainland. It was a very large bird, weighing quite fifty pounds, and so clumsy and slow in its movements, and building its nest upon the ground wherein but a single egg was deposited, that when hogs were introduced on the island they soon devoured the birds and destroyed their nests. Thus it was that the *dodo* became speedily extinct through those ravages.

Pigeons and doves are distinguished for their plumpness of body and strength of wing; though generally they are destitute of brilliant plumage, there are several species in the tropics clothed in the richest raiment, almost rivalling the sun-birds. They usually build their nests on low branches, and lay two eggs of pure white. Doves rear only a single brood each year, but pigeons of the domestic variety lay and hatch every month save March, so that they may be reckoned as being the most prolific of birds. They do not feed their young like other birds, but are provided with a double gullet, in which the food taken is mixed with a secretion that reduces it to a pulpy consistency, which the female has the power to raise again into the mouth,

TURTLE DOVE.

and which she parts with to the young.

The **Turtle Dove** (*Turtur auritus*) is the most familiar of the genera, on account of its prevalence in all parts of the United States, where it is regarded with a kindly feeling, because of the soft and mournful notes it utters during the breeding season.

The **Crested Dove** (*Ocyphaps lophotes*) is a native of Australia, and distinguished alike for its exquisite plumage and the long, pointed crest with which its head is adorned. It is gregarious, going in immense flocks, whose actions seem to be controlled by a leader.

CRESTED DOVE.

The **Bronze Wing** (*O. chalcoptera*) is also a very beautiful Australian species, but rarely more than two are to be seen together, except at watering-places, where, in the evening, these birds congregate in large numbers.

The **Wonga Dove** (*Leucosarcia picata*) is also common to Australia, where, in certain regions, it is quite plentiful. It spends the greater part of its time on the ground, picking up insects, seeds and small gravel. In rising from the ground, the wings vibrate with such rapidity as to produce a whirring noise very pheasant-like.

The **Nicobar**, or **Maned Dove** (*Calænus Nicobarica*), is a native of Java, Sumatra, and neighboring islands, and is a most beautiful bird. It is peculiar in being clothed with resplendent feathers of bright green, bronze, and steel-blue, which, instead of lying close to the body, project loosely, giving it a tousled appearance.

MANED DOVE.

The **Crowned Dove** (*Megapelia coronata*) is, in some respects, the most magnificent species of the genera, being large in size, with splendid plumage, and a royal crest of filamentous feathers that radiate most gracefully from the base of the bill to the back of the crown.

The **Rock Dove** (*Columba livia*) is widely distributed in the north temperate zone, from England to Japan. It affects rocks rather than trees, from which fact the name has been given. It is easily domesticated, and makes a very agreeable pet, though if encouraged by the building of cotes it multiplies so rapidly as to become a nuisance. The color is gray, with hues of purple and green.

CROWNED DOVE.

The **Ring** (*Columba palumbus*) and **Stock Dove** (*C. œnas*) are common domestic species of Europe. The former may be recognized by the ring of white feathers about the neck. The latter is of a dull hue, distinguished only by its nesting habits, which are peculiar. It builds in hollow stumps, or even takes quarters in a deserted rabbit burrow, and is content with only a few sticks in rather aimless arrangement, for the nest seems to have no comfort about it.

The **Parrot Dove** (*Phalacrotéron Abyssinica*), as the name implies, is a native of Abyssinia, though it is also found in other countries of northeast Africa. Except for the color, it bears a close resemblance to the African green parrot, being distinguished from

RING DOVE AND ROCK DOVE.

the parrot family by the *dove* characteristic, viz.: a fleshy protuberance at the base of the bill.

The **Top-knot Pigeon** (*Lepholdimus antarcticus*) is confined to southeastern Australia, where it is found perched in the tallest trees, very rarely condescending to appear on the ground or lower branches. It is gregarious, and is occasionally found in such immense flocks as to break down the tops of large trees by their accumulated weight. In their passage, they fly so closely together that it seems wonderful how they manage to use their wings. It has a very large and thick crest of gray feathers, which lies horizontal and projects some distance back of the head, lending a rather strange appearance to the creature. The neck and breast are hackled, and the tail feathers spotted with white, with some resemblance to a hawk.

TOOTHED PIGEON.

The **Toothed Pigeon** (*Didunculus strigirostris*) is a creature that is supposed to compose the link between pigeons and the dodo, hence the name *didunculus*, or little *dodo*. It is also called *toothed*, because the lower bill is notched like that of the toucan. The upper mandible is very large and sharply curved, by which structure it is able to dig up the soft roots of several plants upon which it feeds. The size is somewhat greater than that of our domestic *pigeons*, and the plumage is attractive, being a raven-black on head, neck, breast and abdomen, and the tail and under coverts a rich chestnut. It is found only in the Navigator Islands of the Pacific.

The **Passenger Pigeon** (*Ectopistes migratorius*) is our best known American bird of the *pigeon* genera, though within the last few years it has so nearly disappeared that it is seldom seen except in the Indian Territory, where one or two large roosts are still visited. When a boy I have seen these *pigeons* flying overhead in such enormous flocks that the sky would be fairly shut out from view by their bodies for hours at a time. These migrations were very frequent, caused by the very great devastations the birds wrought, requiring almost constant change of place to procure food. It is perfectly within the bounds of reason to say, as did Wilson, that as many as a billion *wild pigeons* have been seen to pass over a single course in three days, and that the consumption of food by these birds in the same time was equal to seventeen million bushels. Incredible as their numbers were twenty-five years ago, only a bare remnant now remains, and within a like period they will probably

PASSENGER PIGEON.

become extinct. So quickly do they leave their feeding places and so great is their speed of flight that specimens have been shot in northern New York with crops yet filled with rice taken from the savannas of the far South. As digestion is accomplished in these birds in less than twelve hours, the distance of more than one thousand miles must have been traversed in less than that brief time.

The **Carrier Pigeon**, in which at least six species are included, have been of great service to man, and would no doubt be more generally utilized if the telegraph had not been invented. They are still employed, however, to convey messages, being especially serviceable in time of war.

Allusions to *carrier pigeons* are very frequent in the ancient classic writers, and in the Arabic poets. Anacreon informs us that he held a correspondence with his charming Bathillus by means of a pigeon. And it is related by Ælian that Taurosthenes, a victor in the Olympian games, dispatched a pigeon stained with purple, to announce his triumph to his father, then on an island in Ægina.

Pliny also narrates that a correspondence by means of pigeons was carried on during the siege of Modena, between Decimus, Brutus and Hirtius. "Of what avail," says he, "were sentinels, circumvallations, or nets obstructing the rivers, when intelligence could be conveyed by aerial messengers?" In the crusades, the practice was tried by the besieged inhabitants of Tyre, but with less success. The besiegers had observed pigeons frequently hovering over the city, and began to suspect that these birds were messengers. Having contrived to seize one, they loaded it with false intelligence, in consequence of which they obtained possession of the place. A regular system of posting by means of carrier pigeons was established in the twelfth century by the Sultan Noureddin Mahmoud. It was afterwards improved and extended, and continued till Bagdad fell into the hands of the Mongols in 1258. Sir John Mandeville, who travelled in the fourteenth century, alludes to such a system as practised by the Turkish government. It was described at a somewhat later period as being carried on by means of lofty towers, erected at the distance of about thirty miles apart, and provided with a proper number of pigeons. Sentinels kept watch in these towers to receive the birds and to transmit the intelligence which they had brought by others. The notice was inscribed on a thin slip of paper, enclosed in a gold box of small dimensions and as thin as the paper itself, suspended to the neck of the bird; the hour of arrival and departure was marked at each successive tower, and, for greater security, a duplicate was always dispatched two hours after the first. No such regular system now exists in the Turkish dominions, but carrier pigeons are still much used there. In Aleppo, during the last century, carrier pigeons were in constant employment for the purpose of acquainting the merchants with the arrival of their vessels at Scandaroon. The impatience of the pigeon to see its young was here taken advantage of, as an additional stimulus to procure its quick return. They would travel from Alexandretta in ten hours, and from Bagdad (thirty days' journey) in two days. From Scandaroon, which was distant forty leagues, they required only from two hours and half to four hours. Towards the end of the last century the employment of pigeons from Alexandretta and Bagdad was discontinued on account of the frequent destruction of them by the Curd robbers. The practice was more recently in vogue among the Dutch merchants, for the purpose of anticipating the ordinary means of conveyance in the receipt of stock intelligence, by which they often realized considerable sums.

BURROWING BIRDS.

I have already given a brief description of the burrowing habits of the sand-marten, having to include that bird with the swallow family; but there are several other burrowers which for want of distinct or appropriate classification may be properly noticed under the above head. Indeed, classification at best is but an arbitrary arrangement in many instances, as all naturalists admit, so that it is at least pardonable to ignore arrangements except where the family characteristics are easily distinguishable, and which I shall strive to observe.

The **Kingfisher** (*Alcedo ispida*) is a burrower to the extent of seeking a convenient cover excavated by some more industrious artisan. Having its haunts about the water it does not forsake its familiar range when ready to raise a brood, but goes in quest of a hole near the water's edge, preferring a rabbit burrow, or the abandoned hole of a water-rat. Upon finding a place suited to its rather fastidious desires, the *kingfisher* expends some labor in shaping the aperture, and in digging out the hole anew so that it will have a slope upward towards the rear. This is to prevent the possibility of an invasion from rising water, for the air within the hole will prevent the water from penetrating to the elevated nest, however great may be the rise. The nest which the *kingfisher* constructs is remarkable both for its shape and the material that composes it. Mr. Gould, who has made a study of birds, and is everywhere recognized as an authority, tells us that the nest is composed wholly of fish-bones, minnows furnishing the greater portion. These bones are ejected by the bird when the flesh is digested, just as an owl ejects the pellets on which her eggs are laid. The walls of the nest are about half an inch in thickness, and its form is very flat. The circular shape and slight hollow show that the bird really forms the mass of bones into a nest, and does not merely lay her eggs at random upon the ejecta. The whole of these bones are deposited and arranged in the short space of three weeks.

It may possibly be owing to these bones and the partial decomposition which must take place during the time occupied in drying, that the burrow possesses so exceedingly evil an odor. This unpleasant effluvium, which may indeed be called by the stronger name of stench, is wonderfully enduring, and clings to the bird as well as to its dwelling. The feathers of the *kingfisher* are most lovely to the eye, but the proximity of the bird is by no means agreeable to the nostrils, the "ancient and fish-like smell" being extremely penetrating.

The *kingfisher* is a great egg producer, usually laying eight or ten eggs each season, but if these be removed from the nest with care she will continue to lay throughout the season, like our domestic fowls.

The largest of the species is the **Laughing Jackass** (*Ducelo gigas*) of Australia, its length being about eighteen inches. It is not only larger than its congeners, but differs somewhat from the other species in habits. While not refusing fish as food, it does not confine itself to a fish diet, being known to eat insects, and also rats and even snakes. The name, *laughing jackass*, has been given it because its twittering cry, common to all the several varieties, resembles the guttural call of the striped hyena, and sometimes is a fairly correct imitation of a braying donkey. Though a burrower, this species does not make its nest in the abandoned hole of some earth-dweller, but seeks the hollow trunk of a tree in which the deposit of eggs is made without constructing any nest.

The **Ternate Kingfisher** (*Tanysiptera dea*) is found in New Guinea, and is very remarkable for the extraordinary length of its central tail feathers, which are nearly bare from the junction of the wing tips to within an inch of the extremity, where they broaden out into webbed points.

The **Belted Kingfisher** (*Ceryle alcyon*) is the species common in all parts of the United States, especially along the banks of streams, over which he sits on some bough fishing all day long. When aroused from his vigils he leaps off his perch with a sharp metallic twitter, but soon assumes another position, and proceeds with his fishing. Other species, in no wise differing in habit, are the **Spotted** (*Ceryle guttata*), **Great African** (*C. maxima*), **Black and White**, (*C. rudis*), and the **Speckled** (*Alcedo ispida*).

The *kingfisher* is one of the most voracious of birds, and occasionally pays dearly for his gluttony, as several anecdotes are related to show. Woods writes, in his Natural History :

"Sometimes the bird has been known to meet with a deadly retribution on the part of his prey, and to fall a victim to his voracity. One such example I have seen. A *kingfisher* had caught a common bull-head, or miller's thumb, a well-known large-headed fish, and on attempting to swallow it had been baffled by the large head, which refused to pass through the gullet, and accordingly choked the bird. The *kingfisher* must have been extremely hungry when it attempted to eat so large a morsel, as the fish was evidently of a size that could not possibly have been accommodated in the bird's interior. Several similar examples are known ; but one, which is recorded by Mr. Quekett, is of so remarkable a kind, that it is worthy of notice. The bird had caught and actually attempted to swallow a young dabchick, and, as might be supposed, had miserably failed in the attempt.

FABLED NEST OF THE HALCYON BIRD.

"The most complete instance of poetical justice befalling a *kingfisher*, is one which occurred in Gloucestershire, and was related to me by an eye-witness. The narrator was sitting on the bank of a favorite river, and watching the birds, fish and insects that disport themselves upon and in its waters, when some strange blue object was seen floating down the stream and splashing the water with great vehemence. On a nearer approach it was seen to be a *kingfisher*, from whose mouth protruded the tail and part of the body of a fish. The struggles of the choking bird became more and more faint, and had wellnigh ceased, when a pike protruded his broad nose from the water, seized both *kingfisher* and fish, and disappeared with them in the regions below."

Such a misfortune sometimes befalls other fish-eating birds, for I one time saw a small grebe, in a St. Louis park, meet with a like disaster. It was fishing most industriously for small sun-fishes, quite a number of which it had eaten, when I observed it rise to the surface with a fish double the size of those before eaten; the grebe, with no intention of abandoning its prey, tried for many minutes to gorge the fish, and at length succeeded in swallowing the body so far that only the tail protruded, but was unable to drive it further with all its attempts at deglutition. The bird showed great distress, and gradually its efforts became fainter and fainter, until after some half hour of struggling it dropped its head into the water and expired.

The Jacamars (*Galbulidæ*) of South America, which include four species, viz.: The **Paradise** Jacamar (*Galbula paradisea*), **Green** (*G. viridus*), **Three-toed** (*Jac. tridactyla*), and **Great Jacamar** (*Jac. grandis*), though not nearly so large as the kingfisher, are very similar in form and nesting habits. The small *Galbula paradisea* is a very beautiful bird, rivalling the splendors of the most exquisite birds of New Guinea, but the others, save the *green* species, are of a dull, even sooty color. They are solitary creatures, spending the day on a single perch, watching for passing insects, on which they feed exclusively. In the breeding season, the male and female unite their labors in digging a hole in some sandy bank, inclining the excavation upward for a depth of eight to ten inches, the rear end being made globular in form to receive the nest, in which only two white eggs are laid.

BEE-EATER (*Merops apiaster*).

The **Bee-eaters** (*Meropidæ*) are very closely allied to the jacamars, and by some naturalists are included in the same order. The genera, however, is a considerable one, numbering, as it does, thirty species, and widely distributed. Those peculiar to Europe and Africa are distinguished for their rather brilliant metallic lustre, in which green is the predominant color. The breeding habits of the *bee-eaters* are peculiar. They nest usually in colonies, digging deep tunnels in steep, sandy river banks. This tunnel, which is often as much as ten feet long, opens into a considerably enlarged breeding chamber, where the female deposits usually five white eggs on the bare soil. These tunnels are dug at the expense of extraordinary labor, in which the pair perform an equal part; when the digging is finished, the bills of the birds will be found worn down to nearly one-half their original length.

The **Bee-eater of America** (*Merops Americana*), of which there is a single species, differs from its congeners of the old world in having claws of great muscular strength and of structure like those of the hawk family. It is a small dull-colored bird, and seldom more than male and female are seen together. It sits upon a perch watching for bees or beetles, upon which it darts, uttering at the time a sharp twitter, and at once returns to the perch it left. When wounded the bird will fight most viciously, lying upon its back and offering a stout defence with beak and claws. After wounding a *bee-eater* upon one occasion I thoughtlessly attempted to secure it, when in a moment the bird turned upon its back and as I reached my hand near enough, it seized my finger with a wonderful grip, sinking the claws so deeply into the flesh and holding on so tenaciously that I was some time in breaking its hold loose.

The **Marmots** and **Todies** of the West Indies, which are closely related to the bee-eaters, are also burrowers, digging holes like the sand-marten in sandy banks, but usually selecting dry ravines, and rarely run their tunnels a greater distance than six inches, in which four pearly white eggs are laid.

The **Puffin** or **Mask Bird** (*Fratercula arctica*), found plentifully along the northeast coast and on many islands, is a true burrower, though exercising its power for excavating only when necessity so compels.

As is the custom with most diving birds, the *puffin* lays only one egg, and always deposits it in some deep burrow. If possible, the bird takes advantage of a tunnel already excavated, such as that of the rabbit, and "squats" upon another's territory, just as the Coquimbo owl takes possession of the excavations made by the prairie dog. The rabbit does not allow its dominion to be usurped without remonstrance, and accordingly the bird and the beast engage in fierce conflict before the matter is settled. Almost invariably the *puffin* wins the day, its powerful beak and determined courage being more than a match for the superior size of its antagonist. When it is unable to obtain a ready-made habitation, it sets to work on its own account, and excavates tunnels of considerable dimensions.

The Feroe Islands are notable haunts of the *puffin*, because the soil, which is in many places soft and easily worked, is favorable for its excavations. The male is the principal excavator, though he is assisted by the female; and so intent is the bird upon its work, that it may be captured by hand, by thrusting the arm into the burrow. The average length of the tunnel is about three feet and is seldom straight, taking a more or less curved form, and being furnished with a second entrance. No nest of any kind is used, but the egg is laid on the earth at the end of the burrow, so that, although it is at first beautifully white, it becomes in a short time stained so deeply that it can seldom be restored to its primitive purity.

PUFFIN, OR SEA PARROT (*Alca Arctica*). CRAB-DIVER (*Uria troile*).

So deeply do the burrows run, that when a passenger is walking near the edge of the precipice upon which the *puffins* breed, he can hear the old birds grunting and chattering below his feet, disturbed by the footfalls above them.

The young *puffin* has many foes that endeavor to seize it before the bill has attained its full proportions and defensive powers. The parent birds, however, bravely defend their progeny, and have been known, as a last resource, to grasp the invader in its beak and hurl themselves and foe into the sea. Once among the waves, the *puffin* has the advantage, for it is an excellent swimmer and diver, finding its food among the swift fishes which it catches with facility. Indeed, a *puffin* may be frequently seen with a half-dozen small fishes in its mouth at one time, all arranged in a row with the tails projecting. The bill of the *puffin* is so large and unsightly that the name *mask-bird* is not inappropriate, rendered more so by the singular fact that it is shed each year.

There are many other birds which pass a semi-burrowing life, making their nests in hollows already excavated, and either using them without adaptation, or altering them very slightly for the purpose of depositing their eggs. The **Jackdaw** (*Corvus monedula*) is frequently found making its nest within a deserted rabbit burrow. The **Stock-dove** (*Columba œnas*) is also sometimes found rearing its young within an abandoned tunnel, as is also the **Sheldrake** (*Tadorna vulpanser*). This latter, however, invariably adopts a burrow that is contiguous to water, in order that its young may be more conveniently fed on the insects and crustacea that live in the water-courses near the sea. The *sheldrake* is not fastidious, being content to accept nearly any hole, so that it be suitable for her eggs, which are generally from twelve to fifteen in number, which she carefully covers with down plucked from her breast.

The **Stormy Petrel** (*Thalassidroma pelagica*), more commonly called *Mother Cary's chicken*, is also a member of the burrowing tribe, though its appearance would least suggest such a habit, for it is nearly always met with far out at sea. In fact, so constantly does it seem to be on the wing, and at such remote distances from the shore, that for many years it was supposed to never visit land, but to carry its eggs under its wing and there incubate them. This belief was not disproved until within the last fifty years, and many sailors even yet refuse to discredit the old fancy. It is therefore with much interest that the facts concerning its nesting habits are given.

If the *stormy petrel* can find a burrow already dug it will make use of it, and accordingly is fond of haunting rocky coasts, and of depositing its eggs in some suitable cleft. It will also settle in a deserted rabbit-burrow, if it can find one sufficiently near the sea, and is found breeding in many places which would equally suit the puffin.

Failing, however, all natural or ready-made cavities, the *stormy petrel* is obliged to excavate a tunnel for itself, and even on sandy ground is able to make its own domicile. Off Cape Sable, in Nova Scotia, there are many low-lying islands, the upper parts of which are of a sandy nature, and the lower composed chiefly of mud. Not a hope is there in such localities of already existing cavities, and yet to those islands the *petrels* resort by thousands, for the purpose of breeding. The birds set resolutely to work, and delve little burrows into the sandy soil, seldom digging deeper than a foot, and, in fact, only making the cavity sufficiently large to conceal themselves and their family treasures.

Each bird lays a single egg, which is white and of small dimensions. The young are funny-looking objects, and resemble puffs of white down rather than nestlings. The parent attends to its young with great assiduity, feeding it with the oleaginous fluid which is secreted in such quantities by the digestive organs of this bird. So large indeed is the amount of oil, that in some parts of the world the natives make the *stormy petrel* into a lamp by the simple process of drawing a wick through its body. The oil soon rises in the wick, and burns as freely as in any of the really rude and primitive, though ornamental, lamps of the ancients.

The *petrel* only feeds its young by night, remaining on the wing during the day, and flying to vast distances from the land. Owing to this habit, and its custom of taking to the sea during the fiercest storms, it has long been an object of dread to sailors, whose illogical minds are unable to discriminate between cause and effect, and who fancy that the *petrel*, or *Mother Cary's chicken*, as they call the bird, is the being which, by the exercise of some magic art, calls the storm into existence. They even fancy that the *petrel* never goes ashore nor rests; and will tell you that it does not lay its egg in the ground, but holds it under one wing, and hatches it while engaged in flight. To the vulgar mind, everything incomprehensible is fraught with terrors, and so the harmless, and even useful *petrel*, is hated with strange virulence.

Throughout the breeding season, the *petrel* is indefatigable in search of food, and will follow ships for considerable distances, in hopes of obtaining some of the offal that is thrown overboard by the cook. Even if a cupful of oil be emptied into the water, the *petrel* will scoop it up in its bill and take it home to its young. During the night it mostly remains with its offspring, feeding it and making a curious grunting noise, something like the croaking of frogs. This noise is continued throughout the night, and those who have visited the great nesting-places of the *petrel*, unite in mentioning it as a loud and peculiar sound. The ordinary cry is low and short, something like the quacking of a young duck. By day, however, the birds are silent, and only those who keep nightly watch on the ship's deck, can have an opportunity of hearing their chattering cry.

The burrow in which the young *petrel* is hatched is extremely odoriferous, the oily food on which the bird lives having itself a very rancid and unsavory scent; and in consequence of feeding upon this substance, both the habitation and the inmates are extremely offensive to the nostrils. The young bird is at first very helpless, and remains in its excavated home until it is several weeks of age. One of these birds was seen on the Thames in the month of December, 1823, where it attracted some attention, its peculiar mode of pattering over the water causing it to be taken for a wounded land bird, and inducing many persons to go in vain pursuit of the supposed cripple.

While many different genera rear their brood under ground there is a greater number of birds that are wood-burrowers, but among these there is no similarity in the color of their eggs. Most prominent among the birds with which we are best acquainted, that excavate their nests in the hollows of trees, are the *woodpeckers*, common alike to both the old and the new worlds. Birds of this family are easily distinguishable by the peculiar form of the beak, feet and tail. The powerful sharp-pointed bill enabling them to chip away the bark and wood, while the claws and tail are so formed as to support the bird firmly while this work is being performed.

As is pretty generally known, *woodpeckers* make their nests in a tunnel which they drive through the decayed branches of a tree, never attacking the solid limbs, upon which they could make but small impression. Oftentimes trees which have the appearance of soundness are much decayed beneath the bark, and though this defect is not easily discoverable by man, unless such vegetable parasites as the lichen have made their appearance over the unsound portion, yet a *woodpecker* is able to tell unerringly just how far the decay has progressed. Such places are seized upon by the bird, which drives its lance-like bill into the softened wood, and by industrious hammering soon excavates a hole of proper dimensions and several inches deep, in which four pearly white eggs are deposited without further provision. When incubation begins the female is not easily driven from her nest, for I have more than once climbed, not without much difficulty and noise, up a tree in which a *woodpecker* had her nest, and succeeded in placing my hand over the hole before she would seek flight.

The **Blue-bird** (*Sialia sialis*), so common everywhere in North America, and welcomed as the harbinger of spring, while not a burrower, builds its nest in the hollow of a stump or limb, or will drive away the marten, and take up its domicile in a cote. Every person holds the *blue-bird* in their affections, because of its azure plumage and soft, wooing notes, regardless of its rather vicious habits of making war on other birds.

The **Rollers** (*Coracias garrula*) is a term applied to a family of European birds, found also in northern Africa, because of their curious "tumbling" motions when ascending, in which respect they resemble the "tumbling" pigeons, except that the latter assumes this singular motion while descending, and the former only when ascending. The *roller* nearly equals the raven in size, and possesses an equally discordant voice, which it almost incessantly exercises; the plumage, however, is very pleasing. It is very irregular in its nesting habits, sometimes building in the hollows of trees, sometimes upon the bare ground, and, again, tunnelling deeply into a sandbank, after the manner of the kingfisher.

The **Bell-bird** (*Arapunga alba*), of Guiana, is one of the most singular of the feathered tribe, not only in one, but several particulars. The color is a pure white, and the size and structure that of a pigeon, but in habits it is without example. The most curious features that distinguish it are its notes, crest and nesting habits. Its haunts are among the darkest coverts of South American forests, and usually about damp and most forbidding places, where it remains nearly always on the ground, hunting insects that are found most numerous about sedgy banks. For this reason, notwithstanding its conspicuous color, the *bell-bird* is rarely seen. It may be distinguished, however, at some distance, by the curious horn-like structure which grows from its forehead, and rises to a height of some three inches when elevated. This "horn" is jetty black in color, sprinkled very sparingly with little tufts of snowy-white down, and, as it has a communication with the palate, has probably something to do with the bell-like sound of the voice. The song or cry of this species has been admirably described by Waterton, in his well-known "Wanderings in South America."

"His note is loud and clear, like the sound of a bell, and may be heard at the distance of three miles. In the midst of these extensive wilds, generally on the dried top of an aged mora, almost out of your reach, you see the campanero, or *bell-bird*. No sound or song from any of the winged inhabitants of the forest, not even the clearly-pronounced '*Whip-poor-Will!*' from the goat-sucker, causes such astonishment as the toll of this species.

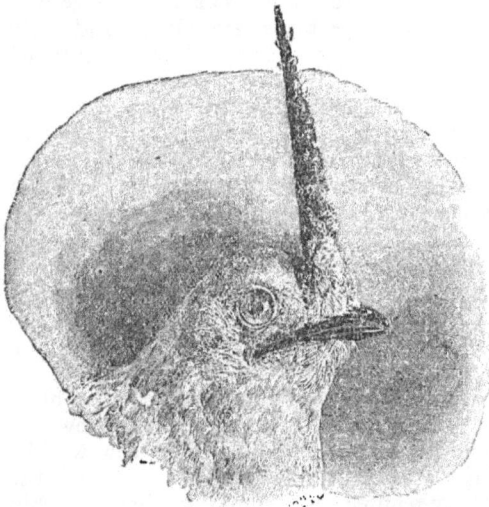

BELL-BIRD.

"With many of the feathered race, he pays the common tribute of a morning and evening song; and, even when the meridian sun has shut in silence the mouths of almost the whole of animated nature, the campanero still cheers the forest. You hear his toll, and then a pause for a minute, then another toll, and then a pause again, and then a toll and again a pause. Then he is silent for six or eight minutes, and then another toll, and so on. Actæon would stop in mid-chase, Maria would defer her evening song, and Orpheus himself would drop his lute to listen to him, so sweet, so novel and romantic is the toll of the pretty snow-white campanero."

The "horn" of the *bell-bird* is only erect while the creature is excited and during the resonant cry, and when the bird is at rest it hangs loosely on the side of the face. Not until within the last few years has anything concerning the nesting habits of this bird been known. Recent travellers through Guiana have at length determined the fact, however, that during the breeding season the *bell-bird* retires to higher ground, and, invariably, to a rocky region, and selects a spot usually within a crevice or between two large rocks; here it digs a hole nearly twelve inches deep, which is lined with grasses rather roughly arranged, on which it lays two white eggs. The nest is so securely hidden from view that only by the greatest accident would it be discovered.

NEST OF THE CANARY.

SINGING BIRDS.

If I were to undertake the labor of describing all the forest warblers of the world, preparation would be necessary for an enormous volume, which, perhaps, no publisher could be urged to undertake. The species are so numerous that every clime has received from God the cheerful blessing of song birds, which vie with the flowers for supremacy in the esteem of mankind. How strange the fact, that among the most gorgeous-plumaged birds of the world we find no songsters; that those of sweetest tongue are almost invariably birds of sombre feathers. The fable of the peacock's cry to Jupiter applies not only to that creature but to all birds quite as well; but while the woods are filled with the trilling melody of sweet singers that charms our ears, our eyes are not slighted, for on every side of sylvan dale, on bush, tree, ground and stem, sit or flit the most brilliant creatures, which represent the scenery of the world's theatre, as the song birds represent nature's orchestra. We cannot all be actors, some must be spectators, others scene shifters, and the bright-plumaged birds play a no less important part than the singers, because they lend an equal charm to the woods, and incite alike our admiration and thankfulness to a Creator so generous of His gifts to please the eye and ear of His children.

Only comparatively few of the songsters can here be described, but those thus introduced are eminently representative of the most interesting species found generally in civilized countries, and especially in North America.

The **Starling** (*Sturnus vulgaris*) is found in nearly all temperate climes, but is most numerous in Britain and Europe. It is gregarious, going in flocks of several hundred, under control of a leader. The nest of this bird is a crude affair, and built in a variety of places, sometimes on the ground, or even in deserted rabbit burrows, or in pigeon cotes, forming a ready affiliation with any other variety. It is an amiable and interesting pet, being easily taught to speak with the facility and distinctness of a parrot, and combines with this faculty a marvellously instinctive judgment. It has a great affection for its offspring, which is admirably illustrated by the following anecdote: A barn, in which a *starling* had her young, caught on fire, an incident which the bird was first to discover, and at once she flew about in mad distraction, as if trying to call attention and help from persons in the vicinity. As the flames grew nearer her young, and perceiving that they were about to perish, she flew through the smoke and seizing one of the fledglings bore it in her beak to a place of safety, returning immediately after another, and in this way she removed her entire brood of five to another spot where they were deposited together, and a new nest formed for their comfort. In its habits the *starling* is very similar to our common cow birds, though its plumage is very handsome.

STARLING.

The **Cross-bill** (*Loxia curvirostris*) is native to both America and Europe, where it is found in considerable numbers, being sparrow-like in both size and habits. This bird is remarkable for the very curious shape of the bill, the upper and lower mandibles completely crossing, giving it an awkward and mal-formed appearance, but which, in fact, is a remarkable adaptation to the feeding habits which nature has designed it to practise. The jaws are very muscular, and by the peculiar form of its bill the bird is able to bite an apple through the centre and extract the seeds therefrom, or shell the seeds out of pine cones, or even to break the shell of hard almonds. The young of the *cross-bill* do not exhibit the peculiar formation of bill noticeable in the parent birds.

CROSS-BILL.

The **Bullfinch** (*Pyrrhula vulgaris*) may hardly be called a wild bird, for it exists in no country in any considerable number, being rather a distinctively cage bird, like the canary. Besides, in its natural state the *bullfinch* is rather an uninteresting bird, against which the charges have been laid of denuding fruit trees of their flowers, through no other than merely mischievous propensities. In domestication the bird is wonderfully interesting, and is the most highly prized of all feathered pets. Before taken into captivity its notes are only a simple chirrup, but in bondage it develops really marvellous powers of imitation, easily learning to whistle even complicated operatic airs, with a soft, flute-like melody. In order to develop this interesting faculty, the bird must be taken when young, and immediately after feeding be given its lessons, which consist in the playing on some soft instrument (the flageolet is preferable) the airs it is desired that the bird should learn.

The *bullfinch* manifests a loving disposition towards some persons, and an inconceivably violent hatred towards others without any apparent reason. Many anecdotes are reported illustrative of both these qualities, one of which relates that a *bullfinch* actually died of love for its beautiful mistress who refused to return its affection. Such stories, however, are unreliable, and not worthy of insertion in a book of this character. This bird, in freedom, builds its nest usually in a thick hazel copse attached to the side of a slender branch, and deposits therein five eggs, beautifully marked with purple and brown streaks, and a ring of greenish white on the larger end. The bird, which is about six inches long, is of a slaty-gray on neck and back, while the head, coverts of tail and wing are black and tips of wing white.

The **Kernel-biter**, or **Grosbeak** (*Coccothraustes vulgaris*), belongs to a family of which our common *red bird* (*Cardinalis virginianus*), also called *crested red bird* and *Virginia night-*

SNOW FINCH (*Plectrophanes nivalis*) AND GOLD-FINCH (*Emberizi citrinella*).

ingale, is a member. The former species, however, is a native of Europe, where it sometimes appears in flocks of twenty or more. It is a very shy bird, with dull plumage and monotonous notes. The name *kernel-bird* has been given on account of its habits of feeding, in which it is so voracious as to swallow the seeds of such fruit as cherries and plums. There are six different species, none of which are very interesting.

The **Chaffinch** (*Fringilla cœlebs*), or **Bachelor Finch**, is a distinctively English bird, and a merry little fellow he is, too. After the breeding season, the males and females separate by voluntary divorcement, and so remain until the next mating time, on which account the name *bachelor* has been applied. His merry notes and *pinck, pinck,* do not serve to establish him in the favor of gardeners, upon whose crops he makes serious invasions, being as bold in his thievery as is the sparrow. The nest which this bird constructs is a marvel of neatness and ingenuity, being composed of moss, wool and hair admirably interwoven and lodged in the main fork of a tree, where it is most difficult to detect, being of the colors of its surroundings, so as to form an almost perfect simulation to the tree branches.

The **Goldfinch** (*F. carduelis*) is much more charming than his congener, both in color and habits, and is a very popular English pet. Its natural haunts are about houses, and captivity destroys none of its graces, but rather promotes its faculties of charming entertainment. No bird can be taught so easily, which accomplishment is turned to good account by bird fanciers, who teach it a great variety of amusing tricks, such as firing a cannon, whistling at the word of command, simulating death, performing on the trapeze, etc. It is a sweet singer, and is beautifully arrayed in a variegated plumage; about the bill is crimson, head and neck a jet black, sides of face, white, back and breast grayish-brown.

The nest of the *goldfinch* is generally located near the tip of a branch bearing a thick foliage, and is so artfully made to assimilate with the surroundings as to be most difficult of detection; nor does the bird fly to or from her nest directly, always creeping some distance along the branch before taking flight.

The **Girlitz** (*Serinus hortulanus*) is a native of southern Europe, and bears a close resemblance to the canary. It is a garden bird with pretty markings of black spots and stripes over a ground color of pale green. The points of the wings and tail are also black. It has a soft note varying from a chirrup to an occasional twitter.

The **Magpie** (*Pica caudata*) is an American bird, and confined largely to the United States. A species called the **Yellow-bellied Magpie** is also found in Europe, an identity of habits being observable in the two. It is a daring bird, of an omnivorous appetite, eating carrion, smaller birds, eggs, the young of quails, rats, frogs, mice, snails, caterpillars, etc., and has such a ravening desire for fresh meat that it will attack the galled places that may be exposed on horses and mules. It is a cunning and intelligent bird, easily tamed and with but little care may be taught to whistle any air, or to talk

GROUP OF FINCHES. CROSS-BEAK, RED BULLFINCH, KERNEL-BITER, CHAFFINCH AND GIRLITZ.

fairly well. Its nesting place is usually a thicket, or close fork of a large tree. The nest is built with great care, of fine and coarse sticks interlaced and then cemented with clay, and of spherical shape, with a hole through the centre from which the head and tail protrude.

Magpies are frequently mentioned in ancient history, and they have a conspicuous place in mythology. Ovid writes of an interesting family of young ladies who were changed into *magpies:*

" And still their tongues went on,
    though changed to birds,
 In endless clack, and vast desire
    of words."

The Greeks and Romans dedicated the *magpie* to Bacchus, because of the garrulity that drink incites.

RAVEN (*Corvus corax*) AND MAGPIE.

The **Blackbird** (*Turdus vulgaris*) is of great variety, and found in both the old and new world. The **Crow Blackbird** (*Quiscalus purpureus*) is found in immense flocks nearly everywhere in the United States and Mexico, where it ravages the fields and is a very pest to farmers. The **Red Shouldered** (*Agelaius phœniceus*) is equally common, though it does not associate in large flocks, and is generally found in the company of other species, such as the *crow blackbird* and *cow-bird.* Its notes are musical but rather melancholy.

The **Cow-bird** (*Molothrus pecoris*) has nothing to recommend it to favor, but much to condemn. The males are polygamous and the females are cuckolds, depositing their eggs in other birds' nests when the rightful owner is temporarily absent, and imposing all the labor of rearing the young on the foster-parents.

The **Cat-bird** (*Mimus carolinensis*) is very numerous in nearly every part of the United States, and is one of the earliest arrivals of spring. It is of a dull lead color, and gives out a very disagreeable note, resembling the mew of a cat, until the breeding season is nearly past, when the male pours forth a rich, warbling melody almost rivalling the mocking-bird.

BLACKBIRD AND NIGHTINGALE.

The **Nightingale** (*Sylvia luscinia*) is confined to the old world, appearing in England in the early spring, and retiring to the south early in the fall. It is about the size of our cat-bird, and of scarcely more attractive plumage, the upper parts of the body being a russet-brown, and the breast and abdomen a grayish-white. The nest is made of grass and leaves, with little regard for appearances or strength, and is usually lodged in the low branches of a tree or on some shrubbery. The eggs are of an olive-brown and usually five in number. The *nightingale* is the sweetest singer of European birds, though at times, like the cat-bird, its cry is very unmusical.

The **Brown Thrush** (*Harporhynchus rufus*) is the American rival of the nightingale, whose notes he can almost equal. This bird comes to us in May and may then be found always in pairs in nearly every thick hazel copse, where it generally builds its nest, though sometimes a spot on the ground is selected, in which usually four eggs mottled with brown specks are deposited. It sings generally during early morn, and sometimes will pour forth its rich melody for hours, making the woods ring with its charming instrumentation. The *thrush* is noted also for its attachment and affection for its young, in whose defence it will brave any danger. The color is a brown on the back, hackled with white on the breast.

The **Mocking Bird** (*Mimus polyglottis*) is beyond compare the most proficient minstrel in all the world's feathered orchestra.

Its home is the new world, being found in North and South America, but rarely above latitude 40°, and being numerous only far south of this line. It is not a bird of bright plumage, but the ease and gracefulness of his motions, the nervous throbbing of his wings, and the joyous

BROWN THRUSH.

animation displayed while warbling and trilling his varied lays, mark him as a creature of surprising intellect and extraordinary accomplishments. He possesses a voice capable of almost inconceivable modulation, ranging a gamut measurable by all the sounds between the gentle "*cheep*" of the sparrow to the harsh scream of the eagle; from the soft *pianissimo* to the *multisonus*, a perfect diapason of harmony. In his native groves among the magnolias at morning's dawn, when the wood is already vocal with a multitude of songsters, his voice rises preeminent above every competitor. His expanded wings and tail, glistening with white, and the buoyant gaiety of his actions arresting the eye as his song most irresistibly does the ear, he sweeps round with enthusiastic ecstasy, and mounts or descends as his song swells or dies away. While thus exerting

himself, a bystander destitute of sight would suppose that the whole feathered tribe had assembled together on a trial of skill, each striving to produce his utmost effect. He often deceives the sportsman and sends him after game birds that remain invisible; even birds themselves are deceived by this marvellous mimic and are decoyed by the fancied calls of their mates, or dive with precipitate haste into the depths of thickets at the scream of what they believe to be the sparrow-hawk. Again, he whistles a call that starts a dog to his master, or chirrups like a young chicken in distress, and causes the hen to hurry with bristling feathers and hanging wings to her injured brood. He imitates the creak of a wheelbarrow, the cluck of a hen, the "*putrack*" of a guinea fowl, or may be taught to whistle an air. In short, the voice of all varieties of the feathered kind is perfectly imitated by this wondrous musician.

The nest of the *mocking bird* is generally built in hedges or thorn bushes, of weeds, sticks, straws and grass, and lined with fine fibrous roots, in which usually four eggs are laid. The bird is very courageous in defence of its nest and young, and has been known to assail, with fatal effect, large black snakes that attempted to devour the brood. Occasionally two broods are hatched in one year, those of the first, however, being much larger.

The **Baltimore Oriole** (*Oriolus baltimore*), (so called from the orange and black of its plumage, those being the heraldic colors of Lord Baltimore), is also peculiar to the United States, distributed as far north as Minnesota, though appearing only in the summer season. As soon as the warm airs

NEST OF THE BALTIMORE ORIOLE.

of April begin to start the verdure the *orioles* come northward in pairs and begin at once preparations for rearing a brood. The nesting habits of this bird are particularly interesting, as its soft warbling notes are charming. In the far south the nest is made penduline of Spanish moss, so loosely woven as to permit the air to circulate freely through it. In the Northern States, it is hung upon the extremities of widespreading branches, invariably at high altitudes, and where it has perfect exposure to the sun. In weaving its nest the *oriole* ties the materials that compose it to the branch, dexterously using its bill and feet for that purpose, and giving to it the shape of a hanging-bag, or old-fashioned purse-bag, leaving a hole near the top for entrance, and the interior lined with the softest texture. In making its nest the *oriole* seizes any material suitable to the purpose. An anecdote is related to illustrate the bird's propensity for selecting soft lining for the nest. A lady in Connecticut was sitting beside an open window sewing, when being called away for a few moments, she returned to find that her spool of silk thread and measuring tape were missing. Diligent search failed to discover the articles until by accident she saw the tape hanging from the nest of an *oriole* that had built in a tree near by. At the expense of much trouble, incited by curiosity, the nest was recovered, when it was found that the silk had been woven into it with such dexterity as made it impossible to disentangle. The female lays four, five or even six eggs, of a light gray color with dark spots and lines, which are hatched in a fortnight.

Wall-Creeper (*Tichodroma muraria*) is the name given to a very common little bird widely distributed over both Europe and America. It seldom perches

WALL-CREEPER.

on a branch, but is continuously circumambulating the trunk of some tree in industrious search for insects, and is most useful to farmers for the destruction it occasions to harmful grubs, caterpillars and noxious flies. It nests in April, building in a crevice or hole after the manner of woodpeckers, and usually lays five eggs of white. Its note is a merry "*chuck-chuck*," but only occasionally uttered, nor does it remain long on any tree, but keeps flitting from one to another in apparently nervous distraction. It is of a dun color on the back, with breast of white. Another species, sometimes called *speckled woodpecker*, and *fly-snapper*, is also plentifully distributed over the Northern States, and is equally useful in ridding trees of harmful insects. It is considerably larger than the *wall-creeper*, and is handsomely clothed in a raiment of white and black, so as to present a speckled appearance. Other species include the *red-head woodpecker*, *yellow-hammer* and *wood hen*, also

called *giant woodpecker*. This latter species is a very large bird, almost equal to a pigeon, but is nowhere plentiful and is very wary. The color is black, with a top-not of scarlet, and has a neck of considerable length which makes its crest appear more prominently. It builds its nest always in the dead branch of a very tall tree, beyond the reach of interruption, for which reason few of its eggs have ever been recovered. The *yellow-hammer* may be classed among our prettiest birds, in a plumage diversified with yellow wings spotted with black, a white breast similarly flecked, a black gorget crescent-shaped, and a spot of red upon the head. Both the *giant woodpecker* and the *yellow-hammer* are esteemed for their flesh, but the other species are rejected as being wholly unfit for food. Not only is the flesh of nearly all the species unpalatable, but they are generally found to be infested with parasites, besides giving off a rather offensive odor. The general habits of all the genera are very similar, and the birds are so well known to my readers as to render more particular description unnecessary.

The **Nut-cracker** (*Nucifraga caryocatactes*) is peculiar to Europe, and though classed by Brockhaus among the singing birds, it rather belongs to the crow genera. The nest of this species is made at the extremity of a long tunnel cut in the wood of a decaying tree, either originally dug by the bird, or altered and adapted to its purpose. The eggs usually number five and are of a grayish color. The body is of a warm brown color flecked with white spots. America also possesses a species of *nut-cracker* scientifically known as *nucifraga columbiana*. It has a rather attractive plumage and formidable claws, though not strictly a carnivorous bird. It is most commonly met with near the sea shore, where it is sometimes seen in large and very noisy flocks.

The **Blue Jay** (*Cyanurus cristatus*) is a noisy inhabitant of the United States and has his congeners in nearly all countries of the globe. It is a beautiful bird but is in great disfavor on account of its predaceous habits. On account of the great number of acorns it consumes, this being a favorite food in Germany, it is called the **Acorn Bird** (*Garrulus glandarius*). He is a regular visitor to gardens and orchards, where he

NUT-CRACKER AND ACORN BIRD.

regales his appetite on cherries, berries, peas, and in fact nearly everything that grows. He also takes the part of a sentry and gives noisy notice of the proximity of an owl, hawk, weasel, marten or rat, and is hated by sportsmen because of his raising a hue-and-cry in the woods at the sight of a hunter, giving alarm to all game that may be within the sound of his voice. The *jay* is as predaceous as the magpie, robbing other birds' nests, not only of their eggs but also devouring their young. The nest of the *jay* is a rude affair constructed of coarse sticks lined with grasses. The eggs are generally five in number, of an olive-brown, marked with dark spots.

SAFFRON FINCH (*Regulus cristatus*).

Finches. There are no less than forty-four species of the *finch* in North America, and almost as great a variety in Europe, and a dozen or more in Asia. There is great similarity in size, but no little diversity in color and habits. The **Siberian Finch**, for example, is a most stupid bird, thrusting its head into the grass at the sight of an enemy and leaving the body exposed, like the ostrich; the plumage, too, is dull and uninviting. The **Bearded Finch** (*Amadina fascinata*), on the other hand, is a bird of extraordinary activity and cunning, of exquisite plumage, and a charming singer. The **Snow Finch** (*Plectrophanes nivalis*) is nearly white, an inhabitant of northern regions, and has a very feeble voice. The **Golden Finch** (*Emberiza citrinella*) is a most charming little bird, with rich mellow notes, but is a great pest to farmers, though extremely shy in its habits. The species in this country are generally familiar about our doors, and are held in high favor for their pretty songs and their usefulness in devouring harmful insects.

The **Titmouse** is represented by twenty-four species in America, and an equally large variety distributed throughout other parts of the world. They, too, bear a striking family resemblance, being distinguishable from other birds by their sharp, strong beaks, boldly defined color of plumage, and a nervous, incessant movement, as if it were impossible for them to remain quiet a moment. The claws are sharp and strong, enabling them to cling to the under part of small branches, or hang suspended from the bottom of an apple while feasting on the fruit. Their appetite, however, is omnivorous, as they feed alike on seeds, fruit, insects, and have a special fondness for fresh meat. They generally build their nests in the hollows of a tree stump, or the crevice of a wall, in which sometimes as many as a dozen whitish-gray eggs are laid. There are noticeable exceptions to this rule, however, as some species rival the oriole or tailor-bird in the wonderful construction of their nests.

GROUP OF TITMICE.

The *great titmouse*, the *blue titmouse*, the *long-tailed titmouse*, the *coal titmouse*, and the *marsh titmouse*, are common in England and other parts of Europe. The *crested titmouse* is found in the northern parts of Europe. The *bearded titmouse*, plentiful in Holland, has tufts of black feathers pendent like whiskers from the sides of its face. The *yellow-checked titmouse* is an Asiatic bird, found chiefly among the Himalayas. It is a very queerly-marked bird, the cheeks and whole under-surface of the body being pale-yellow, the flanks having a greenish hue; the wings gray, mottled with black and white; the tail black, with a slight edging of bottle-green, and the rest of the plumage of a jetty black. The *rufous-bellied titmouse* is found in southern India and Nepaul.

The *cape titmouse* is one of the most ingenious builders of the family, native to South Africa, and is noted for the peculiar shape of its nest, as may be seen in the illustration. The nest is in the shape of a bottle, and is made of cotton. The male constructs for himself a saucer-like pocket on the outside of the neck, while the interior is given to the female and her brood. While the mother is in the nest the father sits in his little sentry-box faithfully guarding the family safety. Instantly the mother leaves her little brood, the father, wishing to accompany her, closes up the narrow entrance of the bottle by beating it with his strong little wings till his end is accomplished.

The **Penduline Titmouse** is also noted for its nest-building skill. This nest, suspended from the branches of trees, has exactly the form of a chemist's retort, but instead of being built of hard material only fine moss and down enter into its composition. The opening is woven with such care that not one fibre projects beyond another. How this bird, while on the wing, enters the inverted neck of the nest it is difficult to tell, as the opening has scarcely the diameter of its body. But it darts in at full flight and without disturbing a single fibre.

In the United States the two most common and interesting

NEST OF CAPE TITMOUSE (*Parus capensis*).

species are the **Black Capped Titmouse**, more commonly called **Chicadee** (*Parus atricapillus*), and the **Crested Titmouse** (*Lophophanes bicolor*) whose merry notes of *peter, peter, peter*, make all the woods cheerful. These two are commonly seen together, associated often with the brown creeper and spotted woodpecker, all of which varieties exhibit wonderful activity in the search for food. The two species named are very like in habits, both building their nests in an abandoned squirrel-hole, or fashioning their habitation, at the expense of great labor, by excavating into decayed tree branches, and both also lay five white eggs. The *crested titmouse* is often called *tomtit*, and though small is a most courageous bird, and fights desperately with both beak and claw. He is also very curious to explore mysteries, and at such a time acts in a truly comical way. Upon one occasion while in the woods I saw a pair of *titmice* flitting from branch to branch and filling the air with their musical calls of *peter, peter, peter*, when out of curiosity I attempted an imitation of their notes. As often as they sang I whistled in response, when presently I saw them drawing gradually nearer. Thus encouraged I continued to answer their calls until they came so near me that I could almost reach them with my hand. All this time they hopped cheerily about looking at me in an exceedingly droll manner, cocking their heads in many positions, as if afraid to trust the vision of either eye. In this manner we continued to watch each other until I became tired of the amusement.

House Wren (*Troglodytes parvulus*). This charming little bird is common throughout the United States, and though a greater number seem to migrate to warmer latitudes in the fall, a few remain and may be seen in mid-winter as far north as Missouri. It ranks next to the humming-bird in size, but though most diminutive of body, it pours forth, at breeding time, very shrill and loud notes, resembling "*chippery, chippery, chippery,*" chip, with a liquid intonation which cannot be imitated by letters. It is very sociable and prefers to build in or near human habitations. They have been known to build in the sleeve of a coat hung against a wall, or even in old hats. If the cavity they select be too large, they fill the unused space with sticks or other convenient material, leaving an entrance barely large enough to admit their wee bodies. In the centre of this mass a hemispherical nest is constructed, compact in its architecture, composed of fine material and warmly lined with feathers and the fur of animals. The eggs are usually seven in number, with a white ground thickly blotched with small spots of reddish-brown. The *wren* is insectivorous, and one of the most useful friends to the farmer.

NEST OF PENDULINE TITMOUSE (*Parus pendulinus*).

Shrikes (*Lanius collurio*) are represented in America by two species, the **Loggerhead** and the **White-rumped,** the former being found chiefly in the Southern States, while the latter extends its range as far north as Wisconsin. The former is much the handsomer bird, so closely resembling the mocking bird that it is frequently mistaken for that species. The latter is distinguished by its darker color on the upper parts, and a white rump. Neither of these are very musical, but the European congeners, especially the **Great Gray Shrike** (*Lanius excubitor*), is noted for its imitative powers and charming notes. All members of the *shrike* family have a remarkable habit of impaling their prey on thorns or other sharp points, for what purpose naturalists have never been able to determine. They subsist chiefly on insects, but sometimes take their prey from among smaller birds, reptiles and mice. Frequently the prey is killed and eaten upon the spot, but as often it is conveyed to a neighboring thorn tree and there impaled, where it is very often suffered to remain until entirely decayed. Even in captivity this same propensity is exhibited, on which account it is commonly called the **Butcher Bird.** The nest is large and coarsely constructed, and usually located in a low tree-top, among vines which offer some concealment.

CRESTED TITMOUSE.

The **Indigo Blue Bird** (*Cyanospiza cyanea*) is found everywhere between the latitudes of Mexico and Nova Scotia, and is one of our prettiest birds, in addition to which charms is a sweet voice, though of brief compass. It delights in perching high up on the small limbs of a lofty tree and singing its many notes, like a praise offering to the Deity. The color is changeable, but during the months of May, June, July and August, the male is clothed in a plumage of bright indigo, but later in the season the color is changed to a blue, then light green, and towards winter to a dull brown. It builds its nest in dense grass, generally between upright stems, suspended on either side, and is constructed of grasses and other fibrous material. The eggs are five in number, of a deep blue, with purple blotch on the larger end. Its food is seeds and insects. Varieties are also found in Europe.

HOUSE, OR JENNY WREN.

Cedar Bird (*Ampelis cedrorum*), also called **Cherry Bird,** is found breeding in every State of the Union. It associates in flocks of considerable size and is partial to cedar trees, off the berries of which it largely feeds. It also strips the mountain ash with such voracity that it sometimes becomes helpless from over fullness. The size is that of the crested titmouse, which it resembles in having a similar crest of erectile feathers, the plumage, however, is more attractive. It also feeds off cherries and other small fruits, but prefers certain insects, the canker-worm being its choice of food. It nests in July and lays five eggs of a light slate color, marked with purple blotches.

22

Dotted Fly-Snapper (*Musicapa grisola*). This is a European species particularly plentiful in England and Ireland, but several representatives of the genus are common in the United States under the names of *gnat catchers* and *warblers*. They are all small birds with no special markings or particularly interesting habits. Their song is generally a simple note, such as *tree, tree, tree*, or a compound, resembling *te de ter-itsca, te derisca*, a striking resemblance to the notes of the chicadee. They are all expert fly-catchers.

SPECKLED TITMOUSE.

Robin Redbreast (*Pipilo erythrophthalmus*), also called *Joree, Red-eyed ground robin, Towhee bunting*, and many other local names, is a familiar bird everywhere east of the Rocky Mountains. It is a bird of attractive plumage, of black, chestnut and white, and of sociable habits, in addition to which features it possesses a sweet voice of great tenderness, so that from the earliest times it has ever been regarded with special favor by man. The story of the Babes in the Woods, and particularly that portion wherein is so pathetically described the tender offices of the robins that covered the bodies of the little children has, no doubt, greatly helped to increase the esteem in which these birds are held, but they have many qualities to commend them not found in fiction.

DOTTED FLY-SNAPPER.   NECKLACE FLY-SNAPPER.

They associate in pairs until the time for migrating, when as many as a score may be seen together. Like the wren, however, some few remain in the Middle States throughout the year. The song it utters may be represented by the words, *towhee, towhee*, twice repeated; but sometimes it is more musical, and trills out, *t'sh'd-witee-te-te-te-te*. The *robin redbreast* builds its nest in a natural depression, even with the surrounding surface, filling the cavity with coarse material, and little or no lining is used. Great

ROBIN REDBREAST.   GARDEN RED-TAIL.

care is taken to conceal the nest, which is usually under an overhanging tuft of grass. Five eggs are laid, of a pale flesh color, speckled with dots of brown.

The **Ground Robin** (*Erythacus rubecula*). This bird is more properly the *robin redbreast*, and is so designated in Europe, for the breast is a pale red, which color is continued about the throat, chin, forehead and around the eyes. This bird is very common in the United States, usually found in fallow fields, but also visits the city parks. In size and habits it is identical with the joree robin, but is more sociable, and quick to adopt the friendship of man. The nest is usually built in the same manner as that of the joree, but a little encouragement will cause it to build in a room, and become the voluntary pet of a family. While insectivorous, it also exhibits queer tastes, being extremely fond of butter, tallow, cream and fat meat, and, to procure these, will not hesitate to act the part of a thief.

The **Garden Red-tail** (*Rucitilla phœnicura*) is native to Europe, as is his congener, the **Black Cap Red Start** (*R. tethys*). They are both excellent singers, with notes sometimes resembling those of the nightingale. The names are derived from the ruddy

BLACK CAP RED START.

chestnut color of the tail-feathers of the former, and the black feathers of the latter. Their flight is rather eccentric, and not unlike our common snowbird and brown wren, starting out of hedges when little expected, and, after a flight of a few feet, darting back again into the thickest parts, where they find concealment. Their nests are usually built in the hollow of a tree or the crevice of an old wall.

The **Ortolan**, or **Wheat Ear** (*Saxicola œnanthe*) also called *Stone* or *Fallow Chat*, is occasionally met with on the Atlantic coast of our country, but only as a stray, its home being in the old world, where it is widely distributed. Though a small bird, and a fair singer, its flesh is so highly esteemed in England, that great numbers are taken and sold in the market. Its habit of taking refuge under a stone, tuft of grass, or any object that seems to offer shelter,

ORTOLAN.

at the least alarm, is taken advantage of by hunters to capture it by laying snares before such places. The *chat* builds its nest in deep crevices of rocks, and otherwise affects a partiality for stones, even his notes bearing a striking resemblance to the sound produced in breaking stones with a hammer. He is believed, by ignorant people, to be a sure precursor of death, on which account he is mercilessly persecuted. One of the principal reasons for this strange superstition, is the fact that the *chat* often builds its nest under stones in burying-grounds, and may be commonly seen in cemeteries, where it utters its doleful notes from the top of a heap, beneath which a dead body reposes.

The **White Wagtail** (*Motacilla alba*) is also a European bird, but a single species of which, the **Yellow Wagtail** (*M. sulphurea*), has been found on the Western Continent, and this in Alaska. The **Pied** (*M. yarrellii*) and **Gray Wagtail** (*M. campestris*) are also natives of Europe, and all the several species are distinguishable alone by variation in color. They haunt the sea-shore, much after the manner of snipes, but they build their nests in the hollow of trees or, like the *chats*, in heaps of stones, and infrequently in brush heaps. They take their name from the constant flitting, or wagging, motion of the tail when not in flight.

WHITE WAGTAIL.

Larks. America is the habitat of several species of the *lark* family, which are recognizable by the great length of the claw of the hind toe, short and conical bill, and the very long tertiary quill feathers of wing, usually being almost equal in length to the primaries. The **Sky-lark** (*Alauda arvensis*) is found in the United States, but its notes are not nearly so musical as the species found in England, nor is it in any respect so interesting a bird. The latter is remarkable not only for the beauty of its song, but also for the manner of its utterance, which is exceedingly strange. It delivers its notes always while on the wing, and at the moment of giving utterance the bird begins to soar upward, continuing its flight skyward, whistling all the while, until it rises entirely beyond the vision of a spectator. It is also credited with extraordinary intelligence, as the following story, related by a lady who claims to have been a witness, and repeated by Woods, will testify:

"A pair of larks had built their nest in a grass field, where they hatched a brood of young. Very soon after the young birds were out of the eggs, the owner of the field was forced to set the mowers to work, the state of the

CRESTED LARK (*Alauda Cristata*).
SKY LARK (*Arvensis*).

weather forcing him to cut his grass sooner than usual. As the laborers approached the nest, the parent bird seemed to take alarm, and at last the mother-bird laid herself flat upon the ground, with outspread wings and tail, while the male bird took one of the young out of the nest, and by dint of pushing and pulling, got it on its mother's back. She then flew away with her young one over the fields, and soon returned for another. This time, the father took his turn to carry one of the offspring, being assisted by the mother in getting it firmly on his back; and in this manner they carried off the whole brood before the mowers had reached their nest." This is not a solitary instance, as I am acquainted with one more example of this ingenious mode of shifting the young, when the parent birds feared that their nest was discovered, and carried the brood into some standing wheat. Mr. Yarrell, moreover, mentions that the lark has been seen in the act of carrying away her young in her claws, but not on her back, as in the previous instance. Perhaps the bird would learn the art of carriage by experience, for the poor little bird was dropped from the claws of its parent, and falling from a height of nearly thirty feet, was killed by the shock. It was a bird some eight or ten days old. The lark has also been known to carry away its eggs when threatened by danger, grasping them with both feet. Many other stories are related illustrative of the lark's intuition, if not reasoning powers, one of which has passed into the fable as "the lark and the mower." This fable, however, is no doubt based upon fact, since the bird has actually been known to remove her young during the time that the cradlers were at work mowing the field in which she had her nest of fledglings. I recall to mind the experience of a gentleman in my native town which he related to me many years ago, substantially as follows: "During the early part of June while passing through a field of ripening grain only a short distance from my country residence, I observed a lark as she swept over my head with a worm in her beak, and dropped down into the wheat some yards away. Seeing the worm in the bird's mouth led me to believe that she had a nest of young where she settled, so out of curiosity I went to the place to discover if my suppositions were correct. As I drew very near the spot the old bird fluttered away as if badly injured, and as I did not immediately pursue her she turned upon her side, opened her mouth and began panting as a wounded bird invariably does. As I started again towards her she held one wing rigid while with the other she continued to flutter apparently in deep distress from a serious injury. Perceiving that she was trying to lead me away I returned and found the nest in which there were five young birds not more than four or five days old. Having heard much concerning the maternal affections of the lark I determined now to test it. Accordingly I took up the young and carrying them to a straw stack quite near my house, made a nest and deposited the brood therein; I then drew off to watch results, being careful to hide within a stable and watch through the cracks. In a short time the mother-bird came to her young and in another moment flew away again. In about fifteen minutes she returned, however, bringing her mate, and the two then proceeded to remove their brood by each taking one at a time upon the back and carrying them to another part of the field, where a new nest was evidently made for them. I followed after the birds, after the removal was completed, and found, as I had expected, an improved nest, which I did not again disturb."

The **Meadow Lark** (*Sturnella magna*), and the **Shore Lark** (*Eremophila alpestris*) are the two prominent species of American birds of the lark family, being found distributed throughout the United States from Texas to Labrador. The nesting habits of the several species are nearly identical, building on the ground, usually under a tuft of grass affording perfect concealment. The number of eggs is usually four, though sometimes five are laid, and the young are affectionately cared for by both parents until fully fledged and with full strength of wing. The *shore-lark* very much resembles the *sky-lark* in its habit of rising almost perpendicularly in the air, wheeling up and up in irregular circles until nearly out of sight, singing at intervals a sweet song, and then descending to the very spot whence he rose. All *larks* are insectivorous, and may be numbered among the farmer's most useful friends; the *shore-lark* occasionally varies its food by visiting the shores of streams, and there feeding on small crustacea.

The **Tree Pipit** (*Antheus arboreus*), also called *Tit-lark*, is quite numerous throughout the United States, as is the **American Pipit** (*A. ludovicianus*) and the **Meadow Pipit** (*A. pratensis*), the habits of which are very similar, but in coloring there is a marked difference. The latter, while classed as a song bird, utters only a single " *tweet* " in a feeble voice, and only while on the wing and at its greatest elevation. It is most common about waste lands, where small flocks assemble, and as these increase towards fall the waste lands are abandoned for cultivated fields. It is, perhaps, strongest of wing of all land birds, having been taken on board a ship nine hundred miles from the nearest shore. The color is an olive-brown, with wash of green on the upper parts; wings and tail a dark-brown sprinkled with white. The breast is a pale white with spots of brown. The *tree pipit* is so called because of its habit of perching

TREE PIPIT.

upon trees, in contrast to all the other species which remain most of the time upon the ground, though it is not nearly so graceful on a perch as when tripping among the grasses.

The song of this bird is sweeter and more powerful than that of the preceding species, and is generally given in a very curious manner. Taking advantage of some convenient tree, it hops from branch to branch, chirping merrily with each hop, and after reaching the summit of the tree perches for a few moments, and then launches itself into the air. Having accomplished this feat, the bird bursts into a triumphant strain of music, and, fluttering downwards as it sings,

MEADOW PIPIT.

alights upon the same tree from which it had started, and by successive leaps again reaches the ground. The color very much resembles that of the former species, but the size is considerably greater. The American *pipit*, like the other species, builds its nest upon the ground under a tuft of grass, and incubation is performed by male and female sitting together. As might be expected from this habit of nesting, the *pipit* is an affectionate bird, devoted with singularly strong attachment to its mate.

The **Black-throated Bunting** (*Euspiza americana*) is abundant in the West, but rarely found east of the Alleghanies. As soon as it arrives from the South in the spring, it immediately begins building, generally lacing together the tops of long grasses and making its nest thereon, though it is sometimes known to build in rose-bushes and other low branches. During the summer this bird destroys immense numbers · of caterpillars, canker-worms, and other harmful insects. It may be easily known by its notes, which are, *chip-chip che-che-che*, the two first syllables being uttered between pauses, and the last three rapidly.

The **Fox-colored Sparrow** (*Passerella iliaca*) is plentiful in the Northern and Western States, though it is not known to breed in this country. It is found in flocks of a dozen, usually haunting the outskirts of thickets and moist woods. They breed in the north of British America, and at this season the male takes on a gaudy plumage of cardinal, and develops a charming voice. It nests upon the ground and lays five eggs of a pale green tint, blotched with brown. Their food is insects and seed, and they imitate our domestic fowls, scratching the ground to uncover insects, seeds or other food.

The **English Sparrow** (*Passer domesticus*) is at once the most numerous and despised bird to be found in America. Though abounding in great numbers in every city and town of the United States, it is an importation from Europe, first brought over about thirty years ago in the belief that it would perform most useful service towards exterminating tree caterpillars. Several pairs were also taken to Australia about the same time, under a similar fallacy. In former years the sparrow was no doubt almost strictly insectivorous, but in the development of nature, which produces many changes, its appetite is no longer that of an insect devourer. No other bird re-

SHRIKE, OR BUTCHER-BIRD.

produces so rapidly, since the sparrows lay from five to six eggs at each sitting, and raise three broods each year. This remarkable fecundity has produced results most annoying, which, added to its predaceous, audacious, and rapacious disposition, make it a proper subject for our animadversions. Australia has tried to destroy or reduce the number of sparrows by offering rewards for their heads and eggs; America will soon have to adopt similar measures, or the depredations of these already innumerable and rapidly multiplying pests will be incalculable. Against the sparrow I must prefer many serious charges: He builds under the eaves of our houses and chokes up our waterspouts with his nests, befouls the roofs, porches, walks and. window-sills, litters up our yards, devours our flower seeds, plucks the ripening oats and wheat, keeps up a perpetual charivari about our doors, and makes himself an intolerable nuisance without exhibiting one redeeming trait. But to these charges I must add another, much more serious: He is not only savagely pugnacious, fighting among his own species, but he is a foe to all of our pretty, useful and singing birds, not only assaulting the parent birds, but destroying their eggs and murdering their young. This latter charge I make upon evidence that cannot be refuted: In the summer of

1885, a robin red-breast built her nest in an apple tree that stood within a few feet of the rear door of my residence, in St. Louis. This familiarity we encouraged by giving her food several times each day, until she came to expect a regular allowance. In due time the robin laid four eggs, and

HEDGE ACCENTOR.

began the interesting process of incubation, not the least alarmed during this period, even when I went within so close as a foot of her. In every respect we were friends. The brood was at length hatched out, and for more than a week my wife and daughters vied with the old birds in giving the young kindly attention. One morning, shortly after daylight, I heard a very great noise in my back yard, and, rising to discover the cause, was astonished to see perhaps a score of sparrows making a united attack upon the young robins, two of which had been dragged from the nest and killed. I hurried down-stairs to the assistance of my pets, but before I could reach them the sparrows had fairly covered the poor little fledglings and bitten their bodies almost into a pulp. I have been told by others that such murderous propensities are often exhibited by these imported pests, which accounts largely for the rapid disappearance of our song birds.

The **Snow-bird** (*Fringilla hyemalis*). A few years ago these birds were so plentiful that flocks of a thousand or more were a common sight, and cruel sportsmen often fired into the swarms, merely out of curiosity to see how many they could kill at a single discharge of a shot-gun. Snow-bird pie became a favorite dish, and this led to the capture, by trapping, of so many that their number rapidly diminished, and now it is a rare sight to see a flock numbering as many as a dozen. The *snow-bird* rarely makes its appearance in the United States before November, and migrates northward early in the spring, for the purpose of breeding. They nest upon the ground, after the manner of larks, except that, in nesting, they do not lose their gregariousness, but continue closely together.

The **Song Warblers** (*Luscininæ*) are represented in America by no less than fifty-seven species. As their name indicates, they are noted for their sweet song, to which is added a graceful form and great activity of movement. In all these birds the beak is strong, straight and sharply pointed, with a notch on the upper mandible near the extremity. The nostrils are placed

SNOW-BIRD.

at the base of the beak, and are pierced through a rather large membrane, and are protected by feathers. They are all insectivorous, and are of the greatest benefit to farmers and horticulturists. Their size is equal to that of a snow-bird.

The **Hedge Accentor** (*Accentor modularius*) is a rather rare bird in America, though occasionally met with in the Western States. In Europe, however, it is quite common. It is a small bird, about five inches long, bluish-gray on the back with brown streaks on the neck. The quill-feathers are of a dark-brown, and the abdomen white with a wash of pale buff. It haunts the habitations of man, and builds its nest at a low elevation, of moss, wool and hair, and lays five eggs of a bluish-green color. The female often lays five sets of eggs but rarely raises more than a single brood.

The **Rice-bird**, or **Bob-o'-link** (*Dolichonyx orizyvorus*), small of body as it is, nevertheless has the distinction of being a game-bird, whose flesh is very highly esteemed, and in season is hunted with a persistency greater than marks the pursuit of any other bird. He has also the further distinction of many names, being known in the South as *Rice-bird*, in the Middle States as *Reed-bird*, in the Northern States as *Bob-o'-link*, and locally called *May-bird*, *Meadow-bird*, *Butter-bird*, *Skunk-bird* (on account of the color of its plumage), and *American Ortolan*. He winters in the south, but comes north early in April, and is most plentiful in New York and New Jersey about the middle of May. They arrive usually in flocks of a dozen, and begin at once the building of their nests, which are located on the ground under a tussock of grass, and composed of small grasses rather deftly arranged, in which five or six eggs are laid, of a dirty-white, splotched with rufous-brown. During the season of courtship the male is dressed in a bright coat of black, crown of head cream color, feathers on the shoulder and rump white, and tip of tail feathers a pale-brownish ash. Towards autumn, however, his raiment seems the worse for wear, and he looks like a member of the shabby genteel. He now has a dull coat, two stripes upon his head and sides, sparsely streaked with brown. Naturalists have generally given him a bad name as the devourer of wheat, barley and corn, when in the milky state, but while this charge may be true, the *bob-o'-link* renders inestimable service in return for his depredations. He is known to feed almost exclusively, during the breeding season, off caterpillars, beetles, grasshoppers, spiders, crickets and the seeds of wild grasses. Recent investigation also proves that they devour immense numbers of the larvæ of the cotton-worm.

The *bob-o'-link* is a charming singer, whether as soloist or in concert, for he sings equally well whether piping to his mate or adding his voice to a hundred others, as they often do. But when busy with courtship, each cock pays court to the hen of his choice, and sings for her his most entrancing melody. There are few things more delightful than on a June morning, when all the earth and sky blend in sweetest harmony, when the scent of apple-blossoms has not been wholly dissipated, to lie in a soft, refreshing bed of luxuriant grass, and listen to the *bob-o'-link's* song of wooing. He sits upon a yielding reed a moment, then rises on trembling wing high into the air, at the same time pouring out a wonderful succession of tinkling, vibrating, ringing, rollicking notes, wheeling and whistling *bob-o'-link, bob-o'-link*, and then jangling off into a succession of the sweetest melody which betrays his rapture.

Though no larger than the English sparrow, beautiful of plumage, charming in manner, an exquisite minstrel, and most useful as an insect destroyer, yet this elegant little bird is most cruelly persecuted, millions being shot and trapped for the eastern markets, where they sell at a most extravagant price. Bryant's poems on this bird under the title of "Robert of Lincoln," and the prose panegyric of Washington Irving in " Wolfert's Roost," have served to immortalize poor little *bob-o'-link*, whose numbers are so rapidly diminishing under the fire and nets of those who fail to appreciate the beauties and charms of nature.

SINGULAR NEST-BUILDING BIRDS.

The beautiful raiment with which God has clothed the birds, and the gift of charming melody with which He has endowed their voices, may well excite our admiration, but to these gifts have been added other characteristics which excite our wonder, until we marvel at the diversity of attributes which distinguish bird life. We have already described some of the curious phases of feathered creation, but it is appropriate here to group together a few of the birds distinguished especially for their ingenuity in nest-building, since by so doing we shall be the better able to comprehend the wonders displayed in bird architecture. We will find many of the trades and professions represented by feathered mechanics, such as masons, weavers, tailors, cooks, sextons, preachers, officers, criers, binders, gardeners, etc.

The **Parti-colored** or **Golden-crested Wren** (*Regulus cristatus*), of England, is scarcely larger than a humming-bird, but wonderfully courageous, and cunning as well. In an aviary, where two of these were kept with a hundred varieties of birds, they made themselves masters of the largest and obtained more than their share of food by adopting many adroit devices, the most curious being to sit upon the head of some larger bird, like the jackdaw, and seize the food at every peck the daw would make. It was certainly a comical sight, but scarcely more so than that of seeing them invariably go to roost by nestling in the feathers of the larger birds.

PARTI-COLORED WREN.

The nest of this beautiful little bird is exquisitely woven of various soft substances, and is generally suspended to a trunk, where it is well sheltered from the weather. Says Woods: " I have often found their nests, and in every instance have noticed that they are shaded by leaves, the projecting portion of a branch, or some such protection. In one case the nest that was suspended to a fir-branch was almost invisible beneath a heavy bunch of large cones that drooped over it, and forced the bird to gain admission by creeping along the branch to which the nest was suspended. The edifice is usually supported by three branches, one above and one at either side. The nest is usually lined with feathers, and contains a considerable number of eggs, generally from six to ten. These eggs are hardly bigger than peas, and extremely delicate.",

The **Australian Jungle-fowl** (*Megapodus*) with its congener, the *bush turkey*, is certainly the most peculiar in its nesting habits of all the birds on our globe. If the descriptions given of their manner of building were not well authenticated by naturalists who have themselves seen the nests, as well as observed the manner of their construction, it would be a great stretch of imagination to credit them.

In several parts of Australia large mounds were discovered which for a long time were thought to be the work of human hands, reared above the remains of departed natives before Australia fell under the civilizing influences of the English. Nor was this belief at once dissipated by the declara-

NEST OF AUSTRALIAN JUNGLE-FOWL.

tions of the natives, who declared the mounds to be artificial ovens, thrown up by *jungle-fowls*, in which their eggs are laid and hatched by the heat of the decaying vegetation which composed them.

The size of these tumuli is truly marvellous, measuring sometimes sixty feet in circumference at the base, and rising to a truncated cone fifteen feet in perpendicular height. These mounds are erected by the industrious *jungle-fowl* of earth, leaves and fallen grasses, which it partly conveys and partly throws with its feet, which are extremely large, for this purpose. Some authorities state that the material is gathered in the grasp of one foot and carried by hopping on the other, while Woods maintains that the bird gathers up the grasses or leaves with its feet and throws the material backwards while it stands on one foot. As this habit was observed while the bird was in captivity, and its range so circumscribed as to prevent the exhibition of all its instincts, Woods' assertion may well be questioned, especially in the light of other observations to the contrary.

The mounds are invariably protected from the full rays of the sun by being located within the shelter of densely leaved trees, otherwise the moisture of the heap would be too rapidly dispelled. When the mound is completed the bird digs a hole in the centre to a depth of six or seven feet, though the excavation is rarely vertical, but rather tortuous, and sometimes, though the hole may be seven feet deep, the eggs are not more than three feet below the surface. Mr. Gilbert, who has made a study of the habits of this bird, says:

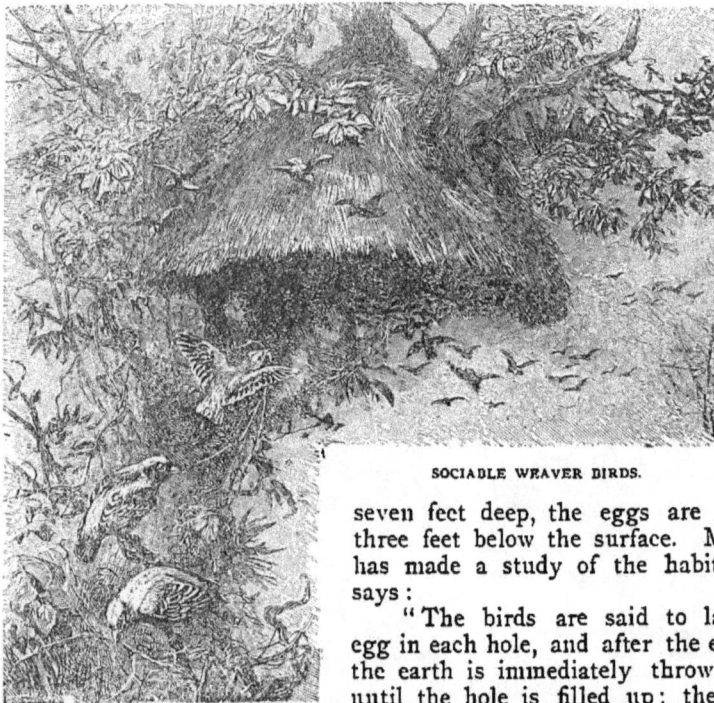

SOCIABLE WEAVER BIRDS.

"The birds are said to lay but a single egg in each hole, and after the egg is deposited, the earth is immediately thrown down lightly until the hole is filled up; the upper part of the mound is then smoothed and rounded over. It is easily known where a *jungle-fowl* has been recently excavating, from the distinct impression of its feet on the top and sides of the mound; and the earth being so lightly thrown over, that with a slender stick the direction of the hole is readily detected, the ease or difficulty of thrusting the stick down indicating the length of time that may have elapsed since the bird's operations.

"Thus far it is easy enough, but to reach the eggs requires no little exertion and perseverance. The natives dig them up with their hands alone, and only make sufficient room to admit their bodies and to throw out the earth between their legs. By grubbing with their fingers alone, they are enabled to feel the direction of the hole with greater certainty, which will sometimes, at a depth of several feet, turn off abruptly at right angles, its direct course being obstructed by a clump of wood or some other impediment."

Mr. Gilbert upon one occasion found a *jungle-fowl's* tumulus in which there was a single young one almost ready to leave the nest. It was the size of a quail, and presented many characteristics of that bird. When put into a box with sand plentifully sprinkled on the floor, even at this early age it exhibited the instincts of the mature bird, and continued gathering sand in its claws and throwing it backward all the day long.

The *jungle-fowl* ranges in the proximity of streams or the sea-beach, but confines itself to such dense shore thickets that it is not often seen. When flying the legs hang down the full length, and when frightened it utters a scream like the peacock. The size is that of a brahma hen, and the coloring a ruddy-brown.

The **Brush Turkey** (*Tellegalla lathami*), also of Australia, where it is known by the name of *New Holland Vulture*, is almost identical with the jungle-fowl in its habits, though greatly differing in appearance. The head and neck are devoid of feathers and are very vulture-like, while the throat is covered with naked fleshy wattles like the turkey, which it also resembles in size and the coloring of the body-feathers. It is a gregarious bird, though rarely appearing in companies of more than a dozen, and in the brushwood has all the characteristics of our wild turkey. Like the jungle-fowl the *brush turkey* con-

TELLEGELLA BUILDING ITS NEST (*T. lathami*).

structs a mound of extraordinary size, of dried grasses and other vegetable fibres and leaves, in which work it employs the feet as represented in the engraving. Nor is the erection of a mound the work of a single bird, but of several, who use the nest for common deposition. After the heap is raised to the required size, a large hole is dug in the centre, about two feet in diameter, in which the hens lay their eggs, two dozen or more in number, which are deftly arranged in a circle with the small ends downward. They are then covered carefully and left to the heat generated by the decaying vegetable matter to hatch, which usually requires a month's time. When the young breaks the shell it forces its way through the loose covering, like a young turtle emerging from its sandy birthplace. It remains then above ground during the daytime, but is covered up again at night by the male parent-bird for three successive nights, when it is sufficiently developed to take wing.

The **Oven-Bird** (*Furndrius fuliginosus*) is a rather small South American bird, specially distinguished for its very curious, oven-like nest, from whence the common name is derived. The North American species is called the **Golden Crowned Thrush** (*Sciurus aurocapillus*), but the size and nesting-habits of both species are very similar, except that the former frequently builds its nest in trees, or on other elevations, of clay and grasses, while the latter invariably, I believe, builds upon the ground and uses no clay in the composition of its nest. They are closely allied to the creepers, about the size of a lark, and are splendid climbers as well as runners. The nest, about which centres the chief interest connected with the bird, is generally shaped like an inverted kettle, or a bee-hive, with entrance at the base. Its walls are very thick, but to this strengthening provision are added other means for increasing its resistance to the rough usage of the wind, rain or violence. If one of the nests be carefully divided, the observer will see that the interior is even more singular than the outside. Crossing the nest from side to side is a wall, or partition, made of the same materials as the outer shell, and reaching nearly to the top of the dome, thus dividing the nest into two chambers, and having also the effect of strengthening the whole structure. The inner chamber is devoted to the work of incubation, and within is a soft bed of feathers on which the eggs are placed. The female sits upon them in this dark chamber, and the outer room is probably used by her mate. The reader will remember instances of such supplementary nests having already been mentioned. The eggs are generally four in number.

SOUTH AMERICAN OVEN BIRD.

Both sexes work at the construction of the nest, and seem to find the labor rather long and severe, as they are continually employed in fetching clay, grass and other materials, or in working them together with their bills. While thus engaged they are very jealous of the presence of other birds, and drive them away fiercely, screaming shrilly as they attack the intruder.

The **Pied Grallina** (*Grallina australis*), of Australia, constructs a nest almost as curious as that of the oven-bird. This bird is a water-loving creature, beautifully colored with white and black, and of an extremely restless movement, whether wading in shallow brooks or hopping among the branches. Its nest is made of clay and vegetable fibres and located on the forks of a horizontal limb that overhangs the water. In shape it very much resembles a low, large-mouthed basket, or a Boston bean-pot.

Black-Headed Synalaxis (*S. melanops*) is found in tropical America. It is a very active bird in the presence of insects, rapidly traversing the trunks of trees, pecking almost constantly, and quick to detect the location of a tree-grub, even beneath the bark. But, like the oven-bird, the *synallaxines* are notable for the curious nests which they construct.

Although these birds are of small dimensions, they all build nests which might easily be attributed to the labors of some hawk or crow. The nest of one species is often from three to four feet in length, and is placed

NEST OF BLACK-HEADED SYNALAXIS.

very openly in some low bush, where it escapes notice on account of its resemblance to a bunch of loose sticks thrown carelessly together by the wind. In its interior, however, the edifice is very carefully made, and, like the nest of the oven-birds, is divided into two recesses, the eggs being laid in the inner apartment, upon a bed of soft feathers.

The **Weaver Birds** are all natives of tropical regions, being confined largely to Africa and India, though a few species are found in South America. So numerous are the *weaver birds* that they have been grouped under a scientific head (*Ploccidæ*), and are usually described in order, though they present many different characteristics, which, if followed, would distribute them among several orders. Generally *weaver birds* suspend their nests to drooping boughs, but there are notable exceptions to this rule, as will be shown. But wherever built, the observer will perceive the intuition that has led the birds to build with a special view to protecting their young from the depredations of monkeys and snakes.

All the pensile birds are remarkable for the eccentricity of shape and design which mark their nests; although they agree in one point, namely, that they dangle at the end of twigs, and dance about merrily at every breeze. Some of them are very long, others are very short; some have their entrance at the side, others from below, and others again from near the top. Some are hung, hammock-like, from one twig to another; others are suspended to the extremity of the twig itself; while others, that are built in the palms, which have no true branches, and no twigs at all, are fastened to the extremities of the leaves.

AUSTRALIAN TRAPPE.

There is a bird found only in Australia and called the *trappe* which possesses not only a singular nesting habit, but is no less curious in other respects. It is so rare that few naturalists have recognized its existence, and none have ever attempted to classify it. This curious creature seems to be a cross between the gallinaceous and wading birds. During the strutting season the male makes himself an object remarkable to behold, as is seen in the engraving. It nests in the high grasses, but quite unlike others of the fowl species, it builds a very large truncated cone leaving only a shallow basin in the top for the four dusky colored eggs which it lays. Its size is equal to that of a guinea hen.

A good example of the last-mentioned description of nest is the **Mahali Weaver Bird**, of South Africa (*Pliopasser mahali*). Although the architect is a small bird measuring only six inches in total length, the nest which it makes is of considerable size, and is formed of substances so stout, that, when the edifice and the builder are compared together, the strength of the bird seems quite inadequate to the management of such materials.

The general shape of the nest is not unlike that of a Florence oil-flask, supposing the neck to be shortened and widened, the body to be lengthened, and the whole flask to be enlarged to treble its dimensions. Instead, however, of being smooth on the exterior, like the flask, it is intentionally made as rough as possible. The ends of all the grass-stalks, which are of very great thickness, project outward, and point towards the mouth of the nest, which hangs downward; so that they serve as eaves whereby the rain is thrown off the nest.

Perhaps the most singular looking nest made by these birds is that of a rather small, yellow-colored species (*Ploceus ocularius*). This nest looks very like a chemist's retort, with the bulb upward—or, to speak more familiarly, like a very large horse-pistol suspended by the butt. The substance of which it is made is a very narrow, stiff and elastic grass, scarcely larger than the ordinary twine used for tying up small parcels, and interwoven with a skill that seems far beyond the capabilities of a mere bird.

If the hand be carefully introduced up the neck of one of these nests, its admirable fitness for the repose of the young birds is at once perceived. When merely viewed from the outside, the nest looks as if it would be a very unsafe cradle, and would permit the young birds to fall

NEST OF THE REDWING (*Turdus iliacus*).

through the neck into the water. A section of the nest, however, shows that no habitation can be safer, and even the hand can detect the wonderfully ingenious manner in which the interior is constructed. Just where the neck is united to the bulb, a kind of wall or partition is made, about two inches in height, which runs completely across the bulb, and effectually prevents the young birds from falling into the neck.

Another of this group is the **Gold-capped Weaver Bird** (*Ploceus ictero-cephalus*). The nest of this bird is notable for the extreme neatness and com-

23

pactness of its structure, for it can endure a vast amount of careless handling, and still retain its beautiful contour. A specimen in the British Museum taken from the banks of a river near Natal, was suspended from two reeds, so as to hang over the water, and at no great distance from the surface.

The entire structure is apparently composed of the same plant, namely, a kind of small reed, but the materials are taken from a different portion of the plant, according to the part of the nest for which they are required. The whole exterior, as well as the walls, are made of the reed-stems, woven very closely together, and being of no trifling thickness. There is a considerable amount of elasticity in the structure, and the complete nest is so strong that it might be kicked down stairs, or be thrown from the top of a monument, without much apparent injury. The interior, however, is constructed after a very different fashion. Instead of the rough, strong workmanship of the exterior, with its reed-stems interlacing each other, as if woven by human art, the inside exhibits a lining of flat leaves, laid artistically over each other, so as to form a smooth resting place, but not interlacing at all. Their color is a blueish gray, and the contrast which they present to the exterior is very strongly marked. In size the nest is about as large as an ordinary cocoanut, not quite so long, but somewhat more oblate.

The **Tailor Bird** (*Orthotomus longicaudus*), though a stranger to Americans, is as popularly known as the most prominent crowned head of Europe, on account of the frequency with which it is described in publications, accompanied by illustrations of its singular nest.

NEST OF THE TAILOR BIRD (*Sylvia sutoria*).

The manner in which it constructs its pensile nest is very singular. Choosing a convenient leaf, generally one which hangs from the end of a slender twig, it pierces a row of holes along each edge, using its beak in the same manner that a shoemaker uses his awl, the two instruments being very similar to each other in shape, though not in material.

When the holes are completed the bird next procures its thread, which is a long fibre of some plant, generally much longer than is needed for the task which it performs. Having found its thread, the feathered tailor begins to pass it through the holes, drawing the sides of the leaf towards each other so as to form a kind of hollow cone, the point downward. Generally a single leaf is used for this purpose, but whenever the bird cannot find one that is sufficiently large, it sews two together, or even fetches another leaf and fastens it with the fibre. Within the hollow thus formed the bird next deposits a quantity of soft white down, like short cotton wool, and thus constructs a warm, light and elegant nest, which is scarcely visible among the leafage of the tree, and which is safe from almost every foe except man.

The *tailor bird* is a native of India, and is tolerably familiar, haunting the habitations of man, and being often seen in the gardens and commons feeding away in conscious security. It seems to care little about lofty situations, and mostly prefers the ground or lower branches of the trees, and flies to and fro with a peculiar undulating flight. Many species of the same genus are known to ornithologists.

The tailor bird is not the only member of the feathered tribe which sews leaves together in order to form a receptacle for its nest. A rather pretty bird, the **Fan-tailed Warbler** (*Salicaria cisticola*) has a similar method of action, though the nest cannot be ranked among the pensiles.

This bird builds among reeds, sewing together a number of their flat blades in order to make a hollow, wherein its nest may be hidden; but the method which it employs is not precisely the same as that which is used by the tailor bird. Instead of passing the t h r e a d continuously through the holes, and thus sewing the leaves together, it has a great number of threads and makes a knot at the end of each, in order to prevent it from being pulled through the hole.

The **Yellow-throated Sericomis** (*S. citreogularis*) constructs its nest in bunches of Louisiana moss that often accumulate at the extremities of d r o o p i n g tree branches.

The **Rock Warbler** (*Origma rubricata*), on the other hand, builds a pensile nest, generally suspended from a shelving rock overhanging a brook, and usually builds in societies, like the fairy martens.

The **Singing Honey Eater** (*Ptilotus sonorus*), the most melodious bird of Australia, attaches its nest to the long, slender, trailing branches of a t r e e called *acacia pendula*, resembling our weeping willows. These birds are

NEST OF THE LONG-TAILED TITMOUSE.

also gregarious, and sometimes a single tree will contain more than a hundred of such nests, rocked by every gentle zephyr.

The **Swallow Dicæum** (*D. hirundinaceum*), a small but very beautiful bird of South Wales, makes an exquisite nest of cotton-wood down, a pure white, which is not permitted to be soiled, and hangs it from the top of a lofty branch, where it looks like a beautiful purse.

The **Lanceolate Honey Eater** (*Plectorhynchus lanceolatus*), of England, a

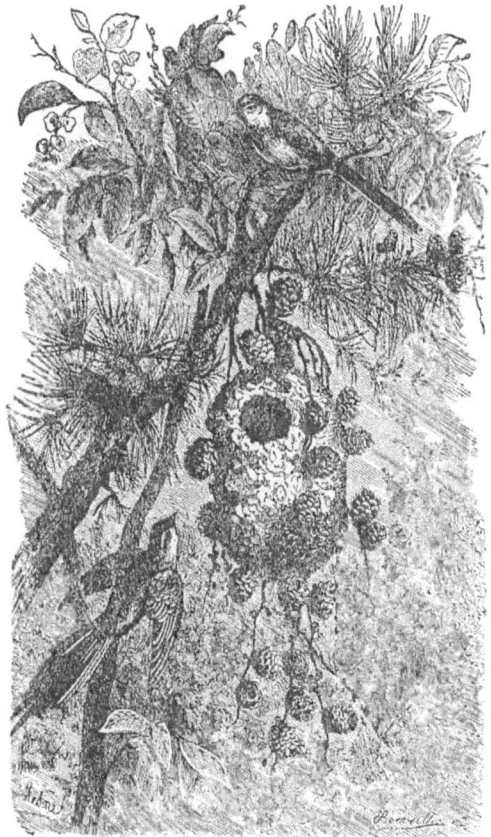

small black and white bird, builds a nest which in appearance is almost the counterpart of one end of a saddle-bag when opened, the ends of which are tied by means of thread to extremities of a longitudinal branch, so that it is a most comfortable pouch in which the bird sets well concealed.

The **Golden Oriole** (*Oriolus galbula*), a near relative of the Baltimore oriole, constructs a nest of equal neatness and ingenuity, though not so long. It is formed of a mesh of leaves finely interlaced in true weaver manner, in the shape of a circular cup, and is attached to the bifurcation of two branches by means of threads usually purloined from some neighboring dwelling.

The **Long-tailed Titmouse** (*Parus caudatus*) constructs a nest quite as curious in appearance as his Cape cousin. It is generally built in some cane-bearing tree, nearly globular in shape, made of moss, with so small an opening as to scarcely admit the body of this little bird. Small as is the nest, scarcely larger than one's fist, it serves to house a numerous brood, there being generally ten or twelve young in each nest.

The **Jupuba Cassicus** (*Cassicus hæmorrhus*), of South America, imitates our oriole in its ingenuity for nest building, as it imitates the domestic fowls of its neighborhood with its wonderfully flexible voice and power of mimicry. The nest is woven of grass fibres into a rather slim bag some two feet in length, with a slit in the upper part for entrance. This purse-bag is generally suspended from the point of a dead limb, usually near a water-course, and sometimes within a few inches of the water.

The **Grass Weaver** (*Fondia erythrops*), though a weaver, constructs her nest at the expense of little labor, and exhibits an indiffer-

NEST OF THE GOLDEN ORIOLE.

ence to appearances. It is made of coarse fibres interlaced, woven into the shape of a cup and attached to a couple of reeds that project above the water or boggy land.

The **Red Wing** (*Turdus iliacus*) is not a weaver, but builds with no less skill than do birds dexterous with the needle-bill. Its nest is wrought from mosses and bits of grass daintily fashioned into a beautiful cup, which is carefully lined with mud and saliva. This lining soon dries and then bears some resemblance to cardboard in evenness and texture. The nest is located usually where two forks start from the trunk of a tree, and is substantial to a degree.

The **Bob-o'-link**, or **Orchard Oriole** (*Xanthornis varius*), is a pretty bird, with cheerful voice and vivacious manners. Its nest, though not so large, is very similar to that which the Baltimore oriole constructs; indeed, there is a striking resemblance to the nests, and habits as well, of all the oriole family.

The **Baya Sparrow**, also called **Toddy Bird** (*Ploceus baya*), of India, is remarkable for the very singular nest it builds, in which respect it has but a single rival, in the species following, which must close our descriptions of the singular nest-building varieties, though the list is not nearly exhausted. The *baya sparrow* is a small species, gregarious in habits, and so extremely sociable that hundreds build their nests in the same tree, and get on together without serious wrangles, unlike their English relatives. The plumage is a bright yellow, with wings, back and tail tipped with brown. The acacia and date-tree are usually selected, from the smaller branches of which, and near the extremity, the nests are fixed, hanging down like so many bottles, the sides of some being provided with a small shelf, upon which the male rests while the female is hatching her brood. The appearance of a tree, with several hundreds of these pensile nests hanging from its branches, is so singular that persons viewing them for the first time invariably suppose the tree to be laden with some strange fruit.

The **Sociable Weaver Bird** (*Philetærus socius*) is a native of Africa, notable only for the surpassingly strange nest it con-

NEST OF GRASS WEAVER (*Fondia erythrops*).

structs, which characteristic has caused it to become one of the best known birds of the world, through the printed matter that has been issued concerning it. Like the baya sparrow, these are gregarious birds, and congregate their habitations in a tree, rearing by their combined labors a structure so large that it may be seen at several miles' distance, and, at times, so weighty as to break down the tree, though it is most commonly built in an acacia called the giraffe thorn, one of the toughest trees known.

The *sociable weaver* is sometimes called the *sociable grosbeak*, being a member of the grosbeak family. It is a desert dweller, and usually selects a tree which, while isolated, is sheltered from the fierce storms common in hot, arid districts. The birds are always found in very large numbers, apparently controlled by a master spirit, who seems to determine the tree upon which the nests of his subjects shall be erected. When this is decided the birds first proceed industriously to make a thatch roof, which they perform by carrying dry grasses, selected from a wiry species known as *booschmanees* grass, which is hung over the branches and ingeniously interwoven until a dome-shaped roof is made, so compact as to turn rain.

On the under sides of this thatch they fasten a number of separate nests, each being inhabited by a single pair of birds, and only divided by its walls from the neighboring habitation. All these nests are placed with their mouths downward, so that when the entire edifice is completed, it reminds the observer very strongly of a common wasp's nest. This curious resemblance is often further strengthened by the manner in which these birds will build one row of nests immediately above or below another, so that the nest groups are arranged in layers precisely similar to those of the wasp or hornet. The number of habitations thus placed under a single roof is often very great. Le Vaillant mentions

ORCHARD ORIOLE.

that in one nest which he examined there were three hundred and twenty inhabited cells, each of which was in the possession of a distinct pair of birds, and would at the close of the breeding season have quadrupled their numbers.

The *sociable weaver* bird will not use the same nest in the following season, but builds a new house, which it fastens to the under side of its previous domicile. As, moreover, the numbers of the nests are always greatly increased year by year, the *weaver birds* are forced to enlarge their thatched covering to a proportionate extent, and in course of years they heap up so enormous a quantity of grass upon the branches that the tree fairly gives way with the weight, and they are forced to build another habitation. So large is this thatch-like covering, that Harris was once deluded by the distant view of one of these large nests with the belief that he was approaching a thatched house, and was only undeceived, to his very great disappointment, on a closer approach.

The object of this remarkable social quality in the bird is very obscure. As in many instances the nests of the *weaver birds* are evidently constructed for the purpose of guarding them from the attacks of snakes and monkeys, the two most terrible foes against which they have to contend, it is not improbable that the *sociable weaver birds* may find in mutual association a safeguard against their adversaries, who might not choose to face the united attacks of so many bold though diminutive antagonists. The shape and general aspect of the nests varies greatly with their age, those of recent construction being comparatively narrow in diameter, while the older nests are often spread in umbrella fashion over the branches, enveloping them in their substance, and are sometimes only to be recognized as a heap of ruins from which the inhabitants have long fled.

In general the *social weaver bird* prefers to build its nest on the branches of some strong and lofty tree, like the giraffe thorn previously mentioned, which also has the advantage of massive and heavy foliage disposed in masses not unlike the general shape of the *weaver birds'* nest. Sometimes, however, and especially near the banks of the Orange river, the bird is obliged to put up with a more lowly seat, and contents itself with the arborescent aloe. The number of eggs in each nest is usually from three to five, and their color is bluish-white, dotted towards the larger end with small brown spots. The food of this bird seems to consist mostly of insects, as when the nests are pulled to pieces wings, legs and other hard portions of various insects are often found in the interior of the cells. It is said that the *sociable weaver birds* have but one enemy to fear, in the persons of the small parrots, who also delight in assembling together in society, and will sometimes make forcible entries into the *weaver birds'* nest and disperse the rightful inhabitants.

The color of the *sociable weaver bird* is brown, taking a pale buff tint on the under surface of the body, and mottled on the back with the same hue. It is quite a small bird, measuring only five inches in length.

NEST OF JUPUBA CASSICUS.

The **Satin Bower Bird** (*Ptilonorhynchus halosericeus*) is a member of the starling family, whose habitat is Australia. In some respects it is the most remarkable of birds, not for the nest it builds, as this yet remains to be discovered, but for the cultivated taste it displays in constructing a bower for lodgment, and the decoration of its surroundings. The bird is believed to nest upon the ground, but because of the superstitious fears of the natives, who hold it in very great awe, they have not discovered or revealed its nesting habits. The bower which this bird constructs is admirably shown in the illustration on page 292, but for a description I am indebted to Mr. Gould, who writes as follows:

"On visiting the Cedar Brushes of the Liverpool range, I discovered several of these bowers or playing-places; they are usually placed under the shelter of the branches of some overhanging tree in the most retired part of the forest; they differ considerably in size, some being larger, while others are much smaller. The base consists of an exterior and rather convex platform of sticks, firmly interwoven, on the centre of which the bower itself is built. This, like the platform on which it is placed and with which it is interwoven, is formed of sticks and twigs, but of a more slender and flexible description, the tips of the twigs being so arranged as to curve inward and nearly meet at the top; in the interior of the bower, the materials are so placed that the forks of the twigs are always presented outward, by which arrangement not the slightest obstruction is offered to the passage of the birds.

"For what purpose these curious bowers are made is not yet, perhaps, fully understood; they are certainly not used as a nest, but as a place of resort for many individuals of both sexes, who, when there assembled, run through and round the bower in a sportive and playful manner, and that so frequently that it is seldom entirely deserted.

"The interest of this curious bower is much enhanced by the manner in which it is decorated, at and near the entrance, with the most gayly colored articles that can be collected, such as the blue tail feathers of the Rose Hill and Lory Parrots, bleached bones, the shells of snails, etc. Some of the feathers are stuck in among the twigs, while others, with the bones and shells, are strewed about near the entrance. The propensity of these birds to fly off with any attractive object is so well known that the blacks always search the runs for any missing article.

"So persevering are these birds in carrying off anything that may strike their fancy that they have been known to steal a stone tomahawk, some blue cotton rags, and an old tobacco-pipe. Two of these bowers are now in the nest room of the British Museum, and at the Zoological Gardens the *bower bird* may be seen hard at work at its surface, fastening the twigs or adorning the entrances, and ever and anon running through the edifice with a curious loud full cry that always attracts the attention of a passer-by. The *satin bower bird* bears confinement well, and although it will not breed in captivity, it is very industrious in building bowers for recreation."

From reports made by visitors to the land where this singular bird is found, it would appear that only a few exercise the singular propensity of bower-building, since, though these birds are rather plentiful, the discovery of their strange structures is an uncommon event. From this fact many persons well-informed have concluded that the bird occasionally exhibits a theatrical ambition, which it gratifies by constructing these bowers as houses of entertainment. to which are invited, at times, such of its neighbors as may be agreeable, though host and guests alike turn players.

It is rather a gregarious bird, assembling in flocks, led by a few adult males in their full plumage, and a great number of young males and females. They are said to migrate from the Murrumbidgee in the summer and to return in the autumn.

The plumage of the adult male is a very glossy satin-like purple, so deep as to appear black in a faint light, but the young males and females are almost entirely of an olive-green.

The **Spotted Bower Bird** (*Chlamydera maculata*) differs from the former species only in the coloring of its plumage, its habits being substantially the same.

GALLINACEOUS BIRDS (*Gallinæ*).

Under the classification technically known as the *gallinæ* (which is a Latin word meaning *hens*) are grouped all varieties of our domestic poultry, and also the *scrapers*, or those species which are in the habit of scratching the ground in search of food, such as the pheasant, grouse, quails, turkeys and many other useful and interesting game-birds.

HOKKO HEN.

The **Crested Curassow** or **Hokko Hen** (*Crax alector*) is one of the most magnificent species of gallinaceous birds, almost rivalling the peacock in brilliancy of plumage, which is of a deep black, with a slight gloss of green upon the head, crest, neck, back, wings and upper part of the tail, and dull white beneath and on the lower tail-coverts. Its crest is from two to three inches in length, and occupies the whole upper surface of the head; it is curled and vel-

GOLDEN PHEASANT.

vety in its appearance, and capable of being raised or depressed at will, in accordance with the temporary feelings by which the bird is actuated. The eyes are surrounded by a naked skin, which extends into the cere and there assumes a bright yellow color. In size the bird is almost equal to a turkey. This species is a native of Mexico, Guiana and Brazil, and probably extends itself over a large portion of the southern division of the American continent. In the woods of Guiana it appears to be so extremely common that M. Sonnini regards it as the most certain resource of a hungry traveller, whose stock of provisions is exhausted, and who has consequently to trust to his gun for furnishing him with a fresh supply. They congregate together in numerous flocks, and appear to be under little or no uneasiness from the intrusion of men into their haunts. Even when a considerable number of them have been shot, the rest remain quietly perched upon the trees, apparently unconscious of the havoc that has been committed among them. This conduct is by no means the result of stupidity, but proceeds rather from the natural tameness and unsuspiciousness of their character. Those, however, which frequent the neighborhood of inhabited places are said to be much wilder and more mistrustful, being kept constantly on the alert to avoid the pursuit of the hunters, who destroy them in great numbers. They build their nests on the trees, forming them externally of branches, interlaced with the stalks of herbaceous plants, and lining them internally with leaves. They generally lay but once a year, during the rainy season; the number of their eggs being, according to Sonnini, five or six, and to D'Azara, as many as eight. They are nearly as large as those of a turkey, but are white and have a thicker shell. Some efforts have been made towards domesticating the *curassow*, chiefly in Holland, and which have had some success.

The Pheasants (*Phasianidæ*) form one of the most interesting groups of the feathered race, whether we regard them for the brilliancy of their plumage or the excellence of their flesh. The *golden pheasant*, of China (*P. pictus*), and another species of that coun-

PHEASANT OF THE HIMALAYAS.

try, known as *P. venesatus*, are reckoned by many to be the most gorgeously attired birds of the world. Asia is believed to be the nativity of all species of the pheasant family, their importation into Europe having been made, it is said, by the Argonauts, who sailed with Jason in quest of the golden fleece. The most splendid species are still to be found in the Himalaya region, of which the *horned tragopan* is specially conspicuous. It is quite a large bird, distinguished for its rich plumage, and for two large wattles that hang from under the eyes and which the bird can inflate at will.

The **Argus Pheasant** (*Argus giganteus*), found principally in Sumatra, is a truly royal bird, the name of which commemorates in a degree the Argus of mythology, who, it is said, never closed his hundred eyes simultaneously until put to sleep by the playing of Mercury on the magic pipe of Pan. As the eyes of Argus were said to be distributed all over his body, so are eye-like spots on every part of the *argus pheasant*, though those on the long tail feathers are much brighter, of white and black. The wings, however, are yet more beautiful, charmingly bedecked with gradations of jetty black, deep brown, and with orange, fawn, olive and white, in exquisite combination. The tail has two extremely long central feathers, sometimes measuring as much as four feet and bewitchingly marked with prismatic colors, re-inforced by secondary feathers of equal gorgeousness. When flying the long tail trails behind with a positively dazzling effect, which is not seen when the bird is on the ground, except it be in the courting season, when the male disports his plumage by expanding both tail and wing feathers and strutting about like the peacock. Its size is equal to that of a guineafowl.

The **Shining Pheasant** (*Lophophorus impeyanus*), of Thibet, is another gorgeous specimen, with plumage reflecting metallic hues of fiery red, green, purple and gold, and rivalling the richest clothed hummingbirds of South America. In the summer season they ascend to great elevations of the Himalayas, but in winter they return in large flocks to the lower altitudes, the two sexes, however, being separated. Great numbers at this time are killed for their skins, and they are fast disappearing by reason of the pernicious demands of fashion in what we wontingly call civilized lands.

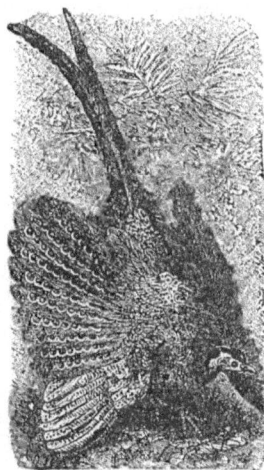

ARGUS PHEASANT.

The **American Pheasant** (*P. colchicus*) is also a pretty bird, or rather the male is, for the female of all the several species has a dull plumage, and much less expanse and length of tail. It is found, but only occasionally now, in the woods east of the Mississippi and north of the Ohio. Like all of the family, it is a ground-loving bird, a splendid runner, and so crafty, as well as swift of flight, that they are most difficult to

SAND PHEASANT (*P. exustus*).

shoot. Their color also, at times, blends so perfectly with the woods, that only a trained eye is quick to perceive them. I once stood for nearly an hour vainly striving to catch sight of a *pheasant* that I heard "drumming" on a log, which I knew to be not fifty yards distant. At last discouraged, I

AMERICAN PHEASANT.

SHINING PHEASANT.

stepped from my hiding place, and at once the bird took flight from the very spot where I supposed it to be, and at which I had been steadfastly gazing so long. In England the *pheasant* is protected in preserves, and poaching is punished by severe penalties. There the bird is quite numerous, and furnishes splendid sport to those whose wealth enables them to indulge in the pleasure. It is a strange fact that the *pheasant*, when proper opportunity offers, will mate as readily with other birds as with members of its own species, so that in England hybrids between it and barn-yard fowls are quite common. Indeed, cock pheasants have been known many times to beat the autocrat of the barn-yard and usurp his place among the hens.

ENGLISH PARTRIDGE.

The **English Partridge** (*Perdix cinereus*) is a bird which forms one of the connecting links between pheasants and quails. By the protection given it in game-kept preserves, the *partridge* is plentiful in England, as well as in a greater part of Europe. Like the pheasant, the *partridge* builds a nest upon the ground, and lays therein a dozen eggs, and sometimes more, which are hatched in about seventeen days. It is a singular fact that the *partridge* and pheasant exchange nests occasionally, so that the eggs found in one nest may be mixed, and a mixed brood is thus hatched, to which the foster-parent is no less devoted than to its own legitimate offspring. In cases of threatening danger to the nest, the female *partridge* has been known to carry her eggs to another spot, though how this is done remains to be explained.

About the middle or end of February, according to the mildness or inclemency of the season, the *partridge* begins to pair; and, as the male birds are very numerous, they fight desperate battles for the object of their love. While engaged in combat, they are so deeply absorbed in battle that they may be approached quite closely, as they whirl round and round, grasping each other by the beak, and have even been taken by hand. So strong, however, is the warlike instinct, that, when released, the furious birds recommence the quarrel.

In nearly all respects the *English partridge* and the *red-legged partridge* differ from the *American bob-white* only in size, the former being nearly as large as our prairie

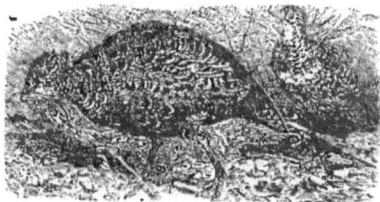

EUROPEAN QUAIL.

chicken, the coloring and habits, being so similar as to show a close relationship.

The **Quail** (*Coturnix communis*) is found very widely distributed over all parts of Europe, and a greater part of Africa and Asia. It will be remembered that during the flight of the Israelites, and while famishing in the Arabian wilderness, a miraculous flight of quails came and covered up the camp. The surprising numbers of these birds in former years would lend probability to the story, even were it related by profane writers. These birds are migratory, and even to this day sometimes pass over Arabia in countless numbers; they chiefly travel by night, no doubt in order to escape the birds of prey that would create sad havoc among them if discovered during the daytime; the females precede the males several days, but why this is so cannot be determined. This species, unlike our American quail,

VIRGINIA QUAIL.

are polygamous, the male adopting the habits of our barnyard cock, and surrounding himself with about a dozen hens. From these habits the inference is unavoidable that the females are very much more numerous than the males. So pugnacious are the males that it has long been a custom among people of Eastern countries to keep large numbers and train them especially for the prize-ring, great sums being staked upon the result of a fight. The breeding habits are identical with those of our American quail.

The **Virginia Quail** (*Ortyx virginiana*) is also called *bob-white* from the sound of its calls in the laying season, and also *partridge*, which latter title is so inappropriate as to belong entirely to another and larger species. It is found throughout the United States east of the Rocky Mountains, and in some places in the West, especially Arkansas and Southern Missouri, it is quite numerous. Like squirrels, *quails* occasionally migrate, apparently assembling at one common rendezvous and departing together, at which time nothing will stay their flight except exhaustion. When I was a boy I saw two migrations of *quails*, both in the same direction, coming from Kentucky and crossing the Ohio into Illinois. The river, at the place where I lived, was fully one mile wide, which was too great a flight for thousands to make without resting, so that they fell into the river near the western shore in great numbers, some reaching the bank by swimming, while others perished, besides those captured by persons in skiffs. But there are now so few compared with their number in former years, while the country is so well settled, that it is not probable similar migrations will be witnessed hereafter.

The *quail* loves cultivated fields where is found an increased food supply of both grain and insects. It builds upon the ground, within the grass or ripening wheat, and lays from a dozen to eighteen eggs. The instant the young are hatched they are ready for energetic action and can run with surprising swiftness. Indeed, the young may sometimes be seen running about with a part of the shell still sticking to the back. In the process of incubation both

AN ENEMY FOILED.

parents perform a part, so, too, when the numerous broods are hatched out do the parents gather the chicks under their wings at night or in bad weather, at which time they sit with their tails together in a rather unfamiliar attitude. Yet they are very affectionate, as well as cunning. The male is courageous in defence of his family, and the mother is very adroit in her efforts to prevent the exposure of her brood. She will feign lameness and flutter along barely out of reach of a person or foe until she has led him a sufficient distance from her chicks which, in the mean time, lie hidden among the grass. When she thinks the danger past she takes wing and flies in a direction opposite to the place where her brood is concealed, and sometime after calls them together. When the young are three weeks old they are able to fly, and rise promptly on an alarm being given by the parent, and thereafter the family will always be found together, whether upon the wing or running along the ground.

Many attempts have been made to domesticate the *quail*, but invariably with poor results. A great many broods have been hatched out by domestic hens, but after a few weeks, or months at most, the young always desert their foster-parents and go to the woods to return no more. The male *quail* is a handsome creature, of a proud bearing, and his notes are loud, cheery and melodious, sounded most frequently at early morn, though in May and June he joyfully pipes his "*bob, bob-white*" from some perch not far distant from the nest of his mate, at all hours of the day.

Quail are so highly esteemed for their flesh that immense numbers are destroyed every open season, by shooting and trapping, the latter being the more destructive. In former years I have seen many quail-drives, by which, during a wet or foggy day, covey after covey were easily driven into a net having wings stretched out several yards and converging toward

BLUE QUAIL.

CALIFORNIA QUAIL.

a centre, where the enclosure was ample to permit the entrance of a hundred or more. When the birds entered the net proper the drivers would rush up, drop the netting and then easily secure the captives.

The **California Quail** (*Laphortyx californicus*) is found only in the South-

THE SUNSET MINUET.

west and west of the Rocky Mountains. In his habits he does not differ essentially from the species just described, but his plumage is very much handsomer, though his flesh is not so palatable. His notes are clear, though not so rich as the Eastern species, bearing some resemblance to "*kuck-kkck-kca-a*." He is beautifully clothed in rich colors: purplish-blue on the neck, black throat, yellow frontlet, two white stripes on the sides of the head, and a delicate crest of three bluish-black feathers that may be raised or lowered at pleasure. The other body-markings are similar to those which distinguish the Eastern species.

The **Plumed Quail** (*Oreortyx pictus*) inhabits the mountain ranges of the Pacific coast. It is about the size of the preceding species, but while the body plumage is not so bright as either, it is provided with two long, black and charming plume-feathers, which may be elevated when the bird is excited.

The **Blue Quail** (*Calipepla squamata*) is found in Mexico and Arizona, and is a specially handsome bird, the body feathers being of a bright sky-blue, the head mottled with rufous, white and black, and crested with a webbed feather that curves gracefully forward. The secondary feathers of the wings are brown, flecked with white.

Next to the pheasant and *quail* in popular esteem as a game or table bird, is the grouse, or prairie chicken, of which there are several species both in this and other countries of a north temperate and arctic climate.

The **Pinnated Grouse** (*Tetrao cupido*) is perhaps the most favorably known of the wild-chicken species, though not so numerous. They are rarely found now east of the Mississippi; nor do any considerable number now exist, so far as I have been able to learn, outside of Dakota, where, in 1884, I found them so plentiful along the Missouri river, some forty miles below Bismarck, that it was an easy matter to "bag" more than one hundred in a day's shooting. These birds are partial to stunted brush-wood, and are particularly fond of a small red and very sour berry, called *bull-berry*, that grows on a low tree, about the size of the hawthorn. When "put up" they nearly always light in a tree, often rising to the topmost branches, from which they are not easily frightened, but will crane their necks and utter their call of *put-put-put-put*, while the sportsman discharges his gun repeatedly and until they are struck. Like others of the species the *pinnated grouse* is a very chivalrous and vain bird during

THE COMBAT.

the mating season. They congregate of an evening in open places and sometimes a covey of several dozen may be seen going through the graceful figures of the minuet, bowing their heads together, turning in a kind of jerky manner, spreading their tails and chassezing in a proud and measured manner. At this time the bird is remarkable for the naked sacculated appendages which hang on either side of the neck, and which can be inflated until they are almost the size and color of an orange. The cocks are also noted for their fighting proclivities, and during the breeding season the most desperate combats take place lasting sometimes for hours, but rarely ever resulting in any great damage to the contestants.

The *pinnated grouse*, like all the species, builds its nest upon the ground, being a simple excavation lined with grass, in which generally about fifteen eggs of a brownish-white are deposited. The plumage of the bird is a mottle of white and chestnut-brown, the latter color predominating. The male is provided with eighteen feathers on either side of the neck, of black and brown, which are distensible, but which generally lie close except during the time of courtship. He has also a small crest and a semi-circular comb of orange-colored skin over each 'eye. The sacculated appendages on the neck are used to produce a drumming noise, or are certainly in some way instrumental in the production of the booming sound which they utter during the strutting season. Naturalists, however, are not united in their opinions respecting the means employed to produce the sound. It has been observed that the *pinnated grouse* inflates the sacs on his neck, and at the instant of their collapse, by a violent expulsion of the air within, a booming noise is heard. But the same sound, though not so loud, is produced by the turkey-gobbler, cock-pheasant, and by other species of prairie-grouse, though none of these are provided with the air sacs. The question, therefore, of just how the noise is produced remains unsettled.

The Hazel Hen (*Tetras bonasia*) is a large species of grouse, resembling the sage hen in habits, but found only in Europe. The body markings are considerably darker than those of our prairie chickens, and it is somewhat larger. The head is distinguished for

PRAIRIE HEN AND HER BROOD.

having a bold crest of feathers, which are balanced, so to speak, by a thick growth of feathers under the throat giving to it a bearded appearance. From base of the bill runs a crescent of white over to the neck and to the breast, while another strip of white crosses on a line with the eyes, which impart a very pleasing appearance to the bird. While found chiefly upon open ground it not infrequently resorts to the brush to prey upon insects that are to be found about rotting wood.

The **Black Grouse** (*Tetras tetrix*) is also a north Europe species, distinguished, as the name implies, for its black feathers which are nowhere relieved except on the inner edges of the wing, which are white. The tail, however, is its glory, being composed of large feathers which curve gracefully from the centre outwardly. The head is also decorated with a small crest of feathers.

24

The **Ruffled Grouse** (*Tetras umbellus*) is more widely distributed than the preceding species, being found in all the Western States, though no longer very plentiful. When I first went to Kansas, in 1870, the prairies everywhere seemed to be fairly animate with them, and early in the mornings of spring so great was the noise created by the drumming of thousands of these birds that little else could be heard. In the winter, though, their numbers seemed to be even greater, for sometimes a heavy snow would drive them in incredible numbers to the shelter of farm houses, and even into dwellings, severe hunger making them so bold. As the West became settled up, the *grouse*, like the Indians, were driven westward in constantly diminishing numbers, until now coveys are only occasionally to be seen. The result of this barbarous destruction of a most useful bird, has been a number of grasshopper invasions that did inestimable damage by the total destruction of growing crops. Had these birds been protected, they would have in turn protected the farmer against the grasshopper, since these insect pests constitute the principal food of *grouse*.

SEEKING SHELTER.

The *ruffled grouse* bears a close resemblance to the preceding species, differing only in having more brown in its plumage, and a ruffle of feathers on the neck, instead of long, wing-like feathers. The **Red Grouse** (*Perdix rufa*) is confined to the British Isles, where, on account of its feathered feet, it is also called the *hare-foot*. The color of this species is variable, but the name *red hen* has been given it because in the winter the male is well clad in a plumage of bright chestnut-brown.

The **Capercaillie**, or **Wood Grouse** (*Tetras urogallus*), is the largest of the grouse species, and is found only in Norway, Sweden, and the extreme northern parts of Europe. Unlike its congeners, this bird spends a greater part of its time among the trees. It also goes under a variety of names such as *cock of the woods*, *mountain cock*, *capercailzie*, etc. Though living among the trees, the *capercaillie* builds its nest upon the ground, and lays from eight to ten eggs. The color of the male is a chestnut, flecked with irregular black streaks, and the breast a black with a gloss of green. It nearly equals the wild turkey in size.

The **Cock of the Plains** (*Tetras urophasianus*) is now a very rare bird, found occasionally in southern California. He differs from other species in having a very long tail, which in the strutting season he spreads in magnificent display, droops his wings like the turkey, erects the silken plumes of his neck and puffs out his crop after the fashion of the pouter pigeon. The flesh is not esteemed.

OPENING OF THE CHICKEN SEASON.

The **Black Grouse** (*Tetrao tetrix*) is a habitant of southern Europe, but it is scarce and extremely wary. This bird differs from other species in several particulars. It crows in a stridulous voice, which has been likened to the sound of whetting a scythe. The male is also polygamous, and rather indifferent in his affections to both females and his young. In the winter the sexes separate, and in the spring, when the courting season opens, the males battle until the authority of the strongest is established. The general color of the male is a glossy black, with a hue of blue. The tail is bifurcated, caused by the greater length of the outer feathers, which curve outward at the tips.

PRAIRIE HEN ON HER NEST.

The **Ptarmigan** (*Lagopus albus*) differs from others of the species, except the red grouse, in having the legs and toes heavily feathered. It is also distinguished for the change of plumage which takes place to accommodate the bird to the rigorous climate where it makes its home. In the summer season the color is a rusty brown, but towards fall the feathers become lighter, fading gradually from brown to gray, and then to a pure white as the severe weather approaches. It dwells among the snow-clad hills, being found as far north as Greenland, and rarely more southerly than British America. In its habits the *ptarmigan* does not differ from the pinnated grouse, which it equals in size.

CAPERCAILLIE.

The **Great Bustard** (*Otis tarda*) was formerly quite plentiful in both Great Britain and Europe, but so few are now to be seen that it will soon be placed in the list of extinct birds. It is about equal to the turkey in size, but has many characteristics of the ostrich. Although the wings are fairly well developed, it rarely exercises them except as aids to its rapid progress over ground, for it is one of the swiftest runners, able to contest with fleetest horse or dog. Its nest is only a shallow basin scooped out of the ground usually in grain fields, in which two or three eggs are laid of an olive-brown, splashed with rufous and having a green tinge.

The **Little Bustard** (*Otis tetrax*) is still found in considerable numbers all over Europe, being particularly plentiful about the Caspian shores, where it ranges in large flocks. It is about the size of a quail and is quite a handsome bird, the flesh of which is held in high esteem.

The **Wild Turkey** (*Meleagris gallopavo*) is beyond compare the most majestic of all game birds, for which reason he is cruelly persecuted, so that the number is rapidly diminishing, though in Missouri and Arkansas it is by no means scarce as yet. It is spread over the whole of America except the extreme north and south, its favorite habitat being Indiana, Illinois, Missouri, Arkansas, Indian Territory and Texas. It begins

BLACK GROUSE (*Tetrao tetrix*).

to mate in February, and then the male puts on his vaunting airs of arrogant pride, gobbling in distracting voice, trailing his wings with a *thud!* that seems to penetrate both earth and air.

The female makes her nest in some secluded place, and is very guarded in her approaches, seldom travelling the same path twice in succession, and if discovered, using various wiles by which to draw the intruder from the spot. As soon as the young are hatched she takes them under her charge, and the whole family go wandering about to great distances, at first returning to the nest at night, but afterwards crouching in any suitable spot. Marshy places are avoided by the *turkey*, as wet is fatal to the young birds until they have attained their second suit of clothes, and wear feathers instead of down. As soon as they are about a fortnight old they are able to get up into trees, and roost in the branches, safe from most of the numerous enemies which beset their path through life.

The great horned owl is, however, still able

PTARMIGAN.

and willing to snatch them from the branches, and would succeed oftener its attempts, were it not baffled by the instinctive movements of the *turkey*. Even the slight rustling of the owl's wings sets the watchful *turkeys* on the alert, and with anxious eyes they note his movements as he sails dark and lethal over them in the moonbeams, his large lambent eyeballs glowing with opalescent light—a feathered Azrael impending over them, and with fearful deliberation selecting his victim. Suddenly the swoop is made, but the intended victim is ready for the assault; as it dips down its head, flattens its tail over its back, and the owl, striking upon this improvised shield, finds no hold for his claws and slides off his prey like water from a duck's back. The whole flock drop from the boughs, and are safely hidden among the dark underwood before their enemy has recovered himself and renewed the attack.

HEATHER SNOW HEN, OR WHITE PTARMIGAN.

The lynx is a terrible foe to the *turkeys*, bounding suddenly among them, and as they hastily rise into the air to seek the shelter of the branches, the lynx leaps upwards and strikes them down with his ready paw, just as a cat knocks down sparrows on the wing. Various other animals and birds persecute the inoffensive *turkey* throughout its existence, but its worst enemy is the featherless biped. Snares of wonderful construction, traps, and "pens," are constantly employed for the capture of this valuable bird; the "pen" being so simple and withal so ingenious that it merits a short description. It is made of logs or fence-rails, laid to a height of about four feet in a square and covered, under which a trench is dug from the centre, leading out some ten or twelve feet from the pen. In this trench corn is scattered to entice the birds. When the *turkeys* discover the grain they proceed quickly to devour it, keeping their heads to the ground and

GREAT BUSTARD.

following up the trench until they pass into the pen. When all the grain is eaten the birds raise their heads and try to get through the interstices between the rails, but never once attempt to escape by the avenue through which they entered, and are thus made captives. Quails are often taken in the same manner.

The **Necklace Hen** (*Pternistes vulgaris*) is an inhabitant of Africa, where it is found in certain parts in great abundance. It is classed with the *francolins*, or between the pheasants and grouse. The males and old females are armed with rather formidable spurs, which they vigorously use. When alarmed they usually rise to the branch of a tree and utter notes somewhat resembling an hysterical laugh.

WADING BIRDS.

The number of species that find their subsistence along the shores of streams, ponds, marshes, or the sea, is very great, and though none of them are endowed with the gift of song, they possess other attributes and curious provisions for adaptation which

LITTLE BUSTARD.

render them no less interesting than the varieties described in the foregoing pages. Most prominent among the wading birds is

The **Flamingo** (*Phœnicopterus roseus*) of the America semi-tropical regions, which is more beautiful than its congeners found in Africa and South America. It has been classed with the *anseres*, or the goose tribe, but by arbitrary assignment rather than discoverable analogy. The *flamingo* is very abundant about

HAZEL HEN (*Tetrao bonasia*).

WILD TURKEY.

the Florida coast, where it is perhaps the most interesting of the many bright plumaged or curious birds of that region. Being very tall, (nearly seven feet) with a long neck and longer legs, its movements are strange, though by no means awkward. When feeding it uses its upper mandible as a scoop, turning the head so as to receive the food within the basin of the upper bill. The tongue is thick and covered with stiff papillæ which point backward so that whatever enters the mouth is not likely to escape. It is a gregarious bird, always feeding in large flocks, over which two or more sentries are placed that keep a vigilant watch for the appearance of any enemy. Though its legs are seemingly abnormally long, they do not prevent the bird from sitting down in a comfortable position, but notwithstanding this ability the *flamingo* constructs a nest of mud and grass in the shape of a truncated pyramid about three feet high, the top of which is hollowed out. In this singular nest the bird lays from two to four eggs, upon which the female sits with her legs hanging awkwardly on either side. The color of this bird is a beautiful scarlet or white, tinted with rose, sometimes almost approaching a light red. Closely allied to the

NECKLACE HEN.

MIRROR HEN
(*Polyplectron chinquis*).

flamingo is the crane, of which there are several varieties and widely distributed, and which are described in the following pages.

The **Bearded Crane** (*Ardia cinerea*) or **Heron** is rarely met with in this country, though its congeners are quite numerous, and the habits of all are nearly identical. This particular species is found in Europe occasionally, but is most numerous in the swamp regions of Africa, though formerly it abounded in England and nearly all the European countries. It is about three feet in height, of a slaty-grayish color, the throat and neck white. It has a long graceful plume of dark blue feathers on the crown, and a beard of white feathers growing from the junction of the neck and breast, from whence the name *bearded crane* is derived. Like all others of the species it is a wader, having very long legs and neck, and extremely light body. It feeds on fish largely, but will also greedily devour young birds, rats, mice and other small creatures. Though not in any sense a swimmer, yet under certain circumstances it has been known to take to deep water to procure its prey, as the following incident reported by Dr. Neill will show: "A large willow tree had fallen down into the pond, and at the extremity, which is partly sunk in the sludge and continues to vegetate, water hens breed. The old cock *heron* swims out to the nest and takes the young if he can. He has to swim ten or twelve feet, where the water is between two and three feet deep. His motion through the water is slow, but his carriage stately. I have seen him fell a rat at one blow on the back of the head, when the rat was munching at his dish of fish." But usually the *heron* procures his food by standing in shallow places and watching, with wonderful patience, for the appearance of minnows or larger fish. He is

NEST OF THE FLAMINGO (*Phœnicopterus ruber*).

a very gourmand, with insatiable hunger, so that the "appetite of a *heron*" has grown into an aphorism to express gluttony. He builds almost invariably in

1. GREAT BITTERN. 2. WHITE CRANE. 3. CURLEW.

BEARDED CRANE.

the tops of high trees, dead branches being preferred, the nest being constructed of sticks, with no attempt at neatness, but the interior is comfortably lined with wool or other soft substance. The number of eggs is four or five, of a pale green color.

The **White Crane**, or **Egret** (*Ardea egretta*), is an American species found in considerable numbers during summer time along the shores of sandbars busily engaged fishing, but like the heron, it will also eat the smaller mammalia. It breeds principally in cedar swamps, on the dead tops of trees, laying generally four eggs. It is an extremely wary bird, instinct probably teaching it that the worst enemy of all creatures—man—covets its beautiful white plumage, and especially the long, delicate train that at certain seasons of the year covers the tail. The height of this pretty bird is between three and four feet.

The **Blue**, or **Sandhill Crane** (*Grus canadensis*), is the largest of the species found in America. Occasional specimens are met with on sandbars of the Ohio river, but their homes seem to be west of the Mississippi, being distributed over the whole territory between 15° longitude and the Pacific, extending north to Alaska. It occasionally makes its nest in the sand, and at other times in the top of tall ferns, in open ground. The eggs are always

BLUE CRANE (*Grus cinerea*).

two in number, of a drab color, and on the greater end are large, irregular blotches of chocolate-brown. The shell is punctulate, with numerous elevations resembling warts distributed over the surface.

The *blue crane*, while certainly not a handsome bird, is by no means ungraceful, especially when feeding. It is extremely wary and rises at the least suspicion, usually giving a hoarse call, as a warning to its mates. Though easily frightened they may be domesticated if taken when young and properly cared for. There is an island in Lake Minnetonka, Minnesota, that has for years been a favorite resort for *blue cranes* as well as for other species. The island is covered with a heavy growth of timber, much of which is destitute of foliage, and in these trees thousands of *cranes* build their nests every year. It is an interesting sight to see them feeding their young, which I have several times witnessed while spending a time at that summer resort. The *cranes* come from every direction, bearing in their strong beaks fishes of various kinds and sizes, including perch, bass, sunfish, croppie and pickerel, some of which latter are occasionally so large as to be quite beyond the ability of even the largest *cranes* to swallow. In such cases, after many vain efforts on the part of the young to gorge the prey,

BEARDED CRANES.

the fish is thrown out of the nest, so that a walk over the grounds reveals thousands of large fish, some alive and others in all stages of decomposition, lying where they have fallen from the nests.

The flight of the *crane* is swift and graceful, and resembles somewhat that of an arrow. The neck is stretched straight out in front, while the legs trail in a direct line with the neck and body behind. At the time of migrating they assemble in immense flocks, choose their leader, and at the word of command leave their perches, and start off on an incline until a great height is reached, when they move in solemn rank, each bird in its assigned place, one behind the other, and the commandant in the lead. The *blue crane* is about six feet in height, with a plumage of unchangeable dun color. It is a vicious bird when unable to retreat, and uses its beak with the effect of a stiletto.

The **Demoiselle Crane** (*Scops virgo*) is a very handsome African species, hardly four feet in height, notable for the eccentric gambols it occasionally indulges in, in which it dances about on the tips of its toes, flaps its wings, and bows its head in a most humorous manner.

The **Crowned Crane** (*Balearica pavonina*) is also a native of northwestern Africa, and like the previous species occasionally indulges in fantastic gambols. It differs from the demoiselle in having a very handsome crest of filamentous feathers of a golden hue, fringed with black barbules. It has an extremely harsh voice, which it frequently uses, and is sometimes called the *trumpet crane*.

The name *trumpeter* is very properly bestowed because the bird gives voice to a call very much resembling trumpet notes, which it repeats very often between sunset and dark. Superstitious people have at-tached the same importance to the cry of the crane as they have to the howling of a dog, and with equal reason, since there is no reason in either. The crane's dancing habits, grotesque almost beyond description, are also made the basis for equally silly prophesy of evil, making of an innocent hilarity which the bird thus expresses, a portent of calamity to persons and even to communities.

CRANE FEEDING ITS YOUNG.

The **Night Heron** (*Nyctiardea gardeni*) abounds in nearly all the marshy districts of America and the British provinces. It breeds about the marshy coasts, building its nests in the densest coverts of swamp lands of sticks that are laid so loosely as to require frequent repairs to retain the young. They are gregarious in building, sometimes as many as half a dozen nests being located in the same tree. The eggs vary in number from three to seven, are extremely thin-shelled considering the size, and are of a pale sea-green color. The young do not leave the tree in which they are hatched for some weeks, but spend little time in the nest after the first week, being usually found clumsily climbing about the branches, and hanging by bill and claws. During the breeding season the male gives utterance to a deep, booming noise, that may be heard for a great distance. Like all the genera, the *heron* is a pugnacious bird, and capable of doing very serious injury with its stiletto-like bill.

The **Blue Heron** (*Ardea cærulea*), also vulgarly called *shite-poke*, is a frequenter of the creeks, marshes and rivers of the Northern States, though it is most numerous about the bayous and lagoons of the South. Its habits are much like the preceding species, except that it does not defend itself so vigorously against foes. It is about two feet in height, and of a dull bluish color.

The **Brown Heron** (*Ardea rufa*), or **Marsh Hen**, is a little larger than the blue heron, and has a dull brown plumage, like the back feathers of a thrush. It is rarely met with except about marshy places, where it feeds off fish, lizards, insects and crustacea. It is a bold bird when wounded, as I can testify by sad experience, using its sharp, strong bill with terrific effect upon man, dog, or any creature that comes within its reach.

EGRET.

The **Great Bittern** (*Botaurus stellaris*) is the largest American species, found everywhere within the United States, and as far north as Hudson's Bay. Its habits are like those of the preceding species, except that its nest is built in sedgy places, in which four eggs are laid, of a pale green. It feeds almost exclusively by night, and when disturbed, rises with a deep, sonorous *kawk*, that is startling to those who hear it for the first time. The plumage is a rusty brown, lighter under the abdomen, and a streak of black underlaid with white, running back from the base of bill to the middle of the neck. The bird stands about three feet high.

The **Golden-breasted Trumpeter** (*Psophia crepitans*) is an extremely handsome bird, with short, velvety feathers on head and neck, and a golden-green lustre on the breast. The body is small, compared with the extremely long neck and legs, but this disproportion serves it extremely well for it is a remarkably swift runner, unlike others of the crane family. The bird is gregarious, inhabiting the heavy forests along the Amazon in large flocks. It is easily domesticated, and is frequently seen among the domestic poultry of the Indians, where it rules, however, with an iron hand. Its height is four feet.

The name *trumpeter* has been given because of the loud and very curious ventriloquous sound the bird produces with the mouth closed. It nests upon the ground.

The **Sun Bittern** (*Eurypyga helias*), also called **Striped-Sun-ray,** is also a South American bird that possesses the characteristics of both heron and rail, but is easily distinguishable from both. It is beautifully variegated with white, brown and black bands and mottlings. It is hardly so large as the brown bittern, has a long, square tail, and a neck of some length, but which is carried drawn back so as to be invisible, except when the bird is excited. The feathers lie close during the life of the bird, but immediately after death they turn up at the ends and appear to be reversed. Though shy in its native haunts, the *sun bittern* is easily tamed, and, like the trumpeter described, it is often found among the poultry of South American Indians. It inhabits the banks of great rivers, and builds its nest at low elevations, of sticks lined with slime and deftly beaten mud. It lays the eggs twice each year, and hatches the young in four weeks. The length is thirty inches.

The **Jacana** (*Parra jacana*) is a very strange bird, found widely distributed in Africa, Asia and Australia. It is distinguished for its extraordinary toes, which are almost wire-like in thinness, yet of the most extraordinary length. But

BITTERN (*Botaurus stellaris*).

we perceive in these a singularly wise provision, showing again the marvellous adaptation of structure to habits. This bird finds its subsistence among the water-lilies, or the floating leaves of other water-plants, upon which, by means of its very long and slender toes, it is able to walk, while picking off the snails and other insects which inhabit water-plants. In South America, cranes may be also seen very often, standing upon the gigantic leaves of the victoria regina, which can support the weight of a body even larger than that

of the greatest crane. The *jacana* is black, with a slightly greenish gloss, running into a rusty red on the back and wing coverts. The wings are furnished at the bend with long, sharp claws, like the bat. At the base of the beak is a leathery appendage, rising to the forehead, and depending to the throat. The claw of the hind toe is of extraordinary length, even exceeding that of the toe itself.

TRUMPET BIRD.

The **Chinese Jacana,** or **Water Pheasant** (*Hydrophasianus sinensis*), is an extremely beautiful bird, with very long, arched tail, like that of the golden pheasant, which it also resembles in other respects. Like that of the previous species, however, it is a good swimmer, and very graceful in all its motions. Both species make their nests of weeds and grasses, upon a support made by weaving together the stems of aquatic plants.

The **Horned Screamer** (*Palameda cornuta*) is found in Central America, inhabiting the morasses of that hot country. Like the jacana, the shoulders of the wings are provided with a bold, sharp spur, which the bird uses with great effect against the snakes which it is often compelled to fight in defence of its young. It is a large bird, almost equal to the turkey, so that a stroke of its wing, armed as it is, may be considered as being more effective than that of the swan. Upon its head is a slender, horn-like growth about four inches long, the use of which has not been determined.

The **Stork** (*Ciconia alba*) is one of the best known and most highly respected birds, found in civilized countries. It is a member of the crane family, distinguished from others of the species, principally, by having eyes surrounded by a naked skin, and partially webbed toes. Its food is garbage, worms, insects, fishes and reptiles of several kinds. The most celebrated of the species is the **White Stork,** which generally passes its winters in the north of Africa, and particularly in Egypt, migrating in the summer to more northern countries of Europe. As the name indicates, the plumage of this species is clear white, with feathers covering the shoulders, and wing

SUN BITTERN.

coverts a glossy black. When the wings are spread, the point of the quill feathers separate, leaving a space between that would seem to interfere with flight, yet few birds are stronger of wing.

These birds have in all ages been regarded with peculiar favor, amounting in some countries almost to veneration, partly on account of the services which they perform in the destruction of noxious animals, and in removing impurities from the surface of the earth, and partly on account of the mildness of their temper, the harmlessness of their habits, and the moral virtues with which imagination has delighted to invest them. Among the ancient Egyptians the _stork_ was regarded with a reverence inferior only to that which, for similar causes, was paid to the sacred ibis, considered, and with some show of reason, as one of the tutelary divinities of the land. The same feeling is still prevalent in many parts of Africa and the East; and even in Switzerland and in Holland something like superstition seems to mingle in the minds of the common people with the hospitable kindness which a strong conviction of its utility disposes them to evince towards this favorite bird. In the latter country more particularly, the protection which is accorded to it is no more than it fairly deserves as the unconscious instrument by which the dikes and marshes are relieved from a large portion of the enormous quantity of reptiles engendered by the humidity and fertility of the soil.

On the other hand, the _white stork_

JACANA.

appears to be influenced by the same friendly feelings towards man. Undismayed by his presence, it builds its nest upon the house-top, or on the summits of the loftiest trees in the immediate neighborhood of the most frequented places. It stalks perfectly at its ease along the busy streets of the most crowded town, and seeks its food on the banks of rivers, or in fens in close vicinity to his abode. In numerous parts of Holland, its nest, built on the chimney-top, remains undisturbed for many succeeding years, and the owners constantly return with unerring sagacity to the well-known spot. The joy which they manifest on again taking possession of their deserted dwelling, and the attachment which they testify towards their benevolent hosts,

HORNED SCREAMER.

are familiar in the mouths of every one. Their affection for their young is one of the most remarkable traits in their character. It is almost superfluous to repeat the history of the female which, at the conflagration of Delft, after repeated and unsuccessful attempts to carry off her young, chose rather to perish with them in the general ruin than to leave them to their fate; and there are many other and well authenticated proofs of a similar disposition. They generally lay from two to four eggs of a dingy yellowish white, rather longer than those of the goose, but not so thick. The incubation lasts for a month, the male sharing in the task during the absence of the female in search of food. When the young birds are hatched, they are carefully fed by their parents, who watch over them with the closest anxiety. As soon as they become capable of flying, the parents exercise them in it by degrees, carrying them at first upon their own wings, and then conducting them in short circular flights around their nest. When in search of food the *stork* is commonly seen in its usual attitude of repose, **standing** upon one leg with its

CRANES ON THE VICTORIA REGIA.

long neck bent backwards, its head resting on its shoulder and its eyes steadily fixed. Its motions are slow and measured, the length of its steps corresponding with that of its legs. In flight its head and neck are directed straight forward and its legs extended backward, an awkward and apparently constrained position, but that which is best calculated for enabling it to cleave the air with rapidity.

In Bagdad, and some other of the more remote cities of Asiatic Turkey, the nests of storks present a very remarkable appearance. The *minars*, or towers of the mosques, at Constantinople and most other parts of Turkey, are tall, round pillars surmounted by a very pointed cone; but at Bagdad the absence of this cone enables these birds to build their nests upon the summit; and as the diameter of the nest generally corresponds with that of the minar, it appears as a part of it and a regular termination to it. This curious effect is not a little increased by the appearance of the bird itself in the nest, which

GERMAN STORK (*Cico*).

thus, as part of the body and its long neck are seen above the edge, appears the crowning object of the pillar. The Turks hold the bird in more than even the usual esteem, which may be partly attributed to its gesticulations, which they suppose to resemble some of their own attitudes of devotion. Their name for the stork is *Hadji lug-lug:* the former word, which is the honorary title of a pilgrim, it owes to its annual migrations, and its apparent attachment to their sacred edifices. The latter portion of the denomination, "*lug-lug*," is an attempt to imitate the noise which the bird makes. The regard of the Turks is so far understood and returned by the intelligent *stork*, that in cities of mixed population it rarely or never builds its nest on any other than a Turkish house.

STORKS ASSEMBLING PREPARATORY TO MIGRATING.

The Rev. J. Hartley, in his "Researches in Greece and the Levant," remarks: "The Greeks have carried their antipathy to the Turks to such a pitch that they have destroyed all the *storks* in the country. On inquiring the reason, I was informed 'The stork is a Turkish bird; it never used to build its nest on the house of a Greek, but always on that of a Turk!' The tenderness which the Turks display towards the feathered tribe is indeed a pleasing trait in their character."

Whale-head Stork (*Balæniceps rex*) is the name given to a species found nowhere, I believe, outside of a small district in Northeast Africa. It does not migrate but spends all its time, in all seasons, about the morasses of Egypt, where it is sometimes seen in pairs and again in flocks of a hundred or more. The most notable feature about this bird is its very curious and very huge bill, the upper mandible being enormously expanded and hanging over the lower.   The point of the upper mandible is a long, hook-like termination, which the bird uses to tear and rip up its prey. Its food is fish, water-snakes and carrion.   It nests upon the ground, laying two eggs in a shallow basin of mud, which is not lined.   The height is about three feet, and the color is a dark slaty gray, with a narrow band of white on the edge of each feather.

STORK'S NEST ON EGYPTIAN MONUMENT.

The Boat-bill Stork (*Cancroma cochlearia*) is another bird distinguished for its singularly large bill, fashioned somewhat in the shape of a canoe, or as the two bills are of an almost identical shape, they resemble two canoes laid together.   It is a South American bird with all the habits peculiar to the heron family, except that it not only wades in the water and watches for its prey, but sometimes sits upon a perch and angles after the manner of the kingfisher.   The male presents an imposing appearance, for though his plumage is by no means bright, he bears a long flowing plume of jet black feathers and a beard of grayish-white. The neck is rather short, and the body about the size of a mallard duck.

The **Spoon-bill** (*Platalea leucorodia*) is perhaps more grotesque in appearance than either of the preceding species, if we choose to judge it by the wonderful bill it supports.   It is found in many parts of Europe, Asia and Africa, South America, and the coast of Florida, always inhabiting marshy regions like its congeners.   The beak is nearly one foot long, flat, and is spatulate at the end and shaped like the bowl of a spoon.   Indeed, the bill is often taken from the bird when dead, scraped very thin, well polished, sometimes set in silver, and used as a spoon.   It builds its nest sometimes in trees, and at other times on the banks of streams in the thick herbage, which it raises above the wet by a plaster of mud.   The color is a pure white, with rose tint about the neck, and a plume of white feathers pendant from the crown.   The height of this bird is nearly three feet.

WHALE-HEADED STORK.
(*Balæniceps rex.*)

The **Adjutant** (*Leptoptilus crumenifer*), of India, is a curious member of the stork family for its several singular features, as will appear. The bill is very large, but uniform in shape, terminating in a very sharp point; it is used as a

MARSH BIRDS.

weapon (and a powerful one it is) as well as for taking its food. In India the bird is so highly regarded for its services as a scavenger that it frequents the streets of towns and cities with all the abandon that distinguishes the English sparrow. So powerful is the bill and so large the œsophagus, that the *adjutant* can easily swallow, as it frequently does, a full-grown cat or a fowl. The carcase of larger animals is easily torn in pieces and devoured with remarkable expedition. It also kills and eats large snakes and other reptiles, in which capacity it performs a very useful service to man.

This creature is so nearly domesticated, by reason of the protection given it, that upon small invitation it will adapt itself to civilization by readily becoming a house-bird, but it never ceases to be an incorrigible thief, on which account it is unpopular as a pet. They have been known to enter a house and seize upon a baked fowl prepared for other palates, and swallow it before an incensed housewife could recover it. Though belonging to the stork family, the *adjutant* has a decidedly vulturine aspect, in that the head and

ADJUTANT.

neck are destitute of feathers, and are covered with fleshy excrescences or wattles. It has also the ostrich's digestive powers, being able to swallow a live terrapin and after digesting the softer parts ejects the shell and bones. The *adjutant* is about five feet in height, nests upon dead trees, and has a very sombre plumage.

The Jabira (*Mycteria australia*) is a close relation of the adjutant, resembling it both in aspect and habits, though presenting a point of difference in having the neck and head covered with feathers of a rich green metallic lustre. It is an extremely rare bird, found only in Australia, and even in its favorite haunts is so uncommon as to lead to the belief that it will soon become extinct. A species of *jabira* is also found in Brazil.

MARSH BIRDS: PELICANS, SPOON-BILLS, CRANES, EGRETS, GODWITS AND PLOVERS.

The **Sacred Ibis** (*Ibis religiosa*) is an Egyptian bird, though it does not breed in that country, migrating further south in the latitude of Khartoum for that purpose. It has long been a matter for dispute as to what species of *ibis* the Egyptians once worshipped, nor can the dispute be settled by the remains that have been recovered embalmed with mummies. Some naturalists even maintain that the *sacred ibis* was never found native to Egypt, but that the specimens which are met with among the mummied remains were imported to serve a religious purpose. Dr. Adams, concerning this opinion, says that while he finds "no reason for considering the *sacred ibis* to have been a native at any time of either Egypt or Nubia," he has "no doubt that it was imported by the ancient Egyptians, and judging from the numbers which are constantly turning up in the tombs and pits of Sakkara and elsewhere in Egypt, and the accounts of Herodotus, Diodorus and Strabo, the *ibis* must have been very numerous and, like the brahmin bull in India, 'did as it chosed.'" Dr. Adams further remarks that every street in Alexandria is full of them. In certain respects they are useful; in others troublesome. They are serviceable because they pick up all sorts of small animals and the offal that is cast into the street, but they are extremely objectionable because of their thieving propensities and their dirty habits.

SACRED IBIS.

Mummied *ibises* are usually found alone, though sometimes they appear with the sacred animals. Hermopolis was the patron city of this bird, but we find its remains also among the ruins of Thebes and Memphis. The White Ibis was also regarded as a sacred bird, having been first imported from Italy and kept in the temple of Isis. It was the emblem of Troth, who was the secretary of Osiris, to whom fell the duty of recounting and perpetuating in writing the deeds of persons deceased.

The *ibis* nests in tall trees, the mimosa being preferred and usually those which stand in the centre of a large morass. Like the herons, these birds are gregarious at breeding times, sometimes fairly covering a tree with their nests, which are very large but carefully made and lined with feathers. Three or four eggs of a greenish white and the size of a mallard's are laid. The so-called *sacred ibis* has a white plumage. The **Scarlet Ibis** is a native of

SCARLET IBIS (*Ibis falcinellus*).

northern South America, but in summer it appears along the Florida coast, and occasionally as far north as North Carolina. It is one of the handsomest birds that is seen within the United States. The **Glossy Ibis** is also a North American visitor, though never seen in any considerable number. Like the scarlet ibis it is a rather small species. The Straw-necked Ibis is a handsome Australian species, never found out of that country, though it is rather common in every part of that continent. Its name is derived from the long straw-colored feathers that adorn the neck. All the several species have their heads destitute of feathers, the body is rather plump, legs short, bills long and curved downward, and all are swift runners. Their general habits are very similar.

The Cobbler's Awl Bird, or Avocet (*Recurvirostra avocetta*) is found in several parts of the United States, its range extending from the Gulf to Labrador. The Mississippi Valley, however, is its favorite haunt, though nowhere is it a common bird. The body is plump and the size of a pigeon, the legs long, the plumage pied, with white preponderating, top of head, back of neck, and wing coverts black. But the one curious feature that distinguishes this bird is the long, slim bill, sharply pointed and curving upward, so that in searching for its food in the mud it uses the bill as a scoop. It nests upon the ground.

EGYPTIAN IBIS AND PAMPAS GRASS.

The Curlew (*Numenius arquata*) is also an American bird, considerably larger than the avocet, and with a bill of equal length but curving downward. Though aquatic, and waders, I have frequently found them in Nebraska a considerable distance from water. Their flight is straight, and while on the

CURLEW.

AVOCET, OR COBBLER'S AWL BILL.

wing they utter a noise that may be imitated by a boy whistling one long, easy note.

The *curlew* builds its nest upon the ground, sometimes upon high hills in the dry vegetation, but more often in marshy localities. Four eggs are usually deposited, of a brownish green color, and the smaller ends of the eggs are always laid together. The general color of the bird is a dark brown, with wing coverts edged with white. The height is about two feet.

WHIMBREL.

The **Whimbrel** (*N. phicopus*), often called the **Jack Curlew**, is found upon the Shetland Islands, but so nearly resembles the *curlew* in appearance and habits that the two are easily confounded. The cry of the latter, however, is very different, resembling in sound the word *titterel*.

The **Godwit** (*Lineosa ægocephala*) is a rather common bird about the fenlands and marshy districts of the southeast, being especially numerous in Florida. They are small, and bear a striking resemblance to the jacksnipe, and their flesh is equally esteemed. Like others of the species, it nests upon the ground, and lays four eggs of light brown color.

The **Jacksnipe** (*Gallinago gallinula*) is a common visitor to the Northern States in the spring and fall, going North to breed, and returning to the South to spend the winter. In the Middle States it makes its appearance, coming from the South, early in April, though sometimes earlier, and remains for about six weeks, when it proceeds northward as far as British America, where it breeds. On the return it again halts during October and November, bringing its full-fledged brood, when it is usually in splendid condition. It generally haunts marshy grounds, though I once saw a large flock feeding on the high sand hills of Dakota. The bird has grown much wilder in latter years, on account of its persistent persecution by hunters. When put up it starts off in a zigzag or eccentric direction, uttering a coarse *schraik*, but soon takes a straight course. The sports-

SNIPE CARRYING HER YOUNG.

man who makes a success hunting this bird must reserve his fire until the snipe assumes a regular flight, which it does before traversing more

than twenty yards. This species of snipe bears a considerable resemblance to
the woodcock, though it is not so large.. The color is light brown, splotched with
white, with stripes of grayish white, alternating with deep brown or black running
over and along the sides
of the head. It builds its
nest under a tuft of grass,
and deposits four eggs of
an olive white, spotted with
brown. No bird is more
attentive to its young, the
mother having been known
to carry away her brood
when threatened by inun-
dation or other danger.
When thus transferring her
brood she takes up one at
a time, bearing them be-
tween her feet, as shown
in the illustration.

GOLDEN PLOVER (*Charadrias apricarius*).

The **Woodcock** (*Scalo-
pax rusticola*) is another
well-known American game bird, found principally in the Western States.
Some years ago it was quite plentiful, but its numbers are rapidly diminishing. It is nearly equal to a quail in size, which it also resembles in color.
The bill, how-
ever, is quite
long, and the
eyes are situated
nearly even with
the crown. It is
found in marshy
places, where it
feeds principally
by boring in the
ground. Like
the jacksnipe, it
is rarely seen
until put up,
when its flight
is generally only
a dash upward
and down again
a few paces away,
hence only a
snap-shot suc-
ceeds in bringing
it down. It flies

WOODCOCK AND YOUNG.

late in the evening, in a straight line, producing a whispering noise. Its
nesting habits and care of its young are very similar to those of the jacksnipe

It is said, though with what truth I have not been able to learn, that the parent often carries its young from the nest to feeding grounds, and returns them to the shelter of the woods again, carrying them between' her feet like the jacksnipe.

The **Ruff** (*Machetes pugnax*) is occasionally met with in the Middle States, but it is more common in Europe. Though frequently haunting marshy places, it is as often seen far inland in dry districts, running on the ground in search of insects. It is gregarious and extremely pugnacious. It derives its name from a prominent collar of long, but closely set, feathers, which distinguishes the male. It is about twelve inches long, and though the coloring is variable, brown always predominates.

The **Spoon-bill Snipe** (*Eurynorhyncus pygmæus*) is a native of the Arctic regions, found principally about the coasts of Nova Zembla, though it is occasionally met with on the northernmost shores of Norway and Russia, being the sole representative of the waders within the boreal zone. In size and color it very much resembles the sand-piper, but is distinguished for its very singularly shaped bill, as will be seen in the illustration, the natural size of which is shown. But, as we have frequently seen, everything is wisely adapted to the purposes of creation, and the wisdom of this seemingly strange provision is manifested in the habits of the creature. It subsists almost entirely off shrimps, which it catches along the ice-bound shores, and which its bill is specially adapted to seize and hold. Its nesting habits are not yet known.

THE RUFF.

SPOON-BILL SNIPE.

The **Golden Plover** (*Charadrius virginicus*) is common throughout the United States, but more plentiful in the Western States. Early in the spring and fall it may be seen in large flocks about the edge of ponds, feeding on insects that burrow in the sand or live about the water, but when spring advances, it moves to freshly ploughed ground, and follows the furrows in quest of grubs and worms turned up by the plow. Flying always in flocks, they offer an easy target to the sportsman, who sometimes brings down as many as a dozen at one shot. In the evening, however, they disperse and roost upon the ground in pairs. During the night they may be occasionally heard piping their thin notes, and quite as ready to take alarm as in the daytime.

The **Kildeer Plover** (*Egialitis vociferus*) is also a common bird, with habits very similar to those of the golden plover, though they do not go in such large flocks. Usually they are found consorting with other species, and making the air resonant with their cries of *kil-deer*, or *te-te-de-dit te-dit*. They are not considered as game birds.

GOLD PIPER (*Charadrias auratus*).

The **Sand-piper** (*Tringoides hypolenca*) is found throughout the Western States in the spring and summer season flying in large flocks, like blackbirds, or running around the shores of ponds tipping their tails in a curious fashion. Again they may be seen, especially towards mid-day, standing as still as so many decoys, in a field, ready to rise at any effort to approach them. Their bodies are too small to admit of them being classed as game birds, though they are frequently sold in markets.

The **Oyster Catcher** (*Hæmatopus ostralegus*) is found all along the Atlantic coast, from Maine to Florida, but it is more common on the Jersey shore. It feeds on crabs, sand-fiddlers and other small shell-fish, and may be seen running rapidly along the water's edge, driving its bill into the sand apparently aimlessly, and again stopping to bore. It will also insert its bill between the open shells of an oyster and eat the mollusk therein. It is a shy bird, and not much hunted. Its length is eighteen inches.

OVSTER CATCHER.

The **Pebble-turner** (*Strepsilas interpres*) is found both in North America and Europe. The name is derived from its movements when feeding, at which times it runs along the beach picking up sand-hoppers, marine worms and other crustacea, and turning over the stones in its course, for the purpose of getting at the little animals that have taken refuge underneath. It also has a curious habit of knocking pebbles from side to side, though evidently done in search for prey. It usually builds its nest under the strong shelter of an overhanging rock, safe from wind and wave. Its color is a bright rust-red, white underneath, with a broad band of black above the tail-coverts.

PEBBLE TURNER.

The **Knot**, or **Irish Sanderling** (*Tringa canutus*) is also a habitant of both hemispheres. It is so called in honor of King Canute, and on account of its habit of running towards the sea at each receding wave, and retreating again at the approach, thus keeping always very near the water. Its habits are, otherwise, similar to those of the sand-piper, which it equals in size.

The **Marsh Swallow** (*Glareola pratincola*) is a singular bird, combining, as it does, the characteristic of both swallow and plover, so that no appropriate classification has been found for it. The tail and wings are like a swallow's, and the mouth resembles that of the goat-suckers, though its habits are those of a marsh plover. It is found chiefly along the Mediterranean. An illustration of this bird will be found in the matter describing the swallow species, over the title of "fallow swallow."

The **Scabbard-bill** (*Chionis minor*) is another bird with conflicting characteristics, so many dissimilarities appearing that it has been variously classed with pigeons, quails, gulls and sandpipers. The latter classification is now generally accepted, though rather by reason of its habits than because of any of its appearances. By some it is called the *short-billed snipe*. The head is certainly that of a pigeon, and the bill is short and thick, presenting a peculiarity, however, in having a sheath issuing from the base, and covering half the length of

IRISH SANDERLING.

the bill. Mr. Darwin is the only person who has given us an account of this bird based on personal observation. It is an inhabitant of Patagonia and Southern Brazil. It nests on the ground, and produces two broods each year. The color of this bird is pure white, the bill is black, and a part of the face is covered by the base of the sheath.

SCABBARD-BILL.

The **Seriema** (*Dicholopus cristatus*) is an Australian bird bearing some resemblance to the crested crane but more to the secretary bird, to which it seems to be related as closely as cousinship, at least. The legs are long and the *seriema* can use them to such good purpose that it may attain the speed of a horse. The head is decorated with a crest composed of a dozen feathers, which the bird raises and lowers rapidly when excited. It rarely enters the water but is never found far from a stream or morass.

The **Sultan Hen** (*Porphyrio smaragdonotus*) is found about the lagoons and marshy districts of northern South America. It is about three feet in height, has very long legs not fitted for speed, however, is of a dun color, thick bill, and a bony crest on the head, which is a provision no doubt to protect its head against the brush through which it creeps, the same natural protection being on the head of the helmeted cassowary. It is a poor flyer, and therefore keeps about the dense sedges. Its food is small fishes, reptiles and insects, in which respects it does not differ from the bitterns. This bird, however, occasionally takes to the water, and though its toes are not webbed it is said to be a fairly good swimmer. In its nesting habits it resembles the water hen, laying usually six eggs of a pale greenish hue in a nest constructed of coarse grass, located generally on the edge of a marsh.

The **Shade Bird** (*Scopus umbretta*) is a native of West Africa. It is usually found alone, except at the nest, in wooded districts, with its head well drawn back watching for fishes or walking with measured stride in search of frogs, worms and snails. It roosts in trees or on its nest, which is a peculiar structure completely vaulted over and shaped like an oven, with entrance from the side. The diameter of this nest is about six feet, which may be accounted for by the fact that it is divided into two or more compartments to provide a resting-place for the parents as well as for the young; as the bird never migrates a single nest serves it for many seasons. The natives regard this bird with great awe, believing it has the power of a witch. The size of the bird is equal to that of a night heron, and its plumage is very sombre. There is a crest of feathers on the head which may be raised or depressed at will.

SHADE BIRD.

The **Rail** (*Crex carolinas*) belongs to a class of which there are thirty-two well-known species distributed over all portions of the globe, a dozen of which are residents of the United States. But the general character of these is everywhere the same. They run swiftly, but their flight is slow, and with the legs hanging down; they become extremely fat, are fond of concealment, and usually prefer running to flying. Most of them are migratory and abound during the summer in temperate regions. The *rail* generally builds its nest in a tussock of grass, and forms it of well-interlaced dry vegetable fibres. The female lays from four to six eggs of a dirty-white, specked with brown or black. The young are covered with a soft velvety black down, and leave the nest almost as soon as they are hatched, taking to the grass where they run about and look like mice.

Rails are seldom met with far from a marshy region. They make their home among the reeds and subsist off snails, bugs and other insects, and also off certain seeds of marsh grasses. Though not specially fitted for the water they can swim well and are excellent divers. If wounded they drop into the water and diving seize hold of the bottom of a reed where they cannot easily be dislodged, and will drown themselves rather than come to the surface if they think it is unsafe to do so.

SERIEMA.

SWIMMING BIRDS.

From the description of the wading birds, all the most interesting species of which having been given, we shall proceed to a consideration of the next order, which, in the natural sequence of progression, must include the numerous species that not only find their subsistence about the water, but which make the water a part of their natural element, swimming, diving, floating and disporting themselves among the waves with the grace and naturalness of fishes. In these birds we observe wonderful provisions, specially adapting them to the life nature has designed them to lead. We find that while the waders have long legs, and toes well spread to give them firm footing, those that swim have short legs and webbed feet, and that the legs are placed well towards the tail, because this affords an increase of leverage in the act of propulsion. We also notice that swimming birds are provided with a very thick coat of feathers, which greatly increases their buoyancy, and that these feathers are covered with an oily substance which exudes from the root of the feathers, and that this natural supply is increased at the pleasure of the bird by an ex-

SULTAN HEN.

WATER RAIL.

traction from the excretory oil-duct that is situated on the bird's rump. In many species this oily exudation is so great as to render the feathers unfit for human use, but in all such cases we find that the supply is necessary because such birds spend nearly all their time in the water. This oil not only renders the feathers impervious to water but also contributes greatly to the warmth of the bird, as well as enabling it to move with greater ease through the water. We shall also find that each species is specially provided with means for taking the

food upon which nature intended that it should subsist. If it be appointed to take its prey during flight the wings are expansive and strong; if it lives upon prey taken beneath the water it is fitted for diving at the expense of flight; if it be designed to fish in the sea or seize its prey in the air, then nature has abundantly qualified it by special provisions of beak, claw and the means for swift progression through the water. All these wonderful attributes will clearly appear in the descriptions following.

The first species we shall consider is the **Goose**, omitting mention of the domestic species, with which every one is supposed to be acquainted, and particularly because it was from the wild varieties that our present domestic geese and ducks descended. There are nearly fifty different species of geese, found very widely distributed, but all essentially have the same habits and a general description may therefore be applicable to all.

PER-WIT (*Vanellus cristatus*).

The *wild goose* is found among the rice fields of the South in winter, though he goes on southward as far as the equator, remaining there until spring when he seeks the breeding-grounds of the far North, extending his flight even to the frozen clime of Baffin's Bay. Their migrations northward are seldom made in flocks, but after broods are hatched and the chill of approaching winter admonishes a change of climate, they assemble in immense flocks under the leadership of a gander who is invariably larger than any of his associates. At the word of command the journey southward is begun. They are thus frequently seen flying at great elevations in flocks of fifty or more, spread out in various shapes, but generally in the form of an acute angle, speeding all day and resting at night. Their cry is frequently heard when they are so high above us as to be imperceptible; and this seems bandied from one to the other, as among hounds in the pursuit. Whether this be the note of mutual encouragement, or the necessary consequence of respiration, is doubtful; but they seldom exert it when they alight in these journeys.

Upon their coming to the ground by day they rang rthemselves in a line, like cranes, and seeme ather to have descended for rest, than for other refreshment. When they have sat in this manner for an hour or two, I have heard one of them, with a loud, long note, sound a kind of charge, to which the rest punctually attended, and they pursued their journey with renewed alacrity. Their flight is very regularly arranged; they either go in a line abreast, or in two lines, joining in an angle in the middle. I doubt whether the form of their flight be thus arranged to cut the air with greater ease, as is commonly believed; I am more apt to think it is to present a smaller mark to fowlers from below, for of all birds the *goose*, despite his reputation for stupidity, is one of the most sagacious. Sometimes they not only alight in the central parts of the United States, but remain in certain localities for a considerable while early in the spring and late in the fall. They are very fond of young wheat and are occasionally very destructive, and especially so in California, where the wheat-growers employ hunters to drive them off their fields. When feeding one or more sentinels are always on watch, with heads well elevated and eyes and ears alert to catch sight or sound of any enemy. The most successful means employed in hunting them may be briefly described as follows: When the geese are found to frequent a sand-bar, as they frequently do, the hunter repairs at night to the spot where they most often assemble, digs a hole in the sand and places a barrel therein sunk to the level of the surface. In this he secretes himself and awaits the dawn of day and the coming of his prey.

Geese often frequent cornfields, in which the corn has been cut and shocked; learning their range the hunter goes before daylight and secretes himself in one of the shocks and then waits the return of the game. Others hunt them by erecting blinds, behind which they lie and take chances of getting a shot as the geese fly over them. Yet others use a gentle horse, behind which the hunter walks as the horse is urged slowly towards a settled flock, but no one

WILD GOOSE (*Bernicla torquata*).

save an experienced hunter is likely to meet with much success putting any of these means into execution, but the novice may find a flock distracted, or less mistrustful, during a heavy snow-storm and make a heavy bag.

The **Cape Barron Goose** (*Cereopsis New Holland*) is one of the largest of the genera. As its name implies, it is chiefly found in Australia. It is sometimes captured and domesticated, while the eggs are often found and hatched out by domestic poultry. But, though it is considered a great acquisition to the poultry-yard, its disposition is very quarrelsome, and it occasionally commits great havoc among the other fowls.

The **Gray Goose** (*Anser cinereus*) is seldom seen in this country, nor is it often met with in Europe now, though formerly it was quite numerous. From this species it is believed our common *domestic goose* is descended.

The **Ring-necked Goose** (*Bernicla torquata*) is our common species of *wild goose*, found plentifully distributed in the western part of the United States, and in the summer time may be found far within the 'Arctic circle. The name *bernicla*, or *barnacle*, has been given to this species by old sailors, who formerly supposed it to be a product of the *barnacle shell*, or so-called *goose-mussel*. This belief was not confined to ignorant sailors, but was affirmed by the early naturalists.

CAPE BARRON GOOSE.

The **Brent Goose** (*Bernicla brenta*), more commonly called **Brant,** is also a frequenter of the United States, usually preceding the other species in the spring and fall by one or two weeks. It is quite numerous, resembles the common wild goose, except that it is smaller in size, and does not have the conspicuous ring of white and black feathers about the jaws.

GRAY LAG GOOSE (*Anser ferus*).

Next to the goose (though it might more properly precede it) is the **Swan,** which is given a subordinate rank because of its scarcity, and because it is very rarely seen in this country outside of a public park. In England the bird has for more than two centuries been under royal protection, very severe penalties being provided against its destruction, except by legal authority. The swan is there regarded as a royal bird, and no one may raise them without a license from the Crown. Each person, so licensed, must file a special "swan mark," by which his birds may be distinguished, the mark being cut in the upper mandible of the bird. All swans which reach a certain age without being marked are called *clear bills*, and become property of the Crown, unless they be kept by special permit, granting the right to seize and keep any adult swan which has not been marked.

The process of marking the swans is termed *swan-upping*, a name which has been corrupted into *swan-hopping*, and is conducted with much ceremony. The technical term of the swan-mark is *cigninota*. Swan-upping of the Thames takes place in the month of August, the first Monday in the month being set aside for the purpose, when the markers of the Crown and the Dyers' and Vintners' companies take count of all swans in the river, and mark the clear-billed birds which have reached maturity. The fishermen who protect the birds and aid them in nesting are entitled to a fee for each young bird.

WHITE, OR HIGH-BACKED SWAN.

Swans are very destructive enemies to fish, devouring the smaller species but doing the largest damage in the spawning season, when they will leave any other kind of food for fish-spawn, and have been known to entirely depopulate ponds of the best food species, such as bass, croppie, pike, and carp.

The swan builds her nest of sticks and straw, and usually locates it beside the water's edge of an unfrequented island. She generally lays six or seven eggs, of a dull or very pale green color, like the duck's, though I have seen the eggs as white as that of a hen's. When incubating, the swan is prompt to resist any invasion of her premises, and becomes a furious fighter in defence of her young. The cygnets—young—are covered with a fluffy down of light blue, no feathers showing until they are two months old. During the first few weeks of their life they mount upon the mother's back, who conveys them from place to place. If on shore she helps them to gain their position by lifting them by one leg, but when in the water she sinks until her back is level with the surface, when they easily scramble into a secure place.

The species are not very numerous, and include only the **Trumpeting Swan** (*Cygnus buccinator*), the **Small Swan** (*Cygnus minor*), the **Whistling Swan** (*C. americanus*), the **White Swan** (*C. olor*) and the **Black Swan** (*C. atratus*), the former being the only species well known in this country, though the *whistling swan* was formerly quite plentiful at certain seasons in Chesapeake Bay.

BLACK, OR MOURNING SWAN.

The **Goose-anger**, or **Goose-sawer** (*Mergus merganser*), also called *Water-pheasant*, *Sheldrake*, *Saw-bill*, and other local names, is a really beautiful bird, combining features of the goose, duck and cormorant. His home is in northern latitudes, extending around the globe near the Arctic circle, and southward to the North temperate. He spends nearly all his time in the water, is an expert swimmer and a wonderful diver, being able to shoot along under the surface with the rapidity of a fish, upon which it feeds. The coloring of the male is a blood-red bill, raven-black head, snowy-white breast, and black-and-dun back and tail-feathers. It nests along the shore usually, and lays from ten to fourteen oblong eggs of a light-green color. The young take to water within an hour after birth, and are so active that they rarely suffer capture by even wild animals that prey upon ducks.

The **Cormorant** (*Phalocrocorax carbo*), like the goose-anger, is an excellent swimmer and diver, both pursuing similar habits in procuring their food. The goose-anger, however, lives chiefly off small fishes while the *cormorant* is

DUCK-GOOSE (*Vulpauser tadorna*).   CORMORANT.

only content with larger prey. I have frequently watched them, while fishing, and noted with what dexterity they would dive and seize a fish of four and five pounds weight, which they would swallow with surprising ease. A more amusing sight is offered by several *cormorants* making a raid on eel-beds, either

when the eels are burrowing in the mud, or ascending streams to breed. It is a most laughable spectacle to see the birds swallow their wriggling, slimy prey, only to have the eel rise again, and be again and again swallowed, until exhaustion prevents the creature from making further effort to escape.

The *cormorant* is such a skilful fisher that it is commonly tamed by the Chinese, and trained to go out with their masters, sitting patiently upon the prow of a boat until given the order to begin the chase. At the word of command, they dash into the water, seize the fish in their beaks, and return with it to their owners. In

GOOSE SAWYER.

beginning to train it an iron ring is put on the bird's neck, which prevents it from swallowing the fish taken, and this is not removed until the *cormorant* fairly forgets that it has the ability to seize and swallow its live prey. After the birds are thoroughly trained, when one of them captures a fish which is too large for it to subdue, one of its companions will come to its assistance, and together they will convey the prize to their master.

26

The *cormorant* builds its nest on the shore, of sticks, sea-weed and grass, in which the female lays usually from four to six rather small eggs, of a pale greenish cast. When the young are first hatched they are destitute of down, being covered with black skin. The general color of the *cormorant* is black, sprinkled with hairy feathers of white, with a crest on the crown of the head. The upper parts of the body are brown, mottled with black, and the front of the throat and under surface is a velvety black. The length of the bird is about three feet.

The **Water Hen** (*Fulica atra*) is common wherever in English waters there are rushes. The *water hen*, or *moor hen*, apportions the brightest plumage to the male, which is dark green in color, except for blackish gray on the head, neck and belly, and red at the base of its bill and on the upper thigh. The *water hen* builds its nests of sticks, sedge and leaves; not content with one

FISHING WITH CORMORANTS.

nest, it makes several, and as its fancy suggests moves its family from one to another. Should the nest be threatened by rising water she continues to add to the height and thus protects it from an overflow. When the *water hen* is to be absent for any length of time from the nest, she leaves it so covered with leaves as to very thoroughly conceal it. The illustration in THE SAVAGE WORLD, however, shows a *water hen* which has just left its nest, and another *hen* watching the sportings of her half-grown chicks. This bird is both a wader and a swimmer, for which reason it is used as a link between these two classes and properly introduces the swimming birds of THE SAVAGE WORLD. Another species technically known as *fulica chloropus* is common to the fresh waters of America under the more familiar name of *mud hen*. It is found in ponds and especially where lily-pads are abundant, the roots and seeds of which it eats with avidity. The toes are only partially webbed, as seen in the small illustration. It is about the size of a teal duck and at certain seasons of the year, notably the fall, its flesh is most toothsome though at other times it is so rank as to be fairly nauseating. The nest of this species is admirably shown in the accompanying illustration. The color of this bird is a dark dun, only a shade lighter on the breast, with sharp pointed bill of pure white and legs and toes a light green.

Sportsmen and epi-cures so often sing the praises of the **Mallard Duck** (*Anas moschus*), which is so fond of frequent-ing the ponds almost about our doors, as to furnish the fullest opportunity for its study. It is regarded as the progenitor of the common domestic duck. Like the eider duck it uses down for the lining of its nest. It requires about a mouth for chang-ing its plumage, shedding the most brilliantly col-ored feathers first. Its head, neck and breast are of glossy green, though

ENGLISH WATER HEN.

the last inclines to brown, except that it wears a narrow white collaret. Its back is russet, deepening into black as it approaches the tail; the tail feathers are brown ash with the exception of four, which look like black velvet, and in the drake are curled. The wings are brown ash till near the extremity where they are black, the two colors being separated by a broad white vertical band.

AMERICA WATER HEN.

The **Bridal Duck**, or American Sum-mer Duck (*Anas sponsa*), is the American species of the *mandarin*, and deserves a more extended notice of its coloring, since it may at any time delight the eyes of such of the readers of THE SAVAGE WORLD as find them-selves awakened to pleasing observation of the animal life around them. The top of its head is green and purple, blended at times but generally distinct; its cheeks are fawn-colored and a cream-colored stripe runs from the back of the neck to the eyes; the neck on its sides is provided with long bright chestnut feathers, while purple prevails on the throat and breast. It has four shoulder-stripes, two of white and two of black, and its russet-colored wings have shining green margins; the underparts make a sharp contrast of white.   As in the case of the Asiatic species the drake wears the fine clothes while the duck dresses sombrely.   It winters at the south, but migrates as far north as Nova Scotia.   It frequents ponds and marshes but builds its nest in the hollows of trees and covers it with small feathers and sticks. Audubon says that when the nest is at some distance from the water, the d u c k will carry in its bill the ducklings, taking them one at a time to their natural element ; on the other hand, if the tree b e i n t h e immediate vicinity the ducklings are allowed to drop from the tree into the water, or are compelled by t h e i r parent to scuffle over the ground on foot.

T h e **Mandarin** (*Dendronessa galericulata*) is a typical mallard, but has its habitat in Eastern Asia, and is distinguished b y wearing ruffs on the sides of the neck, and by fan-like wings.   It is a s p e c i e s m o s t h i g h l y p r i z e d i n China, and rarely seen in other countries even as a naturalized citizen   or a forced sojourner. The plumage of the drake is remarkably beautiful but in the summer ti m̄ e h e exchanges it for the soberest brown a n d gray which are t h e colors uniformly worn by his better h a l f.   During the other seasons the drake luxuriates in the most bewildering combination of the richest purple, russet, green and white.   He is strictly monogamous and with his spouse is to be found perched on trees which border on the ponds in China.

NEST OF THE WATER HEN (*Fulica chloropus*).

The **European Sheldrake**, or **Burrow Duck** (*Tadorna vulpanser*), takes its name from its using rabbit burrows as its nest. It is as cunning as a fox and hence is sometimes called the *sly goose*. To protect its nest from the hunter it will pretend to be lame or maimed, and if he is deceived will finally cause him to believe that he has started in pursuit of an *ignis fatuus*. It does not change its coloring with the seasons, but instead of this moults once a year. The male claims no greater brilliancy of plumage than is allotted to the female. In Jutland the inhabitants have built up quite an industry out of the eggs and downy feathers of the *burrow ducks*, and provided them with

MALLARD DUCK.

burrows which, while satisfying the birds, reduce the labor expended in robbing them, as the duck divests herself of her down with which to line her nest.

WOOD, OR BRIDAL DUCK.

MANDARIN DUCK.

EIDER DUCK.

Many of my readers have seen eider-down, and the next of the ducks of which we are to speak is the **Eider Duck** (*Somateria mollissima*). It belongs to the Arctic zone, and is abundant in Ireland, Norway and Greenland. Norway has recently divided with Africa the palm of interest for pleasure-seeking travellers, and hence any reader of general literature is likely to meet accounts of the *eider-duck*, or at least allusions to its appearance and habits. The eider-down is collected from the nests, and one of these will yield somewhat over a single ounce. It is said that the annual exportation from Greenland and Iceland exceeds six or seven thousand pounds, so that the *eider-duck* feathers not only its own nest but that of the persons who steal the provision it has made for the comfort of its young. In Norway the *eider-duck* has given rise to an industry so considerable, so general, and so important that it is protected by special legislation, as well as by the interested regard of those who derive from it their means of subsistence. The *eider-duck*, like the brahma bulls and the sacred monkeys of India, has learned to improve its opportunities and will in the most fearless manner enter the houses of the peasants and there build its nest, plucking the down from its own breast to insure the comfort of its young. It is allowed to select any location that strikes its fancy, so that at times it will rob the family of the use of their beds, a privation which would be greater if these were any softer than the floor. The *eider-duck* is about two feet and a quarter in length, and its wings when extended will stretch fully three feet. It weighs six or seven pounds, but is rarely eaten because of "an ancient and a fish-like" taste. It is still found in America as far south as the State of Maine, but in earlier days was common even in the vicinity of Boston. It has

SPECTACLED EIDER DUCK.

been discovered that the *eider-duck* if robbed of its nest and eggs will repair its loss by repeating its previous efforts, and that it will continue for several times this attempt to recover from misfortune. Hence the collectors of eider-down avail themselves of this maternal interest, and turn the *duck* into somewhat of a drudge. The *eider-duck* is a good diver and an excellent swimmer, but as a walker it is unusually awkward even for a waddler. The *plumed drake* has a head velvety black on top and green on the back, white side-face, green ear-muffs, white throat and upper neck, buff lower neck,

PIPING DUCK.

black breast and under parts, except for a white spot on the under body back of the legs. Its wings have the first and second series of feathers black, and the unchanging third set long and white; green is the color of its legs. The *duck* is a mottled russet.

The **Piping Duck, Red-headed Poker,** or **Pochard Dun-Bird** (*Fuligula clangula*), appears in northern Europe during the early fall, and migrates in the spring. It is noted as a diver, going deep into the water and swimming long distances under the surface, so that unless entrapped, it generally escapes the hunter. While moving about it keeps uttering a cry which has the monotony of a Scottish bagpipe.

The **Canvas-back Duck** (*Æthia valisneria*) is found at different seasons throughout the United States and Central America, though it is best known through its annual visits to "My Maryland," where, like an epicure, it feasts upon wild celery. It is frequently the victim of the idle curiosity which distinguishes the present period, for whole flocks will swim within short range of the hunter through their interest in the novelty of a dog trained to run up and down the shore while ornamented with a gaily-colored cloth. Epicures differ

TEAL DUCK (*Querquedula crecca*).

about the relative excellence of the various species of wild ducks, but it is certain that the admirers of the *canvas-back* have been most successful in attaining a popular verdict. The coloring above is gray or white, waved with slight touches of black; white beneath; red chestnut on the head and throat; blackish brown on the neck and breast; and black on the rump. It goes in some localities by the names *white-back* and *sheldrake*. Its quickness as a swimmer and diver marks it as an uncertain object for the sportsman.

CRESTED GREBE.

The Grebes (*Colymbi*) are queer looking creatures and suggest the idea that they represent the most conservative and least fully developed species of ducks. In the first place they are destitute of any tail, but then possibly as a compensation their heads are adorned with differently-colored and variously-shaped ruffs, crests, and other similar ornaments. They are unusually and noticeably flat-footed, and the feet are separated into broad lobes. They are found in both hemispheres throughout the temperate zone, and pass their time almost wholly in the water or under the water. They build nests of sea weeds and other vegetation, and fastening them to the grass or rushes, let them float upon the surface of the water. The *grebes* winter in salt water and summer in fresh water. They are remarkable for their expertness and quickness as divers, and though usually swimming well on the surface can, when frightened, so flatten themselves out as to allow nothing but their necks and the ridge of their backs to appear.

GOLDEN-CRESTED GREBE.

The **Golden-crested Grebe** (*Colymbus cristatus*) is European in its habitat. Its fan-shaped cowl and neck valance which, though long in comparison with the size of its head, reaches but a small way on its long neck, the white coloring about its eyes and its beak give to it the appearance of having combined an owl's head, a crane's neck and a grouse's body. The coloring of the *golden-crested grebe* is dark brown on the top of the head, white on the cheeks and around the eyes, and chestnut on the ruff. On the back the undermost feathers are white and the outer ones a chestnut brown and both have a satin sheen. The neck, throat and abdomen are white, and the legs green on the outer side and yellow on the inner part. It swims about with its ducklings now by its side and again on its back.

The **Little Grebe**, or **Dabchick** (*Colymbus minor* or *Podilybus podiceps*), is found on the eastern coast of North America, but lives on inland ponds and rivers. In comparison with other grebes it is a pigmy, resembling a fair sized young gosling. It prefers a roomy dwelling so that its nest looks as if intended for quite a number of ducks fully as large as the mother. These nests will often be found amongst the pond-lilies which serve as material, foundation and landscape for the foreground.

FLOATING NESTS OF THE LITTLE GREBE.

The **Crested Gorfou,** or **Crested Penguin** (*Eudypes chrysocoma*), is found in Patagonia, where its loud, continuous cry and gilded crest make it quite sure of attracting attention. Like the other members of the *penguin* family, it uses its wings not for flight, but for accelerating its movements when running (in which case they are used like fore legs), and for oars when in the water. It has the habits of its family, which are very curious and well worth attention. These birds are gregarious, and the flocks contain seemingly innumerable individuals, reaching quite often forty or fifty thousand. When not in the water, they subject themselves to the strictest martial law, and are told off into corps, battalions, regiments, companies, and squads. The young, those that are moulting, the setting birds, and of every other condition are required to remain with their own kind, and confine themselves to prescribed bounds. The eggs are carried, until hatched, between the thighs of the hen, but the hen does not exert herself seeking food, for her husband is at once devoted and a regular and bountiful " provider." There is but one young one in a brood, and as soon as it can stand upon its own feet the hen joins her husband in going to market, and the young always take their food by inserting their bill into those of their parents. The eccentricity of the *gorfou's* crest, in which each particular hair takes a different direction, suggests the tousled locks of Meg Merriles.

In making its arrangements inland, each *gorfou* is assigned its own square in the regularly laid out encampment, and these *penguins*, sitting bolt upright, produce the impression of an assemblage of Indian huts. The *crested gorfou* is in size

CITY OF THE CRESTED GORFOU.

about as large as a duck, and bears some general resemblance to a trained duck practising its antics.

The **Arctic Parrot, Crab Diver,** or **Guillemot** (*Uria troile*), can walk, fly, swim and dive. The most singular fact in its history is its laying its eggs on the ledges of the sheerest precipices where one would imagine that they would be constantly exposed to destruction from the high winds. The black of its bill runs in a band over the head and, with the exception of the white tips of the wings, prevails upon the upper parts. The bird generally is snowy white, except for a collaret of brownish-gray, and dark gray colorings on the abdomen just below the wings, and again from the feet to the tail. The wing coverts are also of a dark color, sometimes black.

The **Spectacled Auk**, or **Great Auk** (*Alca impennis*), belongs to northern-most Europe. When it is in the water it is almost impossible to pursue it quickly enough to get within shooting range, but like the albatross it can be caught with a hook. It is rapidly becoming extinct and, in spite of the extremely high price which either the bird or its eggs command, the museums of the world contain but thirty-four birds, and but forty-two eggs. Collections of birds' eggs are quite important to naturalists, but the objects sought by THE SAVAGE WORLD forbid any discussion of so large a theme. The *spectacled auk* is black above and white below; around and below the eyes are white markings (which give the *auk* its popular name) and the small wings or flippers are bordered with white on the upper arm.

The **Razor-Bill** (*Alca tarda*) has its habitat in the Arctic sea, and is so called on account of its mandible

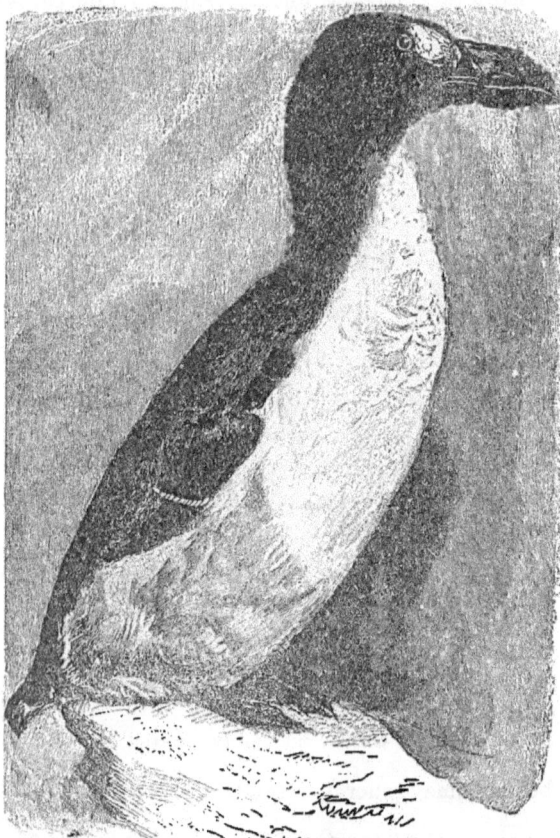

SPECTACLED OR GREAT AUK.

GIANT PENGUIN (*Aptenodytes patagonica*).

bearing some resemblance to the back of a razor-blade. It is very similar in habits and size to the following species.

The **Little Auk**, or **Razor Bird** (*Alle alle*), is called by the greatest variety of names—*sea dove, sea pigeon, rotge*, for example. Its habitat is in the Arctic Atlantic, and it is specially abundant about Spitzbergen. In the frost-made clefts and cavelets of the rocks the *little auk* lays its eggs and raises its family. Its bill, as the illustration on page 412 shows, is altogether unique. The young resemble little fluffy goslings, and are frequently carried on the back of the parent, and left swimming on the water when the parent dives.

The **Giant Penguin,** or **King Penguin** (*Aptenodytes patagonica*), is about three feet in length, bluish-black above and satiny-white below. Its habits are such as have been described when discussing the crested gorfou. It is very numerous in Patagonia and in the Falkland Islands, and is frequently mentioned by explorers and sailors.

The **Pelican** (*Pelicanus*) has, as the reader will have discovered, been praised for qualities which belong quite as much to the eider duck, the whale, and various other creatures, whose affectionate self-sacrifice for their young is quite touching to one's sentiments. To the family belong the *tropic bird*, the *darters* or *snake-birds*, the *gannets*, the *cormorants* and the *true pelicans*. The gregariousness, beautiful plumage and striking appearance of the *pelicans* always attract the attention of those who visit lower Europe and Asia, or Africa, and of those who go to our museums of natural history and to our zoological gardens. Were it not for its head, bill and pouch, the *pelican* might be mistaken for a

GREAT AUK (*Alca impennis*).

PENGUINS (*Aptenodytes patagonica*).

species of goose. The pouch will hold about fifteen pounds of fish, and, when the *pelican* has filled his game-bag, he would be free from the necessity for further effort, except for the fact that he is imposed upon by a species of hawk. This cunning creature, too indolent to catch his own fish, keeps watch upon the *pelican*, and when the latter retires after a day's sport, attacks him and, as he opens his mouth, snatches the fish from it. The **American White Pelican** (*Pelicanus erythrorhyncus*) is common in the Mississippi Valley; the **Brown Pelican** is also found about Southern shores of the United States consorting with the white species. *Pelicans* are strong swimmers and excellent flyers. It is found associated especially with the flamingo, as one often sees a blonde and brunette cultivating companionship. Montgomery is still popular with readers of English poetry, and it will therefore be unnecessary to more than refer to the homes of the *pelicans*, as these are so graphically described in the "Pelican Island." The *pelican* was known long enough ago to receive mention in the Bible, so that its lineage is at least ancient.

The *pelican* has long enjoyed the honor of typifying parental affection and although not undeserving of its wide reputation is not singular in this worthy characteristic, as the readers of THE SAVAGE WORLD will ascertain when reading about the mammalia. The *pelican* is very marked in its domestic virtues and has been known to assume the labor of both birds when a broken leg, a useless wing, or any other physical injury has disabled his companion. In one case which was studied by a competent observer, the male bird constantly brought food to his helpmeet who had her wing broken by the hunter. He manifested the deepest and most sustained solicitude, and seemed to care quite as much that she should not exert herself unnecessarily as that she should have a sufficiency of food. The *pelican* of poetry and fable offers, if need be, its own life's blood for the sustenance of its famished young; the *pelican* of science without going so far, leaves no effort untried to see that those dependent upon it do not suffer from privation. The *pelican* does not put before its young the prey which it has secured, but opening its bill allows the young to forage in the larder which its pouch provides. In the early morning and evening it will gather in companies and start for its fishing-grounds. Having selected a shallow bay or pond the company arranges itself into a crescent some ten feet distant from the shore. They now begin in concert to beat the water with their wings while steadily marching toward the shore and driving before them the frightened fish. When this task has been accomplished they begin to feast and to store their pouches, though they magnanimously permit other birds to enjoy the bountiful provision which by intelligent effort they have thus made.

LITTLE AUK, OR RAZOR BIRD.

The **Tropic Bird** (*Phæthon æthereus*) is a splendid specimen of swift moving creatures, and is very striking in its appearance. Except for the short, stout beak its head and neck suggest the white pigeon. The back immediately over the wings is scalloped brownish-black, and the wings have a black edge from the tips half way their length. The black back markings are separated by a wave-like white band which immediately after expands so as to cover the rest of the body. Its tail consists of several long shaft feathers curved and projecting like an elephant's tusks. Its habitat is Mauritius, and its young in no wise suggest the parent, as they are little round cottony balls. Its body is about two feet and a half long while the tail

TROPIC BIRD (*Phæton æthereus*), FRIGATE BIRD (*Tachypetes aquila*), AND BROWN PELICAN.

will reach a foot and three-quarters. It is very graceful in its motions, seeming to dart through the air, and can remain on the wing without apparent limit of time. It is generally found within two or three hundred miles of the coast, but has been seen at four times this distance. It constantly emits a harsh cry so that it has sometimes been called the *boatswain*, while its peculiar tail has gained for it yet another name,—that of the *star-tail*.

The **Booby** (*Sula fusca*) is a wild gannet which must have given rise to the comparison "as stupid as a goose." It will devote its whole time to fishing, and tamely and wonderingly suffer itself to be robbed of fish after fish by the frigate bird. It is often caught by putting a bait upon a board upon which the *booby* will dash down so violently as to kill itself or else will transfix itself by driving its bill into the wood.

The **Darter** (*Plotus*) exists as an African species as well as an American one. The head and neck look as if they belonged to a long-beaked snake; and in swimming it does not expose its large body, so that the resemblance is thus increased. Add to this that in diving it is quicker than a flash and the propriety of its popular name is evident.

TROPIC BIRD ROBBING THE BOOBY.

The **Sea-Swallow,** or **Tern** (*Sterna hirundo*), is a gull which has been named from the swallow-like, forked-shape of its tail. Jet black is the color of the top of the head and on the back of the neck; the under parts are white, and the rest of the body a gray ashen color; the legs, feet and bill are like red coral though the extremities are perfectly black. It builds the rudest kind of a nest out of sticks, stones and grass. Another species of the *tern* is the **Noddy** (*Anous stolidus*). It flies like a night-hawk, although its habits are not nocturnal. It builds its nest sometimes on the rocks and sometimes in the trees. It is a slovenly housekeeper, and uses the same nest year after year, adding to the structure until it becomes relatively gigantic, but never troubling itself about any house-cleaning. The *noddy* is about a foot and a quarter in length, and its nest is robbed of the eggs for the supply of the table. It dresses in a chocolate color, but varies this by buff on the top of the head and the forehead, by brownish gray on the back of the head, and by black on the bill, on the legs, and on the feet.

The **Scissors-bill Gull** (*Rhyncops nigra*) is met with alike in America and in Africa, and the lower mandible shuts into a groove in the upper in a manner to justify its popular name.

BOOBY.

This beak, black at the tip, constantly grows lighter until at the base it is the color of Guinea gold. It feeds mostly upon mollusks and crustaceans, and is specially fond of oysters in the shell. When the oyster closes the shell before it has been dragged from its home

SEA SWALLOWS HELPING A WOUNDED MATE.

the *scissors-bill* is said to allow its beak to be held by the oyster while it breaks the shell by dashing it violently against the rocks.

GIANT GULL. (*Lestris catarractes*).

The Gulls (*Laridæ*) are divided into true *gulls*, *terns* or *sea-swallows*, and *shearwaters*, which are a frequent and a familiar sight at sea and along the coasts. They are poor swimmers and unable to dive, but they can float with grace and ease, although they pass most of their life on the wing. They are protected by the superstitions of the sea, except when some " summer visitor" at a seaside resort shoots one of them to provide his lady with bird's wings for her hat.

The **Great Black-backed Gull** (*Larus marinus*) is about two and a half feet in length, and is said to be a better swimmer than most of its family. It is said to occur in America, but is found mostly, if not solely, in Europe, particularly in Norway and Sweden and the Shetland Islands. It is a frequenter of the marshes where it always builds its nest like the eider-duck, and will at once lay again if its nest be robbed, and as its eggs are large and well flavored, the natives keep the bird fully occupied. On the Thames river this bird goes by the name of the *cob*. It wears two suits of clothing a year—streaked gray for a winter color, and for summer leaden-gray on the neck and throat, white on the under parts, and black on the upper body, the wings being white tipped; it prefers pink for its stockings and shoes. The third set of eggs is left for the bird's share.

GREAT BLACK-BACKED GULL.

The *gulls*, as well as the hyænas, have their laughing species, the *Larus atricilla*, whose scream sounds like a derisive laugh.

The **Herring Gull** (*Larus argentatus*) is spotlessly white on the head and neck, gray on the body, except with jet-black, white-tipped wings. Its popular name has been derived from its success in the herring fishery.

The **Skua** (*Lestris*) is a professional thief, who, passing his time in idleness, waits until a gull has captured a fish, when the *skua*, like a highwayman, robs the smaller and defenceless congener.

HERRING GULL.

The **Wandering** Albatross (*Diomedea exulans*) is familiar alike to traveller and reader, but even Coleridge's study of the weird, in his matchless "Ancient Mariner," does not discuss the natural history of the bird, and from sailors one learns only the superstitions, which are "as thick as leaves in Vallambrosa." It is the largest of swimming birds, as it very frequently has twelve or fourteen-feet stretch of wing. Its white coloring has the appearance of crested waves, and is broken only by the pink hue of its bill, the green orbit of the eye, the flesh-color tint of the legs, and the tracery

ALBATROSS AND GOLDEN DIVERS.

of black on the edge of its wings, and on the shorter feathers of the back and

BIRDS OF TIRELESS WING.

tail. As the sailor's "home is on the ocean wave," so that of the *albatross* is in mid-air, and its graceful flight is not affected by dead calm or the most

27

furious tempest. It is no uncommon feat for the great ocean steamers to sail at the rate of four or five hundred miles a day, and yet the *albatross*, without apparent exertion, will not only keep them company, but will at least double the distance by the many and eccentric circlings in which it indulges. Whether the *albatross* sleeps on the wing, or taking its rest in the surface of the water makes up for delay by a yet more rapid flight, is a question not yet determined.

The **Stormy Petrels** are so frequently mentioned in accounts of sea-going travellers as to call for at least brief mention. They breed in the crevices of the rocks but build no nests. The young when hatched are left in the nursery during the day, and fed only on the return of

OWL ATTACKED BY A WEASEL.

the parents at night. The *stormy petrels* belong to several species, and we select as their representative the **Fulmar** (*Procellarius glacialis*). In the Island of Saint Kilda, where its nests are numerous, it is preyed upon by the natives, who are fond alike of the eggs and of the young birds. The *stormy petrel* follows whalers and devotes itself to the enjoyment of such parts of the whale as are refuse to the fishermen. This bird is fully described in preceding pages of this work under

LEMMINGS PURSUED BY WHITE OWL AND BUZZARD.

the head of "Singular Nest-Building Birds," where considerable space is devoted to the many curious habits of this little creature.

BIRDS OF PREY.

The **Birds of Prey** (*Accipitres*) naturally represent the next step upward, since for success in the realm assigned to them, in the division of the world's dominion, they require a more complex organism, and a higher structure than the classes of birds heretofore described. As carnivorous creatures, they are provided with strong, hooked beaks, and their claws are adapted for clutching. The short, stout bill has a curved tip and knife-like edges, so that it forms the most perfect contrivance for tearing and cutting, while the muscular power of the jaws fits them for crunching the bones of animals. The upper jaw, like that of the parrot family, is furnished with a sheath (*cere*), which extends as far as the nostrils. The powerful feet are four-toed, and, except in the case of the owl and the fish hawk, are arranged three in front and one behind. The fact that the claws are less fully defined in such birds of prey as feed upon carrion and refuse, is another striking illustration of the wonderful provision in the animal kingdom for each creature having no excuse for not worthily fulfilling the work it has to do.

The **Owl** was sacred to Minerva, for if not possessed of the most remarkable wisdom, it has the appearance of profound sapience.

The **Hawk Owl**, or **Canada Owl** (*Surnia ulala*), belongs to the Polar fauna, and is notable because it can see by day as well as by night.

The **Snowy Owl** (*Nyctea nivea*) is a mighty hunter and fisher, and flies wholly by day. It is specially fond of hares and lemmings. Its orange-colored eyes gleam like gems, and gain in lustre from their contrast with the spotless white of the rest of the body. Its habitat is in the North Polar regions of both continents.

EARED OR HORNED OWL (*Bubo maximus*).

The **Burrowing**, or **Coquimbo Owl** (*Speotyto cunicularia*), is a member of the queer triumvirate, which is composed of the prairie dog, the rattlesnake, and the owl. When the *coquimbo* has no prairie dog to do his mining, he makes his own burrow. This species is small of size, but has very long, stout legs and extremely muscular claws.

The **Boobook** (*Athene boobook*), or **Australian Cuckoo Owl**, though diurnal in its habits, passes the night in song, not unlike that of the cuckoo.

The **Winking Owl** (*Athene connivens*) belongs to Australia, and particularly attacks the bear or kaola, a small quadruped which it frequently makes its prey, though not without having to engage in a fierce battle.

The **Eared Owl** (*Ephialtes scops*) is only about seven inches in length. It is quite common in Italy, but is migratory in its habits, passing its winters in Africa and Asia, and its summers in Europe.

The **Great Owl**, or **Eagle Owl** (*Bubo maximus*), has northern Europe for its habitat. It is extremely fierce, attacking the eagle and the wild dog without considering their formidableness, for which temerity it often forfeits its life. It is very beautiful in its plumage, is susceptible of domestication, and is trained to lure the pugnacious falcon into the hunter's net.

The **Virginian Eared Owl**, or **Horned Owl** (*Bubo virginianus*), utters a cry so weird as to curdle the blood of the traveller. It is very destructive of game birds and displays great fondness for wild turkeys. Like the *scops eared owl* and the *great owl*, it belongs to the horned family, wearing movable, feathered tufts or ears.

FOREST OWL (*Syrnium aluco*). SCREECH OWL (*Otus vulgaris*).

The **Brown Owl** (*Syrnium aluco*), though small, is very bold and strong, and surpasses most species in its fondness for the flesh of young cats. The *brown owl* is perhaps better known as the *barn owl*.

The **Long-eared Owl** (*Asio americanis*) is small but rapacious and is found in many countries. It takes the nest of a squirrel or of a bird, and thus saves the trouble of building for itself. It is very fond of its young, and provides for them most bountifully. It is very common in America and alike varied and beautiful in its coloring. Despite its racial weakness for cat-flesh, a tame *long-eared owl* has been known to live on the most affectionate terms with pussy, who always shared with him her rats and mice. The **Veiled Owl**, or **American Barn Owl** (*Strix flammea*), is light colored and very easily tamed. One of them struck up a friendship for a linnet and used to allow the little bird to ride about seated on its back.

The owl's nocturnal habits and melancholy utterances have resulted in multiplying omens and superstitions associated with him. The French peasants are in the habit of nailing owls to trees or fences and letting them die by the slow torture of hunger. Henry Berthoud, in his "Stories of Bird Life," tells of rescuing one such unfortunate owl and taking it into his household, at Paris. The owl became quite fond of its master and always returned after having been allowed its liberty. Finally it disappeared but was some days afterwards found to have a family, which, together with the male, accepted into their friendship the owner of the mother. If, as sometimes happens, the owl gets abroad in daytime, crows, sparrows, and various small birds, which at night would be the owl's victims, attack and claw and peck at him, against which attacks he is utterly defenceless because unable to see.

The Secretary Bird, or Serpent Falcon (*Gypogeranus serpentarius*), belongs to the fauna of South Africa. Its feats as a serpent-killer outdo the fabled Hercules, and its successful pursuit of snakes has made it a privileged character, protected even by the law. It generally marches along with the most pronounced military dignity, but it

NEST OF BURROWING OWL (*Strix cunicularia*).

BARN OWL AND NEST.

has a curious habit of breaking irregularly into a run as if suddenly irritated or frightened. This bird is about three feet in length, and dark gray in color. Black feathers form a crest and suggested its name, from a fancied resemblance to a clerk with his pen behind his ear. It moves with the greatest rapidity so that it has been called the *Devil's Horse*. As a kicker it rivals the ostrich, and when it attacks a serpent it brings into action feet, wings and beak, but in his household female suffrage is recognized. The American species (*Astur atricapillus*) is specially valuable when trained for hunting.

The **Sparrow Hàwk** (*Asturnisus*) is distinguished for the most reckless courage and pugnacity, which leads him to attack any bird without reference to size. When domesticated, it is affectionate towards its master, and a trustful

SNOWY OWL.

VEILED OWL.

guardian of his fruit. It forms strange friendships, as for example, one formed an intimacy with a puppy, and the two strange companions became such fast friends that the effects of their separation was pitiful to see. At times the swallows and martens will band together and attack the *sparrow hawk*.

The **African Chanting Falcon** (*Meliera musicus*) sings its matins and its vespers, and its notes are said to be very musical.

The **Corn Hawk** (*Circus cyaneus*) is singular among *hawks* because of the difference in coloration of male and female. Its habitat is the flat, open country, from the Isthmus of Panama to the Arctic regions. It is specially serviceable to the planter because it destroys the rats and mice in great numbers.

The **Windhover**, or **Kestril** (*Falco tinunculus*), is a very beautiful European hawk which has the power of remaining poised in the air for a seemingly unlimited time. As will be seen from the illustration presented in THE SAVAGE WORLD, the markings are very beautiful, and the egg is also

SECRETARY BIRD.

prettily spotted with brown. This bird was once used in falconry, for not only were different species trained for special kinds of game, but the distinctions of caste was confined to each class of falcons and hawks, and to certain social ranks, a distinction which was rigidly observed.

FALCONRY.

The **Bengal Falcon** (*Falco cærulescens*) is only a few inches long, but is most expert when trained to hunt the quail. Like all of its family it darts swiftly and directly at its quarry, and losing no time, strikes a second bird as soon as the first one has been disposed of. The **Chicken Hawk,** or **Goshawk** (*Astur palumbarius*), is trained to hunt hares, and being less fleet, is compelled to first stalk and then pounce upon them. The male is much weaker than the female. The training of the *falcon* was a long-continued and arduous labor and deserves mention in order to lend emphasis to the expensiveness of the pleasures of royalty and the nobility, who had to satisfy t h e i r profligate desires by some means of taxing subject and tenant. First the *falcon* was taught to perch, without flapping its wings, lest when it became a hunter it should frighten the quarry. This lesson was repeatedly inculcated by tying the wings, putting a leather band on the legs, and forcing it to occupy the perch as represented in t h e illustration. Then the perch, as will be seen, was made to correspond in general form to the human hand and wrist upon which

THE NIGHT MARAUDER.

the *falcon* was finally to perch. At times, as in the illustration, two *falcons* were put into training, so that when their education was finished they should be prepared to hunt in company. After the *falcon* had learned to perch properly it reviewed the lesson with the substitution of the human wrist for the wooden perch, the falconer protecting himself by a thick leather glove worn on the left hand and wrist, and chaining the *falcon* to his arm. Next in order the falconer walks up and down with the *falcon* on his wrist that it may become accustomed to a moving perch.

THE FIRST LESSON.

FALCON SEIZING A HARE.

While receiving its lessons in perching the *falcon* is hooded with a white leather mask, securely fastened about the neck, and having an opening simply

SPARROW HAWK (*Accipitor nisus*).

EAGLE HAWK PURSUING A HARE.

for the beak. These steps having been taken satisfactorily, the instructions are repeated with the unhooded *falcon*. When the bird has learned its lesson it is required to begin all over again in the presence of numerous spectators, that it may learn not to be diverted by the presence and clamor of human beings.

FISH HAWK.

Next the bird is trained to come at the call of its keeper—an obedience which the *falcon* yields very unwillingly, even with the promptings of pulls upon his chain and of rewards in the form of some agreeable article of food. Gradually the distance between the falconer and the *falcon* is increased until at length it reaches the limit of the vocal powers of the master. Then a leather-covered wooden heron is used as the table from which the *falcon* must feed, so that it learns to associate good living with the back of a heron. Next the bird must learn to answer the call of its keeper, even though it has to leave its feast, after which the counterfeit presentment of a heron is bloodied so that the *falcon* may take another step in its educational progress. Next live birds are substituted for the dummy

WINDHOVER.

GERFALCON.

and all previous lessons reviewed, and when the result proves satisfactory the *falcon* is unchained and compelled to repeat his lessons until he learns his duties. When all this has been accomplished and the *falcon* has been converted into the companion of sportsmen, it is no longer a bird of prey, but when allowed to seek its quarry, kills and drops it, to be picked up by an equerry. Frederick II. was compelled by his *falcons* to raise the siege of Parma; for being unable, during the siege, to deny himself the pleasure of hunting, the Parmese took advantage of the temporary absence of the monarch and his nobility, and sallying forth defeated the leader less soldiery. In France falconry culminated under King Francois I. (sixteenth century), but it continued to be cultivated until interrupted by the French Revolution and the wars of Napoleon. In 1861 the Empress Eugénie undertook, in company with Marshal Bazaine, the Prince of Maskwa and the Baron de Pierre, to revive the institution, but without any permanent results. A French account, belonging to the earlier half of the fourteenth century, relates how a *falcon* brought about a war between England and France. Robert von Artois having been banished from France, sought the Court of England. While there he learned to hunt the heron, and resolved to stir up King Edward to the assertion

FALCON TRAINING.

THE OSPREY, OR FISH HAWK (*Astur palumboris*).

of England's rights in France. Courtier-like, he had a heron served for the royal table, and when it appeared, dilated upon the heron's perfect symbolism of men who, afraid of their own shadow, failed to assert their rights. The monarch made the application and swore by the heron no longer to be diverted from an attack upon France. Robert von Artois next took a solemn vow that he would never return unless victory perched upon Edward's banner. The Earl of Salisbury, catching the infection, swore that he and his would follow and support the king. Then the lady-love of Salisbury, putting her finger upon his eye, prayed that if false to his oath his eye might always remain closed and senseless. Finally the enthusiasm spread to the whole court, and resulted in a French crusade, which nearly proved successful. Nor was this all, for the fact that the English nobility hunted on French soil proved a stronger motive than loyalty to the French nobility, and induced them to lend support to their king, which would have been rendered to no weaker appeal than that of their passion for hunting with the *falcon*.

Falconing was cultivated as early as the fourth century, and continued in Germany till the end of the eighteenth century, while it is still in repute in Persia. Great sums were lavished upon falconries, and the chief falconer ranked fourth from the king. Francis I. of France expended forty thousand florins

CHICKEN HAWK (*Astua palumbarius*). SPECKLED BUZ-
ZARD (*Buteo lagopus*).

KINGLY MILAN (*Milvus regalis*). WANDERING FALCON
(*Falco peregrinus*).

a year, paying his chief falconer a salary of four thousand florins and giving him as assistants fifty gentlemen and fifty falconers. Noblemen and their stately dames devoted much time to hunting with *falcons*, and even when not caring to engage in the sport still kept their *falcons* by them as a symbol of wealth and station. Just as with archery, the terms of falconry formed a vocabulary by themselves, and the historians of the times are filled with allusions to the sport.

Falconry is cultivated ardently by Cossack and Kalmuck, who have inherited love for the sport even though they disregard many of the more stately cere-monies of the nobles who provoked from Froissart his high-wrought descrip-tions of "The Field of the Cloth of Gold."

The **Falcon** (*Falco*) has long been celebrated, and always appears in tales of mediæval knights and their demoiselles. Its beak is thick, stout, curved and

FALCON STRIKING A BITTERN.

supplied with a single tooth, which fits into an empty socket in the lower jaw; its claws are sharply curved and pointed.

The **Gerfalcon** (*Falco gerfalco*) is a native of Greenland, and the very embodiment of courage and fearless activity. In its wild state it hunts hares and rabbits, so that the education which it undergoes when trained is due rather to its wilfulness than to its ignorance. As a rule its plumage is white.

The **Wandering** or **Peregrine Falcon** (*Falco peregrinus*) is the species trained to assist the hunter, for its disposition, without being sweet, is more amiable than that of other *falcons*. The male being a third smaller in size is frequently called the *tercel*. In speaking of the antelope and gazelle mention will be made of the use of the *falcon*.

The **Kingly Milan** (*Milvus regalus*) is frequently trained like the falcon. At one time a number of them had their feet frozen to the boughs upon which they had perched, and fell victims to the peasants. Accidents of this kind may overtake even an animal, as in the case of the eagle which, stepping on an iceberg to take its meal of fish, was compelled to remain a prisoner until it perished from starvation.

GOSHAWK AND NEST (*Astur palumbaris*).

The **Kite** (*Milvus ictinus*) is a very familiar sight in all parts of Europe, but has not, as yet, become naturalized in America. Its name is so often used by British writers that a few words of description will, doubtless, prove serviceable. Its body is about two feet and a quarter in length, but its wings expand more than five feet. Its coloring is reddish-brown, streaked with deep black; its tail is distinctly barred in brown, and forked at the extremity. It is rapidly becoming extinct, but has impressed itself upon the literature of Great Britain, and become familiar to every child through the Chinese kites.

The **Swallow-tailed Kite** (*Elanoïdes forficatus*) is a native of America, where its beautiful black and white plumage, and swallow-tailed coat arrest attention. It is insectivorous, and exhibits the most remarkable grace and swiftness when in pursuit of its prey. The reader, by pausing to reflect upon the difference of wing equipment among the birds already mentioned, will see yet another evidence of design.

The **Mississippi Kite** (*Ictinia mississippiensi*) cultivates the most inexplicable companionship with the turkey buzzard, from which it is distinguished by appearance, flight and strictly insect diet.

The **Little American Eagle**, or **Red-throated Falcon** (*Ibycter americanus*), adds to the many attractions of the South American forests. Blue above and pinkish below, a red-purple for the throat, black for the claws, and yellow for the feet and for the bill-sheath, a diversity of plumage in beautiful combination is such as is seen in no other birds of prey.

The **Buzzard** (*Buteo vulgaris*) is one of the handsomest among the falcons.

FALCON, WITH AND WITHOUT HOOD.

The reader must not confound this bird with what is called the turkey buzzard, for the latter, as will be explained, is not truly a buzzard, but a vulture. It is black or dark brown above, and white below, though spots and streakings add to the richness of its plumage.

The **Honey Buzzard** (*Pernis apivora*) is notable chiefly because it varies its *menu* by the addition of insects, and of the plundered hives of bees.

The **Osprey**, or **Fish Hawk** (*Pandion haliætus*), is always graceful, whether circling in the air, or dashing into the water in pursuit of the finny tribe.

CLAWS OF THE EAGLE.　　SKULL OF AN EAGLE.

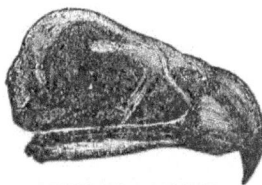

Its claws are long, sharp and curvilinear, and, to complete the perfection of the mechanism, the soles of the feet are roughened, and the outer toe has a flexibility of the human thumb, affording another evidence of adaptation to the demands of nature and the wisdom of the Deity.

The **Sea Eagle** (*Haliætus albicilla*) is European in its habitat, and although a fisherman by calling, is very fond of fawns, hares, sheep and poultry, and has been known to devour the hedge-hog. The **Bald-headed Eagle**, or the **White-headed Eagle** (*Haliætus leucocephalus*), is a well-known symbol to every American. Its head is not at all bald, but as its white is in such sharp contrast with the chocolate-color of the rest of the body, the first impression is that the feathers stop before they reach the head. Franklin objected to the selection of the *bald-headed eagle* as the American symbol, for he dwelt only upon its habit of robbing the fish hawks, but those who overruled his opinion thought rather of the magnificent strength, and aspiring flight of this monarch among the birds, whose form, flight, strength and exulting cry all make him remarkable in the animal world.

Wilson, in his American Ornithology, gives the following spirited description of the *bald* or *white-headed eagle:*

"The celebrated cataract of Niagara is a noted place of resort for those birds, as well on account of the fish pro-

NORTH AMERICAN BALD EAGLE (*Haliætus leucocephalus*).

cured there, as for the numerous carcases of squirrels, deer, bears, and various other animals, that in their attempts to cross the river above the falls have been dragged into the current and precipitated down that tremendous gulf, where, among the rocks that bound the rapids below, they furnish a rich repast for the vulture, the raven, and the *bald eagle*, the subject of the present account.

"This bird has been long known to naturalists, being common to both continents, and occasionally met with from a very high northern latitude to the borders of the torrid zone, but chiefly in the vicinity of the sea, and along the shores and cliffs of our lakes and large rivers. Formed by nature for braving the severest cold; feeding equally on the produce of the sea and of the land;

possessing powers of flight capable of outstripping even the tempests themselves; unawed by anything but man; and from the ethereal heights to which he soars, looking abroad, at one glance, on an immeasurable expanse of forests, fields, lakes and oceans deep below him, he appears indifferent to the little localities of change of seasons, as, in a few minutes, he can pass from summer to winter, from the

lower to the higher regions of the atmosphere, the abode of eternal cold, and from thence descend at will to the torrid or the arctic regions of the earth. He is therefore found at all seasons in the countries he inhabits, but prefers all such places as have been mentioned above, from the great partiality he has for fish. In procuring these, he displays, in a very singular manner, the genius and energy of his character, which is fierce, contemplative, daring and tyrannical; attributes not exerted but on particular occasions; but when put forth, overpowering all opposition. Elevated on the high dead limb of some gigantic tree that commands a wide view of the neighboring shore and ocean, he seems calmly to contemplate the motions of the

EAGLE OF THE ALPS CARRYING OFF MARIE DELAX.

various feathered tribes that pursue their busy avocations below; the snow-white gulls slowly winnowing the air; the busy *tringæ* (sand-pipers) coursing along the sands; trains of ducks streaming over the surface; silent and watchful cranes, intent and wading; clamorous crows, and all the winged multitudes that subsist by the bounty of this vast liquid magazine of nature. High over all these hovers one whose action instantly

28

arrests all his attention. By his wide curvature of wing, and sudden suspension in the air, he knows him to be the fish hawk (*Pandion haliætus,*) settling over some devoted victim of the deep. His eye kindles at the sight, and balancing himself, with half-opened wings, on the branch, he watches the result. Down, rapid as an arrow from heaven descends the distant object of his attention, the roar of its wings reaching the ear as it disappears in the deep, making the surge foam around. At this moment the eager looks of the *eagle* are all ardor, and levelling his neck for flight, he sees the fish hawk once more emerge, struggling with his prey, and mounting in the air with screams of exultation. These are the signal for our hero, who, launching into the air, instantly gives chase, and soon gains on the fish hawk; each exerts his utmost to mount above the other, displaying in the rencontre the most elegant and sublime aerial evolutions. The unincumbered eagle rapidly advances, and is just on the point of reaching his opponent, when with a sudden scream, probably of despair and honest execration, the latter drops his fish; the eagle, poising himself for a moment as if to take a more certain aim, descends like a whirlwind, snatches it in his grasp ere it reaches the water, and bears his ill-gotten booty silently away to the woods.

A FIGHT WITH AN EAGLE.

"These predatory attacks and defensive manœuvres of the eagle and fish-hawk are matters of daily observation along the whole of our sea-board, from Georgia to New England, and frequently excite great interest in the spectators. Sympathy, however, on this as on most other occasions, generally sides with the honest and laborious sufferer, in opposition to the attacks of power, injustice and rapacity, qualities for which our hero is so generally notorious, and which, in his superior, man, are equally detestable. As for the feelings of the poor fish, they seem altogether out of the question."

There is a well autheuticated story of an eagle of this species having descended upon a moun-
taineer's home on the Alps and carrying away a girl six years of age in its cruel talons. This event is said to have oc-curred in 1838, and that though a rescuing party went promptly to the aid of the victim she was not recovered until life had become extinct from wounds received in her breast. Some time in the year 1886 a similar inci-dent occurred in Minne-sota, where a child of four years was carried away by an eagle, but in this case the little one was rescued before it re-ceived fatal injuries. Many stories of a like character, illustrative of the eagle's power and courage, are related, es-pecially of the *bald eagle* of the Alps. A few years ago we read an account of one of these monster birds having swooped down upon a shepherd's house and seized upon a boy nearly twelve years of age who was playing in the yard before his door. In this case the eagle was unable to rise with so great a weight but re-fused to quit its hold until the child's s c r e a m s brought his father to the

EAGLES OF SWITZERLAND.

rescue, who succeeded in killing the eagle with a club that was fortunately at hand.

The **Golden Eagle** (*Aquila chrysœtos*) is a true cosmopolitan, and has his habitat everywhere. Unlike the vultures, it prefers living prey to a carcase, and will successfully attack animals as large as the kangaroo. It is free from the distinctive weakness of the vulture tribe, as it never disgorges. Its head and neck are of red and gold, the legs and tail grayish brown or brownish gray.

The **Brazilian Eagle** (*Morphus urubitinga*) feeds mostly upon reptiles. It is strictly monogamous, but in case of accident, never continues to be a widow or widower. Its strength is enormous, for even when heavy from overfeeding, it cannot be held by the arms of a man, and is notable for its diminutive size, which is not greater than that of a crow. The female eagle is much the larger and stronger of the · pair, and is fully able to manage her helpmeet.

The **Crested Eagle**, or **Harpy Eagle** (*Thrasœtus harpyia*), is a terror to the sloth, the deer, and the opossum. It is the most muscular of the eagle family, and though less notable in power of flight, is remarkable for its ability to secure prey. It wears on its head a fan-shaped crest, which it erects when aroused from its lethargy. It is met with as far north as Texas.

GOLDEN EAGLE (*Aquila chrysætos*).

The solitary habits of the eagle, and its limited fecundity seem to be a provision against any multiplication beyond the needs which they serve in the economy of nature.

The **Eagle of the Steppes** (*Aquila bifasciata*) is found throughout Tartary, and frequently fastens upon and destroys the antelope. The Abbé Hoe, like

Bruce in southern Africa, lost a kid which was being boiled for supper, by an eagle which swooped down upon it and carried it safely away.

The eagle was the emblem of "all-seeing Jove" and no member of the animal world could so fitly symbolize a keen-ness of sight which allowed nothing to escape it. Again, the high altitudes at which the eagle flies, and the easy sweep of his magnificent wings, together with the suddenness with which he appears in the lower realm of earth, all unite to increase the excellence of the symbolism. The Roman armies bore insignia in which the eagle had great part, for he typified mastery, fearless courage, resistless might, seem-ing omnipresence. So, too, when Jove would please himself with the possession of the beautiful boy Ganymede, he employed the

EAGLE OF THE STEPPES.

eagle as his minister; and poet and artist have made familiar to us the soaring eagle and the surprised but fearless child sailing away into the blue empyrean.

1. CATHARTES URUBU.   2. KING OF THE VULTURES.

As has been said the eagle has more than once been known to seize and bear away not merely infants, but children as large as the one in our illus-tration, for whose rescue the maddened father is represented struggling so valiantly.

The **Black Vulture,** or **Carrion Crow** (*Catharista atrata*), is notable for its de-struction of the eggs of the alligator. Its sight is keen, and its appetite insatiable, but still in accordance with the won-derful laws prescribed for the natural world, it does not ren-der extinct the alligator species, but simply prevents it from multiplying too rapidly and thus occupying the earth.

The **King Vulture,** or **White Crow** (*Sarcorhamphus papa*), is specially beautiful in its coloration. Cream white is the prevailing tint, but the lemon color of the throat and back of neck change into scarlet on the sides of the neck and on the crown of the head. The bill is orange and black with an orange colored *cere*. In his presence other vultures are submissive and will never attack a carcase until the *king vulture* has claimed his regal precedence.

The **South American Condor** (*Sarcorhamphus gryphus*) is the type of the vulture family. It is about the same size as the bearded vulture, but has the extraordinary wing expanse of ten feet, while its strength is surprisingly great in proportion. Its general color is a gray-black though the markings are not always the same. Its head and neck (as if to insure greater ease and neatness in its work as a scavenger) are destitute of feathers; the wings grow white towards the ends, and around the neck is a beautiful, white, fluffy ruff. The male is crested, and as the *condor* when at rest conceals its beak in its ruff, it then presents the semblance of a curious freak of nature. Its vision is even more remarkable than that of the telescopic eyes of its congeners, and while itself at an altitude so great as to render it invisible to man, will espy its quarry and steal into its presence like a phantom. All accounts of travel in South America, Africa and Asia, are full of the constant, sudden and mysterious appearance of vultures when game has been shot.

GREAT VULTURE OF THE ANDES (*Vultur gryphus*). .

The *condor* ordinarily contents himself with carrion, but at times it will attack the antelope or even the puma ; it will sometimes join forces with a mate and successfully attack cattle. Its two-feet-and-a-quarter body hardly lead one to expect the enormous muscular strength which it exhibits, and its tenacity of life renders it substantially invulnerable to the ordinary bullet. The *condor* is a glutton, and this leads to his easy capture by the natives, who, baiting with a carcase, wait until the *condors* have gorged themselves and then easily lasso them. The natives despise the *condor* as they do a poisonous snake, or as the average man does a Norway rat. Hence, after capturing the creature, they spare no torments which their ingenuity can devise, so that the *condor* may be regarded as the frequent victim of a modern inquisition.

Some naturalists believe that the *condor* is directed by scent, instead of by sight, but this belief is hardly as well supported by investigation as is the other. It is stated that four *condors* dragged the carcase of a grizzly bear several hundred yards although it weighed over a hundred pounds. The *condor*

EGYPTIAN VULTURE.

VULTURES.

1. BEARDED VULTURE (*Gypaëtos*). 2. SOCIABLE VULTURE. 3. CATHARTES PERCNOPTERUS.

and his tribe exhibit considerable ingenuity in conducting their feasts. It does not find palatable or nutritious the leathery skin, hairy covering, and bony skeleton of its prey, and therefore with a skill which a taxidermist would envy, opens the stomach and entering into the cave of flesh eats away everything but the bone and the integument. It is said by travellers to be a common experience to find the effigies of various animals of which nothing continues but the outward semblance and an empty name. In the desert the conditions are favorable for a long-continued preservation of these empty body-cases, so that the multitude of lifeless forms have all the appearance of a great natural museum whose taxidermist possessed the most perfect skill.

The *condor* would seem to establish the fact that in spite of its seeming provision for quickness of scent, it resembles the greyhound in hunting by sight. Repeated experiments have been made by offering the *condor* favorite articles of food concealed merely by a paper wrapping, and it uniformly happened that he remained wholly indifferent, while if his eye was allowed to fall upon food otherwise wrapped up, he would at once tear to pieces the wrappings and feast upon the *bonne-bouche* which these concealed. Sixteen thousand feet above the earth is no uncommon altitude for the *condor*, who seems to believe in pure air and free exercise, even though his natural instincts make him fond of the most "gamey" food, and lead him always to gorge, even though, like Launcelot Gobbo, he does not "rend apparel out."

The **Bearded Vulture, Sociable Vulture**, or **Lammergeyer** (*Gypætusbarbatus*), is supreme on the Alps, the Pyrenees, and the Indian Himalayas. It is magnificent in size, and terrible in appearance, but ordinarily is harmless and subsists upon carrion. If, however, it be pushed by hunger, it will successfully attack lambs, hares, mountain goats, and chamois; when still more desperate it will carry off small children, and in extreme cases has been known to enter into contests with the mountain hunters. Its method of capturing the goat and chamois is ingenious, for it will suddenly rush upon them and push them over the brink of the precipice. It has learned how to enjoy even the tortoise, for having seized one in its talons it will fly to a great height and break the shell by dashing it down upon the rocks. In flying around the cliffs the *bearded vulture* keeps at a uniform level and within the shadow. It builds its nest high up on the cliff and, like the more destructive birds, lays but a single egg, so that by the accidental destruction of the eggs, and by the usual infant mortality there is an equivalent for what has been called the Malthusian doctrine. An African traveller tells of the loss of part of his dinner by the open and fearless appropriation of it by a *bearded vulture* which manifested the greatest contempt for the attempted protests of the servants. It has given rise to many a thrilling adventure, and is supposed by many to have been the Roc which carried Sinbad, the sailor, to the Valley of

CONDOR.

Diamonds. The illustrations in THE SAVAGE WORLD admirably represent the several species in life-like attitudes, and the manner of taking their prey. The coloring of this bird is found as a distinction of the male. The under parts are yellow, the back, wings and tail black with white shaftings, the forepart of the head cream-colored, and the sides of the head black. It wears a long black, bristly chin-beard, and the orange iris is surrounded by a blood-red coat. Its claws represent the weaker variety belonging to the scavenger family, and hence the greater necessity for the ruses by which the *bearded vulture* converts a living animal into a defenceless carcase.

The **Egyptian Vulture** (*Neophron perenopterus*) is found also in Europe and Asia. It frequently appears in Egyptian symbolism and is called *Pharaoh's chicken*. Its usefulness as a scavenger assures it protection, although it quite frequently desecrates the graveyards.

The **Arabian Vulture** (*Vultur monachus*) is not, as its name might suggest, confined to Arabia, but is common throughout Europe, Asia and Africa. Parts of its neck and head are blue, though its prevailing tint is chocolate. It has the unique ornament of a tuft springing from the point where the wing joins the body.

BEARDED VULTURE (*Gypætos barbatus*). MONK'S-GOWN VULTURE (*Vultur monachus*).

Everything connected with the land of the Pharoahs has a strange fascination for many persons, and hence this bird has been made the theme of many animated descriptions. It is protected alike by law and superstition, and therefore walks the streets of Egyptian cities as though they had been made for its convenience. In a country where the heat is so great, and clean linens so rare, the services of the *Egyptian vulture* are inestimable.

The **Turkey Buzzard** (*Cathartes aura*) is a common sight in the West and

South. Its serviceableness is incalculable for the carrion it devours, but its repulsive appearance and fœtid odor make it especially repulsive.

THE OSTRICH FAMILY.

In the introductory remarks describing the general characteristics of birds, I endeavored to briefly explain the analogies that seem to exist between reptiles and birds, which serve to connect the two species by a well-defined link. In this closing department of birds I will therefore introduce The **Ostrich Family** which supply a link that binds together birds and mammals. With the explanations given we find incontestable proofs of the universal chain that binds all nature in a well connected whole, and the beautiful harmony that exists in all the orders of creation, exhibiting a gradual development that excites our admiration no less than our reverence for the Master-hand of such wise and beneficent bestowal, who d o e t h all things so generously and with such marvellous perfection.

The most conspicuous differences that serve to distinguish birds from mammals may be briefly stated as follows, viz.: the former have two legs and

ALPINE CONDORS (*Sarcorhamphus gryphus*).

are covered with feathers, while the latter have four legs and are provided with a covering of fur. Some quadrupeds, however, are destitute of fur, being supplied instead with bristles, like the hedge-hog, or quills like the porcupine, or a scaly armor, like the armadillos, etc. Again, there are mammals that have no legs, the cetaceans for example, and others that have only rudimentary legs, or flippers, such as the *phoca*, or seal family. Hence, we find that the differences between mammals and quadrupeds is no less marked than is the distinction between birds and mammals.

But if we pursue our investigations with critical care we will be sure to discover a striking analogy subsisting between birds and mammals, though this connection is confined to a single family numbering a half dozen species; the most prominent characteristics we will now proceed to briefly note.

The *ostrich* is a descendant of a gigantic bird that once had its home in Australia and New Zealand. It is very properly known as the **Dinornis**, meaning *terrible bird*, because in life it stood no less than eighteen feet in height, and possessed of proportionate strength, though the skull indicates that it was a stupid bird and less fleet of foot than the *ostrich*. There were several species of this monster bird, all of which are now extinct, though it is probable that its disappearance occurred some time during the present century. Allied to the *dinornis* was the Apteryx, or **Wingless Bird**, which was supposed to be extinct, but is now known to exist in the swamp regions of Australia. Like the

GREAT APTERYX OF AUSTRALIA RESTORED.

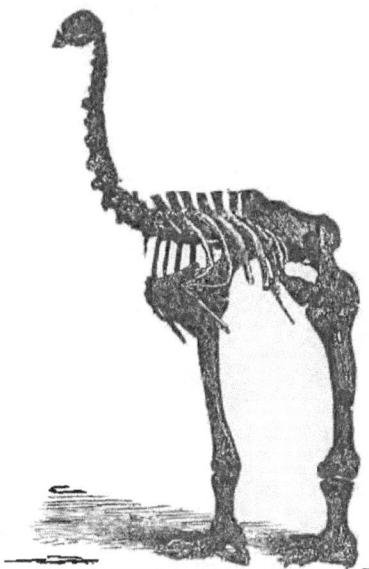

SKELETON OF DINORNIS.

dinornis, it is destitute of wings, being provided with simple rudimentary appendages which are useless even to increase its speed when running. The largest species, which measured some three feet in height, is no longer met with and is supposed to have entirely disappeared. Though both the *dinornis* and *apteryx* are classed among the struthious (*ostrich* kind) birds, their appearance and habits were very different. The former was very *ostrich* like, having long legs and neck, and short bill. Its toes were made for scratching, and its food was vegetable, such as grain and grasses. The *apteryx*, on the other hand, is a short, thick bird, with rather short legs and medium neck, but the bill is very long and snipe-like, and is used for boring into the ground for worms and small mollusks, which compose its diet. When sleeping, the bill rests upon the ground to balance the bird, so that it stands upon a tripod, composed of the bill and two legs. The two are common, however, in respect to their plumage which is more like hair than like feathers. The skin of the *apteryx* is also very similar to that of a mammal, and is used by the natives for making dresses, for which purpose it is very highly valued. The *dinornis* differed from the *ostrich* in having four toes, like those peculiar to gallinaceous birds, in being covered with filamentary feathers, if a license for such an expression is allowable. The plumage, however, certainly resembled hair more than it did feathers, but the internal organism of the *ostrich* is identical with that of the *dinornis* and *apteryx*. Concerning these two wingless birds a writer in the *American Cyclopædia* observes: "The occurrence of these gigantic birds in New Zealand adds much to the evidence that similar apterous and low-organized reptilian birds existed in America during the red sandstone epoch (the age of reptiles) when the cold-blooded and slow-breathing *ovifera* (egg bearing) exhibited such various forms and so great a number of species."

The *ostrich* has been likened to a camel, though the resemblance is certainly not pronounced, if we except the one single characteristic of a common omnivorous appetite and marvellous digestive powers. The plumage is hair-like and coarse, save the wing-feathers which compose the ostrich-plumes of commerce. The creature is entirely bare on the sides and thighs, leaving the dun-colored and sometimes livid flesh-colored skin exposed, which is wrinkled, and resembles the meshes of a net. The legs are covered with large scales, as are the feet, which are cloven to form two toes of unequal length, the inside one being the longer. The neck is long and swan-like, covered with short, white hairs, and the head is comparatively bare, being surmounted with tufts of white bristles. At the joint of each wing, there is a spur or sharp quill, an inch in length, and of a horny substance. The eyes are provided with lids and lashes like those of a man, in which respect it has a mammalian characteristic. The internal parts of this animal are formed with no less surprising peculiarity. At the top of the breast, under the skin, the fat is two inches thick, and on the fore part of the belly it is as hard as suet, and about two inches and a half thick in some places. It has two distinct stomachs. The first, which is lowermost, in its natural situation, somewhat resembles the crop in other birds; but it is considerably larger than the other stomach, and is furnished with strong muscular fibres, as well circular as longitudinal. The second stomach, or gizzard, has outwardly the shape of the stomach of a man; and upon opening is always found filled with a variety of discordant substances: hay, grass, barley, beans, bones and stones, some of which exceed in size a pullet's egg. The kidneys are eight inches long, and two broad, and differ from those of other birds in not being divided into lobes. The heart and lungs are separated by a midriff, as in quadrupeds, and the parts of generation also bear a very strong resemblance to those of the mammals.

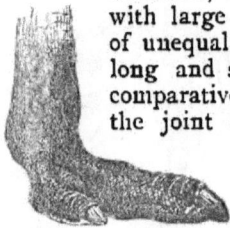

FOOT OF THE OSTRICH.

The voracity of the *ostrich* has become proverbial, nor can any statement, it would seem, exaggerate the fact. It will swallow indiscriminately glass, stones, pieces of metal, or any other substance. This habit is, no doubt, prompted largely by necessity, since we know that all birds swallow pebbles, which serve the double purpose of keeping the coats of the gizzard apart, and to assist the process of grinding up the natural food.

The *ostrich* is gregarious, associating together at times in flocks of a hundred or more, though they are frequently seen in pairs or even solitary. Their food is chiefly vegetable, to obtain which they travel great distances, which they easily accomplish, since no other creature is so swift of foot. When running they raise their rudimentary wings to a right angle with the body and by vigorously flapping, fairly sail over the burning sands. The laying season is early in July in northern Africa and towards the last of December in the south. The eggs are about five inches in diameter and weigh nearly fifteen pounds, being provided with a shell of great strength, to serve as a protection against the animals that would devour them if they were covered less securely. But even strong as is the shell-covering, hyenas and jackals contrive to break them, though by what means naturalists have not yet been able to determine. The *ostrich*, while a timid creature generally, is most courageous during the nesting period, as well as a dangerous combatant. To its powers of defence is added a voice that may well inspire awe, especially if raised during the stillness of night, as it frequently does at

AFRICAN OSTRICH (*Struthio camelus*).

the breeding season. The cry is so nearly like that of the lion that persons most familiar with the king of beasts are readily deceived, and all animals within hearing distance retreat with precipitation in the full belief that it is a lion abroad.

The male *ostrich* (upon the authority of Thomas Pringle), at the time of breeding, usually associates to himself from two to six females. The hens lay all their eggs together in one nest; the nest being merely a shallow cavity scraped in the ground, of such dimensions as to be conveniently covered by one of these gigantic birds in incubation. A most ingenious device is employed to save space, and give at the same time to all the eggs their due share of warmth. The eggs are made to stand each with the narrow end on the bottom of the nest and the broad end upward; and the earth which has been scraped out to form the cavity is employed to confine the outer circle, and keep the whole in the proper position. The hens relieve each other in the office of incubation during the day, and the male takes his turn at night, when his superior strength is required to protect the eggs or the new-fledged young from the jackals, tiger-cats and other enemies. Some of these animals, it is said, are not unfrequently found lying dead near the nest, destroyed by a stroke from the foot of this powerful bird.

OSTRICH RUNNING.

As many as sixty eggs are sometimes found in and around an *ostrich's* nest; but a smaller number is more common, and incubation is occasionally performed by a single pair of *ostriches*. Each female lays from twelve to sixteen eggs. They continue to lay during incubation, and even after the young brood are hatched; the supernumerary eggs are not placed in the nest, but around it, being designed to assist in the nourishment of the young birds, which, though as large as a pullet when first hatched, are probably unable at once to digest the hard and acrid food on which the old ones subsist. The period of incubation is from thirty-six to forty days. In the middle of the day the nest is occasionally left by all the birds, the heat of the sun being then sufficient to keep the eggs at the proper temperature.

The *ostrich* of South Africa is a prudent and wary animal, and displays little of that stupidity ascribed to this bird by some naturalists. On the borders of Cape Colony, at least, where it is eagerly pursued for the sake of its valuable plumage, the *ostrich* displays no want of sagacity in providing for its own safety or the security of its offspring. It adopts every possible precaution to conceal the place of its nest, and uniformly abandons it, after destroying the eggs, if it perceives that the eggs have been disturbed or the footsteps of man are discovered near it. In relieving each other in hatching, the birds are said to be careful not to be seen together at the nest, and are never observed to approach it in a direct line.

The food of the *ostrich* consists of the tops of the various shrubby plants which even the most arid parts of South Africa produce in abundance. This bird is so easily satisfied in regard to water that he is constantly to be found in the most parched and desolate tracts which even the antelopes and the beasts of prey have deserted.

When not hatching they are frequently seen in troops of thirty or forty together, or amicably associated with herds of zebras or quaggas, their fellow-tenants of the wilderness. If caught young the *ostrich* is easily tamed; but it does not appear that any attempt has been made to apply his great strength and swiftness to any purpose of practical utility.

The *ostrich* is valued not only for the incomparable plumage of its wings, but also for its flesh, especially the young, which are said to be most palatable. In the time of Rome's grandeur the *ostrich* was highly esteemed as a rarely rich dish. Apicius has left us a recipe for making sauce for the bird, and Heliogabalus is distinguished in history for having dressed the brains of six hundred *ostriches* in one dish. Among all people now the eggs are greatly prized though too expensive a luxury to be often indulged in, notwithstanding the fact that one egg would suffice to satisfy the hunger of a dozen men.

There are several ways of hunting the *ostrich*, but the two most popular with the Hottentots is by riding the bird down, and by stalking it, the hunter being covered with an *ostrich's* skin and passing a stick up through the neck with a handle at the lower end, by which he is able to counterfeit every motion of the bird.

APTERYX, OR KIWI (*Apteryx australis*).

Although as before stated the *ostrich* is the swiftest runner of all creatures, he is stupid, which weakness is taken advantage of, as will be seen. When the mounted hunter discovers an *ostrich* he sets out after his quarry in a slow gallop, so as not to give any unnecessary alarm. The bird does not take immediately to flight but canters off apparently conscious of his ability to distance pursuit. As the hunter draws gradually nearer, the *ostrich* increases its speed, but instead of running directly away from the enemy it moves in a zig-zag direction, so that the experienced hunter saves his horse and, by cutting across the tracks, maintains his distance until the *ostrich* becomes fatigued and abandoning further effort at escape stops and burrows its head in the sand, presenting its body an easy target for the persistent sportsman.

Occasionally in later years the *ostrich* is taken by means of nets, into which the birds are driven by dogs specially trained for that purpose.

When the Arabians capture an *ostrich*, they cut its throat, and making a ligature below the opening, shake the bird thoroughly. After so doing the ligature is removed, when there pours from the wound a quantity of blood that is mixed with the fat, which is considered the most palatable of dainties. After feasting they flay the bird, using the skin for many useful purposes, such as making nets, cuirasses and bucklers.

The inhabitants of Lybia breed *ostriches* extensively both for the feathers which they yield, in which there is a very great profit, and also, as is told, for riding animals. Several travellers tell us that the natives of South Africa break *ostriches* to the saddle and bridle, in which capacity it is affirmed they are tractable and most serviceable.

The Emeu (*Dromaius australis*) is peculiar to the central plains of Australia, where in former years it was quite plentiful, but is now so seldom seen that we may expect it soon to become extinct. Specimens may be preserved, however, in aviaries, as it is easily domesticated and breeds in captivity. The *emeu* resembles the ostrich in many particulars; enough, indeed, to give to it the relation of a half-brother. The head is very similar, except that the crown is covered with a tuft of coarse feathers; the neck, too, is not so bare as the ostrich's. The most noticeable difference is found in the body covering, which in the *emeu* is of long, dark and hemp-like feathers, destitute of beauty or usefulness; the toes, too, are three in number, instead of two, as in the ostrich. Though the coarse feathers have no commercial value, the bird is hunted for its flesh, and for the oil that may be rendered out of its fat, which commands a high price. It is very swift of foot, but can be run down by horses and dogs without much difficulty. The dogs are trained to reserve the attack until the bird is thoroughly tired out, and then spring upon the throat in such manner as to escape the violent kicks which the *emeu* deals fiercely around, and which are sufficiently powerful to disable an assailant. The *emeu* does not kick forwards like the ostrich, but delivers the blow sideways and backwards like a cow.

The food of the *emeu* consists of grass and various fruits. Its voice is a curious hollow, booming or drumming kind of note, produced by the peculiar

EMEU (*Dromaius*).

construction of the windpipe. The legs of this bird are shorter and stouter in proportion than those of the ostrich, and the wings are very short, and so small that when they lie closely against the body they can hardly be distinguished from the general plumage.

The nest of the *emeu* is made by scooping a shallow hole in the ground in some scrubby spot, and in this depression a variable number of eggs are laid. Dr. Bennett remarks that "there is always an odd number, some nests having been discovered with nine, others with eleven, and others, again, with thirteen." The color of the eggs is, while fresh, a rich green, of varying quality, but after the shells are emptied and exposed to the light, the beautiful green hue fades into an unwholesome greenish brown. The parent birds sit

RHEA, OR SOUTH AMERICAN OSTRICH AND HER YOUNG.

upon their eggs, as has been related of the ostrich. The *emeu* is not polygamous, one male being apportioned to a single female.

The **Rhea** (*Rhea americanus*) is a species of struthious birds found only in South America, being particularly abundant in the La Plata region, and south as far as Patagonia. Like the ostrich, it is gregarious and wonderfully swift of foot, but its fleetness is of little advantage because of its habit of running in circles, so that expert hunters, when well mounted, easily come within shooting range, or ride the bird down. The Patagonians hunt the *rhea* on horseback, and with no other weapon than the "bolas," which is made by sewing a ball of lead into each end of a leather cord several feet long, though sometimes the cord is short, not exceeding three feet. Mr. Barrows has thus described a hunt of the *rhea*, or South American ostrich, of which he was a spectator:

"During our stay at Puau, about three hundred Indians united in a **two**

29

days' ostrich hunt, resulting in the capture of about sixty birds of all sizes, from the full grown adult to the two months' old chick. They began by beating over a large track of the plain, and, then closing in around, the game started. Stout greyhounds are used to good purpose, usually pulling down the swiftest birds within two miles at farthest. The Indians use the 'bolas' with much skill, the ones used for ostriches consisting of two half pound leaden balls, connected by eight feet of twisted rawhide twine. Whirling this about the head and letting fly at the running bird, they often entangle his legs at a distance of thirty to fifty yards, and I was *told* that it was frequently done at one hundred. Single hunters sometimes stalk ostriches in the following way: Getting to windward of the bird, the latter soon scents him and lies down, only sticking up his head above the grass. The hunter may then creep directly up within shot, if the grass be long enough to shelter him."

Mr. Darwin, to whom we are indebted for most of our knowledge respecting this bird, among many other interesting things, thus writes:

"This bird is well-known to abound on the plains of La Plata. To the north it is found, according to Azara, in Paraguay, where, however, it is not common; to the south, its limit appears to have been from 42° to 43°. It has not crossed the Cordilleras, but I have seen it within the first range of mountains in the Uspallata plain, elevated between six and seven thousand feet.

CASSOWARY.

They generally prefer running against the wind, yet at the instant they expand their wings, and, like a vessel, make all sail. On one fine hot day I saw several ostriches enter a bed of tall rocks, where they squatted concealed till nearly approached.

"It is not generally known that ostriches readily take to the water. Mr. King informs me that at Patagonia, in the Bay of St. Blas, and at Port Valdez, he saw these birds swimming several times from island to island. They ran into the water both when driven down to a point, and likewise of their own accord, when not frightened. The distance crossed was about two hundred yards. When swimming, very little of their bodies appear above water, and their necks are stretched a little forward; their progress is slow. On two occasions I saw some ostriches swimming across the Santa Cruz river, where it was about four hundred yards wide and the stream rapid.

"The inhabitants who live in the country readily distinguish, even at a distance, the male bird from the female. The former is larger and darker colored, and has a larger head. The ostrich, I believe the cock, emits a singular deep-toned hissing note. When first I heard it, while standing in the midst of some hillocks, I thought it was made by some wild beast, for it is such a sound that one cannot tell from whence it comes, or from how far distant.

"When we were at Bahia Blanca, in the months of September and October, the eggs were found in extraordinary numbers all over the country. They either lie scattered singly, in which case they are never hatched, or they are collected together into a hollow excavation, which forms the nest. Out of the four nests, which I saw, three contained twenty-two eggs each, and the fourth twenty-seven. The Gauchos (natives) unanimously affirm, and there is no reason to doubt their statement, that the male bird alone hatches the eggs, and that he, for some time afterwards, accompanies the young. The cock, while on the nest, lies very close; I have myself almost ridden over one. It is asserted that at such times they are occasionally fierce, and even dangerous, and that they have been known to attack a man on horse-back, trying to kick and leap on him. My informant pointed out to me an old man whom he had seen much terrified by one of these birds chasing him."

The *rhea* is darkish-gray, taking a blackish-hue above, and being rather lighter below. The plumes of the wings are white, and a black band runs round the neck and passes into a semi-lunar patch on the breast. The neck is completely feathered. The average height of the *rhea* is about five feet, about the same as that of the emeu.

HELMETED CASSOWARY (*Casuarius galeatus*).

Three species of *rhea* are, however, all inhabitants of South America, namely, the common *rhea* just described, Darwin's *rhea* (*Rhea darwinii*) and the *large-billed rhea* (*Rhea macrorhyncha*).

The **Cassowary** (*Casuarius emeu*) is found only in the Malaccas, nor has it been often seen anywhere except in the northern part of Australia. The appearance between the emeu and *cassowary* is very similar, as are their habits. The principal difference is observable in the head, which in the latter is bare of feathers but surmounted by a helmet, or bony protuberance, which enables the creature to rush through a thicket with ease and without danger of injury, as the branches or thorns striking this helmet pass over the head and thus oppose no obstruction. Unlike the ostrich, emeu and rhea which live upon the open plain, the *cassowary* is a strictly bush bird, making its home in the thickest copses, through which it is able to run with surprising speed by reason of the head protection described. The plumage is very hair-like, in some respects resembling the long hair of the yak, except that it is of a glossy-black and flat at the ends.

The eye of the *cassowary* is fierce and resolute, and its expression is carried out by the character of the bird, which is tetchy of disposition, and apt to take offence without apparent provocation. Like the bull, it is excited to unreasoning ire at the sight of a scarlet cloth, and, like the dog or cat, has a great antipathy towards ragged or unclean persons, attacking such individuals with some acerbity merely because their garments or general aspect do not please its refined taste. It is a determined and rather formidable antagonist, turning rapidly about and launching a shower of kicks which can do no small damage, their effect being considerably heightened by the sharp claws with which the toes are armed. In the countries which it inhabits, the native warriors are accustomed to use the innermost claw of the *cassowary's* foot as a head for their spears. The bird stands about five feet in height, and is about equal to the rhea in size of body. The nesting habits of the two are identical.

The **Mooruk** (*Casuarius bennettii*) was supposed by Dr. Bennett to be a distinct bird, differing from the cassowary in the one single respect of having its head protected by a horny plate, resembling blackened mother-of-pearl, instead of the helmet-like protuberance. But later investigation proves that this slight difference is due to age, the older cassowaries having the helmet worn down until it appears as a flat, bony plate. The cassowary and *mooruk* are but different names for the same creature.

There is a remarkable resemblance among all of the struthious or ostrich birds, not alone in habits, but in appearance and organization as well. Among no other family are the several characteristics so nearly identical, and yet we find the species widely distributed, as just described. Like the human race, they must have sprung from one common source, and like man, became scattered, and in the separation the few peculiarities which distinguish them apart have become developed. The causes of this distribution are not given us to know, but the fact that one species is peculiar to Africa, another to Australia and another to South America, excites in us anew the ever-recurring thought, "How wondrous are the ways of Providence!"

BATS.

HE reptilian aspect of the flying pterodactyl, as described in my introduction to birds, formed a no more distinctly pronounced link connecting reptiles and birds than the *bat* furnishes between birds and mammals, and is but another of the many proofs furnished incontestably demonstrating the evolution of species by structural differences to accommodate the new births to changed conditions of the earth, as already explained. Progression is not always evidenced by these new creations, so to speak; on the other hand there appear signs of retrogression, or deterioration of species, the primitive births generally presenting less gracefulness of form and less adaptability of structure. Thus we observe in the *bat* a creature which, if regarded as a bird, is unsightly, malformed as a flyer and destitute of every economic feature that makes the bird a marvel of perfection in the field to which it has been assigned by nature. If we consider the *bat* as a mammal its imperfections are no less pronounced, for which reason the combination of its bird-like and mammalian characteristics compels its assignment to an intermediate class, or a link which binds the two orders together.

With all the seeming faults of structure we are able to see in the *bat*, it is a living illustration of the attempts or processes of nature towards completing a radical change of organization and the development of distinct and more highly organized species. The *bat* may therefore be regarded as the result of a modification of the bird by nature which, in the effort to develop into a mammal first divests itself of feathers but retains the phal-

HEADS OF EGYPTIAN BATS.

anges and wing structure. For the feathers, soft fur, corresponding to down, is substituted; or, if we chose to proceed more critically, the quill feathers are first removed but the creature is permitted to retain the down.

The next process is probably conspicuous in the bones of the bird thus undergoing metamorphosis. The hollowness of the bones which distinguishes the bird gradually changes by the softening of the osseous ducts ramifying the cavities until marrow appears, for though the bones of *bats* are not nearly so

solid as those of mammals, certainly they are not so light, nor is the hollow so great, as are the bones in birds. The next change is to be noted in the mouth, which has lost the beak destitute of teeth, and instead the *bat* has received a provision that enables it to seize its prey in a different manner. The molar teeth, however, do not appear, as the creature has not yet developed into a mammal so highly organized. The keeled breast of the bird has disappeared and the fulness which the breast-bone supplied is partially retained by the substitution of ribs, to which fore-legs are attached, followed by the appearance of mammæ, the only distinct feature which binds this singular creature to the mammalian order.

MOSS-NOSED BAT (*Synotus barbastellus*).

I do not mean by the above reasoning that the changes, as described, have actually taken place, but rather to indicate the probabilities, and what seems to be justified by the analogy that exists, showing the correspondence the *bat* occupies between birds and mammals.

The *bat* is a common and by no means pleasant sight, for even when harmless there is something uncanny about its appearance and restless flight. A wise man once said that "all that is unknown is accepted as great or terrible," and this is certainly true of the harmless species of *bat* which flies about our rooms. The *bat*, to enable it to fly, has a long fore-arm and fingers united by a membrane which extends back to the hind feet, which are so constructed as to bend backwards and thus keep its sail unfurled. More than four hundred species of *bats* have been distinguished, so it is easy to see that there is room for the greatest variety of habits. Doubtless the popular fear of the *bat* has been increased by the hoary antiquity of the superstitions which have gathered about it and which like a great snow-ball, ever increase as they roll along. The Jews classed

EARLY MORNING BAT (*Vesperugo noctula*).

the *bat* among unclean animals (and this with no reference to the parasites which constantly attend it), and the sanction of religion thus lent to popular superstition must, of itself, have ruined the reputation of the creature. The Greek mythology made the *bat* one of the symbols of the Queen of the Plutonian World, possibly from its choosing the dusk for an appearance so mysterious that the creature seemed to spring out of empty space. The Greek mythology has been the honey-cup of all imaginative writers of succeeding peoples, and hence again human teachings would lead one to associate with the *bat* peculiarities none the less alarming because preposterous. Still farther the rapid motions and ever-changing directions of flight, the darkness amidst which it exerts its activity, and the isolation of the places which it selects for its abode, all unite to deepen the impression with which one's early education leaves him. Artists, likewise,—those men who address the eye and the imagination so effectively,— have so associated the *bat* with the horrible and the repulsive, that even the

LONG-EARED BAT (*Plecotus auritus*).

witches, whose attendants the *bats* are so often made, seem human and beautiful in comparison. Demonology has interest of its own, but the evil demon is represented as having a bat-like appearance, so that the poor creature seems to have been cursed on all sides. Among the Orientals, however, the *bat* is no harbinger of misfortune, suffering, or crime, but the accepted precursor of good luck, happiness, and virtue;

LONG WINGED BAT (*Miniopteris*).

and in some countries the *bat* is even converted into a cherished household favorite. It has been said that the *bats* differed among themselves greatly in the matter of habits. Many of them, though giving birth to twins, raise but one of the young, thus intuitively accepting the Spartan method of limiting population and destroying the weak and unpromising. In some species, the male is the one provided with breasts for the nourishment of the young—a circumstance less surprising to those who, like myself, have known men whose mammary glands secreted quite copious supplies of milk. In other species, again, the males use their

GRAY CLAP-NOSED BAT (*Rhinopoma microphyllum*).

superior strength in carrying the young while these are too weak to care for themselves. Thus does the animal creation suggest many an analogue or teach many a lesson to its highest order—Man. There are men, and their number is

constantly increasing, who believe that while it is "excellent to have a giant's strength, it is tyrannous to use it like a giant," and the day is dawning when mankind will not excuse each other for the non-use or the abuse of the power which each has, nor use woman's greater physical weakness for excusing unnecessary abstinence from effort, or for adding unfairly to her burdens because she can make no successful resistance.

It may not be known to all that though *bats* are gregarious and companionable with each other, their polity is like that of the Shakers, and forbids co-education, flirting or the commingling of the sexes. The young ordinarily fasten themselves to the body of the parent, and are carried about wherever the mother flies, though scarcely perceptible, so closely do they cling to her body.

FLYING FOX (*Pteropus edulus*)

Bats are divided, first, into the fruit-eating *bats* and the insect-eating *bats*. The **Fruit-eaters** (*Frugivora*) are inhabitants of the tropics, and are not found in America, so that their depredations may not be added to the fears of those who always assert that "the fruit is all killed." The *fruit-eaters* will take entire possession of trees, so that to say they are "as thick as leaves in Vallambrosa," would be no exaggeration, and they produce the impression of a new and curious and luxuriant foliation. They do not seem to like the cocoanut or the mango, but do not show the same aversion to the juice of the cocoanut, of which they will drink till intoxicated, and then exhibit all the amusing and all the disgusting effects of inebriety. Those who have not succeeded in "getting dead-drunk," fly in a veering sort of way, and in no respect differ from the "antic disposition"

HORSE SHOE BAT (*Rhinolophus ferrum equinum*).

which the farmer assumes when, "having sold a load of wheat, he takes with him too heavy a load of old rye." An Indian species (*Cynopterus marginatus*) are specially destructive to orchards, and, as they will travel forty miles in a single night, it is difficult to exterminate them. The insect or carnivorous *bats*, however, are the more numerous.

The **Horse-shoe Nose Bat** (*Rhinonycteris ferrum equinum*) seems to feed upon beetles and to seek "the dark, unfathomed caves of ocean" in their hatred of the light. Its wing covers an expanse of a foot and a quarter, so that its size might terrify even though we are not prejudiced against the family.

The **Lyre-like Bat** (*Megaderma lyra*) is a blood-sucker, for which office the conformation of its snout qualifies it. It has been known to seize a smaller bat of

AMERICAN BAT (*Vespertilio murinus*).

a different species and after sucking its blood, to then proceed to devour everything but the bones and head. To these smaller bats the nursery fables of ogres are only too true, and unfortunately for their comfort the ogres are only too numerous. One is made to think of Ulysses and his companion in the care of Outis and of his superior good fortune in being able to escape the dreadful fate which threatened him.

The **Red Bat** (*Atalalapha noveboracensis*) is exceedingly common in the Atlantic States, and disturbs one's comfort by invading the sanctity of one's domicile though simply in pursuit of the insects attracted by the lamplight. It is harmless unless attacked, when it is both pugnacious and able to inflict slight

THE NYCTURUS OF UPPER EGYPT.

but painful wounds. When excited to anger it is not contented with action, but like a human being raises its shrill voice, not believing that "words to the heat

of deeds too cold breath give." One of its devices when worsted is to feign
death as we have found some insects and the opossum also to do. This *bat* has
been known to follow the captor of its young just as the mother-bird will do,
or as a cat will follow in the hope of rescuing its kittens.

The **Long-eared Bat** (*Pleocotus macrotis*) is asinine in its provision of ears,
which in length, breadth and uprightness are wonderful to behold, and possibly,
as in the case of Little Red Riding Hood's tormentor, enable it "to hear the
better, my dear." It is not uncommon in the south or southwest and is likewise
found among the other remarkable products of the Pacific slope.

The **Collar Bat** (*Chiromeles torquatus*) has, irrespective of sex, pouches for
carrying the young, so that evidently, under its form of civilization, both male
and female take charge of
the nursery.

SOUTH AMERICAN VAMPIRE (*V. spectrum*).

The **Hare-bat** (*Noctilio
leporinus*) is notable, because in eating it stuffs its
cheek-pouches until these
can hold no more, then
swallows their contents, re-
jects and again swallows
them before digestion be-
gins. An old gentleman,
disabled by the gout, was
listening to his little girl
read, and when she nar-
rated the possession of four
stomachs by the camel, he
could contain himself no
longer, but full of the bliss
of unlimited eating, ex-
claimed: "My dear, what
a wonderful dispensation of
Providence!" Had he but
heard of the double-gor-

mandizing *hare-bat*, how much greater might have been his envy. The
hare-bat climbs backward, thus reproducing the terrestrial locomotion of the
crab, and, finally, he is like the fish-hawk, an angler darting upon the succulent shrimp.

The **American Rose-leafed Bat** (*Macrotus waterhousii*) flourishes in California, Mexico and the West Indies. It is a singular looking creature, whose
thick nose protrudes from a space which resembles an inverted **V**.

The **South American Vampire** (*Vampyrus spectrum*) has, after many days,
had his character vindicated, for science, although admitting that he loves blood,
has ascertained that he confines his attentions to animals not human, and even
then any injury proceeds not from the loss of a little blood but from the inflammation of the bitten part. It is admitted that in some cases this much traduced
creature will fasten itself to the exposed toe of a restless sleeper, and will draw
blood, but the damage done will not be great.

The **Kalong** (*Pteropus edulis*) is a black bat found in Borneo and Java, whose body is as large as that of a small dog and which is hunted as an article of food.

The smallest of bats is the **Pteropus minimus** of Java, but its diminutive size does not interfere with its having a tongue which it can protrude to the length of two inches.

The **Egyptian Wingfoot** (*Pteropus aegyptiacus*) has eighteen inches of expanse to its wings.

The **Glossophaga**, or **Tongue-using Bats** are found in Brazil and Guinea, and belong to the vampires.

The **Wart-lipped Vampires** (*Stenoderma*) have their peculiarity indicated by their name. There is also a **Tailless Vampire** (*Desmodes rufus*).

ATTACKED BY VAMPIRES.

Of the horse-shoe family of bats, mention may be made of the Egyptian species (*Rhinopoma microphyllus*); the pouched species belonging to Egypt (*Nycteris thebaica*), and the one found in Java (*Nicteris javanica*).

THE COLUGO, WITH YOUNG.

Of the common bat family we may add at least the names of **Vespertilio alecto**, the **Mozambique Green Bat** (*Vespertilio viridis*), the **Arboreal Bat** of Europe (*Vespertilio serotinus*) which lives in pairs, the **Senegal Flying-bat** (*Molossus daubentonii*) and the **Flying Marmot** (*Vespertilio nigrita*).

Doubtless the fact that bats frequented altars and feasted upon the remains of sacrifices added them at once to the symbolism in Egypt, and led to their presence at scenes of witchcraft. Shakespeare, whose knowledge of folk-lore was very great, makes one of Macbeth's witches say "But in a sieve I'll thither sail, and, like a rat without a tail, I'll do, I'll do, I'll do." And again when the witches make their demoniac broth, they say "Wool of bat and tongue of dog." Some eight or ten more bats deserve mention because of one peculiarity or another.

There is the **Pouched Bat** (*Cynonycteris grandidieri*) which seems to have been fitted out so that the male shall help the female in carrying their papooses.

The **Common European Flying Bat** (*Pteropus medius*) may be encountered while our restless travellers are "doing the old world."

The **African Hammer-head Bat** (*Sphyrocephalus monstrosus*) is remarkable for his cranium.

The **Armor-bearing Bat** (*Phyllorhina armigera*) is an Indian species, notable for being possibly the smallest of the horse-shoe bats.

The most common **American Bat** (*Vespertilio sublatus*) is with us so frequently that we may care to know his name.

The **Fruit-scenting Bat** (*Rhinopoma nicrophyllum*) is of scientific interest as a connecting link between the fruit-eating and the flesh-eating bats, as he does somewhat in both directions.

The **Nyctinome** (*Nyctinomus nasutus*) is peculiar from the facility with which it can walk upon the ground.

The **Mexican Tongued Bat** (*Phyllonycteris sezecorni*) uses its feet squirrel-like while it extracts flavor and succulence from berries.

The **Fruit-eater** (*Artibeus perspicillatus*) is not a favorite in Jamaica, because of its destruction of fruit; but, with all the efforts made to destroy it, the species continue numerous.

The **Colugo** (*Galeopithecus volans*) is found in Sumatra, Borneo and Malacca. It is arboreal, and its membranous parachute enables it to pass with ease and rapidity from tree to tree, although it is not a true flyer. It has been known to cover the distance of seventy yards, but in doing this descended forty feet. This creature bears a striking resemblance to the flying fox, and also possesses many characteristics peculiar to the monkey family, with which it is classed by many naturalists, but as a compromise it may be properly regarded as a transition link between bats and monkeys, being more nearly related to the marmoset. Its membranous wings—so-called—resemble those of the flying squirrel, but its powers of flight are much greater. They have been seen to leave the top of a tree and sail to another quite as much as seventy-five feet distant, and that, too, without descending more than a dozen feet. By a peculiar structure of the tail, which controls a membranous expansion extending from the rear, the creature is able, in a measure, to direct its course and also to rise over inconsiderable obstructions which may chance to be in its path.

The *colugo* never stirs about in the daylight, but issues forth at early twilight and breaks its fast upon winged insects, but retires after a few hours of activity to certain trees off the leaves of which it feeds as a dessert to its insect food. It is also said to eat fruits of various kinds.

MAMMALS.

UR delightful labor is now approaching a close, for we have arrived at the last division of our subject, and whether we look backward or forward we are ready to recognize the marvellousness of creation and to feel with new force the thoughtfulness, power and mercy of a Creator whose wisdom is equally apparent whether exhibited in the most minute or in the largest of creatures; whether studied in the simplest form of *protozoa*, or in the delicate and complicated mechanism of the highest of animals—Man. Think of the harmony of adjustment required for the exercise of such endowments as enabled earth's most gifted sons to live and do and suffer; to transform mere inert matter into the subtlest thought or the most ennobling deed. Consider all ·that is implied in the mere fact of human existence, and we shall realize that no inspiration of poet or seer approaches the strangeness, the unsuspected harmony which lies all about us awaiting only our awakening from a life in the senses, and a life which is dead in comparison with its possibilities. I have endeavored to approach the story of creation in an humble and reverent spirit. I have sought to escape the folly of preconceived ideas, and to substitute for carelessness and indifference an acquaintance with a life that is all about us, and which illustrates each instant the wisdom and mercy of the omniscient and bounteous Creator. Matter is to my mind but the sensible form assumed by spirit; the web which' spirit weaves into many a pattern and many a design. Matter acted upon by the power of the Creator becomes the visible expression of the infinitely varied tones which unite to form the world's diapason which has marked the rise and setting of each day's sun through the countless æons of the past. It is not alone the rocks that tell the story to such a reverent mind as that of Hugh Miller. It is not simply the plant life which has such interest for Asa Gray that tells the story of the never-ceasing processes which illustrate the care of the Creator, his infinite wisdom and his method of governing the·world of matter by law, which shall be the natural expression of the harmonious relationship of each object to its work in life. Our story has now brought us to the threshold of the highest realm in the animal kingdom, and yet we shall not find greater wonders than were furnished by the worm, the fish, the reptile and the bird.

.By the process which is termed *evolution*, the Lord hath seen fit to unroll one form from another, ever adjusting the writing on the new page to the fullest harmony with what has gone before and with what is to come after. If the creature is, like the *tarpan*, to inhabit dreary wastes and serve the needs of a barbaric people, then will it be found to have evolved those characteristics which will best enable it to fulfil the conditions of the life which it is to lead. If my attempt has been at all successful, I have shown the succession in life-forms and how this succession is as necessary as the link to the chain. It is not alone in the church that there is a doctrine of apostolic succession, but all life is but an illustration of God's mighty wisdom and providence, and his creatures are constantly giving him the praise which consists not in mere petition, but in the proper conduct of the life assigned to them, and the discharge of such offices as have been entailed upon them. The animal world as such recognizes, through instinct, the will of its Creator, and yields an unreluctant obedience to the laws prescribed for its being. To an animal, "*laborare est orare*" and by night and by day the animal world is faithful to the laws of its being, and thus justifies the opening of the anthem called Genesis: "And God said, let the waters under the heaven be gathered together unto one place, and let the dry land appear; and it was so. And God said, let the earth bring forth grass, the herb yielding seed and the fruit tree yielding fruit after his kind, whose seed is in itself, upon the earth; and it was so. And God said, let the waters bring forth abundantly the moving creature that hath life, and fowl that may fly above the earth in the open firmament of heaven. And God created great whales, and every living creature that moveth, which the waters brought forth abundantly, after their kind, and every winged fowl after his kind; and God saw that it was good. And God said, let the earth bring forth the living creature after his kind, cattle and creeping things, and beasts of the earth after his kind; and it was so." Or as the non-inspired poet hath sung:

> Thou from primeval nothingness did'st call,
>   First chaos, then existence; Lord! on Thee
> Eternity had its foundation; all
>   Spring forth from thee; of light, joy, harmony,
> Sole origin; all life, all beauty, Thine.
>   Thy word created all, and doth create;
> Thy splendor fills all space with rays divine;
>   Thou art, and wert, and shalt be! Glorious
>   Light-giving, life-sustaining Potentate!
>
> Thy chains the unmeasured universe around;
>   Upheld by Thee, by Thee inspired with breath!
> Thou the beginning with the end hast bound.
>   And beautifully mingled life and death!
> As sparks mount upward from the fiery blaze,
>   So suns are born, so worlds spring forth from Thee;
> And as the spangles in the sunny rays
>   Shine round the silver snow, the pageantry
>   Of heaven's bright army glitters in Thy praise.

With the end of the reign of birds, the world was ready for the appearance of the mammal and a new order entered upon existence, beginning with the duck-mole and echidnæ, and progressing upward till it reached the *quadrumana* or *primate*, the highest class below man and supposed by some to have been man's ancestors. Such theorists however seem to overlook the fact that the quadrumana while developing even structural differences of moment, are wholly

destitute of those intellectual and moral attributes which alone fit man for terrestrial sovereignty. If the quadrumana are to be likened to man at all, it
would seem as though it should be only to the debased and wholly animal
man whose representative in fiction is Shakespeare's Caliban.

The readers of THE SAVAGE WORLD will have found out for themselves that
I have in the introduction and in the chapters preliminary to the various
orders, at least endeavored to submit the evidence which, to my mind, proves
conclusively that science is not like the dead languages, a subject understood and exhausted, and that there seems to be with expanding knowledge a
fuller appreciation of the necessary agreement between intelligent science and
the Genetic account of the material world. I have regarded evolution as a
new term, not as a new idea. I accept the fact that the finite mind is
prone to make many mistakes in regard to the purposes and methods of the Infinite Mind, and therefore suggest that, regarding evolution as a method of the Creator, and not as one of the gods of the Pantheist, there is everything to convince the student of its reality and truth. An effort has been made to trace transitions without at all detracting from the inter-est or popular

DUCK MOLE AND NEST.

value of THE SAVAGE WORLD. In the various subordinate introductory chapters,
the fossil forms have been referred to as likely to be of interest to the intelligent
reader, and as bearing out the conclusions reached by me claiming
nothing for myself but honesty of purpose and strict adherence to all promises
made to the reader. The experience of breeders throws light upon and adds
strength to the doctrine of "natural selection," and sufficiently shows *how*
animals are adapted to their environment. Fossil life has been sufficiently
dwelt upon to open the field to the ingenious, and enough for the purpose of
illustration. Classes and orders have been considered so far as they could
have popular interest, or as they serve for types, and it is believed that no

work on the subject treated by THE SAVAGE WORLD will be found more complete or more satisfactory for any one desiring the real knowledge which persons at large wish in regard to subjects which do not directly form a part of their daily life. No one more than myself can appreciate so fully the distance between desire and achievement; but many may justify my belief that there is need for information which a person in active life

THE URCHIN (*Echidna hystrix*) AND DUCK MOLE SWIMMING.

has not time to extract from mere data, especially when these are obscured by technical terms. The changes of animals when domesticated have been sufficiently used in illustration of the fact that it lies within the opportunities of any one to watch the method which the Creator has prescribed as the law of animal life, and which will be further exhibited in the descriptions which follow.

MONOTREMES.

The **Duck-billed Platypus** (*Ornithorhyncus anatinus*) is aquatic, and may well serve as a link between birds and mammals. It is about a foot and three-quarters in length, and its long, flat, otter-like body is clad in soft, thick fur, brown above and whitish beneath. Its toes are united by a membrane, its tail is flat and obtuse, and its broad, elongated muzzle looks as if it had been borrowed from the duck. It burrows in the banks of streams, and is therefore called *river mole* or *mulligong*. The animal is, as has been said, properly aquatic, but it can move about on land, and even climb, though in the latter exercise it braces its body against two opposite walls and wriggles itself up in a way quite well known to the fertile-minded "small boy." It is

PORCUPINE ECHIDNA.

Australian in its habitat, and may well serve the purpose of marking the transition from the highest types of birds to the lowest forms of the mammals.

The **Echidna**, or **Porcupine Ant-eater** (*Echidna hystrix*, or *aculeata*, or

Tachyglossus aculeatus), is about ten or twelve inches long, has only a rudimentary tail, and a snout noticeable for the way in which it tapers to a point. Its dark brown body is covered with white spines, which are black on the tips. Its fore feet, employed in digging, are armed with long, stout nails, while the hind feet are used as hand-barrows (or feet barrows), with the unpatented addition of a shovel. The mouth is a mere aperture, sufficient, however, for protruding and retracting its much-celebrated tongue—its weapon of offence. This long and much-used member is constantly moistened with a viscid fluid, which enables the animal to hold securely whatever its tongue can reach. In the absence of teeth, it is provided with spines on the roof of its mouth and on its tongue. Among its other peculiarities of structure is the depression or seeming disappearance of the mammary glands when these serve no purpose for the suckling of the young. When the mammary glands are distended, there are formed two pouches, which serve as cosy apartments for the young, but which disappear altogether with the distension of the mammary glands. The flesh of the *echidna* is said to taste like that of young pigs. Like the preceding species it is peculiar to Australia, where so many singular forms of life are to be found.

ECHIDNA.   DUCK MOLE.

The link which seems to bind birds with mammals is found in the duck-billed mole and the *echidna*, in a direction where we would be least likely to expect it, and exhibits a fact which compels us to pause with profound astonishment at the marvellous provisions and apparent eccentricities of nature. The surprise to which I refer proceeds from the fact that these two creatures, instead of producing their young like all other mammalians, are bird or reptilian in respect to the laying of eggs, from which the young are brought forth by incubation. This assertion was long ridiculed as an idle tale, worthy to rank with that which represents the cock as laying an egg, or a mare's building a nest; but it is no longer so regarded, since the truth of the claim has been well established, and is now accepted by all naturalists. The eggs which these two quadrupeds lay are very much alike, that of the duck mole being only a little larger. They are oblong, and covered with a leathery integument like those of many reptiles. Whether the eggs are really incubated is still a question for some dispute, though the weight of authority is in the affirmative. When the young issue from the egg they proceed at once to extract nourishment from the mammæ of the mother after the manner of other quadrupeds, and are therefore true mammalians, with only the differences above explained.

30

MARSUPIALS.

The **Marsupials**, or **Pouched Animals**, are curious from their appearance and habits, highly considered by hunters, and of interest to naturalists. Their organization is higher than that of the preceding class and they require for the support of life conditions less favorable than do those of the group last described. The *marsupials*, as we know them, like the monotremes, are not typical forms but varieties of these, so that succession in the order of animal life requires that these two families be studied by means of fossil remains. This study, unpromising at first sight, becomes of interest so soon as those who visit our museums have been supplied with an intelligent object and sufficient stimulus to cause their curiosity to lead them to a personal examination of specimens. The **Metatheria**, or **Fossil Marsupials** have pubic bones, double system of teeth, five-toed · feet, besides one or two other characteristics not of popular interest. The **Dromotherium** (an American species represented solely by its lower jaw and teeth) was judged by the naturalists' safest test, the teeth, to have been a pre-historic ant-eater. The **Dryolestes** seems to have been a small opossum whose progeny were to become relatively "sons of Anak." The subsidence of the continent of Australia is comparatively recent and its fossil remains, though gigantic in size, evidently belong to species still existent. Still they serve to show that the founder of the *marsupial* family was at once carnivorous and herbivorous, and to suggest that its successors developed along distinct lines, some of them favoring the kangaroo type and becoming frugivorous, and others developing the other side of their possibilities and becoming *dasyures* and carnivora. The typical fossil is called **Thybacoleo**; the earliest kangaroo the **Diprotodon** or **Nototheria**; and the ancestors of the *dasyures* being named **Phascolotherium** (*amphilestes*), **Amphitherium, Spalacotherium,** or **Triconodon.** The manufactured names of the systematic zoologists have no great significance when translated, but as a name is only used for identification, any one about to visit a museum can copy the technical names and study such specimens as the museum possesses. In some of these forms the joining of the toes has proceeded so far as to give four toes instead of five, and to suggest the process by which a toed animal may in the lapse of time grow into one which has a solid hoof. With the *marsupials*, as with the monotremes and all other classes, we shall

MERLIN'S (EXTINCT) OPOSSUM.

hardly expect to trace every step of an insensible gradation or to fail to find varieties which will, because of the peculiarities of their structure, be classed differently by various naturalists, as these are most impressed by one or another structural peculiarity. This very complication of structure seems to our minds most conclusive evidence of the gradual evolution of the animal kingdom, as well as for the reversions and freaks which might reasonably be expected. Some of the *marsupials* have not even a rudimentary pouch, and yet the other features of their structure relate them closely to the pouch-bearing family. The existing forms of *marsupials* do not directly represent succession to the monotremes, but a study of their fossil ancestors has led naturalists to assign to the family the second place in the ascending scale.

THE OPOSSUM.

Many things have united to lend reputation to the **American Opossum**. For hunting this animal is a familiar sport to those who live in the Southern States; its flesh is highly prized by the negroes; its form, its habits and its characteristics are all such as to attract attention and excite comment. The **Common Opossum** (*Didelphis virginiana*), as the name implies, is common to nearly all the States of America, though most persistently hunted in the South, yet the flesh of those found in the Northern States, where the cold is severe, supplies a much more palatable dish. This species is about three feet in length, and half as much in thickness, is grayish-white, and its prehensile tail is allowed about a foot and a quarter of length, and supplied with scales instead of fur. Being a remarkably good climber, and uniformly in a state of ravenous hunger, the *opossum* is very troublesome to the fowls and eggs, which farmers do not raise for the

CRAB-EATING OPOSSUM.

COMMON OPOSSUM.

pleasure of such tramps as he. Still, it is to the *opossum* that we owe our symbol for feigning successfully; for, when no other course is open before him, the *opossum* will roll itself into a ball, and "play dead," so that no abuse can make it betray its continued vitality. The young, as soon as born, are retired to the ventral pouches of the mother where they remain until they have attained their growth.

BROAD-NOSED WOMBAT.

The **Crab-eating Opossum** or **Didelphis Cancrivora** (*Philander cancrivorus*) is smaller and darker-hued than the common *opossum*. Even more than in the case of other species is the tail of the *crab-eating opossum* his dependence for clinging to branches of a tree while feeding on the fruit. It is specially fond of crabs, and quite successful in fishing for them in Brazilian waters. The flesh of the *opossum* is too rich for the untrained palate, but by those who have acquired the taste, it is said to resemble the English hare. **Merlin's Opossum** (*Philander dorsigerus*, or *Didelphis dorsigera*) has such a poorly-developed pouch that very soon does it transfer its young to its back, where they hold on partly by their feet, but much more by twisting their little tails around that of their parent. This is a Surinam species, and is quite handsomely colored in gray-brown, or whitish-yellow, with darker brown markings around the eyes and forehead. The **Yapock** (*Cheirionectes yapock*, or *variegatus*) is entirely aquatic, cannot climb, but is a remarkable swimmer and diver, has webbed-feet (the front ones used for burrowing being less so), and squirrel-like cheek-pouches. It is found in Brazil and in Guiana. There are many species belonging to the marsupial

LARGE-BROWED WOMBAT.

genus, some of which may be briefly mentioned as follows:

The **Didelphis Quirca Opossum** is a Brazilian species differing mainly in smallness of size, from the Virginia opossum. They are here referred to because one reading accounts of life in Brazil is likely to see them mentioned. The **Didelphis Elegans** and **Murina** are pouchless and carry their young as do the Indian women. The **Didelphis Imaavida** and the **Noctivaga** belong to the Peruvian forms which are strictly vegetarian. The **Broad-nosed Wombat** (*Phascolomys platyrhinus*) is notable for the broadness of its nose, its size and its yellow coat of fur. The **Broad-fronted Wombat** (*Phascolomys latifrons*) has a hairy muzzle and goes clothed in silken hair. The **Kaponne** (*Cuscus orientalis*) is considered by the natives a delicacy. The **Vulpine Phalanger** (*Trichosurus vulpinus*) eats like a squirrel and looks like a fox. **Albert's Petaurus**, and **Bernstein's Petaurus** (*Petaurus albertisii*, and *bernsteinii*), after carrying the young about for a long period safely housed in their pouches transfer them to their backs like the s p e c i e s known as Merlin's opossum above noted. The **Dromiciæ** (*Dromicia concinna*, and *neillii*) are very small, have the skin of the body extending down the legs as far as the ankle, and suggest a connecting-link between the parachute-carriers and the species which lack

AUSTRALIAN BEAR.

this appendage. The **Tailed Dromicia** (*Dromicia caudata*) is so named from its vast superiority over the other species in the matter of a caudal appendage.

PHALANGER (*Phalangista vulpina*).

The **Great-browed Wombat** (*Phascolomys latifrons*) has the general appearance of a bear of small size. It is a burrower, nocturnal in its habits and feeds upon roots and herbs. It is a tailless animal and its large head seems to have been purchased at the expense of the rest of his anatomy. It has its strongest likeness to the bear, possibly, in its awkwardness, as it is pouched like the marsupalia, toothed like the rodentia, has many anatomical features in common with the badger, and is as ridiculous in appearance, movement and amiability as a trained bear. As a mining engineer it a b s o l u t e l y threatens the permanency of human structures, by communicating subterranean passage ways which suggest the difficulties encountered and the success achieved at Mt. Cenis.

The **Australian Bear**, or **Kaola** (*Phascolarctos cinereus*), is tailless but keeps the ventral pouch; it is arboreal, nocturnal, and produces the same general impression as a bushy, little puppy, whose furzy ears would occupy one's attention but for the quizzical expression of its face. It is specially interesting to the naturalist because it seems to be a connecting link between the opossum, the kangaroo, the bear and the sloth.

The **Spotted Cuscus** (*Cuscus maculatus*) is found in New Guinea, and on the Molucca Islands, but not elsewhere in the same sea. It is hunted alike for the table and for its fur, which is generally some shade of white, with black spots arranged as if forming a regular pattern. It is arboreal, and nature seems to have gifted its tail with such tenacity as to avoid any occasion for conferring locomotive power on its feet. It is another of the animals which will "play possum," and as it will never move while conscious of observation, children sometimes capture it by wearing out the muscles even of so prehensile a tail, for the children will sit and watch the animal until its muscular force is exhausted; naturally the natives would find more expeditious ways of bagging their game.

FLYING ARIEL.

The **Great Flying Phalanger** (*Petaurus australis*, or *flaviventer*,) belongs to the family which is supplied with a parachute and a prehensile tail. Its color is so variable that a description given from one animal might be contradicted by the next specimen subjected to examination. The species called **Sugar Squirrel** (*Petaurus*, or *Belideus sciureus*,) possibly because it is no squirrel at all, is very beautiful as an illustration of what an animal Beau Brummel can achieve in the combination of furs. It is nocturnal, arboreal, and makes nothing of a flight of thirty or forty feet when exercising as a performer on the trapeze. It is playful, gregarious and not unable to endure captivity. Its fur is a soft brownish-gray, the under parts whitish, a broad black stripe down the spine, and a tail whose length and thickness ought to satisfy the most exacting wearer of fur.

TASMANIAN WOLF. (*Thylacinus cynocephalus*).

The **Ariel Flying Petaurus** (*Petaurus ariel*) is so graceful in its progress through the air as to have deserved, as well as to have received, the name of the daintiest imagination of the world's greatest poet, who seems to have lavished upon the "Tempest" all the beautiful fancies which could find suitable place, even when the scene was to be cast in a wholly unreal world, and the poet was to be limited solely by the quality and fecundity of his fancy.

The **Taguan** (*Petaurista taguanoides*) is mentioned because travellers seem frequently to forget that what is familiar and commonplace to them, may be wholly meaningless to their readers. The *taguan* is found in New South Wales, and its more than three feet of body and tail render it eminent in the matter of size among the flying phalangers. Its habits are nocturnal, so that by day it is easily captured, and it is greatly esteemed by the natives for food.

The **Australian Ariel** (*Petauristus*, or *Belideus breviceps*,) is smaller, but otherwise is very much the same as the ariel.

MOUNTAIN KANGAROO (*Petrogale xanthopus*).

The **Yellow-bellied Petaurist** (*Belideus flaviventer*) is arboreal, and awkward and helpless when on the ground. Its soft short fur is chinchilla-like, and added to the savoriness of its flesh subjects it to many an attack from the natives of New South Wales.

The **Flying Mouse, Little Petaurist,** or **Opossum Mouse** (*Acrobata pygmœa*), is about the size of our common mouse, and when at rest its white-trimmed umbrella is folded away so as to result only in undulations of white fur. It is not a real flyer, the parachute lending temporary support and not being usable as wings; the peculiar feathering of its tail increases its

BANDED BANDICOOT.

resemblance to that of a bird. It enriches the fauna of New South Wales.

The **Beaked Tarsipes** (*Tarsipes rostratus*), of Australia, is noticeable as an insect-feeder. Its coat of gray is striped like Magruder's pantaloons, save that it wears three black stripes, not one.

The most common species of *bandicoots* in Australia are **Perameles Macroura** and **Perameles Obesula,** and they require no description since the mere name is sufficient for those who read of life in that Oceanic continent. The species in New Guinea are named **Perameles Doreyanus** and **Perameles Longicaudatus.**

The **Pig-footed Bandicoot** (*Chœrops castanotis*) belongs to South Australia, and differs only in the skill with which it builds its nest in the underbrush.

The **Banded Bandicoot** (*Perameles fasciata*) is Australian in habitat and very peculiar in its appearance. Its arched, mouse-like body is covered with alternate yellow and black hairs, and its hind-quarters are accentuated by tolerably broad bands of black. The tail is mouse-like except for the black hair that covers the upper portion. Its legs and feet look as if they had been borrowed from some one of the tribe of birds; its ears are large and erect, and its head reaches a sharp point in its descent into a snout.

The **Long-nosed Bandicoot** (*Perameles nasuta*) is prevailingly brown in color, although the upper parts are shaded with black, and its sides exhibit a tendency to purple. The *bandicoots* move about with a gait which seems to be

LONG-NOSED BANDICOOT.

a compromise between a walk and a jump, but which carries the animal along quite rapidly, although not very gracefully.

Hunting the kangaroo is an exciting sport because of the remarkable fleetness and endurance of the animal—a run of eighteen miles not being considered unusual. Among hunters the kangaroo is pursued by kangaroo-dogs, a special breed trained for the purpose. These do not escape unharmed, for although the kangaroo is timid by nature, it lends emphasis to the warning "Beware the fury of a patient man." If near water, it will plunge in and, if followed by the dogs, deliberately seize them and drown them by holding them under water with its hind feet. If on land it will get a tree at its back and fence most dangerously with its hind feet. If the hunter be on foot, it will disregard the dogs and make an attack upon the man. By the Dingos it is captured not only by means of pitfalls, nets, and snares, but in a manner having some resemblance to an ambuscade. Forming themselves into large parties, the natives will surround a herd of kangaroos, and then some of them advancing and throwing their spears will drive the frightened animals almost into the arms of another party lying in concealment, which in turn will drive them in some other direction, until finally the whole of the herd has been killed. Often when hotly pursued, the kangaroo has been known to deliberately throw its young into the bushes, so that they at least might escape.

The **Bridled Kangaroo** (*Macropus frœnatus*) and the **Crescent Kangaroo** (*Macropus lunatus*) belong to the long-nailed kind and differ from each other and from the species which will be described later, as the type *unguifer*, simply in coloring.

The **Papuan Kangaroo** (*Macropus papauna*) is of interest to the naturalist because its discovery upset the conviction that the fauna of New Guinea included no representative of the family.

The **Antelope Kangaroo** (*Halmaturus antilopinus*) is large but takes its name rather from its deer-like skin. It belongs to Australia.

The **Brush Kangaroo** (*Helmaturus bennettii*) is Tasmanian, and valuable alike for its flesh and its covering.

BRUSH KANGAROOS.

The **Nailed Kangaroo** (*Macropus unguifer*) is a typical variety of the smaller kangaroos, whose tail seems to be supplied with at least a rudimentary nail. It is found in the Austro-Malayan Islands and illustrates transitions of form.

GIANT KANGAROO.

The **Great Kangaroo**, or **Giant Kangaroo** (*Macropus major*), belongs to New South Wales, and is quite graceful when making its immense leaps of as much as twenty feet. It is warmly clad in brown yellow hair, and attains the height of six or seven feet. Like the rest of its tribe, when leaping, it does not use its fore legs, but jumps like men in a sack-race. It has a special adaptation of its incisor teeth which enables it to clip the thinnest blades of grass as though they were cut by shears. The natives in hunting it use no weapon but a club, although the *kangaroo*, if desperate, will sometimes seize dog or man in a bear-like embrace, while it uses its hind feet as feed-cutters.

HUNTING THE KANGAROO WITH BOOMERANG.

The **Whallabee** (*Macropus ualabatus*) is quite numerous in New South Wales, and attains very great size. Like the **Tasmanian Whallabee** (*Halmaturus billardieri*) and the **Padlemon** of New South Wales (*Halmaturus thetidis*), it is gregarious and lives in large herds. The **Woolly Kangaroo** (*Macropus laniger*) is often called the **Red Kangaroo**. This species uses its tail simply as the third leg of a tripod, but when jumping seems to depend entirely upon its legs.

The **Rock Kangaroo** (*Petrogale penicillatus*) can run up and down the precipitous rocks as if he were a monkey, but sometimes sacrifices himself by basking in the sun, possibly occupied with fancies compared to which "The Reveries of a Bachelor" would be the baldest prose, and the idyls of celebrated poets but unsuccessful attempts to give to airy nothings a habitation and a name.

ZEBRA WOLVES PURSUING A KANGAROO.

The **Hare Kangaroo** (*Lagorchestis leporoides*) has a short muzzle, close

curly hair and very handsome markings. The upper coat is some hue resulting from combining white and black and cream color; the feet are variegated,
and each eye, as well as the neck, is circled with some shade of yellow. It

"squats in a form,"
and from its rapid-
ity when moving,
furnishes as much
sport as the "regu-
lation British
hare." It is very
successful as a
leaper, whether the
jump' be a stand-
ing one, a running
one — on a level,
or straight up into
the air.

The Hare
Kangaroo of New
Guinea (*Dorcopsis
luctuosus*) is a mel-
ancholy little crea-
ture, about twenty
inches in length.

BANDED ANT-EATER.

It is of an ashen color, having the hair on its throat combed forward, frill-like.

ZEBRA WOLF.

The Burned Tree-
Kangaroo (*Dendrologus
inustus*) is noticeable for an
appearance of having been
partially singed, and of having survived to tell the tale,
as well as to wear a tail.

The Great Red Kan-
garoo Rat (*Æpriprymnus
rubescens*) is hunted in New
South Wales for its flesh.

Mueller's Hare Kan-
garoo (*Darcopsis muelleri*)
is small, chocolate-colored,
neck hair pointing forward,
and an inhabitant of Mysol.

The Kangaroo Rat
(*Hypsiprymnus murinus*) in
temper and appearance re-
sembles the smaller kangaroos, but its movements are
those of a rat whose legs have not been built on a uniform pattern. It
would be harmless except for its inability to resist the temptations of a potato
hill or a melon patch.

The **Tree Kangaroo** (*Dendrologus ursinus*) has fur so dark as to take its name from the black bear, and a tail so long and stout as to insure his remaining on a tree so long as he cares to do so, all that is necessary is to retain his balance.

The **Banded Ant-Eater** (*Myrmecobius fasciatus*) is pouchless, but has long hair which serves as a coverlet for the young. Its fur is mostly brown, growing black as it approaches the haunches, yellowish-white on the under portions of the body, and fawn-colored on the shoulders; the back is ornamented by numerous white stripes running crosswise. It has the tongue as well as the palate of the ant-eater; is easily domesticated, and makes quite a pretty-looking pet. Its habitat is the western coast of Australia.

Phascogale, the Tapoa, Tafa, Brush-tailed Phascogale (*Phascogale penicillatus*), is a beautiful but deceptive little creature. It dresses in long, soft, woolen fur, gray above and white below. The tail is half the length of the body and the greater part of it is covered

TASMANIAN DEVIL. (*Dasyurus ursinus*).

by black hair, which at the tip culminates into a tuft. The *phascogale* can climb anything but a smooth wall, is fearless, bloodthirsty, and always rapacious. It is arboreal in its habits, and is somewhat "handy with its flippers" in its own defence. It belongs to Australia.

The **Pouched Mouse** (*Antechinus apicatus*) has a rudimentary pouch, and a rudimentary tail. It is insectivorous, and like the **Yellow-footed Pouched Mouse** (*Antechinus flavipes*) is very common in New South Wales and in Australia. Both of these *pouched mice* are about the size of the common house-mouse, and are arboreal. The *yellow-footed* species is dark gray above, chestnut on the sides, and white on the under parts of the body. The **Woolly Dasyure** (*Antechinomys lanigera*) belongs to Central Australia where its three-inch body, and five-inch tail are frequently seen. It is mouse-colored above and white below, and has a tuft of wool as the crowning glory of its tail. The pouch is lacking, and the mammæ rudimentary.

The **Tasmanian Devil** (*Diabolus*, or *Dasyurus ursinus*,) has finally been greatly reduced in number to the gratification of the farmers of Van Dieman's Land. Although smaller than the Tasmanian wolf, its stoutness, pugnacity and unreasoning ferocity render it worthy of its name. It dresses in black; relieving any sombreness by stripes, bands, or spots of white. It is doubtless the most uniformly ill-tempered creature in the animal world, as rage seems to be its normal condition. The fact that in the differentiations of species it went one way while the kangaroo went another is to be noted.

ZEBRA WOLVES HUNTING.

The **Zebra Wolf**, or **Dog-headed Opossum** (*Thylacynus cynocephalus*), is not strictly an *opossum*, for its hind feet lack thumbs, the tail is hairy and non-prehensile, and it has too few incisor teeth in each jaw. It is called the *Tasmanian wolf*, the *Australian tiger*, the *Zebra wolf*, and the *Australian hyena*. It is carnivorous like the wolf, to which it has many other resemblances. Its body slopes forward in consequence of its hind legs being longer than the fore legs; its elongated thick muzzle is almost cylindrical, its tail broad at the base tapers to a point, and it dresses itself in gray indulging, however, in black stripes across its back and hind legs. It is very destructive to flocks and is hence anything but a favorite with farmers. Its

DASYURE.

digestion is sufficiently remarkable to admit of its competing with the goat, or the ostrich, since it has been known to eat the porcupine, quills and all. The animal is nocturnal in its habits and specially particular about making its home wherever the light of day cannot penetrate. The animal has only rudimentary marsupial organs.

The **Dasyure** (*Dasyurus viverrinus*) is regularly dark-brown, inclining to black with irregular white spottings. It is a nocturnal animal, lives in the hollows of trees, and takes its name from the fact that its tail is hairless as well as non-prehensile.

EDENTATES.

The *Edentata* show a great advance in organization and have an intestinal placenta, a distinct nervous system, and a larger sized brain. They are destitute of front teeth as well as of other teeth. The wide differences between the animals belonging scientifically to this class has special interest because it illustrates the many accidental variations of an evolutionary scheme.

The **Edentata**, or **Toothless Animals**, include the armadillos, the sloths and the ant-eaters. The **Megatherium** is the fossil representative of the family and exhibits characteristics which fit it for being the common ancestor of progeny whose appearance, structure and habits render them as unlike as blonde and brunette coloring, or the differences of temperament exhibited by different members of a human household. In number and structure the teeth of the *megatherium* will be found identical with those of the sloth, but they are four-sided or prism-like in form. The long legs ended in feet whose size might suit the mastodon. This fact is understood that the family had not yet been reduced in size and in appetite sufficiently for it to live in a tree.

PEBA, OR NINE-BANDED ARMADILLO (*Tatusia novemcincta*).

The fossil armadillo, **Glyptodon**, has been named **Macrotherium, Limognitherium** and **Ancylotherius**.

The **Peba** (*Tatusia novemcincta*) is nine-banded (sometimes eight-banded) and is the only armadillo in the United States. The **Eight-banded Peba** (*Tatusia octocincta*) differs only in the number of its bands.

The **Tatons**, or **Giant Armadillos** (*Priodonta gigas*), are represented in South America by five species: the *Cachecames*, the *Cabassous*, the *Apars*, the *Priodonta* and the *Enconbettes*. Like the other members of his family, he never takes his armor off, nor doffs his helmet. The mail on his body consists of bands of shell which admit of the greatest freedom of motion; he carries a triangular headpiece, or a large buckler on his shoulders, and a carapace over his haunches, while his tail is supplied with a series of bony outside rings. His feet are built upon the principle of a steam-plough so that when the *armadillo* starts for the subterranean regions nothing but smoke is fast enough to catch him. Its body is upward of three feet long, and the tail adds another foot and a half. The tail is used by the natives as a trumpet, but whether as a symbol of "the last trump" is doubtful, as the *taton* will disinter bodies unless the graves are walled. As a burrower the *taton* would "put to the blush" Shakespeare's mole-like ghost.

The ordinary **Armadillo**, or **Poyou** (*Dasypus sexcinctus*), is relatively small, not exceeding a foot and three-quarters. It is tireless as a scavenger, and of great service in the hot countries where it lives. The **Armadillo of the Pampa** (*Tatusia tricincta, Apara*, or *Mataco*,) is diurnal in its habits, and is protected by armor, of which only three bands on the body are solid, so that the *armadillo* can readily roll itself into a ball when needing protection. The **Tatouhon** (*Tatusia septemcincta*) or seven-banded *Peba*, is in size intermediate

PICHICIAGO—TWO-THIRDS LIFE SIZE (*Chlamydophorus truncatus*).

between the common armadillo and the "giant." The **Pichey Armadillo** (*Tatusia minuta*), can live for long periods without requiring water. The **Tatoway** (*Xenurus unicinctus*) is singular from wearing no armor on its tail. The **Pichiciago** (*Chlamydophorus truncatus*) is the Chilian armadillo. The tail-piece suggests that the original plan was suddenly given up, and a piece of armor used to conceal the incompleteness. The body, except the head, back and haunches is furred like that of the mole.

The **Collared Sloth** (*Bradypus torquatus*, or *triadactylus*,) is, in point of size, the leading member of the three-toed sloths. It is covered with long, orange-colored hair, which changes to yellow on the top of the head, to red or reddish gray on the breast, and to black in the wide collar which graces its neck. Its head, like that of its family, is relatively small and quite round, and its black face is clear-shaven. His long claw-armed legs are so joined to his skeleton as to render it easy for him to hang without effort with his face to the sky and his back to the ground; his coloring, likewise, so increases his invisibility as to suggest another adaptation of the creature to its environment. Its home is in the tree-top, and it will never descend unless starvation drives it to a new source of supplies. Its paws have great strength, and if the animal is attacked by a dog while it is on the ground it will turn upon its back, and if it once succeeds in getting hold of the dog it will squeeze and claw it to death. The sloth is nocturnal, and while not preternaturally frisky even then, manifests more signs of life than during the day. When struck, it moans, but gives no other evidences of resentment. The sloth has become a symbol for sheer laziness, which

AI (*Two-toed sloth*).   UNAN (*Three toed sloth*).

it undoubtedly deserves because of its indifference under all circumstances, scarcely exhibiting enough spirit to retreat when danger threatens it. I believe no specimen has ever been brought to America, and in my travels abroad I have seen only a single one. This I happened upon in the beautiful zoölogical garden at Rotterdam, and which the keeper declared to me never moved except when it was feeding. The primitive sloth, whose remains are found in South America, was one

THREE-BANDED ARMADILLO (*Tolypeutes tricinctus*).

of the largest of land animals, measuring some twenty-five feet in length.

The Ai (*Arctopithecus ai*, or *flaccidus*,) disturbs the silence and increases the weirdness of the Brazilian forests by its oft-repeated cry of *ai, ai, ai.* The *ai* is a three-toed sloth; it is grayish-brown in color and is streaked on the back. Its coloring, together with the algæ which abound upon it, serves as an assistance to that power of making itself invisible, though present, which is a characteristic of the whole sloth family. Thus identified in color with the tree, which it has selected, there is required an experienced eye to discern the *ai*. If missiles are thrown at it, it will patiently endure its punishment,

having learned that " it is better to bear the ills we know, than suffer those we know not of." The hunters shoot off the branch, and as the *ai* clutches like Macbeth at the empty air, its whereabouts is no longer an unbetrayed secret. As the *ai* and other sloths feed upon leaves they can well afford to pass their time "up a tree" and let more grasping creatures energize for a living. It is known that a family of sloths passed several years without once coming down from the tree of which they had taken possession. This close confinement seems the more wonderful, if we bear in

GIANT ARMADILLO.

mind that the sloth does not roost, or, like the marten, live in nests, but that it hangs by its claws to the branches, letting its back and the full weight of its

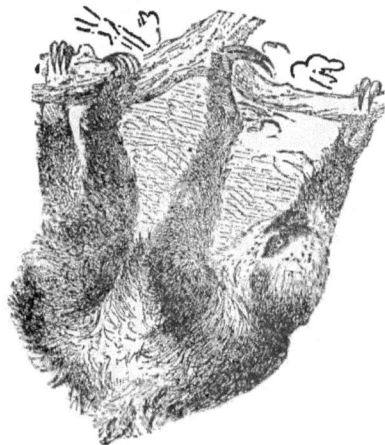

TWO-TOED SLOTH.

body remain without support. It has been said that the more miserable the conditions of life the more strongly do creatures cling to it, and this truth is strikingly illustrated in the case of the sloth. Its existence seems sufficiently limited to furnish no exciting pleasures, and yet the creature displays not simply the greatest tenacity of life, but the greatest power of resisting all efforts of the grim reaper—Death. Possibly the Mohammedan heaven, to which alone it can look forward, seems to promise nothing that cannot be obtained on earth. The sloth, as a rule, lives alone—at times, in a family group; he is apathetic in his passions and affections, as well as in his individual desires. It is too phlegmatic to be combative, but if forced to act in self-defence throws itself upon its back and endeavors to embrace and strangle its adversary. He is the very type of the ennuyed, monosyllabic, empty-pated " blarsted Britisher;" nothing can excite in him any surprise or interest, and no suffering seems to exert more than a moan of

31

astonishment at so unseemly an interruption of the *dolce far niente* of so harmless a being. Dicken's fat boy, "Joe," was not so unruffled, so wholly limited to the pleasures of eating and sleeping as is the sloth. In this rushing, bustling, unresting age of ours, one can almost find relief by giving himself to a few moments of comradeship with the sloth, even though the sloth be in the pages of a book—for the sloth is quite as much alive there as in his natural condition. Walt Whitman wanted to "lean and loaf," but it would be much more restful to buy a sloth and watch its entire contentment with a masterly inactivity. As has been said, the *ai* makes night musical by uttering from time to time the most hair-raising and supernatural moans or cries, which sound as if a lost spirit was writhing in extreme agony, or as if some demented creature,

GREAT ANT-BEAR (*Myrmecophaga jubata*)

lost in the pathless forest, was otherwise uttering the most lugubrious lamentations.

The **Two-toed Sloth,** called the **Unau** (*Cholœpus didactylus*) when of the Brazilian species, and the **Central American Two-toed Sloth** (*Cholœpus hoffmanni*), save the waste of energy required even in uttering a moan. The *unau* is about two and a half feet in length, and is grayish-brown in coloring. It is quite a remarkable climber, wearing the soles of its feet upon the inside. Locomotion on the ground is very laborious, as the *unau* has to lie on its back and pull itself along in a hand-over-hand fashion.

The **Great Ant-eater,** or **Ant-bear** (*Myrmecophaga jubata*), is a queer-looking creature

GREAT ANT-EATER AND ARMADILLO LORICATA.

covered with long brown hair which at times is sprinkled with white or gray. From the throat a singular blank triangular band runs across the shoulders. Its tail is so liberally provided with long hair that the *ant-bear* uses it for an umbrella whenever it wishes to sleep under cover. Its great size (five to seven feet), its long, curved beak, its feet so little suited for rapid locomotion, all unite to render the animal noticeable even though not handsome. It is not unfrequently called the *tamanoir*.

The **Tamandu** (*Tamandua tetradactyla*) is smaller, its hair is not allowed to be so long and unkempt, and the fore-shortening of its head makes it a much more handsome creature. The **Little Ant-eater** (*Coclothurus didactylus*) is arboreal and in many ways suggests the squirrel. Its tail is prehensile, it sits on its haunches and uses its front paws; its foot-and-three-quarters of length is provided with a very fine, silken fur; and like all its tribe it has a somewhat more than ample provision of tongue.

Among the *edentates* is the **Aard-vark, or Earth-hog** (*Orycteropus capensis*), which though lacking the armor of the armadillo, is nevertheless an ant-eater who displays the same taste in diet, and the family skill as a grave-digger, a miner, and a burrower. Southern Africa, that new wonder-land, is the habitat of the *aard-vark*. It is about five feet in length, the tail representing a foot and three-quarters. Its fore legs are very powerful and furnished with hoofed claws, or clawed hoofs. For the world of the ants, the *aard-vark* is a terror more real and dangerous than the prowling burglar and safe-blower whose praises are celebrated in the daily papers. For with the coming of nightfall, the *aard-vark* sallies forth and with great directness and singleness of purpose takes his way

LONG-TAILED MANIS.

to some ant-hill. Once there he demolishes the walls of the structure, and with his viscid tongue sweeps the ants by quantities into his gluttonous stomach.

The **Long-tailed Manis, or Phatagin** (*Manis longicauda*), is an inhabitant of Africa, and his armor is very beautiful while at the same time it can be made to serve as a weapon of offence. The fact that the long tail claims three-fifths of a length of five feet makes the effect of the scales all the more striking. The termite is the special quarry of the *manis* whose methods of swallowing the ants has already been explained when we were considering the insects. Another species of *manis*, the **Short-tailed** (*pentadactyla*), is found in India and Ceylon. It is gentle and can be converted into a pet. Both species of *manis* can burrow to the most surprising depths, which they are able to do by reason of their long and formidable claws. They are thus provisioned by nature to enable them to ravage ants' nests, and they are armored as a protection against the sharp jaws of such ants as are able to destroy much larger creatures than the *manis*.

The **Tricuspid Pangolin** (*Manis tricuspis*) is African in its habitat, and has three rows, each seven-banded. The **Giant Pangolin** (*Pholidotus giganteus*) is about five feet in length, is armored throughout, and flourishes greatly on African soil. The **Indian Pangolin** (*Pholidotus indicus*) is, likewise, a giant, and is represented abroad as well as at home, by the **Chinese Pangolin** (*Pholidotus dalmanni*).

RODENTIA.

The **Rodents,** or **Gnawing Animals** (*Rodentia*), form a very numerous class of mammals, ranking, as judged by their organization, next above the *edentata* or *toothless animals.* Though predominantly frugivorous, or at least insectivorous, they, in some species, have learned to become omnivorous. This statement furnishes the most suitable occasion for the remark, that the habits of animals, as well as of human beings, are greatly modified by long-continued civilization, so that none but a naturalist would at all times suspect a common ancestry. Hence, the reader must, so far as he can, divest himself of prejudices based upon an acquaintance with an animal in its domesticated state only, and leave his mind free and unbiased for the consideration of evidence, which it is the office of THE SAVAGE WORLD to submit. If one stops to reflect he will at once appreciate the many and marked differences between the best type of the American and the representatives of the Caucasian race who inhabit northern Africa; nay, he will at once appreciate the wide differences which distinguish the dweller on the northern Atlantic coast and the native of the Gulf States. Hence, he will be ready to see that the habitudes, resulting from long-continued domestication, will utterly separate an animal from his congeners, whose

ANT-EATER IN ATTITUDE OF DEFENCE.

lives and hereditary traits know nothing beyond the savage freedom of their natural state. But, still again, the *rodents* emphasize the fact that variety in the midst of unity lends strength to the scientific hypothesis of evolution, while relating this to the Genetic account of the creation. Finally, the succession, in order of time, strengthens the position that each class appears when the earth is ready for its services, and becomes extinct when the work which it had to perform has been completed. This extinction, be it noted, is like the creation of a new class or species, not accomplished by some sudden suspension of the laws which a beneficent and all-wise Creator has assigned to the inhabitants of our earth as the normal manifestations of their life. Hence, while we can even now find fossil progenitors of the *rodents*, they differ from their progeny, and the descendants represent the greatest variety of development. Those forms of the *rodents*, which still serve a useful purpose in rendering the earth more habitable for the classes above them, and for the successful dominion of man, have shown an adaptability which has enabled them to change with the conditions surrounding them, and to continue their really beneficent work, even though its execution involved new responsibilities, and demanded changes in the organization of the creature itself. The *rodent* is able to live everywhere and anywhere; to adjust himself to all climates, and to any regimen in the matter of food; its smallness of size and disproportionate courage, persistence and strength fit it to take care of itself under conditions which would prove fatal to the largest mammals, and its duty is limited to no one locality, no one geological, geographical or historical period. Dentition, or the structure of the teeth, is, as has been said, the accepted scientific standard of classification, and the *rodents* are evidently named, solely with reference to their dental peculiarities. The *rodents* have no canine teeth, for these would be useless to them. Hence, between the incisors and the canine teeth there is an unfilled gap. No living *rodent* has more than two developed inci-

AARD-VARK, OR EARTH HOG.

sors in each jaw. The typical number of molar teeth is six for each jaw, subject to increase in certain animals.

Furthermore, all the teeth are solidly rooted, and yet grow without limit, as the tooth-making supplies are perennial. The arrangement of the dentine (or soft material of the tooth), and of the enamel is such that, when we add the way in which the teeth come together, we have every condition for the constant presence of teeth always in good condition, and constantly sharpened by contact with each other. The molars, however, grow less readily, and in form are specially suited to the work which they are to perform. The *rodents* develop the cerebellum at the expense of the cerebrum, whose convolutions and complexities are few. It follows, therefore, that, while their intelligence is striking, it is an intelligence whose limits are very narrow. The stomach is simple in its structure, and the intestinal canal proportionately favored. The provision for acute hearing, clear sight, quickness of scent, and delicacy of touch is as notably great as one would expect from the lives which the *rodents* are to live. To illuminate all this verbal description, let the reader think of an example of each of the four families—the hares, the porcupines, the rats and mice, and the squirrels. The one fossil *rodent*, (if, as many naturalists believe, it is a *rodent*) is the **Mesotherium Cristatum**, belonged to the Pliocene Age, and dwelt in the South American Pampas. In the lower jaw, it had two small supernumerary incisors, the teeth were enamelled on all sides, and hence wore down more like those of the *equidæ* than like those of living *rodents*. The molars are ten above and eight below, four in the upper jaw and two in the lower, being simply pre-molars; they have no roots. The skull is large and stout; and the feet, though five-clawed, are hoof-like; in the latter particular, as well as in the gigantic size of the animal, the height of whose skull was a full foot, seem to relate the *mesotherium* to the *toxodontia*. Whether or not the *mesotherium* was the Adam of the *ro-*

EARTH HOG (*Orycteropus capensis*).

dents, it is quite certain that it represents a less differentiated type than is found in existing species. Its remains, like those of the *monotremes*, *marsupials* and *edentates*, will interest such visitors to our museums as care to make real to themselves the resemblances and differences between earlier forms and the varieties which have been evolved from these.

Passing now to existing rodents, we shall consider first the **Leporidæ**, or **Hare Family**, which has the two-fossil forms of **Palæologus** and **Titanomys**, corresponding to the two branches of the *hare family*—the *hares* proper and the *hare*-like animals. There is an extra pair of "milk" incisors in the upper jaw, and the enamelling is such that the teeth do not get the edge which is found in rodents, other than the *hares*. Externally, the *hares* are distinguished by length and mode of articulation of the legs, which unite to specially adapt them to their leaping mode of locomotion, and to their no less customary squatting for concealment, or for rest. The front paws are five-fingered; the hind paws, four-fingered. The palms and soles are furred, and the paws are not used as hands. The temperate and frigid zones are, as one would expect, the favored habitat of the family, although species are found in southern Africa and India.

HARES.

The genus *Lepus*, which includes the animals designated in common language by the names of *hares* and *rabbits*, is characterized by a peculiarity in the incisors of the upper jaw, which have behind them two smaller teeth of nearly the same form; by the flattened summits and transversely disposed plates of enamel of the grinders, of which there are six above and five below on each side; by a tuft of hairs on the inside of the cheeks; by the elongation of the ears, and by the abbreviation and recurvature of the tail.

The **Common Hare**, of England and the Continent of Europe (*Lepus timidus*), presents the following characteristics: the body is large, compressed and deep; the neck very short; the head of moderate size, convex above, broad and obtuse in front, the nose being depressed, the lips tumid or swollen and separated by a deep incision; the eyes very large, prominent and inserted laterally; the ears of the same length as the head, narrow, deeply concave and with the tips rounded. The hind legs are much the longer and have only four toes, while there are five on the fore feet; the soles of all the feet are covered with hair; the claws are rather long, slightly arched, compressed, somewhat acute, but those of the hind feet blunted in older individuals. The tail is very short and recurved. The fur is of two kinds, as in all the species of this genus; the longer hairs are very slender at the base, enlarged towards the end, recurved and intermixed with still longer straight hairs; the shorter, extremely fine and tortuous. On the feet the longer hairs predominate, and are straightish and rather stiff, on the ears they are short; on the nape of the neck they are wanting. The mystachial bristles are long, faintly undulated on two opposite sides, disposed in several series, the lower forming a tuft; five or six long bristles arise over the eye and some shorter ones beneath it. The hair on the lower parts is longer, on the tail soft and woolly. The fore part of the mouth within is covered with stiffish woolly hairs.

The upper parts are light yellowish-brown, mingled with a dusky color on the back and sides, and with gray on the hind quarters; the fore part of the neck and a portion of the breast are a dull, light yellowish-red, as are the feet and part of the flanks; the abdomen, inside of the thighs, and a large patch on the throat, are white; there is a whitish line over the eye, and a patch of grayish-white before it, the ears are pale yellowish-red on their anterior margin externally, dusky intermixed with yellowish-red on their anterior half, whitish on the posterior, with a patch of black at the end; internally with whitish hairs at the base, dusky at the middle of their posterior margin, reddish-white in the rest of their extent, except the margin of the tip, which is black. The tail is black above and white beneath, or rather behind, as it is recurved. On the upper parts the hair is grayish-white, sometimes pure white at the base, dusky beyond the middle, and yellowish-brown at the ends; the elongated slender hairs are black, but on the sides of the body and the lower parts, reddish or white. The fur or fine hairs are white, with the extremity dusky. The length to the ends of the tail is about twenty-six inches, the head five, the tail three and a half, and the ears about four and three-fourths inches long.

The *common hare* feeds entirely on vegetable substances, such as grass, clover, corn, turnips and the bark of young trees, sometimes inflicting great injury on the latter, especially in winter. Towards evening it comes abroad in quest of food, and continues to search for it during the night, in conformity with which habit, the pupils of the eyes are large and of an oblong form. It advances by leaps, and as its hind legs are much longer than its fore legs, it runs with more ease up an inclined plane than down a declivity, especially if this be steep. During the day it reposes in a crouching or half-sitting position in its form, which is a selected spot usually resorted to by it, amidst grass or ferns, or in the midst of shrubs.

Its sense of seeing and of hearing are extremely acute; its eyes, being directly on the sides of the head, take in a wide range, and its large ears can be readily turned in any direction, forward, outward or backward, so as to catch the slightest sound. Being in a manner defenceless, and having no burrow or fastness to which it may retreat, it must trust to its vigilance and great speed to enable it to elude its numerous enemies. The excellence of its flesh makes it liable to be destroyed. Moreover, its frequent occurrence, extreme timidity and great speed, render it a favorite object of the chase.

It is chiefly to the lower and more cultivated districts that the *common hare* resorts; but it is also found in the upland valleys, and on the slopes of hills at a considerable height. Timid and gentle as it is, it is by no means innocuous, for the injury it occasions to the young corn is often considerable. In winter it finds an abundant supply of food in the turnip fields, and sometimes visits gardens at night, especially when urged by hunger during a continued frost. It has been observed to cross rivers by swimming, and even to enter the sea for the purpose of gaining an island or point of land on which food was more abundant.

The female goes with young thirty days, and several times in the season produces a litter of from three to five young ones, which are born covered with hair, having their eyes open and capable of running. The young squat in the fields, remaining motionless, and are with difficulty perceived. Even the old *hares* are not readily driven from their forms, in which they will sometimes remain until a person is quite close to them, when they at length start off, exhibiting in their motions the haste and perturbation of extreme fear. The timidity of the *hare* is indeed proverbial, as is its propensity to return when wounded or even when hunted to its usual place of repose. Besides being pursued with hounds and shot for pastime, it is snared in its form or in the paths which it has made in the herbage. Its flesh is superior in flavor to that of the white *hare* or rabbit; and its fur is used for various purposes, especially in the manufacture of hats. Large individuals weigh from nine to twelve pounds; the ordinary weight is about eight pounds.

The **White Hare, Variable Hare** or **Changing Hare** (*Lepus variabilis*), is considerably smaller than the common hare, which, however, it resembles in form, although it has the ears and hind legs proportionally shorter. The number of toes, and the nature of the fur are the same as in the other species; but the latter is softer and more woolly. In the middle of summer the head is reddish-brown, the lips and chin brownish-white, the ears dusky on their anterior half, grayish on the posterior with the tip black, the nape whitish; the general color of the upper parts is dull grayish-black, intermixed with reddish. The long hairs are gray at the base, then black, with a small portion of a yellowish tint, the tip black. Some of the longer hairs are entirely black. The fine hair or under fur is light-gray at the base, pale reddish-brown at the end. The tail is grayish-white. The lower parts are grayish-white, as are the legs over the greater part of their extent; the toes brown, the claws dusky.

In September the colors begin to assume a paler tint, many of the dusky hairs having disappeared. In October the change is farther advanced, and towards the end of the month the muzzle, hind neck and feet are white, though there are spots and patches dispersed here and there. In December the fur seems entirely white, but has an intermixture of long blackish . hairs on the back; the anterior external part of the ear is brownish and its tip black. The under-fur is light bluish-gray at the base, pale yellowish or cream color towards the end. In this species the hair is almost always changing; in April and May there is a general but gradual shedding, after which the summer colors are seen in perfection; towards. the middle of autumn many new white hairs have been substituted for colored ones, and by degrees all the hairs and fur are shed, and renewed before the end of December, when the fur is in the perfection of its winter condition, being closer, fuller and longer than in summer.

The *white hare* is not uncommon in the middle and northern divisions of Scotland, residing in the valleys of the Grampian and other mountainous tracts, but not ascending the hills to their summits, although in summer it keeps on the ferny slopes. This species also occurs in some of the northern parts of England. In winter, it descends to the bottom of the valleys, but never visits the lower districts. It does not burrow, but conceals itself among the ferns or heath, often in stony or rocky places. Its flesh is whiter than that of the common hare, generally leaner and therefore less esteemed. It is said to be easily domesticated if taken young, and to exhibit less timidity and more playfulness than the common hare.

The **California Hare** (*Lepus californicus*) is long and slender, it has a long tail and ears, it is reddish-brown above, but the under parts are yellowish-white. The average length is less than two feet. It differs from its relatives in its degree of fleetness and in its timidity, in which qualities it is their successful rival. This hare is to be found in California and the southern part of Oregon.

The **Sage Rabbit** (*Lepus artemisia*) from the west and from the plains of Mexico and Texas, cannot be satisfactorily distinguished from the last species.

The **Jackass Rabbit**, or **Texan Hare** (*Lepus callotis*), is so named from its very long ears, which measure about five inches, though the animal is rather smaller than the European hare. It is yellowish-gray above, waved irregularly with black; the upper part of its tail is black, sides gray, and it is a dull whitish below; its nap is a sooty-black. It is found in Mexico, Texas and Oregon. The long and slender legs indicate rapid locomotion and a capacity for making long leaps; it is a solitary and not very common species, and has not been found in California.

Rabbits at home will sit in their doorways even when they have been driven thither by the hunter. The dead are always removed from the warren by their companions. In spite of their timorousness, one of them has been known to kill two large rats which invaded the burrow with no good intent.

The **Tapeti**, or **South American Hare** (*Lepus brasiliensis*), is happy in the possession of both hare-like and rabbit-like qualities. It is peculiar in its small ears, short stumpy tail, and in its littleness. It is found in Brazil and the adjacent countries.

The **European Rabbit**, or **Cony** (*Lepus cuniculus*), the lapin of the French, is about sixteen and one-half inches long, with the tail three inches in addition, and the ears also are about three inches long; the tarsus is shorter than in the hare; the general color grayish-brown above, white below, the back of the neck rufous, tail white below, blackish above, but pencilled with dirty-white; ears not tipped with black. Compared with that of the hare, the skull has the muzzle, inter-orbital space, and incisive openings narrower; the mammæ are five pairs, two pectoral, three ventral. In the wild state the *rabbit* inhabits Europe except the more northern portion, and northern Africa. It is thought to have been originally from Spain, but being hardy, to have been carried to most parts of the world. It is easily distinguished from the hare by its smaller size, grayish color, and short feet and ears; it also differs from the hare in its habit of burrowing.

Unable to escape from its enemies by speed, it seeks safety in deep holes dug in dry sandy places; living gregariously in what are called warrens, with an ample supply of food at hand, in places suitable for burrows, such as sandy heaths covered by a prickly furze. Remaining concealed during the day, they come out at twilight in search of food, and often do considerable mischief by digging up the newly-sprouted corn and gnawing the bark from young trees. The warrens are often of large extent, and a source of great profit from the flesh and skins of the animals which are caught in snares and traps, dug or drowned out, or hunted with dogs and ferrets.

They are very prolific, beginning to breed at the age of six months, and having several litters in a year, producing from five to eight at a time; the period of gestation is about three weeks, but as the uterus is double, there may be two distinct litters at an interval of a few days. The young are born blind and naked, in a nest lined with the mother's soft fur. Rabbits are said to live eight or nine years. They seem to have social laws, the same burrow being transmitted from parents to children, and enlarged as the family increases. It is estimated that a single pair of rabbits would, if unmolested, become the progenitors of more than one million two hundred and fifty thousand of their kind. To check this increase there is the persecution of man, and of carnivorous beasts and of birds. The ravages of the rabbits are more than counterbalanced by their flesh, which forms a nutritious and easily digested article of food, and by their skins which are used in making hats and are dyed to imitate more expensive furs. White and gray are the prevailing colors; in the silver-gray variety the hairs are white and black.

The **American Gray Rabbit**, or **Cotton-tail** (*Lepus sylvaticus*), is about sixteen and one-half inches to the root of the tail and twenty-six and one-half to the end of the outstretched legs; the fur and pads of the feet full and soft; on the back light yellowish lined with black, grayer on the sides; on the rump mixed ash, gray and black, pure white below; upper surface of tail like the back, below pure cottony-white; posterior edge of ears whitish, edges of dorsal surface toward the tip black, the rest of ashy-brown; fur lead-colored at the base. This is among the largest of the short-eared leporidæ of America, being the largest in the west and the smallest and coarsest-haired south; it is found almost throughout the United States, from the southern parts of New Hampshire to Florida, and west of the upper Missouri, being most abundant in sandy regions covered with pines. It also frequents woods and thickets, concealing itself in its form, in thick bushes, or in holes in trees or under stones by day, coming out at night to feed. It is fond of visiting clover and corn-fields, vegetable gardens and nurseries of young trees, where. it does much mischief. It does not burrow like the European rabbit, and comes rather in the class of hares; when pursued it will run with great swiftness and with doublings to its hole in a tree or rock. Though it will breed in enclosed warrens, it does not become tame, and has not been domesticated.

It is very prolific, or else it would be exterminated by its numerous enemies. It often runs into the hole of the woodchuck, skunk, fox or weasel; in the last three cases frequently being a victim to the carnivorous inhabitant of the burrow. It is hunted with dogs, shot from its form and caught in snares and traps; its flesh is much esteemed. It somewhat resembles the European rabbit in its gray color, but it does not change its dress like the latter; it is furthermore smaller and more slender.

The **Pikas** constitute the genus of the family leporidæ, which includes the tailless hares. The *pikas* have no visible tail, the ears are short and rounded, the hind legs short, the molars five in each corner of the mouth, the feet densely clothed in fur, except small naked pads at the end of the toes. The *pikas* are of small size, the largest not surpassing a Guinea-pig. They are found only in Alpine or sub-Alpine districts, where they live in burrows or among loose stones remaining quiet by day and feeding at night. The food consists of herbage of different kinds, which they store up in little piles in autumn for winter consumption. When feeding they often utter a chirping or whistling noise.

The **Alpine Pika** (*Lagomys alpinus*) is about nine and one half inches long, with long and soft fur grayish next the skin; general color above grayish-brown, below yellowish-gray; feet pale with a yellowish tinge; the ears margined with white. It inhabits Siberia from the river Irtysch to Kamtchatka.

The **Pigmy Pika** (*Lagomys pusillus*), from southern Siberia and the Ural mountains, is six and three-fourth inches long, of a general brownish tint pencilled with black and brownish-yellow; feet and under parts yellowish-white.

Other species are found in Hindoostan (*Lagomys ogotona*), some of them 6000 or 8000 feet above the level of the sea.

The **Rocky Mountain Pika,** or **Little Chief Hare** (*Lagomys princeps*), is about seven inches long; the general color is grayish above, yellowish-brown on the sides, and yellowish-white below. It is found along the Rocky Mountains from latitude 42° to 60° north. It frequents heaps of loose stones, coming out after sunset.

The **Polar Pika** (*Lagomys hyperboreus*) is the smallest known species; it is only five and one-fourth inches long, grayish-brown above, tinged with red on the head and sides; it lives in northeast Siberia.

The **Common Squirrel** (*Sciurus vulgaris*), the only species of the genus *sciurus* to be found in Great Britain, is one of the most beautiful and lively of the

British quadrupeds. Its form is compact, its body being of moderate length, rather full, with the back usually arched; the neck short and thick; the head of moderate size, with the fore part flattened; the nose prominent; the lips broad; the ears of moderate size; the eyes large and prominent; the anterior limbs of ordinary length and muscular, with four slender toes, furnished with large, arched, and much compressed and acute claws, and a rudimentary inner toe reduced to a mere knob; the hind limbs proportionally longer, with five toes, and similar but shorter claws; the tail nearly as long as the body and head. On the fore feet the lateral toes are nearly equal and much shorter than the middle toes, which also are equal. On the hind feet the first toe is very short, the second longer than the fifth, the third and fourth longest and about equal. The incisors of the upper jaw are rather short and chisel-shaped; those of the lower much longer, narrower and pointed. There are five grinders in the upper and four in the lower jaw, on each side.

When the fur is complete in autumn it is rather long, dense and soft, the ears fringed at the end with longish hairs; the tail bushy. The general color of the upper parts is brownish-red, minutely dotted with yellowish-gray, the hairs being whitish and marked with brown; the tail of a darker brown with a very small portion of the tip whitish; the lower parts pure white; the feet and a band along the side light red. The mystachial bristles dark-brown. The female is smaller than the male, and generally of a lighter color. In younger individuals the color is redder than in adults, in which it is seldom destitute of a gray tinge, owing to the minute markings above described; and I have seen some in which the gray predominated over the red. In April and May the hair of the upper parts assumes a singularly faded appearance, losing its gloss, and assuming a light yellowish tint. In the latter month the process of shedding begins, to be completed by the end of June, when the ears are destitute of tufts. It appears that the long hairs which fringe the ears are not proportionally longer than the rest until November, that then they gradually elongate, attain their greatest length in spring, and remain unshed until June. In the northern regions of Europe the gray color in winter is more decided, and the fur denser and of finer texture.

ALPINE MARMOT.

The agility of the *squirrel*, its lively disposition and beautiful form, make it a general favorite. It is amusing to watch it in its arboreal excursions, when one sees it ascending the trunk and branches with surprising speed, running out even on the slender twigs, always when in motion keeping its tail depressed, occasionally leaping from one branch to another, and when alarmed scampering away at such a rate that one almost expects to see it miss its footing and fall down headlong.

It feeds on nuts, beech-mast, acorns, buds and the bark of young branches; generally, while eating, sitting on its haunches with its tail elevated, holding the object between its paws, and dexterously unshelling the kernel, from which it removes the outer pellicle before munching it. It does not live all the time in trees, but frequently resorts to the ground, where it moves with nearly equal agility, leaping like a rabbit. The female produces three or four young ones about midsummer, which are deposited in a nest, formed of moss, fibrous roots, grass and leaves, curiously interwoven, and placed in a hole or in the fork between two branches. In autumn it lays up a store of provisions for winter,

BOBAC (*Arctomys bobac*).

but usually in an irregular manner, depositing nuts in different places in the ground and in holes of trees. When the cold weather commences, it becomes less active, and often dozes for days in its retreat; but it does not become completely torpid; and it has been seen abroad in the midst of a most severe snow-storm. If the weather be comparatively mild, it exhibits its usual activity, feeding on barks and twigs. The squirrel may be domesticated if taken young, and becomes an agreeable, playful and gentle pet. It is generally distributed through the wooded parts of the country.

The **Flying Squirrel**, or **Red Yaguan** (*Sciuropterus volucella*), is about a foot in length, half of this belonging to the body, and is provided with a membrane on each side, which extends as far as the wrist and ankle. It is not truly a flyer, for the membrane acts rather as a parachute, and when the squirrel wishes to rise it is compelled to leap upward so as to fall downward at the right angle. There is also a European species (*Sciuropterus volans*). The largest species is the **Taguan** (*Pteromys petaurista*).

The **Scale-tailed Squirrel** (*Anomalurus fraseri*) is a flying squirrel, but the scales on its tail serve the uses of a fifth foot. It is African in its habitat, as is also another species, the **Shining Scale-tailed Squirrel** (*Anomalurus fulgens*).

The **Red Squirrel** (*Xeres hudsonius*) is the American representative of the common squirrel of Europe. It is frequent in the Northern States, and is sometimes called the *hackee*, or *chickaree*.

COLORADO MARMOT (*Spermophilus citillus*).

The **Bobac** (*Arctomys bobac*) is yellowish-gray, with dark mottlings, the under parts, throat and tail being russet, and the muzzle silver-gray. Its habitat is northern Europe and Asia, but it seeks lower altitudes than the alpine species.

The **Woodchuck**, or **Ground Hog** (*Arctomys empetra*), has a great variety of names, not only its numerous popular nicknames, but also more than its share of scientific titles. One reason of his various names is no doubt found in the fact that he is very widely distributed, for he is found in the Canadas, as far south as the Carolinas, and as far west as the Rocky Mountains The body is thick and squatty, and the legs peculiarly short. The body is about eighteen

ARCTIC MARMOT.

inches long, the tail about four. The color varies in different individuals. The *woodchuck* does not go far from its burrow during daylight, but after sundown comes out to forage, hunting for grass, fruits and vegetables, often doing great damage to clover fields and kitchen gardens. They are solitary in their habits, not forming communities of more than a single family. They hibernate where the winter is cold, and are believed to fast during the entire period of their hibernation. Their burrows are on the slopes of hillocks, often near the roots of a tree. They extend from twenty to thirty feet from an opening, rising inside into a large room, which serves as the dormitory for the entire family, and as a nursery for the young ones. Farmers in New England sometimes flood them out, sometimes kill them with rifles, but more frequently catch them in steel-traps hidden by grass and leaves. In defending himself the *woodchuck* bites most severely, and is no mean adversary. Although his walk is plantigrade, that is, done on the soles of his feet, he occasionally climbs up trees or bushes to the height of a few feet, taking a sun-bath upon an outstretched limb. He cleans his face and smooths down his fur. His fur is of no value, and his flesh eaten only when one is pressed by hunger, or in search of a new sensation. Its taste is said to resemble very much that of pork, but is more decided. The *wood-*

EUROPEAN MARMOT (*Aretomys marmota*).

chuck is called, sometimes, the *Maryland*, the *European*, or the *Alpine marmot.*

The **Prairie Marmot**, or **Prairie Dog** (*Spermophilus* or *Cynomys ludovicianus*), sometimes called the **Wish-ton-Wish,** is about thirteen inches long; the upper parts are reddish-brown, mixed with gray and black; the under parts

PRAIRIE DOGS.

are of a dull whitish tint; it has cheek-pouches about three-fourths of an inch deep, and its body is short, thick and clumsy. It lives on the prairies of the Missouri and Platte rivers. It is found as far south, also, as Texas, New Mexico and the borders of California. In the colder parts of this region they hibernate, but remain active all the year around in the warmer localities. The *prairie dogs* are gregarious in their habits. They live in burrows. Before the entrance to each of these there is a little mound. The front hall slopes downward and inward, at an angle of about forty degrees. From this passages diverge sideways or upwards. At the end is a bed of dry grass. Several hundred of these burrows are often congregated into a village, called by the hunters and travellers a dog-town. The inhabitants of such a town often sit upon their haunches on these little mounds that form a sort of front door to their burrows, surveying the landscape o'er and uttering a sharp, short sound, called barking. As they bark they wiggle their tails, as though to say,

"What a great boy am I." If a person approaches they immediately dive into their holes, being notified of his arrival by the animals nearest his pathway. Presently a head here and there pops out, as though to discover whether the coast is again clear. The *prairie dogs* are happy-go-lucky little fellows, and spend much of their time visiting and gossiping. One very curious thing about these dog-towns is that rattlesnakes and burrowing owls are to be found in the houses with the rightful owners. Though these snakes and owls seem to live on good terms with the *prairie dogs*, there is evidence that they, at least occasionally, devour their entertainers. This has been explained by supposing that they eat only such animals as die in their holes, thus performing the office of scavengers. The *prairie dog* receives its name not from its appearance but from the resemblance between the sound it utters and the barking of a dog.

The **Beaver** (*Castor fiber*) is rapidly becoming extinct, as mankind has been too unrelenting in the warfare which it has waged upon the creature. The *beaver* is so clever a builder that he has received at least his full share of praise. His carpenter's outfit consists of chisel-shaped teeth, a long, scaled, convex, trowel-shaped tail, and webbed hind feet. After having cut down a tree, and used the same principles as those of a Maine lumberman, he cuts the wood into long, pointed timbers of small size, and upon these

BEAVERS BUILDING THEIR DAMS.

puts layers of stone and of mud; his dam, likewise, takes cognizance of the resistance with which it may be expected to meet, and has its form determined by the character of the stream. Having provided his dam (whose walls are not unfrequently several feet in thickness), he tunnels himself a residence which shall lie quite a distance back from the water, and which shall be provided with a double entrance to the water—one for egress and the other for entrance to the house, and unerringly digs deep enough to suffer no inconvenience from the ice. Having built its house it lives upon the "dormitory plan," making its rooms of the most generous size, and arranging its beds along the walls.

The *beaver* is common in the northern and north temperate latitudes of both Europe and America, but is very rare in the middle latitudes, and unknown in the south. They formerly abounded in England as far south as Berkshire, and some persons suppose that oral tradition still survives, relating to their existence in that country. Their bones are found, in some districts, in the accumulation of peat in the fens, and on marshy river borders. Until recently they were abundant in the Northern, Middle and Western States of the United States, as the large number of their dams, and of the beautiful level beaver-meadows, caused by the accumulation of soil and filling up of their ponds by alluvial matter, sufficiently indicate. The gradual clearing up and cultivation of the country has, however, banished them, mile after mile and day after day, from the haunts of intrusive and encroaching man, until the *beaver* is scarcely to be found at all on this side of the streams which have their springs among the roots of the Rocky Mountains. Even there, also such unwearied war do the wild trappers of the various fur companies wage against them, and so largely tempting have been the

TREE PORCUPINE (*Cercolabes prehensilis*).

sums paid for their spoils, that they are rapidly decreasing, and may ere long become extinct. It has been said, however, that the application of silk to the manufacture of hats, and the large use made in late years of plain felt, by causing a very material fall in the price of *beaver* has procured them such a respite, that they are again becoming numerous in places where they were a few years since almost extinct.

CAPYBARA.

The *beaver* colonizes like the ancient Greeks and the modern Europeans. The old move up stream where living is more plentiful, and the young go down stream, but the original home always retains tenants. When it builds an island lodge, it makes one straight, ascending pathway, and a second sinuous and abruptly descending. The bank lodges have already been described. The *California beaver* makes no dams: the *Missouri beaver* constructs slides on the banks. When building dams the *beaver* walks on its hind legs and carries the stone pressed against its breast. When the wood is at any distance it builds canals twenty-five feet in width and three in depth and "rafts its lumber."

32

The **Boomer**, or **Mountain Beaver** (*Haplodon rufus*), is substantially tailless, lives inland, and burrows in the dry land. Its habitat is Oregon, Washington and California. It adds to our illustrations of the changes which animals undergo to adapt them to their environment.

PORCUPINES, CAVIES, AGOUTIS, AND OTHER RODENTS.

The two sub-families of the *hystricidæ*, or *porcupines*, are the *cercolabina* and *hystricina*. The *cercolabina* is confined to America, and the latter is spread over the old world. In both sub-families, the collar bones are nearly perfect, attached to the sternum but not to the shoulder blade; the eye cavities are very large; the forehead very broad; the cheek bones destitute of an angular process on the lower margin; the molars four in each side of the upper jaw, and four in each side of the lower jaw; the dorsal vertebræ usually fourteen, and the lumbar, four; the feet short, body more or less armed with spines or quills, capable of being raised by muscles under the skin. The *cercolabina* live almost entirely in trees, and their feet have generally only four equal toes with long compressed and curved claws; there are sometimes five toes on the hind feet; the soles are thickly studded with small flattened warts; the skull short and broad, with a minute lachrymal bone forming no part of the lachrymal canal; the palate between the molars is on a lower level than the anterior portion; the molars converge in front, and are distinctly rooted, each having a fold of enamel on either side, the worn crown presenting two deep transverse cavities surrounded by enamel; incisors small; anterior and posterior clinoid processes wanting. This sub-family contains the genera *erethizon*, *cercolabes* and *chætomys*. The genus

EUROPEAN HEDGEHOG (*Erinaceus europæus*).

erethizon has a non-prehensile tail, short, thick, flattened, covered at the base above with hairs and spines, and on the under side and at the apex with stiff bristles; nostrils close together; feet short and broad; toes four or five, with long curved claws; hind feet with a distinct inner toe with claw, without any projecting semicircular lobe on the inner side; upper lip slightly notched, but with no naked mesial line; body stout and covered with a long and dense fur from which the spines project, limbs short and strong.

The **Capybara** (*Hydrochœrus capybara*) abounds in South America; it is killed for its flesh, is persecuted in the water by the crocodile, and hunted upon land by the jaguar, so that it must often wonder whether life is worth living. It is about the size of a sheep, has a large mouse-like head, small, round ears, great black eyes, a nose set off with formidable whiskers, a short neck, short legs, a coarse covering of russet hair. When in motion it has the appearance of a prize hog, but when seated on its haunches it looks like nothing but itself.

The **Patagonian Cavy** (*Dolichotis patagonica*) is a burrowing animal, but sometimes takes possession of the excavations made by other creatures, instead of constructing its own home. It is found, sometimes, a long way from its retreat; two or three generally go together on these rambles. In its manner of running it imitates the rabbit; but in spite of its long limbs it is not particularly fleet. Once in a while, though not very often, it squats down to rest. It is ever on the lookout for enemies, yet bold enough to seek its living by daylight. It is easily distinguishable from other *cavies* by its long and well-developed eyelashes. It generally produces twins, whose cradle is the parental burrow. When its flesh is cooked, its whiteness makes it look very inviting,

PATAGONIAN CAVY.

but experiment diminishes its attractiveness, by proving the flesh dry and almost flavorless. A careless observer might mistake this animal for a hare, on account of its long legs, long cocked-up hairs and diminutive tail. It is much larger than the hare and weighs from twenty to thirty pounds. Its fur is pleasant to the touch, brown on the back and fawn-colored upon the sides. It is found in such parts of Patagonia as have a desert character, and these rodents hopping one after another give life to a rather dreary landscape.

The **Guinea Pig** is no pig at all, neither does it come from Guinea, but from Guiana. When eating it generally sits upon its haunches and uses its fore paws as if it was a squirrel. It is pretty to the eye, and readily domesticated, but as a pet it is dull and uninteresting. Its usual coloring is red, white and black (not blue as one might expect from the combination); these colors are distributed in irregular patches.

The **Restless Cavy** (*Cavia aperea*, or *Cobaye aperea*,) the Cochon d'Inde of Buffon; the Ferkel-maus of the Germans, is sometimes, from the peculiar sound it utters, called the *coui-coui*. About the size of a rat, it is far less gracefully proportioned, being thick, clumsy, short-legged and tailless. The fur is long and somewhat coarse, the pencilling of that on the upper portions of the body is black and dull yellow. It inhabits the banks of the Rio de la Plata and is common in the vicinity of some of the towns located upon that river. It is found in Paraguay, Bolivia and Brazil. It prefers to dwell in marshy places, beautified by the green leaves of aquatic plants; but occasionally lives in sand hillocks or in the hedge-rows. Wherever the soil is dry it makes burrows, but where the herbage is luxurious it lives concealed among the leaves. It is gregarious in its habits, forming little communities of from six to fifteen individuals, who breakfast very easily, do without dinner, eat their supper at sun-down, and never stray far from home. It breeds but once a year and then has but one or two young.

AGOUTI AND CAPYBARA.

The **Bolivian Mountain Cavy** (*Cavia boliviensis*) is always found at a considerable distance above the sea-level. In this respect it is totally different from the *cavia aperea* which is to be found only in low lands. The *Bolivian cavy* called the *cavia flavidens* is very abundant in the plains around Lake Titicaca.

The **Southern Cavy** (*Cavia australis*) may be found in Paraguay to the Straits of Magellan. It is a lively little creature, but its timidity saves it from too much self-assertion. It makes itself a home near the habitations of man, but spends most of its time in deep burrows, made in sandy places, protected by shrubs.

The **Paca** (*Celogenys paca*), often called the *brown paca*, once lived in the West Indies, but now seems to be found only in Brazil and the neighboring countries. It looks thick-set and clumsy, but surprises the spectator by its prompt and sudden movements. Its general color is blackish-brown, variegated by four rows of parallel spots beginning at the shoulders and reaching back to the haunches. The tail is greatly abbreviated. The length of the body is about twenty-one inches; the height about a foot. The *paca* likes to live in damp forests, and where he can conveniently reach the water. He digs a burrow, which has so thin a roof as to form a pitfall for the unwary traveller, whose feet crushing through it drive the unprepared host from house and home. These burrows generally have the entrances closed by doors of dry leaves and branches. The hunters who hope to catch their prey alive, stop up two of these exits and dig into the third. When the citadel is reached, the hapless animal fights the besiegers with teeth and claws, only to be captured at the last. Like the cat it washes its face and whiskers with its fore-paws, but unlike pussy it does not hate to enter the water, for it swims and dives with great skill. Its food consists of fruits and tender plants. It is nocturnal in its habits, for its eyes are not able to bear the full effulgence of the king of day. It has a sweet tooth, and sugar-cane plantations suffer from its love of good things. When it is not asleep it is eating, and when asleep it lies in a soft little bed which it spares no pains to make comfortable, even luxurious. In spite of its laziness it is cleanly and well-bred. Its flesh is prepared in the same way, and tastes like that of a suckling-pig. The female, at a birth, produces but one, which remains with her for a long time. The breeding-time falls in the rainy season. The *paca* is said to be possessed of small intelligence but favored with fine instincts. In captivity it is quiet and contented.

The **Agouti** is a deer, bearing some slight resemblance to a pig. Its head and form are moose-like, its legs slender, its ears short, but open, its tail lacking, or represented by the merest stump. The **Yellow-rump Agouti** (*Dasyprocta aguti*) is quite abundant. The **Agouti Acouchy** (*Dasyprocta acouchy*) is distinguished by quite a long tail. The Yellow-rump Agouti is brown, which grows lighter in shade, and even approaches white on the breast and belly, and becomes yellow or almost golden upon the rump. The hair is long and, as the animal has no tail, it falls over the hind quarters like a carefully adjusted bed-spread or valance. It is nocturnal in its habits, and its motions are marked by grace and quickness. It can swim, but it cannot dive; it is sometimes called the *South American rabbit*, but really belongs to the Guinea or Guiana pigs. It has the family appetite, which is distinguished by voracity rather than delicacy. In eating it uses its front paws, after the fashion of the squirrel, and plays such havoc with the crops that the planter is always waging war against it.

The **Crested**, or **European Porcupine** (*Hystrix cristata*), is found in southern Europe, where it has come from northern and western Africa; it is about twenty-eight inches long, the tail about eight more. The muzzle is large and obtuse, sparingly clothed with small, dusky hairs, with scattered longer and coarser ones on the upper lip; the anterior and under parts and limbs with spines not more than two inches long, with which are mixed some coarse hairs; it has a crest of numerous very long bristles, extending from the crown to the back and curving backward; the hind parts of the body and tail are covered with quills, some slender and flexible, others shorter, stouter and very sharp. There are a few on the tip of the tail. The prevailing color is brownish-black, with a white band on the fore part of the neck. This is the *porc epic* of the French, the spiny pig, so-called from its heavy, pig-like look and its grunting voice. It lives in rocky crevices or in burrows, becoming torpid in winter; the food consists of various vegetable substances, and its flesh is well flavored. It can erect its quills at pleasure, but cannot discharge them. Besides its grunts it makes a rattling noise by shaking the tuft of hollow quills on its tail; when angered it also strikes the ground with its feet.

The **Nepaul Porcupine** (*Hystrix hodgsoni*) has no crest, and is covered chiefly with spiny bristles with long, hair-like points, and the quills are rather black than white. It is very abundant in the sub-Himalayan region, and very mischievous, digging up potatoes and other root crops. It is monogamous and has two young at a birth. Its flesh is very delicate and is eaten by all classes, even by the high caste Hindoos. It is easily tamed, and breeds in captivity, and it is considered lucky to have a family of them about the stables.

AGOUTI.

The **Brush-tail**, or **Fasciculated Porcupine** (*Atherurus fasciculatus*), sometimes called the *Malacca porcupine*, is found in the Celebes Islands, and the Isles of the Indian Archipelago. The most peculiar thing about it is its tail. The body is nearly covered with spines, mostly white at the base and black toward the extremity, pointing backward upon all ordinary occasions; the beginning of the tail is also decorated with spines; then comes a portion thereof which is bare even of hairs, and the tail ends in a tuft or bundle of long flat bristles, bearing a close resemblance to narrow slips of parchment, slit in an irregular manner. This tuft is nearly white and about two inches long. The entire tail is about five inches, the body a little more than a foot long. The eyes are small and black, the ears short, round and naked. It sleeps all day, is cross and fretful, and when irritated or disturbed stamps with rage, erects its spines, and swells itself to its utmost size. Its intelligence is very limited.

The **Yellow-haired**, or **Prairie Porcupine** (*Erethizon epixanthus*), is smaller than the preceding; the color is blackish-brown, the long hairs of the body tipped with greenish-yellow; the anterior molar is considerably larger than the rest; it is found west of the Missouri to the Pacific Ocean.

The **Canada Porcupine** (*Erethizon dorsatus*) is about two and a half feet long, and weighs from twenty to thirty pounds. It appears larger than it really is from the length of the hair and spines; the fur is generally dark brown, soft, woolly, and grayish next the skin, coarse and bristly in some parts, six or seven inches long on the back, coarse hairs usually with dirty white points, giving to the whole a hoary tint; the spines, more or less hidden by the fur, and abundant on the upper surface of the head, body and tail, are two or three inches long, white with dark points; the tail is about ten inches additional to the above length; the incisor teeth are of a deep orange color.

It is a very clumsy animal, with much-arched back, snout thick and tumid, ears short and rounded, and tongue rough with scales. It is found between northern Pennsylvania and latitude 67° N., and to the east of the upper Missouri river. It is an excellent though a slow climber; is not able to escape its enemies by flight, but cannot be attacked even by the largest carnivora with impunity; dogs, wolves, the lynx, and the cougar have been known to die from the inflammation produced by its quills; these are loosely attached to the skin and barbed at the point, so that they easily penetrate, retain their hold, and tend continually to become more deeply inserted; when irritated it erects its quills, and by a quick lateral movement of the tail strikes its enemy, leaving the nose, mouth and tongue beset with its darts; it has no power of shooting the quills. Its food consists of vegetable substances, especially the inner bark and tender twigs of the elm, basswood and hemlock; it seldom quits a tree while the bark is uneaten, except in cold weather, when it descends to sleep in a hollow stump or cave; as it kills the trees which it ascends, its depredations are often serious. The nest is made in a hollow tree, and the young, generally two, are born in April or May.

It is almost as large as a beaver, and is eagerly hunted by the Indians, who eat the flesh and use the quills to ornament their moccasins, belts, pouches, bags, baskets and canoes, for which purpose they are often dyed with bright colors; it is very tenacious of life; it does not hibernate as the European *porcupine* is said to do. This animal shows plainly that the quills are only modified hairs, as it presents quills on the back, spiny hairs on the sides, and coarse, bristly hairs on the under surface, passing into each other in regular gradations. These quills, or more properly spines, vary in length, those of the greatest length being so soft and flexible as to offer little resistance, but beneath these are the shorter spines which constitute the animal's real armament. Their length is from five to ten inches and they are both stiff and very sharp-pointed. In making an attack or resisting its enemy the *porcupine* moves backward with all its spines spread. When it strikes a foe these shorter spines are left in the wounds, being so slightly attached to the skin of the animal. They are so pointed that if not quickly withdrawn they work deeper into the flesh and will cause death. In Africa and India, where the *porcupine* abounds, it is a rather common circumstance to find a leopard or tiger that has been killed by the penetration of its flesh by the *porcupine* spines. In one instance, a tiger was found dead whose head, paws and ears were filled with the spines of a *porcupine* which it had vainly tried to kill. The *porcupine*, though conscious of its power, is by no means aggressive, and will escape by flight rather than risk a conflict, but when set upon there are few more dangerous adversaries.

COMMON PORCUPINE.

The Couiy (*Sphiggurus insidiosus*) is a great climber, and is assisted in his arboreal excursions by his partly-naked prehensile tail. They are covered by short, sharp spines concealed by the hair. Any one who attempts to smooth its hair is apt to be severely wounded by this hidden arsenal. It is about a foot and a half long, and brown in color. It is' decidedly lethargic, being very deliberate in its movements and sometimes indolent enough to remain motionless in one spot for twenty-four hours. It lives upon flowers

KANGAROO RAT.

and fruits which it eats while comfortably seated upon its haunches. It lives in South America, where it is also called the *sphiggure*.

In the genus **Cercolabes**, which includes the tree porcupines, the body is similarly armed with spines and spiny-hairs; the tail is long and prehensile; all the feet are four-toed, with long and curved nails, the hind feet having each a rudimentary inner one, a small nailless tubercle, and being with the palm much expanded by a semicircular lobe on the inner side; the soles are rough and naked, the claws long, and the hind feet so articulated that the soles are directed inward; the lobe can be bent inward, being supported by several bones, some supernumerary; the tail is thick and muscular at the base, slender and bare above, and prehensile at the end, the upper surface being applied to the branches, and the tail coiled in a direction just opposite to that of the monkeys of the same country. The muzzle is very movable, hairy, thick and obliquely truncated; the eyes small but prominent; ears small and sparingly clothed with

SKUNK (*Mephitis mephitica*).

hairs; the incisors are narrow. The animal emits a disagreeable odor, somewhat like that of garlic. The food consists of fruit, leaves and tender bark. They are usually seen singly, and sleep during the heat of the day, feeding at morning and evening. They are harmless, easily reconciled to captivity, but have very little intelligence. They inhabit America from Mexico to Paraguay, living on trees, on which they are expert though slow climbers.

The **Brazilian Tree Porcupine** (*Sphiggurus prehensilis*) is sixteen to twenty inches long to the base of the tail, the latter nearly as much more. It is abundant in Guiana, Brazil and Bolivia, and feeds on the fruit of the palms.

The **Mexican Tree Porcupine** (*Sphiggurus mexicanus*) is mostly black; the spines are nearly all hidden by the fur, yellowish or whitish with black points; it is about eighteen inches long, its tail about fourteen; it inhabits the temperate mountain regions of eastern Mexico, between 2000 and 4000 feet above the sea.

The **Ground Pig**, or **Ground Rat** (*Aulocodus swinderianus*), is a beaver-porcupine diminutive in size, found in South America and chiefly interesting because fossil bones of the same species have been discovered in the Eocene strata.

The **Cuypu** (*Myopotamus coypu*) inhabits the banks of a great many South American streams, and is found on both sides of the Andes. On the eastern side its habitat extends from Peru to 43° south latitude, on the west from Central Chili to Terra del Fuego. It is found also along the bays and channels between the little islands of the Chonos Archipelago. The specie peculiar to river-banks, subsist on vegetable matter, while it is said that those found near the sea add shell-fish to their bill of fare. It has the size and general appearance of a beaver, in fact it is sometimes called the *La Plata beaver*. Its general color is a dusky brownish-yellow. The hair is fine and silky, and at its base is a fur similar to that of the beaver, which has become a considerable article of commerce. The animal is nocturnal and is hunted only at night with dogs, with which it fights ferociously. Its flesh is white and agreeably flavored, though it is not generally eaten. Buenos Ayres is the headquarters of the trade in this sort of fur.

The **Viscacha**, or **Biscacha** (*Lagostomus trichodactylus*), is somewhat like the rabbit. It has smaller but wider-spreading ears; its tail is about one-fourth of the length of the body and its tip turns upward. Its fur is close and fine, brownish-gray above and shading into white below. It flourishes on the pampas of Buenos Ayres. Here they are eagerly hunted and promptly put to death, not because their flesh is good for food, but because they dig up the soil and damage the crops. They serve one useful purpose after death, since their fur is made into caps. They live in companies, seldom venture far from home, are vegetarian in diet, move by leaps, like to pose upon their haunches, and carry food to their mouths with their fore paws very much as squirrels do. They express their feelings by a variety of short cries. The female produces four or five at a birth. The favorite resort of the *viscacha* is that part of the plains which during one-half of the year is covered with immense thistles, to the almost entire exclusion of other plants. Like the magpie, the *viscacha* carries off objects for which it has no possible use. Every hard object it finds, it drags forward to the mouth of its burrow, but for what purpose no one knows, as it does not use them as weapons of either attack or defence. One dark night, as he was riding along, a traveller dropped his watch. Missing it at his journey's end, he retraced his steps the next morning, examining the entrance of each *viscacha* burrow on his route, when at one of these lay the lost property.

Cuvier's Lagotis, or the **Alpine Chinchilla** (*Lagidium cuvieri*), resembles the viscacha, and is itself sometimes called by that name, yet its body is more slender, the thickest part being near the tail; the ears are longer and stand straight up from its head, and the tail is long and carried straight out. The body above is a greenish-yellow in parts slightly mottled with black, while below it is auburn and the mustache is black. It has beautiful fur, and its flesh is edible. It lives on the western slopes of the Andes from 18° to 30° south, in rocky and stony places, where it digs a two-storied burrow, and dines on herbs and shrubs. The Indians prize it for its fur and its flesh.

The **Pale-footed Chinchilla** (*Lagidium pallipes*) belongs to the same genus and is like the Alpine chinchilla in most of its characteristics, but it makes its home only in the rock valleys of Chili.

The **Woolly Chinchilla** (*Chinchilla lanigera*) is the one that gives us the well-known chinchilla fur. It is a small creature, being but six inches in length from the nose to the root of the tail, which is only moderately long and covered with soft fur. It looks very much like a rat; its rich fur is so thick and fine as to resemble wool, and some of its threads so long that they may be spun. It is very mild and gentle, never bites, and likes to be petted. It is very cleanly, and free from any offensive odor. It digs burrows and is sociable in its habits. It is to be found in Chili and Peru. Its fur makes a valuable article of commerce, and has thus occasioned great slaughter of these animals. They are hunted by boys with dogs and sold to the traders.

The **Short-tailed Chinchilla** (*C. brevicaudata*) is larger than the preceding species, and lives in Peru, but exhibits few differences in either habits or appearances.

The **Camas Rat-pouched Gopher** (*Geomys borealis*) has a stout, thick body seven or eight inches long, and a tail about two and a quarter. It is of a reddish-brown above, a darker shade below. It has a cheek pouch upon each side, about three inches deep and lined with hairs. These it uses as its larder. They open externally so that the food has to be taken out of these receptacles and carried round to its introduction into the mouth. They subsist upon a vegetable diet and take their meals sitting upon their haunches, using their fore paws as hands. The pouches are used also as baskets to convey their marketing home to their burrows, which they dig in sandy places, and which are deeper than those dug by the mole. The nest is rounded, made of soft substances, is lined with hair which the mother plucks from her own body, and is located at a place where several of these underground galleries converge. In this nest the female deposits from five to seven young ones, during the month of March or April. These animals not only fight human beings when attacked, screaming and seeking to bite the intruder, but they fight among themselves with their snouts, as hogs do. They are to some degree nocturnal in their habits, and in the colder parts of the district which they inhabit, become dormant during the winter. It can travel nearly as fast backward as forward. It is trapped and destroyed to prevent the damage that it does to gardens and orchards. The ears are scarcely noticeable, and the eyes very small; its fur is soft and thick. It is to be found in Canada, and is spread westward to the Pacific Ocean, and in some localities as far southward as Arkansas.

The **Southern Pouched Rat**, or **Gopher** (*G. tuza*), differs from the Canada pouched rat in the fact that its cheek-pouches open into its mouth, but in many other respects resembles it. It is about eight or nine inches long, brownish-yellow above and gray beneath. It lives in Georgia, Alabama and Florida.

The **Chestnut-cheeked Pouched Rat**, or **Gopher** (*G. castanops*), is of a pale yellowish-brown color, and is about eight inches long, and has its home on our southwestern prairies.

The **Mexican Pouched Rat,** or **Gopher** (*G. mexicanus*), is larger than the chestnut-cheeked gopher, being about eleven inches long. It is darker, too, its color approaching black.

The **Jerboa,** or **Bush-tailed Kangaroo** (*Bettongia penicillata*), is about the size of an English hare, is nocturnal in its habits, and specially clever in trying, like a true Briton, to hedge himself off from everything but "his majesty, himself." When grass is short and scanty, or the hillside devoid of herbage, the *jerboa* confronts the architectural problem of how, with insufficient material, to build a domicile, such as its instinct teaches it that it should have. He meets the difficulty by going elsewhere in search of building material, and having found it, collects the grass into little hay-ricks or sheaves, which he grasps with his prehensile tail, and then skips along to the site selected. As long as the young occupy the nest, the fond and careful parent never leaves or comes into her house without carefully concealing the entrance-way. It is to be borne in mind that, in the case of the *jerboa*, as in so many others, the popular appellation is no indication of family, as determined by anatomical structure; for the *jerboa* is no kangaroo at all, but a member of the *rodentia*.

The **African Petromys** (*Petromys typicus*), or **Rock-rat,** is found in south-western Africa, in the valley of the Orange river. It is about seven inches in length with a tail nearly as long as the body. It is of a reddish color, and though the hind legs are not disproportion-

JERBOA (*Dipus ægypticus*).

ately developed, walks upon them, instead of going upon all four. It builds its nest among piles of stones, or in the crevices of rocks.

The **Dormouse** (*Muscardinus avellanarius*), or **Mus Avellanarius** (*Linnæus*), a beautiful little animal, is very intimately allied to the squirrels, among which it has been placed by some authors, from which it differs chiefly in wanting the anterior, small molar in the upper jaw; in having the tail less bushy, and the hind legs less elongated. Its form is compact and full, the neck short, the head rather large, the nose prominent, the eyes of moderate size, the ears rather short and broadly rounded, the feet of delicate structure. The general color is light, yellowish-red, gradually becoming paler beneath, the fore part of the neck nearly white, the tail dull red. It is about five inches long.

In its habits the *dormouse* resembles the squirrel, inasmuch as it climbs with facility, and exhibits great liveliness and agility; but it is also allied to the mouse, and passes a great part of its time on the ground, feeding on grass, corn and various small fruits. It resides in thickets, generally remote from human habitations, placing its nest in bushes, and forming it of grass and leaves, intricately interlaced, and disposed in a roundish form, with a narrow aperture at the top. Having laid up a store of food, and, like other hibernating animals, having become very fat towards the end of autumn, it betakes itself to its retreat, and rolling itself up into a ball, falls into a state of torpidity, from which it is now and then aroused by an unusually mild day, when it partakes of its provisions, and relapses into its usual condition. The young are of a brownish-gray color, four or five in number.

Mr. Salmon gives the following account of a *dormouse :* "As I was pushing my way amidst the briars and brambles, I chanced to stumble upon an interesting incident, in the shape of a little ball of grass curiously interwoven, lying on the ground. It was about eight inches in circumference, and on taking it up I soon ascertained, by the faint sound emitted from its interior on my handling it, that it contained a prisoner. I bore my prize homeward for examination, and on

LEMMINGS.

my making a small opening, immediately issued forth one of those beautiful little creatures, a *dormouse.* The heat of my hand and the warmth of the room had completely revived it from its torpor; it appeared to enjoy its transition by nimbly scaling every part of the furniture in all directions. It experienced no difficulty in either ascending or descending the polished backs of the chairs, and, when I attempted to secure it, it leaped from chair to chair with astonishing agility for so small a creature. On taking it into my hand, it showed not the least disposition to resent the liberty; on the contrary, it was very docile. On being set at liberty, it sprang at least two yards on to a table. I was much gratified at witnessing its agile movements. In the evening I placed my little stranger, with its original domicile, in a box, of which, on the following morning, I found it had taken possession, and again relapsed into a state of torpidity."

The **Rat** is a well-known rodent, and is the type of the *muridæ* or mouse-family, and the common house-*rat* is a gift to America from the old world.

The *rat* is distinguished by unfailing presence of mind, and by never-flagging readiness of resource. It is very sympathetic with its kind, and has been known to lead about a blind companion. A tame *rat* suffered so from separation from its companions, as to die seemingly from melancholy. In carrying off eggs, two methods seem to be in vogue. First, a *rat* will lie on its back, and holding an egg in its feet will allow itself to be used as a handcart by its companions. The second plan is to form in line, and pass the egg from *rat* to *rat*. If a stairway is to be ascended, the lower *rat* in effect turns a semi-somersault, and presents the egg to the *rat* on the step above. When a bottle, containing liquid is to be plundered the *rat* uses its tail. In attempting to get at a jar, which could not be scaled, *rats* have been known to throw up mounds of plaster to the required height.

The **Bay Bamboo Rat** (*Rhizomys sumatrensis*) is a reddish-brown, and even its incisors are enamelled in red. It is about the size of a half-grown rabbit and sports an egg-shaped head. It looks like a mole and is very destructive to the bamboo crops. Its scientific name was given under a false impression for its habitat is Malacca. It digs deep holes near streams where grass and the bamboo are found, but in its fondness for cultivated crops it will stray far from its subterranean home, and thus exposes itself to the retribution which the farmers love to inflict upon it. The hind feet being webbed to adapt them to life in wet and marshy regions, the grace and agility which distinguishes the *bamboo rat* as a wader or swimmer is lost when it has to run over ploughed fields or dry ground. Hence, while feasting upon the fruit of others' labors, there comes the abhorred planter and cuts its thread of life, for although the creature is a foot and a half in length, it is powerless when deftly seized by its hind feet. Companies of men engage in the hunt, and, like the Indian of tradition, dash out the animal's brains against the rocks.

BAMBOO RAT (*Aulacodes swinderianus*).

The **Zemmi** (*Spalax typhlus*) belongs to southern Russia. It is tailless, eyeless (at least externally), earless and yet it is keen of hearing. It is about three-quarters of a foot in length, has a head broader than its body, end of the nose clothed with a skin-case, and the nostrils underneath, short legs, and short, stout claws. Its hair is furry, dark brown, with ashen gray extremities, and it is a remarkable mining engineer. It keeps its tunnels connected, and every few yards makes an opening to the surface, building hillocks one or two feet in circumference and of corresponding height. Whether regarded as a "freak of nature," or as a wonderful adaptation for the life which it is required to live, it is equally noteworthy.

The **Lemmings**, or **Arctic Musk-rats**, are so numerous in hyperborean regions as to have given rise to serious discussions as to whether they do not fall in showers from the clouds, as Jupiter visited Io.

The **Snowy Lemming** (*Cuniculus torquatus*) turns white in the winter season, as though to secure greater protection by identifying itself with the fancies and freaks of the blithesome snow. In the summer time, on the contrary, it is white only on the tail and on the feet; its back is black or blackish, and the rest of its body beautifully mottled in buff, chestnut or gray, while about its neck it seems to wear a white collar, from which fact it is sometimes called the *collared lemur*. Its plump little body is wrapped in plentiful and thick fur mantle, and like Kriss-Kingle, it wears furred boots and gloves. It has but a poor apology for a tail—a mere suspicion of hairs—but it wears sharp claws and manages to hear without external ears.

The **Myodes Lemmings** do not change color with the seasons, they often lack claws, and have a different dental structure. The orange-colored species (*Myodes obenses*) is short-clawed and a strikingly brilliant creature.

The **European Lemming** (*Myodes lemmi*) is specially notable for the migrations in which all *lemmings* indulge. They reproduce the process of the historical colonizations of Greece or the invasion of Rome by the Goths and Vandals. Whenever the *lemmings* are troubled by a Malthusian excess of population, then some must move, bag and baggage, and once started on their way they persist in "fighting it out on their chase-line, even though it take all summer." Rivers, lakes, towns, forests, mountains, valleys, plains or precipices seem to them not only not insurmountable but alike matters of indifference. From Browning's "Pied Piper of Hamelin" one may

HAMSTER AND LEMMINGS (*Cricetus vulgaris, Myodes lemmus*).

derive the most adequate idea of the irresistible progress of their great hordes —not, be it understood, that Browning speaks of the *lemming*.

The **Short-tailed Field Mouse, Field Vole**, or **Campagnol** (*Arvicola arvalis*), is red above and gray beneath. It is a good climber; in the winter it builds subterranean nests, but in the summer it prefers the surface of the ground. This is the little creature to which the poet Burns devoted one of his most characteristic and popular poems :

> " Wee, sleekit, cow'rin, tim'rous beastie,
> Oh, what a panic in thy breastie !
> Thou need na start awa sa hasty,
> Wi' bickering brattle !
> I wad be laith to rin an' chase thee,
> Wi' murd'ring pattle ! "

An extraordinary instance of the rapid increase of *mice*, and of the injury they sometimes do, occurred a few years ago in the plantations made by order of the Crown in Dean Forest, Gloucestershire, and in the New Forest, Hampshire. Soon after the planting of these forests, a sudden and rapid increase of *mice* took place in them, threatening destruction to all the young plants. Vast numbers of young trees were killed, the *mice* having eaten through the roots of five-year old oaks and chestnuts, generally just below the surface of the ground. Hollies, also five or six feet high, were barked around the bottom; and in some instances the *mice* had crawled up the tree and were seen feeding on the bark of the upper branches. In the reports made to the Government on the subject it appeared that the roots had been eaten through, wherever they obstructed the runs of the *mice*. Various plans were devised for their destruction; traps were set, poison laid, cats turned out, in vain; nothing appeared to lessen their numbers. At last is was suggested that if holes were dug, into which the *mice* might be enticed, their destruction might be effected. Holes, therefore, were made, about twenty yards apart, being about twelve in each acre. These holes were from eighteen to twenty inches deep, and two feet one way by a foot and a half the other, and they were wider at the bottom than at the top, being excavated or hollowed under, so that an animal once in, could not easily get out again. In these holes at least thirty thousand *mice* were caught in the course of three or four months, that number having been actually counted and paid for by the officers of the forest. It was, however, calculated that a much greater number than this was taken out of the holes by weasels, hawks, owls and other birds. Cats, also, and dogs resorted to the holes to feed upon the *mice*, and many were destroyed by traps and by poison. In Dean Forest alone, the number killed was calculated at not much less than one hundred thousand. In the New Forest, from the weekly reports of the deputy-surveyor, about the same number were destroyed; in addition to which it should be mentioned that these *mice* are found to eat each

SHORT-TAILED MOUSE.

other, when other food falls short. Hence the total destruction of *mice* in these two forests would probably amount to more than two hundred thousand.

Mice desiring to cross a stream make rafts of dried mushroom sacks, and embarking as a family party boldly sail the main, after the manner of squirrels.

The **Brown Water Vole** (*Arvicola amphibius*), commonly named the *water rat*, has the body full; the neck very short; the head short, broad, rounded and convex above; the limbs small; the tail rather long and slender, and the snout is small. The ears are short and entirely concealed in the fur. It has five toes on each foot. The general color of the fur above is dark-brown, the under parts and sides of the head, light brownish-red; the teeth are brownish-yellow, the eyes black, the nose dusky, the soles of the feet pale flesh-color, the claws pale yellowish-gray.

The residence of the *brown water vole* is in the banks of rivers, brooks, canals, mill-dams and ponds, in which it forms long and tortuous burrows. It frequently betakes itself to the water, where it swims and lives with ease, and generally has an entrance to its retreat beneath the surface, so that in cases of danger it may effect its escape without appearing on land. In fine weather, especially in the morning and evening, it may often be seen sitting at the mouth of its hole, nibbling the grass or roots there; but in the middle of the day it usually remains under ground.

It feeds entirely on vegetable substances, chiefly roots, and has been known to deposit a store even of potatoes for winter use, for it does not appear to become torpid in the cold season, although in time of snow it does not come abroad.

Five or six young are produced early in summer, and deposited in a nest composed of dry grass and other vegetable matters. This animal never makes its appearance in houses; nor is it injurious to man, otherwise than by perforating the banks of canals.

The **Wood Mouse**, or **Long-tailed Field Mouse** (*Mus sylvaticus*), is generally distributed throughout Great Britain, and lives not so much in woods as its name implies, as in thickets, hedges, cornfields and gardens. It resem-

FIELD MOUSE (*Mus sylvaticus*).

bles the domestic mouse, and is of nearly the same size, but is easily distinguished by its reddish color, and its more elongated tail. It produces from five to eight young at a time, and is supposed to litter several times in the year. It is, in consequence, very abundant in many districts, and frequently commits considerable ravages in the cornfields and gardens. Its food consists of seeds, especially those of grasses, acorns, nuts and insects. Like the squirrel, it lays up a store for the winter, depositing great quantities of vegetable substances in its holes, which are formed in banks, or under the roots of trees, or in the open fields. Sometimes it takes possession of the deserted runs or nests of moles. It does not become torpid in winter, at least it has been seen in the midst of snow, when it had come abroad to search for food. Its more formidable enemies are kestrels, owls, ermines and weasels. Although extremely timid it may be easily tamed.

The **Musk-rat, Musquash**, or **Ondatra** (*Fiber zibethicus*), is abundant in North America and not at all an unfamiliar sight. The blackish color of the young turns later into a dark brown, inclining to gray on the under parts, and its glossy fur is water-proof. The tail, half as long as the body, is scaled and thin, and serves the *musk-rat* as a rudder when in the water, and as a spade when it wishes to excavate its underground abode. The hind feet are bent at an angle and the toes webbed, so that they serve the uses of a pair of oars. The nose is covered with fur, and the small ears are likewise almost buried. The *musk-rat* is timid, and hence is not so readily secured as it is easy to see. It will excavate as far as fifty feet, and builds its nest of dry reeds and grasses. At one time the *musk-rat* played a prominent part in our foreign commerce, but more acceptable furs have driven his from the market. Were it not for the damage which they do to dams and embankments (which they mistakenly look upon as natural advantages not to be neglected), the *musk-rat* might now pass his life in peace, so far as fear of man is concerned, but owing to a want of harmony of effort between man and himself, he is slain, lest he innocently cause great damage.

The **India Musk-rat, Sondeli**, or **Monjourou** (*Sorex murinus*), though not hunted for the musk of commerce, is so strongly impregnated with strong odors that its mere presence is as baneful as the secretions used by the skunk. Its hair is short, mouse-colored, and growing whitish on the under parts.

The **European Hamster** (*Cricetus frumentarius*) is very much like the true rat, but has large cheek-pouches. Its range is of wide extent embracing most of Europe and Asia. The *hamster* is about nine inches long; its tail three. The color of the upper parts is a reddish-brown, below it is black and the feet are white. It has one white spot on the throat, another on the breast, and three light spots on each of its sides but different varieties are of different colors, one species being black. They are a great pest to the farmers, for they not only have a voracious appetite and an enormous capacity for stuffing themselves, but having eaten all they possibly can, they crowd into their pouches all the wheat, peas, and beans these will hold, and carry this food off to their burrows for winter use. Here the forage is carefully cleaned and the husks and chaff thrown away. The peasant who goes during the winter to hunt the *hamster* for its skin, opens the burrow and pos-

MUSK-RAT.

sesses himself of the edible contents, sometimes finding in a single storehouse as much as two bushels of grain. The animals' mode of constructing these combined magazines and dwellings is very elaborate. First, they form the vestibule by digging down obliquely. At the back end of this, the male sinks a single perpendicular shaft, the female several. At the end of these passages several rooms are formed, for each young one is said to have its private apartment, and some are used for pantries. Except during a very short season of mating the male and female occupy separate apartments and see little or nothing of each other. The *hamsters* fight, kill and devour animals of their own species, and lesser animals of other sorts. During the entire cold season they hibernate, yet so great is the damage they do during the summer, that famines have been caused by them, and governments have had to set a price upon their heads.

.33

The **Jerbillies** (*Gerbillinæ*) suffer constant depredation from a horned viper (*vipera cerastes*), whose appearance alone might well affright the timid little creatures, even though the snake's designs were not known to be murderous. The *jerbillies* vary in length from two to three feet, and are clad in a tawny, yellow, shell-like armor, marked with brown to increase the resemblance to the sandy soil, in which it burrows. It is distinguished for having very short fore feet and hind limbs of very great length, so that its locomotion is identically like that of the kangaroo. The tail is equal to the length of the body.

The **Harvest Mouse** (*Mus minutus*) is somewhat smaller than our common house-mouse, which it resembles. Its hair is brown but tipped with red, and it it white on the under parts. It hangs its nests to stems of straw and stubble, and weaves them into perfect spheres of fine grass. It leaves no opening, and naturalists have not yet decided how it provides for its young.

The **Brown**, or **Norway Rat** (*Mus decumanus*), has a body eight to ten inches long, and the tail six to eight inches, scantily covered with hair and about two hundred rings. The color above is grayish-brown mixed with rusty, grayer on the sides and ashy white below; the upper surface of the feet dirty white. This species, originally from India and Persia, entered Europe through Russia, appearing in the central countries about the middle of the eighteenth century; it was brought to America about 1775, and has since greatly increased in numbers, driving out here, as in Europe, the black rat, which had been previously introduced. It is now generally distributed over the world, having been transported thither in ships, and is most abundant near the sea-coasts. Its haunts are well-known to be cellars, sewers, canals, docks and similar dirty places, wherever it can make a burrow or find sufficient food. It is a great household pest, and so prolific that its devastations are sometimes very great. It breeds from three to five times in a year, having twelve to fifteen at a birth, the males always being the most numerous. Not only the black rat, but other species indigenous to the old world, are driven off or destroyed by it; the dead and even living persons are attacked by it when hard pressed; it is not only pursued by man, dogs and cats, but the stronger will kill and devour the weaker of its own species.

The **Black Rat** (*Mus rattus*) is seven or eight inches long, with a tail of eight and a half inches. The color is very dark, often nearly black, with numerous long hairs projecting from the short and soft fur, lead-colored beneath and the feet brown. It has a slighter form than the brown rat, with the upper jaw more projecting, the ears larger, and the tail much longer in proportion. It is not very strong, but exceedingly active; being rather timid, it is exter-minated by the larger and fiercer brown rat; the habits of the two species are much the same, but the *black rat* is less a burrowing animal, and prefers the upper parts of the houses to the cellars and dirty places. It used to be the common house-rat of Europe and warm countries until driven off by its con-gener; it appear to have been brought to the new world about the middle of the sixteenth century; it came originally from Central Asia and, like the preced-ing species, it is omnivorous.

The **Roof**, or **White-bellied Rat** (*M. tectorum*), is about six and a half inches long, and the tail about eight, with two hundred and forty rings; it is colored above like the brown rat, the lower parts and upper surface of feet yellowish white; the head is rather blunt, the eyes large, whiskers long and black, ears very large, and the thumb rudimentary. It came originally from Egypt and Nubia, thence passed to Italy and Spain, and from the last to America in the fifteenth century; it is common in Mexico and Brazil, and in the Southern States, but is rarely found above North Carolina; it is fond of inhabiting the thatched roofs of houses, whence its name; it is the same as the *mus alexandrinus* and *mus americanus*.

The **Giant Rat** of Bengal and the **Coromandel Coast Rats** (*M. giganteus, Raffles*) have bodies thirteen inches long and tails as much more; they are very destructive in gardens and granaries, devouring chickens and ducks, undermining houses, and piercing the mud walls; they are the largest of the sub-family, a male weighing as much as three pounds; it is often eaten by the lower caste Hindoos. All these rats are fond of fighting, and with their omnivorous habits are decided cannibals, eating not only their conquered brethren, but also their own young. Though living in the filthiest places and in the foulest air, they always have a sleek coat, and take the greatest pains to clean themselves, licking their paws in the manner of a cat; during mastication the jaws move very rapidly; they drink by lapping; when asleep the body is coiled in a ball, with the nose between the hind legs and the tail curled around the outside, leaving only the ears out ready to catch the least sound of danger; as food fails they migrate in companies from one place to another.

ANIMALS OF BORNEO.
MUSK DEER, MACQUE, FLYING FOX AND HORNED OR RHINOCEROS BIRD.

There are more muscles in a rat's tail than in the human hand; this most useful appendage with its chain of movable bones and its numerous muscles is covered with minute scales and short stiff hairs, making it prehensile and capable of being used as a hand, a balancer or a projecting spring. The teeth are long and sharp, but there is nothing specially dangerous in a wound made by them; their strength enables them to gnaw ivory, as dealers in this article well know; in fact, even in Africa, elephants' tusks are found gnawed by rats, squirrels, porcupines, and perhaps by other rodents, as long as any gelatine can be found in them. These animals are greatly subject to tumors of the skin, which often end fatally; they also soon perish without water. Persecuted as these animals are, they have their uses, especially as scavengers for devouring refuse matter, which would otherwise engender disease in tropical climates or in large cities, in whose sewers they live in legions. Their skins are employed for various purposes, as in the manufacture of the thumbs of gloves, but are too delicate for any article requiring much strength. The Chinese and other Asiatic nations, and many African tribes consider the flesh of rats a great delicacy, and Arctic travellers have often found them a welcome addition to their bill of fare.

The **Florida**, or **Wood Rat** (*M. floridana*), is about eight inches long and the tail six inches, the short stiff hairs of the latter not concealing the scaly rings; the color above is dun, mixed with dark and yellowish-brown, lighter on the sides beneath, and the feet white; tail dusky above, below white; the head is sharp. It is abundant in the Southern Atlantic and Gulf States, and is found occasionally in the West; its habits vary much in different localities, living in some places in the woods, in others under stones or in the ruins of buildings; in swampy districts it heaps up mounds two or three feet high of grasses, leaves and sticks, cemented by mud; sometimes the nest is made in the fork or the hollow of a tree. It moves about only at twilight, is very active and an excellent climber; the food consists of corn, nuts, cacti, crustaceans, mollusks and various roots and fruits; the disposition is mild and docile; from three to six young are produced twice a year.

The **Bush Rat** (*M. mexicana*) is rather small, light brown above, fulvous on the sides, under parts and feet white; tail hairy. Larger species are found west of the Rocky Mountains, very destructive to the furs, blankets and stores of the trappers; for an account of these see Vol. VIII., of the "Reports of the Pacific Railroad Expedition." In the bone caves of Pennsylvania have been found the remains of a species whose body must have been at least twelve inches long. In the genus *sigmodon* the general appearance is that of a large field mouse; the body is stout, the hair long, the muzzle blunt and hairy, except on the point of the nose; the upper lip slightly notched; thumb rudimentary; soles naked, with six granular tubercles; incisors stout, the upper much rounded; ears and tail moderate; molars rooted with a plane surface, the last two lower with enamel in the form of an S, whence the name. The genus is confined to the southern parts of the United States.

The best known species is the **Cotton Rat** (*Sigmodon hispidus*), about five inches long with a tail of four; the color above is reddish-brown, brightest on the sides, lined with dark-brown, and under parts grayish-white; the hair is long and coarse, and the claws very strong. It is more abundant in the Southern States than the meadow mice in the North, living in hedges, ditches and deserted fields, and consequently doing but little damage to the planter. It is gregarious, feeding on seeds of grasses and leguminous plants, and also on flesh; it picks up wounded birds and small mammals, crawfish and crabs; it is very fierce and pugnacious, the stronger killing and devouring the weaker, and the males often eat the young; it is also very fond of sucking eggs. Nocturnal in habit, it is seen by day in retired places; it digs very extensive galleries not far from the surface, a family in each hole; very prolific, it breeds several times a year, having four to eight in a litter; it swims and dives well. It received its name not from any injury it does to the cotton plant, but from its lining the nest with this substance, which it is said to collect in large quantities. It is preyed upon by foxes, wild cats, hawks and owls; it is not found north of Virginia.

In Africa is found the **Mixed-colored Tree Rat** (*Dendromys mesomelas*), a dweller in the trees, and colored gray, with a black stripe in its beak.

The **Skull Cap** (*Lophiomys imhausi*) was discovered some seventeen years ago, and until quite recently was represented by a single individual specimen. Even now but four of the animals are known to naturalists, and these come from Abyssinia. It is stout in body, short in legs, long-tailed, small clawed. It is as badly compounded of the peculiarities of different animals as was the artificial "humbug," submitted to the entomological professor. It is in skull like a reptile, in feet like the opossum, in coloring like the skunk, in size and form like the guinea pig, in having an opposable thumb like the monkey, in tail like a fox, and in mane like the fossil elephant. Its own peculiar manifestations are an ability to erect its mane at pleasure; furrows between the mane and the hair on the sides of the body; hair in the furrows which is spongy and unlike any other known hair, fur or bristles, and an extravagant length to the hair of the body, as in the case of the sheep of the Dinkas, the buffalo and some other animals.

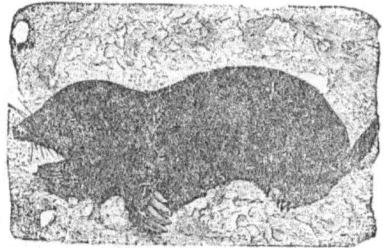

MOLE (*Talpa europæus*).

The **Azar's Agouti** (*Dasyprocta azaræ*) is South American, and for the purposes of THE SAVAGE WORLD needs no further particular description beyond its name.

The **Hutiaconga** (*Capromys pilorides*) is Cuban in its habitat, about two and a half feet in length (of which the tail claims a third), black and yellow in coloring, short-tailed, naked and scaly, and is, in short, a tree rat.

The **Hutia-carabali** (*Capromys prehensilis*) is also Cuban, is smaller in size, and longer of tail, and its tail is to some extent prehensile.

SECTION OF MOLE'S NEST.

The **Tuko-tuko** (*Ctenomys brasiliensis*) is a stout, little gray or brown creature, about a foot in length from the end of its nose to the tip of its tail. Its ears are rudimentary, and its feet have the singular appendage of a comb of bristles. It is frugivorous, nocturnal and subterranean. Its habitat is South America, from Brazil to Patagonia.

The **Jumping Hare** (*Pedetes caffer*) belongs to South Africa, and derives its name from its method of progression, which is by kangaroo-like leaps of twenty or thirty feet. As it is gregarious a sight of a troop is much more comical than any "sack race." The tail, equal to the body in length, is very bushy, and the ears, in size, arrangement and coloring, resemble those of the hare family. Its hind feet, although four-toed, are provided with claws, which have advanced a great ways in their effort to convert themselves into unmistakable hoofs.

The **African Briste-toe Gundi** (*Ctenodactylus massoni*) is of the size of a small rabbit. Its tail is short, its prevailing color gray. It is to be met with in Southern Africa.

The **Jumping Mouse** (*Zapus hudsonius*) is North American. It is built on the plan of the house-mouse, but enlarged behind; has very long hind legs, adapted for leaping; internal cheek-pouches, and altogether suggests the kangaroo undergoing a transition into a mouse.

The **Mole-like Pocket Rat** (*Thomomys talpoïdes*) is about three-quarters of a foot in length, and varies in color from gray to russet, the mouth, feet and tail generally being white. It takes its name from being pouched.

The **Rocky Mountain** species (*Thomomys clusius*) is noticeable only as a member of the pigmy family.

THE INSECTIVORA.

With the subsidence of the waters and the perfection and riotous variations of plant-life, sea and fen, land and air, alike required and were fitted for the dominion of the insects, and we should naturally expect as the next stage in progression to find a class of animals which will prevent what had become the disastrous results of the fecundity of the insects. Nor are we to be disappointed, for we find that with the rodents to restrain the extravagance of vegetable growth, we have the **Insectivora, or Insect-eating Animals**, who limit the otherwise absolute and tyrannical sway of lower animal forms. Here belong the *moles*, the *hedgehogs* and the *shrews*, some species of which will be known to all of our readers.

The *insectivora* add canine teeth, which are not perennial. The legs are intermediate between those of the rodents and the winged-legs of the bats; the brain is less convoluted than that of the rodents, but more highly developed in its well-defined direction.

Perhaps in none of the animal creation, so much as in the **Mole** (*Talpa europea*), is adaptation to the necessities of life so evident. It is insectivorous, and, to forage to the best advantage, must do its work underground. Its nose and fore paws enable it to burrow with a rapidity which makes its speedy disappearance almost spectral. Its rich, thick fur lies equally well in any direction, and is absolutely proof against defilement from the earth and other substances with which it comes in contact. Its muscles are abnormally developed in the shoulders and fore legs, and rest upon a frame-work stout in proportion. Its little eyes are hidden beneath its dense fur and are protruded when under unusual circumstances it needs their service. Its sense of hearing and its sense of smell are remarkably acute, and its sensibility of touch most exquisite. It is a fierce, rapacious little creature, giving the freest play to passions and appetites, which one might rather expect in the case of the leopard and the tiger. It is a skilful and ingenious architect, so much so as to justify a description of its home. It first constructs quite a large mound or hill. It then makes two galleries one running around the hill, near the top, and the other near the bottom; these galleries it connects by five passages. It now digs a circular hole in the centre of the floor, and unites this to the lower gallery by means of three hallways. A large passage is then run from the central pit under the lower gallery to the ground outside, and a large but varying number of passages from the centre to the lower gallery. The pit is the bedroom of the *mole*, although in the warm months it does not use it. When the female builds its nest it does not construct it in the house, as described, but in a separate mound, connected with the other. The black, soft fur of the *mole* is sometimes used for making purses and pouches, but thus far has not played any large part in commerce.

The **Changeable Mole, Gilded Mole, Shining Mole, or Cape Chryso-chlore** (*Chrysochloris holosericea*), is among the notabilities of the Cape of Good Hope. Its silken coat is bronze-green or bronze-red, constantly changing in the light like a variable silk. Its teeth do not meet, but the ones below fit into vacant spaces beside the ones above.

The **Star-nosed Mole, Long-tailed Mole, Knotty-tailed Mole,** or **Radiated Mole** (*Condylura cristata*), has a proboscis like a nose which terminates in what looks like a rose-colored sea-anemone, and this singular addition to his olfactory organ can be expanded or contracted at pleasure. It has North America for its habitat.

The **African Mole, or Golden Mole** (*Chrysochloris aurea*), has a peculiar arrangement for securing the leverage that such a constant and deep-digging miner requires, while at the same time the length of limb shall not increase the resistance with which he is to meet. The limbs then are exteriorly very short, but there are deep cavities in the walls of the thorax which admit of the presence and use of what may be called leg-pistons.

The **Hairy African Mole** (*Chrysochloris villosa*), and **Trevelyan's African Mole** (*Chrysochloris trevelyani*), have skull-ridges which give the hair of the head the appearance of a necklace, or a cap.

The **Rice Mole** (*Oryzoryctes hovi*) is a small animal whose powers of destruction are a matter of no small concern to the farmers of Madagascar.

The **Long-tailed Microgale** (*Microgale longicaudata*) likewise belongs to Madagascar, and is notable among mammals for the possession of an extremely large number of caudal vertebræ.

The **Golden Geogale** is a soft-furred *mole-shrew* whose habitat is the western part of Madagascar. Its main importance is a connecting-link between the preceding and following species.

INDIAN TUPAIA.

The **Java Tupaia** (*Tupaia javanicus*) is smaller than its Indian congener, but has no other popular distinctions.

The **Borneo Feather-tailed Tupaia** (*Ptilocercus lowi*) is only about five inches in length, but its long tail is more strikingly ornamented than those of its congener.

The **Indian Tupaia** (*Tupaia ferruginea*) is found in India and Sumatra. Its coat is silken, brown and yellow in coloring, and its bushy tail is very like that of a squirrel. Its head is prolonged into a whiskered snout, of which the pointed upper jaw is much the longer. It is arboreal, and its sharp claws render it quite successful as a climber.

The **Oregon Mole** (*Scapanus townsendii*) is notable for having a very dark purple color.

The **Hairy-tailed Mole** (*Scapanus breweri*) is found in the Western States, and is, for the purposes of THE SAVAGE WORLD, distinguished by the hairy covering of its tail.

The **American Mole** (*Scalops aquaticus*) has a fringed muzzle, and is a common sight in the eastern parts of the United States. Its muzzle is longer than that of most of the family, and its claws are especially sharp and adapted to rapid burrowing.

The **Haytian Agouti** (*Solenodon paradoxus*) is unusually large-sized, being about eight or ten inches in length, exclusive of its tail. It is soft-furred and passes its life in the mountains.

The **Almiqui, or Cuban Agouti** (*Solenodon cubanus*), belongs to the same family as the Haytian agouti, but has certain peculiar habits. For example, when pursued it will, like a young child, hide its head in the first crevice or hole, and consider itself securely concealed. It also unwisely utters a shrill cry while robbing the poultry yards. The length of its body is about a foot, and the tail continues the length for about three-quarters of a foot.

ELEPHANT SHREW.

The **Swift Potamagale** (*Potomagale velox*) is a water animal belonging to equatorial Africa. It is built upon the plan of Winan's cigar-shaped ocean steamer, as the long snout and oar-like tail seem to be merely elongated continuations of its stout body. The hind feet have the toes connected by a membrane, and the animal is an exceptionally rapid swimmer. It is said by some travellers and naturalists to vary its natural diet of insects by courses of fish. It is undoubtedly true that it is an enthusiastic and successful fisherman, even though its carnivorous tastes be yet unauthenticated.

The **Oared Shrew, or Black Water Shrew** (*Crossopus fodiens*), has its black coat mixed with white hairs, and its under parts are grayish. The tail for some distance is somewhat cylindrical, and the remaining portion is flattened like the blade of an oar. Its thick, soft fur is a sufficient protection against the dampness of the water. It is sportive, graceful and incessantly active. Its ears are provided with a set of valves, which close automatically whenever the creature dives.

The **Rustic Shrew, Etruscan Shrew, or Italian Shrew** (*Corsira rustica*), is exceedingly minute, being but one and a half inches in length, with possibly another inch of tail. It is generally believed to be the smallest of the mammal species.

EUROPEAN SHREW.

The **Elephant Shrew** (*Macroscelides typicus*) is South African, and carries a thin proboscis of disproportionate size. It is dark brown in coloring, although sometimes tinged with red. It is a burrower, rapid in its movements, and frequently sits upon its haunches.

The **Jumping Shrew** (*Macroscelides typus*) belongs to the Cape of Good Hope, and is a burrower. It takes its name from the fact that it has been selected as a typical *jumping shrew*. The **Algerian Jumping Shrew** (*Macroscelides rozeti*) does not differ except in habitat, progressing by leaps like the jerbillies.

The **Mozambique Jumping Shrew** (*Petrodromus tetradactylus*) does not burrow but lives amidst the rocks, and his anatomy is varied to meet this change of condition.

The **Mozambique Beak-toothed Shrew** (*Rhyncocyon cirnei*) is nocturnal, subterranean and about three-quarters of a foot in length. The Zanzibar variety of the *beak-toothed shrew* (*Rhyncocyon petersi*), as well as the **Long-legged Beak-tooth** (*Rhyncocyon macrurus*) and the **Golden-rump Beak-tooth** (*Rhyncocyon chrysophagus*), of the Mombaco, are recent discoveries and therefore likely to meet the attention of the readers of current travels.

The **Pig-like Wood Mouse** (*Hylomys suillus*) belongs to northern India and has short legs, clawed feet, and although properly a hedgehog is very rat-like in appearance.

The **European Shrew** (*Sorex vulgaris*) lives either in the fields or the fringe of the woods, though it is sometimes to be found in bogs and fens. This variety of habitat is quite consistent with the fact that they are not strictly insectivorous, but find birds and mollusks quite palatable. It is preternaturally

SHREW MOUSE.

pugnacious, and its remarkable fecundity (annual families of six or seven being the rule) seems to require a corresponding liability to a decimation of its numbers. It is furthermore subject to a yearly plague whose ravages are quite extensive. It flesh is so tainted with a disagreeable secretion that no creature but the owl will try to fatten on its carcase. Those familiar with the popular superstitions of Europe (through such books as those of Baring-Gould) will call to mind the wealth of stories relating to vodooism in which this poor little *shrew* plays a conspicuous part.

The **House Shrew** (*Crocidura araneus*) is found in northern Africa, in Russia and in Siberia.

The **Ciliated Shrew** (*Crocidura suavolens*) has its habitat about the shores of the Mediterranean and is one of the mammalian pigmies.

The **Broad-nosed Shrew** (*Crocidura platyrhinus*) belongs to the North Atlantic coast States.

The **Mexican Shrew** (*Crocidura craw-*

TANREC HEDGEHOG.

fordii) has no popular peculiarities, yet its habitat renders it likely to form the acquaintance of American readers.

The **Mole Shrew** (*Blarina brevicauda*) is common in the United States. Its ears are entirely concealed, and its tail is so disproportionately short and thin as to suggest that it must, by great exertion, have escaped from a more than ordinarily muscular grasp.

The **Ash-colored Mole Shrew** (*Blarina cinerea*) has fewer teeth (thirty) and a peculiar coloring.

The **Marsh Shrew** (*Neosorex palustris*) is aquatic, has valved ears, and is found from the extreme Eastern States as far west as the Rocky Mountains. The corresponding species found on the west of the mountains, the **Pacific**

Marsh Shrew (*Neosorex navigator*) is substantially identical with the last-named species.

The **Spanish Desman** (*Myogale pyrenaico*), like the other *desmans*, does not burrow, but tunnels passages which, meeting in a common centre, form the radii of a circle. Its peculiar secretory glands alike protect it from attack, and increase the length of the tail, at the same time better fitting it for use in the water.

The **American Shrew Mole** (*Neurotrichus gibbsii*) is like the mole in its system of dentition, although the teeth themselves are patterned upon those of the shrew. It is subterranean and tunnels like the Spanish desman. It is found on the prairies in the recent addition to the galaxy of stars and stripes, which but recently was not the State but the Territory of Washington.

The **Shrew Mouse, Fetid Shrew, Ranny,** or **Erd Shrew** (*Corsira vulgaris*), is a pugnacious little subterranean creature, protected against attacks from animals by its exceedingly offensive odor. In seasons when the worms and insects dig too deeply for the *shrew mouse*, the latter dies by hundreds, but finds no animal except the owl to offer its stomach as a sepulchre. Around the *shrew mouse* has gathered the most generous abundance of superstitions, mostly turning, however, upon its supposed venom.

The **Hedgehog** (*Erinaceus europæus*) is well-known because of his prickly armor, and his ability, when curled into a ball, to resist all attempts to uncoil him. Its spines are about an inch long, lying in a horizontal position, pointing backward, and furnished with a head which holds them fast in the skin. This armament enables the *hedgehog* to fall from considerable heights without injury. The young are born with unopened eyes and to a careless observer would seem to be young birds. It is readily domesticated and is superior to any powder in its ability of destroying cockroaches and other insects. It seems to be poison-proof and is especially fond of killing and eating snakes. It successfully attacks poultry, grouse, and even hares and rabbits, and is not averse to fish. When combative it uses its teeth and claws. It has been known to show a fondness for strong drink.

The **Siberian Hedgehog,** or **Long-eared Hedgehog** (*Erinaceus auritus*), is smaller than the common hedgehog, and its head is very pig-like in appearance. Its spines are tri-colored—white, brown and yellow.

The **Tanrec,** or **Madagascar Hedgehog** (*Centetes ecaudatus*) is seemingly longer in body and limbs. It is tailless, but the length of its muzzle seems to be a compensation for this privation. Its quills are shorter and yellow with black tips. It passes its winters in a burrow, and while hedgehogs are not, in the strictest sense, hibernating animals, they for the most part pass the season in a state of lethargy or torpor. It is nocturnal in its habits and has the color of the mole.

The **Spiny Tanrec** (*Centetes spinosus*) is smaller, and its quills are white, with mahogany-colored tips. Its habitat is Madagascar and it is captured for its flesh.

The **Banded Tanrec** (*Centetes madagascarensis*) is almost black but has three broad stripes of whitish-yellow running longitudinally. It is a native also of Madagascar.

The **Sumatran Gymnura** (*Gymnura rafflesii*) is an opossum-like animal, prevailingly black with white head, neck, flanks and tail (in the latter half). Its white face is made more striking by a pronounced black stripe over each eye.

The **Colugo** (*Galeopithecus volans*) is not a true flyer, but can sustain itself in air for a short time, while passing from tree to tree. It uses its tail as a rudder, but perhaps more frequently flies down than up. It prefers climbing to flying where it has any choice. It is generally quiet during the daylight and the darkness, preferring the hours claimed by Aurora, and those distinguished by the term crepuscular (as belonging to the twilight). Passing its life amidst the thick foliage of the forest trees, it lives upon leaves as one might expect, although it varies its fare by occasional dishes of insects. It is often classed among the bats and has been described in that department of THE SAVAGE WORLD.

CETACEANS

As bats are the flying mammalia, so cetaceans are their antithetical, or swimming congeners, which strain from the water of the sea the smaller animal life which supplies them with food. Some of the **Cetaceans, or Whales,** have already been considered in connection with the fishes, since, regarded otherwise than structurally, the *cetaceans* have many external resemblances to the order of fishes. The sea has thus far been needed by man only as furnishing channels for his commerce, and as supplying the materials for certain industries. Hence, as should be expected, if the author's theory and conclusions have seemed to be sound and well-supported, the larger forms of life have persisted longer in the waters than upon dry land. Still, the *whales,* narwhals, grampuses, porpoises and dolphins, are rapidly growing fewer, and doubtless making way for a new succession of life, which will be found better suited, and more directly contributive to the needs of man's higher civilization. The *cetaceans* are protected by their bulk, strength, and thick skins against attacks from dangerous enemies other than man—irresistible man. Being mammals, and therefore requiring oxygenated air, they not only rise to the surface, but find their abundant blubber useful, not alone as securing buoyancy, but as protecting them against the inclemency of the deep sea. Regarded anatomically, the *cetaceans* rank high or low, as we regard the organism of the brain, or that of the skeleton. Their dentition is peculiar in that the teeth are not incisors, canines or molars, but special cone-like forms, suited to the uses to which they are to be put. The stomach of the *cetacean,* like that of the camel and other of the cud-chewing animals, is divided into chambers, and while the design of this structure is not certainly known, it would seem to furnish an illustration of the wonderful means by which all animals are adapted to the normal conditions of their life. The circulatory system, likewise, is specialized, and the veins and arteries form reservoirs, so as to provide, seemingly, for the frequent and long-continued descents which the *cetacean* makes into the deep sea. The *cetaceans,* as fossils, make their earliest appearance in the Eocene period, where we find the *phocodontia,* the *squaladon,* and the *zeuglodon.* These fossil-forms give evidence of having differentiated into the *cetacea* and the succeeding class, *sirenia.*

The **Australian Two-toothed Whale** (*Ziphius australis*) is ash-colored, lighter above and darker below, and in connection with the **Nova Zembla Two-toothed Whale** (*Ziphius novæzælandiæ*), the **European Two-toothed Whale** (*Ziphius cavirostris*), the **Two-toothed Cow-fish** (*Mesoplodon bidens*), the **New Zealand Cow-fish** (*Mesoplodon grayi*), the **Bottled-nose Whale** (*Hyperoodon butzkopf*), are interesting as belonging to a species regarded, until quite recently, as wholly extinct, and as illustrating the change of habitat, as man takes possession of what was before the dominion of the *cetacean*.

The **Beluga**, or **White Whale** (*Delphinapterus leucus*) begins life as lead-colored, becomes mottled as it approaches maturity, and finally assumes a garb of the most beautiful cream-color. It "roams the wide seas over," has been found as large as fifteen to seventeen feet in length, is well supplied with teeth, and prefers to feast upon the larger fishes, which cannot compete with its rapid and powerful swimming.

The **Gladiator Whale** (*Orca gladiator*) is round-headed, stout-toothed and heavy-jawed, as it needs to be, since, contrary to the conventions of the *cetaceans*, it is pugnacious, and directs its attacks particularly against the whalebone whales. It is common in the waters of the north Atlantic.

The **Black Gladiator**, or **Straight-finned Killer** (*Orca rectipinna*), is the corresponding species found in the Pacific ocean.

The **Caing Whale**, or **Black Globe-headed Whale** (*Globiocephalus melas*), is gregarious, and a common sight to travellers upon the broad Atlantic.

The **Short-finned Whale**, or **Southern Black-fish** (*Globiocephalus brachypterus*), is toothed, black in color, and found from the latitude of New York southward.

The **Pigmy Sperm Whale** (*Kogia floweri*) belongs to India and Australia, and, though a pigmy in comparison with the mammoth forms of the sperm whale, yet reaches the not inconsiderable length of eighteen or twenty feet.

The **Rorqual of the Atlantic** (*Agelaphus gibbosus*), or **Scrag Whale**, appeals to our patriotism, as it was well known to the early New England fishermen, and led to the possibilities of those ancestral traditions which are now so highly prized by some. Its back is supplied with rough, bunch-like protuberances, which have given it its second name. Its baleen is white.

The **Humpbacked Whale** consists of some seven species, of which we mention the **North Sea Humpback** (*Megaptera longimanus*), and the **Pacific Humpback** (*Megaptera versabilis*).

The **California Gray Whale** (*Rachianectes glaucus*) is an oil whale, noticeable for its periodical migrations. It passes part of the year in the Arctic regions, but prefers to spend the early winter and spring on the California coast. It was selected for mention in our discussion of the fishes.

The **Atlantic Fin-back** (*Balænoptera rostrata*) is small in size, as is also the **Pacific Fin-back** (*Balænoptera davidsoni*).

One of the swiftest of the cetaceans is the **Pacific Razor-back** (*Balænoptera velifer*) and it carries the immense length of upwards of sixty feet.

New Zealand plays so important a part in our modern life that mention should be made of the **New Zealand Right Whale** (*Balæna australiensis*), and the **Biscayan Whale** (*Balæna cisarctica*) deserves notice for the adventures to which, in times past, it gave rise.

The **Cape Whale** (*Balæna antipodarum*) furnishes sport and brings profit to the South Africans.

The **Pigmy Right Whale** (*Neobalæna marginata*) is, like our other pigmy whale, only small when compared with the other members of its family, as it is frequently fifteen feet in length. Its whalebone is its point of superiority, and excels that of all other whales. It dwells in the waters about Australia and New Zealand.

The **Pacific Black-fish** (*Globiocephalus scammoni*) swims near the coast, and though its oil is inferior in quality and less in quantity than that of the sperm whale (which is generally found in the same neighborhood), it still has sufficient value to make man its mortal enemy.

The **Mediterranean Grampus** (*Grampus rissoanus*) is naturally the best known of the grampuses, since the pen of the historian, the tale of the traveller, and the song of the poet, have delighted in celebrating the glories of this world-historical region.

The **North Sea Grampus**, or **Gray Grampus** (*Grampus griseus*), one meets with in tales of Arctic exploration.

The **Cape Grampus** (*Grampus richardsonii*) is found usually haunting the southern coast of Africa, from which fact the name is given.

The **White-headed Grampus** (*Grampus stearnsii*) is smaller in size, but more striking in appearance, since its head and front present, in their whiteness, the sharpest contrast to the black which elsewhere prevails.

The **Banded Porpoise** (*Phocæna lineata*) is an inhabitant of the Atlantic, and the contrasts in its coloring are very effective. Its body is white below and black above, but these colors are separated on the side by a rosy-hued band or stripe.

The **Indian Dolphin** (*Orcella fluminalis*) is a fresh-water animal, relatively small in size, and colored like the unwashed linen of Queen Isabella, whose abstinence from soap and water endowed mankind with a new hue (*isabeau*), to which, with unusual gratitude, they have given her name. No wonder, since a monarch can achieve such great deeds, that some faint-hearted Americans should long for the coming of monarchial government, under which they seem to assume that they would naturally and necessarily be the ones who sat upon the throne.

The **Short-nosed Dolphin** (*Orcella brevirostris*) grows to the length of eighteen or twenty feet, and although passing most of its time in salt water, quite frequently ascends the fresh-water streams to what, at least for his purposes, is the head of navigation.

The **White-beaked Dolphin** (*Leucorhampus peronii*) is South American in its habitat, the upper and lower halves of the body are colored, the one black and the other white.

The **Right Whale Porpoise** (*Leucorhampus borealis*) is a species similar to the last, but going no further south than San Diego.

The **Spectacled Porpoise** (*Leucorhampus perspicillatus*) belongs to the same family, but has black rims like spectacles above its eyes. It is tri-colored, black above, white below and lead-colored between.

The **Pacific Dolphin** (*Leucorhampus obliquidens*) is well known to those who visit the Pacific slope, for it surrounds the ships in large schools, and entertains the passengers by its gambols. It has seemingly an appreciation for music, even though lacking the power to create music for itself and for others, and amidst the limited amusements of a sea-voyage its peculiarities often form the theme of the story-teller, whether he be content with his own experiences and imaginings, or mingle with these the plentiful stores of Greek mythology. Its under parts are pearl-colored, the upper parts dark bottle-green, and its sides are striped alternately with black and gray.

The **Pacific Cow-fish** (*Tursiops gillii*) is a solitary dolphin, which frequents the coast of Lower California. It is black throughout, paling, however, on the under parts.

The **Mediterranean Dolphin** (*Delphinus delphis*) is naturally the dolphin of dolphins, since it has had so many Homers to celebrate its praises as an Achilles. As an offset to its glories, as described by the poets, fabulists and sentimentalists, it is a great destroyer of mackerel, which it pursues even to the coasts of England and of France. As " no one is a hero to his own valet," the *Mediterranean dolphin* receives but scant praise from the fishermen.

The **Pacific Dolphin** (*Delphinus bairdii*) is slender in form, and has a long slim muzzle or snout. It is parti-colored, being dark-green above, gray on the sides and below, and with white streaks above the mouth, and others running from the corners of the mouth to the fins, while the belly is provided with a large, lance-shaped white patch.

The **Ganges Dolphin** (*Platanista gangetica*) has narrow jaws well supplied with fanged teeth, and, as its name suggests, prefers the river to the ocean.

The species found in the Indus is known as the **Indus Dolphin** (*Platanista indi*), and that of Bolivia as the **Dolphin of the Amazon** (*Inia geoffroyi*).

Partaking of the same nature as the dolphin, with which it is frequently confounded, is the **Coryphene**, a somewhat larger fish and also more beautiful; in fact, our description of the changing colors and magnificent splendor of the dolphin applies more appropriately to the *coryphene*. The porpoise is also quite frequently mistaken for the dolphin, owing to the fact that their gamboling movements in the water are almost identical. One particular difference between the two is found in the fact that while the dolphin is purely carnivorous and, we may say, cleanly in its habits, the porpoise is something of a scavenger, and roots in the mud like a hog, feeding at times on worms, snails and burrowing mollusks, though it also commits great ravages among the fish. They are very sociable, and are the most familiar objects one beholds at sea. During a stay of some weeks at the mouth of the Mississippi, I made several short voyages out on the Gulf of Mexico; on each trip I met great schools of porpoises and amused myself with the somewhat cruel sport of lying in the prow of the boat and shooting the animals as they rolled up within a few feet of me. But in every instance where I succeeded in hitting one—I used only a small pistol—the creature would give voice to a kind of grunt and immediately the herd would cease rising and not another would be seen until a new school appeared. The grunt was evidently a note of warning.

The **Manatus, Dugong** and **Lamantin** are all herbivorous and their flesh is excellent food, not inferior to beef or veal. In the Malay Archipelago the *dugong* principally abounds, and on account of the affection which the mother bears for its young, it is there called the *water-mother*. Not only does it exhibit rare maternal devotion, but while suckling its young, the mother holds it to her breast by means of her flippers in the most loving way. When attacked she covers her young with her body and will invariably sacrifice her life in its defence rather than abandon it. The three species are quite similar in their habits, though confined to widely-separated districts—the *lamantins* being peculiar to the South American coast, the *manatus* to Africa, and the *dugong* to the Malay coasts.

The **Stellar**, (also called *sea-calf, sea-cow* and *sea-bull*), found chiefly in the Kamtchatka seas, resembles the dugong, but its habits are little known.

ELEPHANTS.

The Elephant (*Elephas*) by its wide distribution, if, as is proper, we include fossil forms, is one of the most striking exemplifications of the sequence of animal life, and of its various evolutions in accordance with changes in the conditions to which it succeeded. The *elephant* makes its appearance first during the period of gigantic forests and the rankest growths in the vegetable world. The fossil remains are found throughout the globe, with the exception of its insular portions, and demonstrate that the now arctic and temperate regions were once tropical, and that the theory of the order of the earth's development is correct. So, too, the fact that, their mission accomplished, the *elephants* gave way to orders qualified to succeed them, illustrates alike the methods and wise guidance of the Creator.

The **Maltese Pigmy** (*Elephas melittensis*) once existed in abundance in the Island of Malta. It was about as large as a calf, and although the explanation

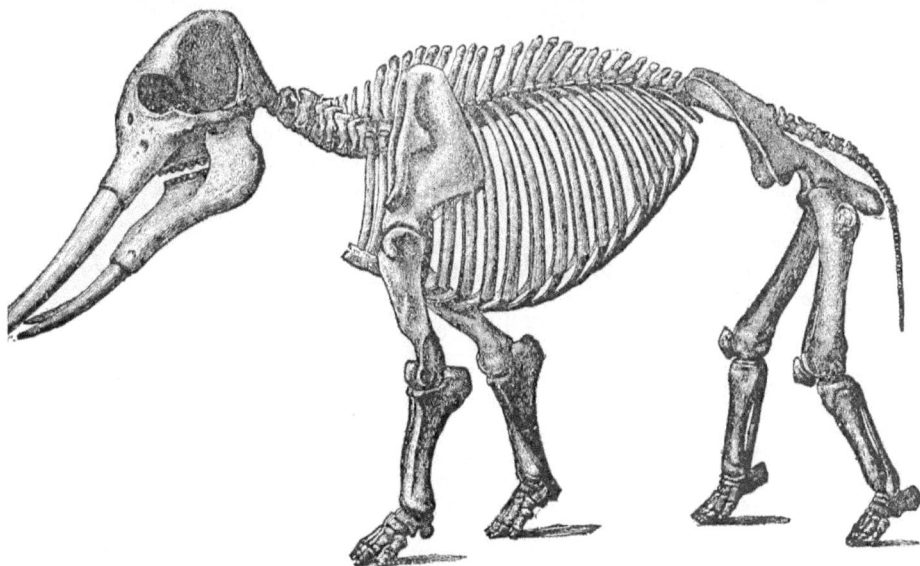

MASTODON ANGUSTIDENS.

of its diminutive size has not yet been certainly found, the presence of the fossil remains is very significant in its bearing upon the geological changes which scientists are convinced have taken place. The presence of this now extinct species in a region so far removed from the present habitat of its kind is likewise significant as a factor in the theory of genetic evolution, as announced in THE SAVAGE WORLD. Another **Dwarf Fossil Elephant** (*Elephas falconeri*), while found in Malta, extends over into Italy. Although the *elephant* can swim, it is adverse to marine adventures, and hence it is reasonable to conclude that in the palmy days of these two fossil forms the land was not, as now, separated by bodies of water, and the volcanic character of the Mediterranean region still further strengthens this conclusion. The new world,

America, being geologically more ancient than the old world, it is not surprising that its fossils represent earlier forms.

The **Alaskan Mammoth** (*Elephas columbi*) has furnished at least specimens of its teeth, and to the trained naturalist a knowledge of dentition is quite as much as the scale of a fish is said to have been to Agassiz, for he feels that with a tooth he can safely proceed to construct the entire animal. The *Alaskan mammoth* must, it is concluded, have borne a close resemblance to the Indian elephant of to-day.

MAMMOTH DISCOVERED IN SIBERIA.

The **Mastodon, or Mammal-toothed Elephant** (*Mastodon americanus*), seems to mark the time when the elephants and the classes immediately succeeding had not yet become differentiated. It was called in Indian tradition, "The Father of all Oxen," or "The Original Ox," and from the fossil remains having been discovered first in Ohio, it is sometimes spoken of as "The Ohio Beast." A **Nebraska Mastodon** (*Mastodon mirificus*), a **Chilian Mastodon** (*Mastodon humboldtii*), the **Lower Tusked Mastodon** (*Mastodon productus*) have still further contributed to enrich our American museums. Europe and Asia have furnished examples of the *mastodon*, and the **European Mastodon** may well have its technical name supplied as likely at any moment to confront the reader of works on Natural History; it is *Mastodon angustidens*.

The **Asiatic Mastodon** (*Dinotherium*) was first found in India, but remains have since been discovered in Germany; it is quite distinct as a species. Its teeth in many ways suggest a pre-historic form of the tapir, and its remarkable tusks run towards the ground, as though designed for shovelling. From its dentition it is concluded that it lived upon soft food, of which there must have been an abundance while yet the waters were contesting the sovereignty of the earth.

The **Siberian Mammoth** (*Elephas primigenius*) had brown hair nearly a foot in

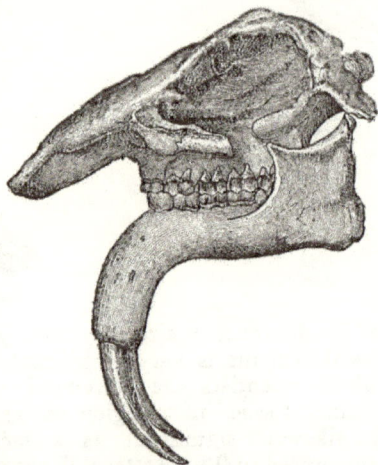

SKULL OF DINOTHERIUM.

length, and shaggy mane, and would appear to have wandered over Great Britain, France, Central and Northern Europe, Siberia, Alaska and even Oregon. The skeleton preserved in the Museum of St. Petersburg has a body sixteen feet in length and nine and a half feet high, and the weight of the entire animal is estimated at twenty thousand pounds.

Of existing members of the family the Asiatic elephant and the African elephant are the representatives. The African elephants differ from their Asiatic congeners in having extraordinarily large ears, a forehead convex and prominent, a head hanging down, tusks for both male and female, four horny hoofs on the fore feet and three on the hind feet, and callosities on the front knees. Until comparatively recently it was unknown in Europe. Of the Asiatic elephants we may name: The **Indian Elephant** (*Elephas indicus*), the **Ceylon Elephant** (*Elephas cingalensis*), the **Sumatran Elephant** (*Elephas sumatranus*), and the **Siamese Elephant** (*Elephas indicus albino*). The *indian elephant* is fifteen feet in height, and in color brown, spotted with gray. They live in moist localities

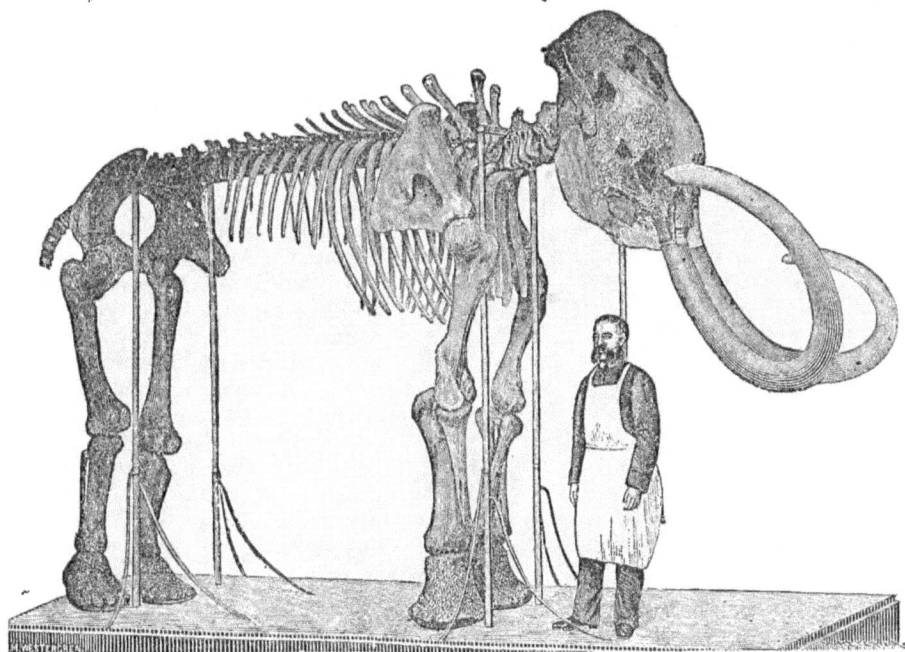

SKELETON OF THE MAMMOTH IN ST. PETERSBURG MUSEUM.

where they can find a vigorous and abundant vegetation. Their immense size justifies their large appetite which requires them frequently to change to new feeding grounds. They live in herds which are under the patriarchal government of an old male. They cover the ground with great quickness, but find it difficult to turn or to descend declivities. Their size is about eight feet in height, by ten or fifteen in length. They finish their period of adolescence in about twenty-five years and the average subsequent duration of life is about fifty years. Some are used by royalty and called the *koomareah;* some as hunters, *merghee;* and the so-called white elephants are regarded as sacred and as being protected by the spirits of the ancient kings. Of course any positive statement is always met by equally positive denial, and therefore many excellent travellers deny that the ele-

34

phant is ever worshipped. The white elephants, rose-tinted as they are, are as strik-ing as they are relatively rare. They are albinos and simply "freaks of nature."

Though the people of Burmah worship the white elephant, this observance is not so general or so much a matter of faith with them as with the *Siamese*. Longing as the Hindoo does for the relief from earthly toil and care known as Nirvana, he regards the elephant as the symbol of this life in death to come. To him the elephant seems to have achieved all that is possible for mere body, and to have attained a patience which is an evidence of progress towards the envi-able state when, losing one's identity, a person becomes a part of the mighty power which crushes mankind, instead of being an isolated being created appa-rently solely that he may be crushed. The palace or temple of the white ele-phant adjoins that of royalty, and the furnishing of his abode is no less regal and foolish. Sakyamuni, the priest and prophet, was himself at one time the occupant of the body of the white elephant. Endurance, patience, submission being the cardinal principles of his teachings, no wonder that religious enthusiasts make of themselves a tessellated pavement over which the white elephant is to walk.

MAMMOTH RESTORED.

The *Ceylon elephant* is small, tuskless (or almost so), gentle and docile. The *Sumatra elephant* is more slender, delicate and intelligent. Albinos are found in all these species, and in Siam are called royal elephants. It has been said that the average lon-gevity of the Asiatic elephant is seventy-five years, but in at least one instance, an elephant, though in captivity, lived one hundred and fifty years. The elephant plays quite an important part in Asiatic history and is often hunted for its hide and its flesh, as well as for mere sport or adventure. When Alexander the Great was engaged in his conquest of the then known world, he met among other enemies King Porus, of India, and for the first time had to contend against the *elephant* in war.

HOW ELEPHANTS ARE CAPTURED.

When the Indians need a fresh supply of elephants they proceed in one of two ways. First they may use as decoys females trained for that purpose, and who occupy the attention of the males until they have been lassoed about the legs and securely fastened to trees. After this they are left to wear out their strength and grow enfeebled by hunger and thirst, until they are not wholly intractable. Then they are fed first upon the food which they consider least palatable, and grad-ually upon what they prefer, until they learn that their entire dependence is upon their captors. The other method is to build a stout corral toward which the herd is driven, and which it is induced to enter by the Delilah-like behavior of female elephants trained for that service. Once in the corral, they are tempted one at a time by the female elephants into a smaller enclosure where they are made captive. When hunted for profit or pleasure, they are sometimes driven into pits floored with sharpened stakes; sometimes their feet are spiked, and sometimes they are shot. The vulnerable spot in the Asiatic elephant is the head. The elephant is grateful only in the sense of the cynic who defined gratitude as "a lively sense of favors yet to be received." The behavior of the domesticated elephant, like that of some persons, is better in company than when at home. Jumbo, whose fame is known to all in Great Britain or in America, was, it will be remembered, amiable to excess towards the outside world, and always tractable to his keepers at home. The oldest records exhibit the elephant as an important feature of all oriental pageants. They appeared in the Roman triumphal processions at least two thousand years ago, and were used in a war against the Gallic tribes.

The elephant is well known alike from accounts of travel or sport in the East, and as a popular member of all travelling menageries. It was well known in Egypt, and no one will have forgotten its having been employed by the Carthaginians, more especially by Hannibal

HEAD OF AFRICAN ELEPHANT.

when he crossed the Alps for the invasion of Italy. Its average weight is from seven thousand to eight thousand pounds, and it can carry burdens equal to the united efforts of from seven to twenty yoke of oxen. Its proboscis is a combination of its nose and its upper lip, and is a singular illustration of adaptation to use. To insure the most perfect strength and flexibility, it is muscular and membranous, not cartilaginous. To give perfect control it has at its command not less than forty thousand muscles. To fit it for its many and varied functions, it has the utmost delicacy and the greatest toughness. To the elephant the proboscis has to serve the uses of the prehensile tail of the monkey tribe, the beak of the bird, the air-bladder of the fish, the nose of the hound, the tongue of the ant-eater, the

tactile service of the human hand, the suction tube of the bee, the arms of the ape.
To the elephant the proboscis represents an organ for seizing, a means of respiration, scent, taste, touch, suction and power of grasping. Great is its need for a union of sensibility and power; for it must deal with the grass as well as with the full grown tree; must be able to pick up a nail as well as to lift an engine; must brush off a fly as well as smite a formidable foe. The elephant believes in the political principle of clanship, and will admit to companionship no elephant not born within the fold, nor if born, if it has ever strayed from home. It recognizes no return for prodigal sons, and once an elephant strays from the family circle, a vagrant is he to remain forever, and as if he were an Ishmael; every elephant's hand is to be against him, and his against all his fellows! These estrays are called rogues, and wandering about without other company than their own, develop abnormally a spirit of mischief and malice.

The **African Elephant** (*Elephas africanus*) is generally supposed to be smaller than his Asiatic congener, but authorities differ upon this point. Its head is rounded, and it has three instead of four nails on its hind feet. It has not in modern times been domesticated, although Roman coins make it clear that it

A BULL ELEPHANT DEFENDING ITS YOUNG.

was used as a beast of burden during the continuation of Rome as a government. Some authorities assert that the elephants used by Hannibal, by the Carthaginians, by the Egyptians and by the Romans, were from Africa, others claim that they were from Asia. Herds as large as eight hundred in number have been met, and three hundred is no unusual size. The male far exceeds in size the female. His tusks are arched, long and tapering, from six to eight feet in length, and weighing from fifty to a hundred pounds. Estimating by the annual exports of ivory (and making no allowance for the animals diverted to other uses) thirty thousand elephants a year are now being slaughtered. The ivory brings in England about one dollar and a quarter a pound, and is of much greater value than that furnished by the Asiatic elephant. The elephant, whether spraying his back in the streams of Asia and Africa, or exhibiting in a zoological garden his wonderful control of his proboscis, or carrying children about on its back, is always an object of interest. Stories about elephants are numerous and always exciting. The *African elephant* is, as travellers tell us, naturally amiable, and will attack man only when infuriated. But if the elephant does make a charge, it well behooves the hunter and his steed to make up in agility what they will find themselves to lack in speed. Unlike the Asiatic elephant, the African species is not vulnerable in the head, and he who fails to know or to remember this will not be exposed to the same forgetfulness another time, even though he escape the consequences of his mistaken course of action.

The natives insist that the elephant is naturally jealous of the

HUNTING THE TIGER BY MEANS OF THE ELEPHANT.

rhinoceros, and that this feeling extends so far that upon the mere sight of the rhinoceros the elephant breaks heavy branches from the trees and proceeds to belabor the rhinoceros and drive it hither and thither in the style of the lion and the unicorn, so familiar in nursery rhymes.

The mother elephant will protect her calf at all expense, and it is quite affecting to see her take her calf between her fore legs and herself stand the brunt of all harm. A herd of elephants having been discovered swimming down the river and not yet having learned the possible danger of allowing the approach of a steam vessel, finally discovered their mistake, and all escaped but a calf, which was captured and towed along by its proboscis. A frenzied sportman slashed the proboscis with his knife, and though the wound was immediately sewed up the calf died—possibly from surgical malpractice. As a rule the elephant, like most animals in their natural state, is temperate and abstemious, but there are temptations which even good elephants cannot withstand. If they can find the right kind of tree the elephant will get drunk on unguana. Then there ensues scenes of revelry similar to those which occur among men. Antics of all kinds; playfulness, sullenness, amiability, quarrelsomeness—all these are exhibited. The great creatures go reeling about until the scene looks almost like a Walpurgis night. Many have been the occasions when a domesticated elephant, escaping from a menagerie and its keeper, has spread terror among quiet people who neither expected nor desired their strange visitor. On one occasion an elephant undertook to run a foundry. Entering and putting to flight the merely human mechanics, it began the conduct of the shop upon a plan of its own. It tried hammers and other tools, and was quite happy in its work of devastation until at last it was tempted by a vaulting ambition to fool with the heated forge, whereupon it desisted, and with a roar of anger and of anguish rushed forth into the street, where it was captured and led back to a more orderly and familiar life. The wild elephant, as has been said, is frequently caught in pit-falls, so that if he escapes he retains a lively recollection of their possible danger. When a herd is moving, the old male, with parental solicitude, moves on in advance, and uses his trunk in testing every place that seems to him the possible trap laid by his enemies for him and his. The elephant loves seclusion and prefers to pass his time in the depths of the forest, coming forth only to get water, and this at times only once in several days. He never approaches nor retires from his watering place by the same routes, seeming never to forget that "eternal vigilance is the price of liberty." Elephants, as a rule, sleep standing, but bulls will sometimes lie down on their side. They take but little sleep and always feed upon awakening. They are destructive not solely because of their enormous appetites, but also because they will wantonly destroy branches which they leave untasted.

ANECDOTES OF THE ELEPHANT.

The fact that the elephant's feet are padded renders his step noiseless in spite of his great weight, and his extraordinary nimbleness seems almost incompatible with his great bulk. Livingstone in his "South Africa" tells of a party of natives who hunted an elephant with spears. When first seen the cow was suckling her calf, and upon discovering danger at hand, immediately put herself between danger and her young. The natives began with a triumphant song, and then threw a volley of spears which fastened themselves in the body of the elephant. She retreated, all bleeding as she was, keeping guard, however, over her calf. In crossing a stream the young was killed and the mother shot so full of spears as to resemble a mammoth hedgehog. Upon discovering the loss of her calf, the cow became furious, and charged again and again upon the natives, who escaped only by taking advantage of the elephant's inability to turn quickly. Finally, weak with loss of blood, the fond mother fell to the ground and, with a last roar, died and became the prize of the hunters.

The elephant if not specially inclined to remember favors, seems never to forget or forgive an injury. The experience of the tailor is possibly so old as to bear repetition. "The Discomfited Tailor" might properly be given for a title. It seems that a domesticated elephant, in going to water, passed this tailor's shop every day and that he fell into the habit of giving it, each time, some elephantine tidbit. But one day, worried by tardy debtors, irritated by the annoyances of life, or possessed by a sudden spirit of malice, he gave the elephant nothing but a jab of his needle in its nose. The next day the elephant did not stop as it went by, but on its return drew up in front of the tailor and deluged him with the enormous volume of water with which he had filled his trunk.

At times the calf will reciprocate the affection of the dam, although usually, like children, they ar too self-engrossed to be thoughtful. The grief of a calf over its mother's death is thus told by a traveller: A party had been out elephant-hunting and had succeeded in shooting several before the others took to flight. Returning the next day to collect their

HERD OF ASIATIC ELEPHANTS.

booty, they were met by a young calf elephant which ran up to them, twisted its trunk about their arms and legs, and seemed very much interested in securing their co-operation in some enterprise. Presently they approached the spot where lay a dead elephant, and the calf at once began running round it, attempted to raise it to its feet, and all the time manifested the liveliest grief and uttered the most pitiful moans. Finding at length that all its efforts were vain, the calf rejoined the travellers, as if to say, "Now that you have slain my mother you must take care of the orphan." I have spoken of the fearful destruction of elephants for the sake of the ivory of their tusks, but it should be added that the wanton wastefulness of hunters—sportsmen they can hardly be called—multiplies this greatly. The tusks have no roots, but grow out of perennial pulp, so that as necessity requires they are renewed. Bullets have been found imbedded in the tusks which must have lodged there while the place of their deposit was still soft as the tusk is at the base, and which must subsequently have been carried along as the new ivory below pushed along the older ivory above. The elephant, like man, has milk-teeth, which in due course of time drop out and are succeeded by the second and permanent teeth. These teeth are peculiar, inasmuch as each one has three different structures. On the out-side the tooth is constituted for crushing, the next layer is adapted to tearing, and the last layer for grinding. Perhaps, as an illustration of the inter-depen-dence of all terrestrial life, it may be worth while to mention that the invention of celluloid, and the discovery of vegetable ivory, has, by furnishing a substi-tute for ivory, done much to postpone the time when the elephant will be an

ELEPHANTS IN SERVICE.

extinct animal. Elephants have been known to have three tusks, but these were mere *lusus naturæ*.

Indian elephants when domesticated are made to earn their living, not simply as companions to the hunter, beasts of burden on a journey, or as ele-ments of a triumphal procession of royalty. By the British they have been made to act as animated gun-carriages, the howitzers being securely strapped to their backs and fired while in that position. By the natives they are used in " logging " and in piling timber, and their great strength and patient intelli-gence render them the most useful, industrious, well-behaved and uncomplaining of lumbermen.

The natives transport the ivory either in the shape of long, heavy strips, or cut up into squares and regular figures, and we present an illustration of natives thus carrying their spoils to market. The elephant is troubled with a highly emotional nature, and not unfrequently falls dead from excessive excitement. His memory is tenacious and he does not forget a kindness, even though he sometimes offsets it with recollection of subsequent ill-treatment. On more than one occasion a traveller who has cured the wounds of an elephant has been held in the liveliest remembrance and recognized affectionately after prolonged absences. At times an elephant will escape and, joining a herd of wild ones, will become feral in its nature. And yet repeated experience has proved that it recognizes the voice of its former master, and speedily yields to the habit of obedient submission. Still the stories about tame elephants turn mostly upon their avenging slights and wrongs. On one occasion a human brute, after having fed an elephant in the menagerie, suddenly stuck a large pin into its outstretched trunk. A year or two later the same person happened to revisit the menagerie, and while wholly unsuspicious of any ill-will on the part of the elephant, was surprised by its seizing his new silk hat, tearing it to tatters, and throwing the pieces at him. A traveller tells of the pursuit of a small boat by an elephant whose dignity had been disturbed by the boatmen, and of

AN ELEPHANT FIGHT.

the elephant being satisfied merely to pursue and constantly drench them with water for the distance of a mile or two.

Though the elephant's intelligence is limited in range, it is certainly very great in degree. His nature is extremely emotional, and manifests itself in a sense of humiliation, in admiration of its own cleverness, in affectionateness as well as in the most terrible anger and the most abject fear. With elephants and the rest of the animal kingdom it is unsafe to build upon experiences with a single individual. The individual elephant as well as the individual human being is, to some extent at least, the creature of its temperament, surrounding and training. Many stories are related to illustrate the affection and kindness manifested by wild elephants by surrounding and protecting the bull whose tusks exposed him to the cupidity of the hunter; instances, too, are numerous in which members of a herd have supported and assisted one of their number which had been wounded.

The elephant's knowledge of its own bulkiness, and its prudence in testing the solidity of stairways, bridges and similar structures, have been spoken of frequently

ASIATIC ELEPHANTS SPORTING.

" in the books."   I may add an instance in which it being necessary for the draught elephants to ascend a hill by a temporary stairway, the pioneer refused to tread upon any plank until he had tested its security, and in case of doubt, compelled the builders to reconstruct their work. Upon safe arrival on the cliff, the elephant gambolled about as if indicating that he was well through with a hazardous undertaking. A smaller elephant followed, and the pioneer watched his progress with the most unflagging interest, and as the climber approached the top, reached over its trunk and lent it sympathy and support. When the second elephant had achieved the summit, the two linked trunks and executed a short dance of joy. In proceeding to the banks of a stream, the commander-in-chief will halt his family at some distance, while he assumes the dangers and responsibilities of a scout. Having satisfied himself that there is no danger from pitfalls or from any enemy, he will return to the edge of the jungle and call forth a number of elephants which he posts as sentinels. After another and final scouting expedition he will give a signal, and the herd will rush tumultuously to the water's brink. The elephant shows great fortitude under suffering caused by its driver, or under necessary pain inflicted by a doctor. It has been known

when being treated with nitrate of silver for an affection of the eyes to lie unresisting, contenting itself with suppressed moans, and to voluntarily submit itself to further treatment. In at least one instance a cow caught its badly wounded and crazy calf and held it firmly while its wounds were being dressed. In another case the elephant stood patiently while a surgeon cut out an ulcerated spot on its back. The possible longevity of the elephant has been ascertained only approximately, for the East Indian government had in its daily employ one elephant whose services it had used for a full century, and another which had been on its pay-roll for fifty years. The decoy elephants seem to learn from man a love of cunning, and appear to enjoy the sport of capturing their kind. As in India the capture of elephants is a regular and fully-organized in-

A HEAD SHOT.

dustry, the devices of the decoys are infinitely varied: Two decoy female elephants having selected the most magnificent tusker approached him, Delilah-like, and once having separated him from the herd, guarded him on each side until the natives had slipped the noosed rope over one of his hind legs. They now divided their efforts, one of them keeping off the rest of the herd while the other wound the free end of the rope around a tree so as to get the advantage of a capstan. The captive having himself circled about the tree so as to prevent the failure of the plot, the elephant which was no longer occupied with the herd, drove it back while the other decoy carried out its original scheme. The captive finally resisting any attempt to pull it nearer to the tree, the elephant not busy with the rope deliberately butted it back inch by inch, and foot by foot, while its companion "hauled in the slack." If captive elephants refuse to lift their front feet so that they can be noosed, the decoys will risk their own safety in attempting the enterprise, sometimes holding up the foot by putting their own leg under the great weight. The elephants trained as lumbermen have the weakness of other slaves who receive for their hard labor "more kicks than pence." If not kept in sight the elephant will quit work and use "his master's time" in the lounging which he enjoys. An example of the politeness of which an elephant is capable, is furnished by the case of one found carrying an ebony log through the jungle to its master's lumber pile. Meeting a mounted traveller where the jungle was thick and the road narrow, it threw down its load and backed itself into the thorns, uttering a sound as if to say "After you, your honor." The traveller not being quick to take the hint, and to accept the civility, the elephant repeated its salute and its retiring again and again, until at last the rider passed by, when it returned to the road, picked up its burden and resumed the drudgery of life. I have spoken of the extreme delicacy as well as the power of an elephant's trunk. This is well illustrated when the elephant is allowed to pluck a fan, and to use its long handled brush in keeping itself free from flies. The lightness and grace of its motions are then beyond those of the fairest daughters of the most enervating clime. An elephant will generally betray his wrong-doings by a certain air of sheepishness. On one occasion the master having arranged his oven and chained his elephant, departed on

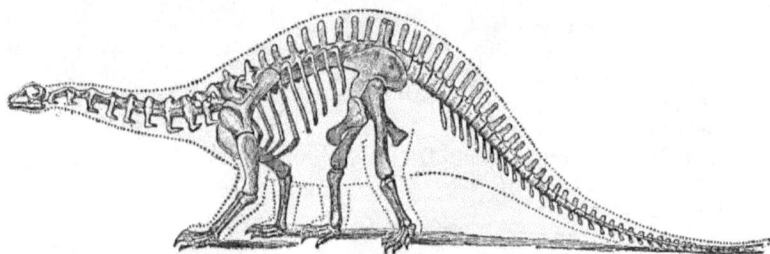

SKELETON OF BRONTOSAURUS, FROM AMERICAN JURA.

some errand. As soon as he was safely out of sight, the elephant unchained himself, robbed the oven, and again chained himself up, being, however, unable to do more than to wrap the chain about his leg. The owner on his return, found himself dinnerless, but also found in the preternaturally innocent behavior of his elephant, a clue to the depredator. Of experiences in hunting there is no lack. A hunter was saved by falling from his horse, after which the elephant continued to charge in entire disregard of the man over whose body it passed without doing him any further injury than that of paralyzing him with fear, and forcing him to conclude, measurements to the contrary notwithstanding, that the mammoth must have had a successor in the mountain of flesh which, for a few moments, loomed above him.

It is among the singular facts in elephant lore, that while the forehead is the vulnerable spot of the Asiatic species the heaviest balls do not seem fatal if fired into the forehead of the African elephant. Sir Samuel Baker relates that he fired three heavy bullets into the forehead of an African elephant, and that though the three wounds were within the space of three inches, they apparently were harmless. The native aggageers, when hunting the elephant, hamstring it, if it be awake, and cut off its trunk, if it be asleep. Three of them, mounted upon their trained horses, will provoke a tusker to pursue one of their number while the other two follow close upon the heels of the elephant, changing offices if the elephant concludes to pursue one of the followers. As soon as the vanguard catches up to the elephant, one of them, throwing the reins to his companion, slips off of his horse and by a dexterous movement cuts the sinews of the elephant's leg.

While three aggageers were hunting an elephant, one of them dismounted just in time to find himself and horse both knocked down. The elephant stepped on the man's thigh, but continued to pursue the white horse. A second aggageer, though on foot, successfully hamstrung the bull before it reached the *cul de sac* in the jungle, where his own horse and the one mounted aggageer were penned up without possiblity of escape. A German wounded an elephant, which he found the next day partially devoured by a lion. Tracking the live game, he succeeded in mortally wounding the lion, but it did not die until it had charged upon the hunter, struck him with its paw on his head, seized him by the throat, and witnessed his dying agonies. A companion succeeded in blowing out the lion's brains, but not until the first hunter was beyond the need of mortal aid.

The feral elephant, or domesticated elephant, which has returned to a savage

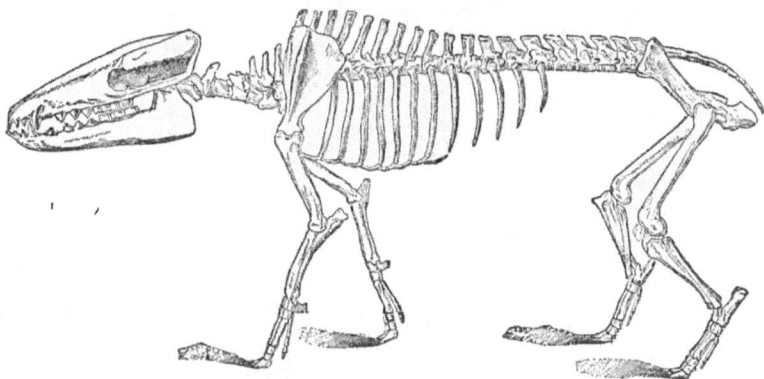

GREAT BEAST OF THE COAL PERIOD (*Anthracotherium magnum*).

life, will still obey the orders of its former mahout or keeper. As another illustration of the elephant's recollection of indignities, may be mentioned the fate of a practical joker, who, having given an elephant a cayenne sandwich, was, six weeks afterwards, deluged with dirty water. The elephant's loyal service may be illustrated by an incident in the lives of two elephants employed in bringing buckets of water from a stream. The larger animal robbed the other of his bucket, and while filling it was butted by the smaller elephant into mid-stream, and dropping the bucket, the rightful owner reclaimed it, and trotted away with his burden. A herd shows its knowledge of the aim of the hunters by placing the tuskers in the centre of a circle into which they form themselves. A performing elephant, which had changed owners, was grievously wounded because it refused to trust an insecure platform. It stood the misunderstanding with only a groan of protest, and when its former owner was summoned, and bound up its wounds, it manifested the liveliest appreciation, embracing him with its trunk, and as soon as the platform had been strengthened, ascending it without waiting for the order. A rogue elephant which had treed a hunter, deliberately constructed a platform in the hope of reaching its adversary.

An African hunter got lost in an elephant jungle, and was scented by the herd, but for the time being the hunter remained undiscovered. He was making the best of his way along an elephant path, when he suddenly came upon a hippopotamus, to which he conceded the right of way, although immediately thereafter he shot it as a proper punishment. After three days he came directly upon one of the elephants, and although the bullet knocked it over, it got up and made off. He found it again, and the elephant charged upon him, only to receive a fresh bullet. Again tracked, shot while charging and brought to his knees, although still able to stumble to its feet it was powerless to do more than to stand still while the hunter finished it. On one occasion a hunter ensconced himself in a tree, and while twenty elephants were defiling beneath him, broke a bull's shoulder with a single shot.

One of the most remarkable things about an elephant is the entire noiselessness with which it is able to move its vast bulk, even when pushing its way through the densest thorn. Once a hunter shot an elephant and was horrified to hear the shrill trumpeting of a whole herd by which, without his knowledge, he was surrounded. He was com-

BATTLE BETWEEN ELEPHANT AND RHINOCEROS.

pelled to seek safety in the hollow of a fallen tree, and the elephants after vain attempts to extract him from his hive, finally relieved his mind by their unwilling departure.

CONIES (Hyrocoidea).

Though the Conies were known to the great Greek naturalist, Aristotle, and are mentioned in the Bible in Proverbs: "The high hills are a refuge for the wild goats, and the rocks for the *conies*," it was reserved for Agassiz to re-discover the *cony*, so to speak, and bring it within the boundary of well-understood animals. This creature we now know quite well, though from its singular characteristics we are not able to classify it, since it exhibits the peculiarities of rodents, ungulates and the probiscidea. This little animal is now regarded as a transitional form from the elephant to the hoofed-animals, the fossil forms of *toxodontia* alone intervening.

The **Daman of Syria** (*Hyrax syriacus*) is undoubtedly the creature referred to in the Bible, and mistakenly called a coney or cony. Another species is found at the Cape of Good Hope, and is called by the natives the **Klipdach** (*Hyrax capensis*). Still another species, the **Mozambique Daman** (*Hyrax arboreus*) is found in South Africa, while the **Daman of Guinea** (*Hyrax sylvestris*) is represented in West Africa. There are yet other species, but these are the ones best known.

The **Daman of Syria**, or coney of the Bible, is about the size of a rabbit, and is dressed in coarse fur of a brown color. It feeds upon plants and shrubs, makes its nests in the rocks, is susceptible of domestication, and possesses neither value nor interest, except in so far as it vindicates the accuracy, as well as the effectiveness, of Bible imagery, the mistakes to which well-meaning and careful students are exposed, and from the endless vexation which it has occasioned both Biblical students and naturalists. It is said always while feeding to station sentinels' and scouts, who announce by a shrill cry the approach of anything to be feared. The **Klipdach** has, so far as known, no characteristics different from the *daman of Syria*. The *Mozambique daman* is spotted along the back and wears longer hair, while the *daman of New Guinea* is arboreal, living in the hollows of trees. The *damans* have the appearance of a furred or

CONIES (*Hyrax syriacus*).

woolly pig which has lost its tail, or at least so much of it as to leave nothing but a rudimentary, fleshy, round root. Its face is not unlike that of a cat or monkey, while the longer, light-colored hair which, extending from the under jaw to its breast and including the cheeks, has all the appearance of a short fur bib or tucker. The paws are padded, hoofed and clawed, and instead of being uniformly four-toed, its hind feet have but three.

TOXODONS.

The **Toxodon, or Bow-toothed Fossil**, is of interest as an apparently connecting-link between the daman and the ungulate. In size it is about the stature of the hippopotamus, and exhibits structural resemblances to the hoofed-animals, the rodents, the ant-eaters and armadillos, and to the dugongs and manatees.   The skull slopes forward and upward, and is flattened.   The upper jaw contains four incisors and seven molars; the lower jaw, six incisors, six molars and two canines; the molar teeth are rootless and exhibit the power of continuous growth.   The trunk and limbs are like those of the elephant and of hoofed-animals.   The shoulder blade is like a tapir's, and the bone of the upper part of the arm like that of the rhinoceros; the foot resembles most nearly that of an elephant.   The *toxodon* has been discovered only in pieces, and some of his members are still missing, but as the naturalist was able to reconstruct the fish from a single scale, so he has filled out in plaster what would seem to be the necessary substitutes for the missing parts.   To go into a large and carefully arranged museum (of which we have several in this country, as that in New York or the one in Boston), is interesting only to one who brings with himself knowledge enough to cause him to take an active and intelligent interest. Doubtless many a one has, like myself, in my earlier days, stared ignorantly at the re- mains of the mega- therium or of the mas- todon, and wondered what possi- ble object there could be for such expendi- ture of time, labor and money. But once get the idea of the evolution, progres- sion, and

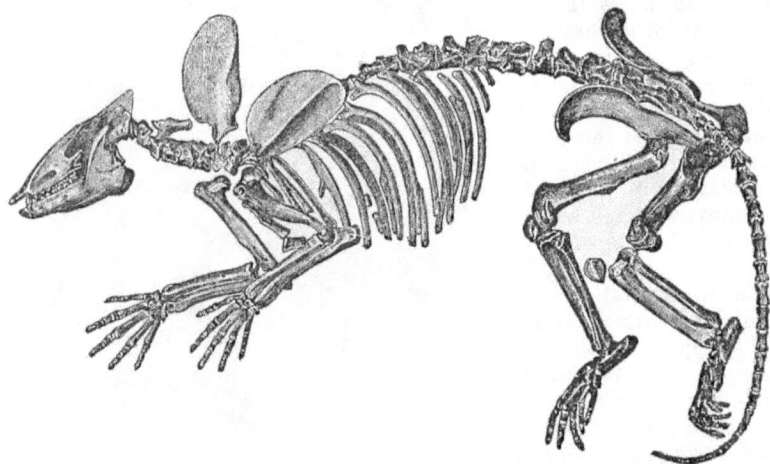

SKELETON OF THE PHENACODUS PRIMÆVUS.

precession of animal life, and all that is changed.   It is my hope to make only such mention of extinct forms as shall illustrate the principle of modern scientific investigation, and possibly to induce the reader to make real the illustrations by visits to museums, where the representative types can be found.   Who, in a tour of observation, ever realizes the wealth of enjoyment, information and stimulus which can be derived from the museums of our great cities?   Rather, do not most travellers go through these repositories with even greater weariness and ignorance than they do the art galleries?   Does the typical traveller (and the Americans are a nation of travellers) ever think of concerning himself deeply about museums of natural history, academies of fine arts, the great organized industries of manu- facture and commerce, or the libraries whose shelves contain wonders to them unknown?   Or does he return with nothing better than a hasty view at parks, buildings and cemeteries; or than a knowledge of the superficial differences in dress, and furniture and viands?   When Addison visited Italy he took with him knowledge only of such scenes and objects as the poets had touched upon, and returned in entire ignorance of what had been described in prose. Yet Addison was an intelligent man, whose outfit, partial as it was, far exceeded that of most modern travellers. What can be the significance of a visit to the field of Waterloo for one who does not know the great events to which this battle was the close? What possible interest can Westminster Abbey have to a visitor ignorant of all that makes it so significant as the resting-place of England's vanished heroes? But with guidance no greater than I can hope to furnish here, the reader will find that, having something to see with, as well as to look at, that which before was wearisome will become the source of active pleasure.

The **Nesodon** as yet exists for us only in the shape of skull and teeth, but it seems that he must have been a smaller species, more nearly approaching the hoofed-animals.

UNGULATES.—TAPIRS AND RHINOCEROSES.

We have now reached the order of **Hoof Animals**, or **Ungulates** (*Ungulata*). The hoof is hardened and of modified skin, which forms a case for the last joint, and serves as a substitute for the soles or pads found in the elephant. The *ungulates* walk upon the ends of their toes, and hence the usefulness, if not the cause, of their conformation. The *ungulates* are sub-divided according as they have one toe or two: those which live on dry ground having one toe, and those which live in marshes, two. The *ungulates* illustrate the method by which these changes are brought about in nature. First, in locomotion upon ground, the middle toe received most of the

TAPIR.

burden and developed at the expense of the others, which, as they became more and more useless, grew more and more rudimentary.

The **Odd-toed Ungulates** (*Perissodactyla*) have the axis passing through the third toe (if there be so many as three or more than three). They are large-sized animals, thick skinned, have sparse if any hair, and their skulls are elongated. The earliest fossil form is called *phenacodon*, which was succeeded by the *lophiodontidæ* and the *calicotheriidæ*. These animals show the reduction

35

to three toes, and the beginning of the change from tubercular teeth to the crescent pattern of the horse. Then come the *palæotheridæ* and the original rhinoceroses. These have but three toes, which indicate an approaching diminution of number.

The **Tapir** stands between the elephant and the hog. The **American Tapir**, or **Mborebi** (*Tapirus americanus*, or *terrestris*), is found abundantly in the South American tropical forests. It keeps near the water, of which it is very fond, and contrary to the popular idea in regard to the swine family, it is an excellent diver and swimmer. Its height is about four feet, and its build is proportionately strong. Its hide is a protection, as it rushes through thorns and brambles. It is naturally peaceful, but if wounded becomes aggressive

HUNTER ATTACKED BY A WHITE RHINOCEROS.

and dangerous. It whistles instead of grunting, but unlike the hog exercises its vocal powers but seldom. Its color is brown, and it wears a short, standing, black mane. While young it is generally spotted and striped with yellow. It is susceptible of domestication, but its size and active curiosity are adverse to its becoming a favorite and a pet. Its habits are nocturnal, and it remains true to the single mate which it has selected.

The **Central American Tapir** (*Elasmognathus bairdi*) is black or blackish-brown; the cheeks and sides of the neck, red; the chest, throat and chin, gray. It exceeds in size the American tapir.

The **European Tapir, Malayan Tapir**, or **Kuda-ayer** (*Tapirus, indicus* or *malayanus*), has its body so clothed in white as to suggest its having run off with some one's clothes-line, or its being engaged for a sheet and pillow-case party. The deep black of the rest of the body renders the contrast quite startling. It has no mane, but triumphs over its American congener by possessing a much longer proboscis, as well as in size. It is not a swimmer.

The **White Rhinoceros** (*Rhinoceros simus*) is nocturnal in its habits, about six and a half feet in length, acute of hearing, keen-scented, near-sighted. Its speed, when hurried, is greater than a man's and less than a horse's. It wears two horns, of which the front one is straight, flat, and from a foot and a half to four feet in length, while the posterior one is much shorter. It is one of two white species, is larger than the black species, has an elongated head, the muzzle of which, however, is square. It is patient even when attacked, unless it has young to protect. On one such occasion a rhinoceros turned, thrust its horns into the belly of a horse, and having thrown it off of its feet retreated without attacking the hunter. It has Africa as its habitat.

INDIAN RHINOCEROS.

The Long-horned **White Rhinoceros** (*Rhinoceros oswellii*) is rare, and is found only far in the interior of Africa. The front horn is curved forward, so as to enable it the better to tear up the ground, an exercise in which it often indulges. It is sufficiently long for manufacture into various weapons and walking canes. Though the rhinoceros may be found in company with others, this association is purely accidental, as it is in no sense gregarious.

The **Keitloa, Equal-horned,** or **Blue Rhinoceros** (*Rhinoceros keitloa*), though smaller than the white, charges an enemy without waiting to be attacked.

In spite of its name it is pale yellow. Its horns are of equal length. It is often called the *black rhinoceros.*

The **Two-horned Borneo Rhinoceros** (*Ceratorhinus sumatrensis*) has its skin folded, not into shields, but into capes on the shoulders and haunches. It is of a dark slate color, and measures about eight feet in length.

The **Rhinoceros of Assam** (*Ceratorhinus lasiotis*) is taller, smoother and paler than the preceding and is light brown in color.

The **Indian Rhinoceros** (*Rhinoceros unicornis,* or *indicus*), is one-horned, about ten feet in length, and its skin is formed into a number of shields, which are covered with tubercles. The folds of the collar come off into a dewlap, and the shields over the withers and each fore leg is triangular; the hindquarters are shielded as far as the knee-joint. It is about eight or nine feet in length and weighs several thousand pounds. This is the rhinoceros which is read of in Roman history and which is most frequently seen in our zoological gardens and museums. Its habitat is Hindostan.

The **Rhinoceros of Java** (*Rhinoceros javanus*) is smaller, has a larger upper lip, and a larger neck shield, which is saddle-shaped. It is nocturnal in its habits and does great damage to the coffee and pepper plantations, and to cultivated plants. It is more amiable than most of its kind.

The **Rhinoster** (*Rhinoceros bicornis*) has two horns, which are brown, tinted with green. It is

RHINOCEROSES FIGHTING.

about eleven or twelve feet in length, and yellowish-brown in color. It is found at the Cape of Good Hope, and is called the *borele,* or *little black rhinoceros.* The front horn is long and bent backwards; the other, short and conical. It is very active, frequently aggressive, very fierce, and very dangerous. It sometimes quarrels with its own kind. It is nocturnal, and sleeps soundly during the day. Like the other rhinoceroses, it uses its horn to dig up roots. Its tendon Achilles or most vulnerable part is back of the shoulder.

The fossil forms are the *aceratherium,* which has four toes and no horn; the *canopos,* which reduces still further the toes of the front foot to three; the *aphelops,* which had achieved the teeth and horn; the *diceratherium,* which exhibits two rudimentary horns; the **Siberian Rhinoceros** (*Rhinoceros trichorhinus*) which persisted longer than the mammoth and which was covered with long hair.

The young of the rhinoceros have no horn, and the development of this weapon extends over many years. It is not set in the skull, but is held only by the skin, from which it may easily be separated. As an illustration of the wounds which they inflict upon each other mention may be made of the shooting of one animal which had had a handsbreadth of hide and flesh torn from him. His own horns consisted of one two feet in length, and a second which, though but three-quarters of a foot, was as sharp as the finest dagger. They are sometimes trapped by the natives whose device is quite ingenious. As the animal is very wary, a pit is dug in the path by which he returns to his resting-place, and spikes, like the spokes of a wheel, are attached to a rope and placed therein—the other end of the rope is then made fast to a heavy log, and every precaution is taken in covering the pit-fall to give it the appearance of having previously been walked over by the rhinoceros. If he does not suspect the device he steps upon the mere covering of earth, falls upon the spikes, which tangle him up in the rope, and though he escapes and drags the log some distance, he is easily followed and dispatched. His horn is no part of his skull, and can be removed by cutting away the skin. It rests upon

A RUNNING FIGHT.

an arch, formed by the bones of the face, and thus protects the brain from concussion. This horn was formerly regarded as a discoverer of the presence of poison, and was therefore manufactured into drinking cups for the nobility. The fact that the rhinoceros seems, as a rule, not to attack when it has a fair view of its object, would seem to indicate that its apparently wanton fury towards logs and trees and other senseless things, arises from its imperfect vision and its distrust of novelties. The thickness of the skin makes it proof against insects, a protection much needed by a creature of its habits.

Although *R. simus* is generally spoken of as the white rhinoceros, there is not much difference of color between it and *R. bicornis*. It is a huge ungainly beast, with a disproportionately large head, a large male standing six feet six inches at the shoulder. Like elephants and buffaloes they lie asleep during the heat of the day, and feed during the night and in the cool hours of early morning and evening. Their sight is very bad, but they are quick of hearing and their scent is very keen; they are, too, often accompanied by rhinoceros birds (*Buphaga africana*), which, by running about their heads, flapping their wings and screeching at the same time, frequently give them notice of the approach of danger, and are further of service in ridding them of parasites. When disturbed, they go off at a swift trot, easily distancing a man on foot, but they are no match for a good horse.

The anterior horn of a full-grown animal is from eighteen inches to over four feet in length, a cow having a thinner and usually a longer horn than a bull. Occasionally they are curved backward, but generally straight and flattened by friction on the anterior surface. The posterior horn may vary from three or four inches to two feet, and there appears to be as much variation in relative length as in individuals, which fact has led to no little confusion in fixing the species definitely.

The rhinoceros is said to be easy to kill if one shoots him in the neck in the region of the withers. But on the other hand, the animal is not possessed of such amiability and weakness as to render him a desirable foe. Of course most hunters are led by their pride to describe their own skill and prowess, rather than the exciting adventures which terminated in their favor, and the reader is less affected by the boasted skill of a hunter unknown to him personally than by the dangers of the sportsman which so appeal to the imagination. An African traveller relates an amusing incident by which he came near losing his camp equipage, while discovering that a red blanket was as objectionable to a rhinoceros as to a mad bull. The creature having gored the blanket and having thus acquired an unexpected and unlooked for ornament which not only blinded his eyes but also interfered with the natural use of his front legs, started off, blanket and all, and ran a race like that of Cowper's celebrated John Gilpin, only having to "carry weight" in the shape of a fatal ball; the rhinoceros ran in only one direction until his life's blood had ebbed away. It not unfrequently happens that unskilful native-hunters will wound the rhinoceros only superficially and then the creature will at once "carry the war into Africa," and if Africa is not both spry and lucky he will not be let off with being chased hither and thither. Presently it will be found as unfor-

HEAD OF INDIAN RHINOCEROS.

tunate to supply the native African with fire-arms as a similar experiment has always proved in the case of the North American Indians, for they will speedily render extinct animals which will yet be needed for the support of a by no means limited population. Of course the costliness of the trip keeps the number of European sportsmen within limits, but re-enforced by the natives, no game preserves can be expected to hold out. So far as the rhinoceros is concerned, it must not be forgotten that as a food-animal it is held in high esteem. Naturally, when accompanied by its cub, the rhinoceros is specially dangerous; the cub, however, if taken away from "its deceased parent," will not object to consorting with domestic oxen and cows. An exciting ride on a rhinoceros, which, though not resulting fatally to the rider, discouraged further experiments, is related by a traveller. While hunting, a kaffir used his opportunity to jump upon the back of a rhinoceros. The creature, mad from so unaccustomed a burden, ran wildly but swiftly and every moment bore the rider further and further from hope of relief. To tumble off was fatal, for then the rhinoceros would charge upon him; to stay on was not only uncomfortable and dangerous, but was rapidly growing impossible. Finally the native took off his blanket and threw it upon a bush in front of the rhinoceros. Seeming now to ·have discovered its tormentor, the animal charged into the blanket while the kaffir slipped off and beat a hasty retreat. In a case where there was a double charge, the shooting of one seemed only to further infuriate the other which charged and charged again, until it fell a victim to its own invincible courage. On another occasion a surprised hunter was twice knocked down, and owed his final escape to the contemptuous magnanimity of the rhinoceros. The rhinoceros and hunter coming suddenly upon each other, the animal was ready but the hunter was not. The creature charged so fiercely as to carry the man off his feet, while it passed clear over him and dug its horns into the

KEITLOA, OR BLACK RHINOCEROS.

ground. But before the hunter could scramble up the rhinoceros recovered itself and charged again, this time ripping the flesh the full length of the hunter's leg. Having thus avenged its wounded honor, the creature stalked away with the greatest indifference to the recent object of its wrath.

The gladiatorial shows are best known to us through the extravagant public spectacles provided by Roman politicians, as we rarely think of the Spanish bull-fights in the same ·light. But among the barbarian natives of Asia, and specially of Africa, the value of human prowess leads to frequent royal amusements of this kind. Seated in an amphitheatre, carefully protected against the intrusion of the animals in the arena, the savage monarch, his courtiers and attendants, watch the contests between all kinds of pugnacious animals. At one time two ugly-tempered rhinoceroses will be irritated to frenzy and then pitted against each other. One will be painted red, or yellow, or blue, and the other white, or red, or green, in order that one may follow every movement of each combatant. They will fight until exhausted, trying to oppose head and jaw to each other's horns, and thus to escape the throat thrust which finally gives the victory to the conqueror. Native attendants will from time to time throw water upon them that their energies may be taxed to the utmost.

Again, two male tusked elephants will be driven to combat with each other, and will shriek and struggle, and charge again and again until success falls to the share of one of them.

Still again, a man will be pitted against an excited elephant, and the

RHINOCEROS HUNTERS.

spectators watch the contest between power and subtlety until either the elephant seizes the man and pins him to the ground with his knees, while it viciously gores him with his tusks, or the native, using the opportunity, succeeds in hamstringing his bulky foe.

Yet again, rhinoceros and elephant will be joined in a death struggle; the one endeavoring to rip the other to pieces, and the other to put his enemy under his feet and grind him to powder. Two buffalo bulls will lock horns and butt each other, until at last one is successful in pushing back its foe, and forcing it to resign the contest; and finally, natives, with their hands clad in the skin and claws of the lion, the leopard, the tiger, or the panther, will fight like wild beasts until one or both fall dead.

While a party of hunters were occupied in cooking an antelope, a rhinoceros intruded upon the camp, scattered all the camp equipage, and after having been wounded by an elephant-gun, calmly walked away. The next day the hunters tracked him and shot him anew, but even then he compelled them all to seek refuge in the trees, while it galloped off to a jungle and began mixing up its tracks. It was again tracked and shot, and a second time the hunters were obliged to climb a tree. Finally its immense power of vitality began to be exhausted, and it at last succumbed to the succession of bullets buried in its flesh. On another occasion six rhinoceroses charged upon and treed the hunters, though one of them was driven, torn and bleeding, for a long distance through the thorns. A tracking of one of the creatures resulted in a new exhibition of rapid climbing, and thus the pursuit went on with varying fortune until the rhinoceros had received nine mortal wounds.

Two sleeping rhinoceroses were surprised by hunters, and charging one of them, although receiving the contents of a gun directly down its throat, was able to first trot away and then to gallop off. Seven mounted hunters chased the pair for two miles, running neck and neck, but were never able to get nearer than two yards distant. On another occasion a *keitloa* or two-horned rhinoceros escaped on the run, after receiving a sabre cut from the aggageer—a small feat for an animal which, apparently, can run as fast and as long on three feet as on four.

Rhinoceroses are trapped as follows: A two-foot hole is dug in the pathway which he frequents, and a running-noose is laid on top of a hubless wheel whose sharpened spokes overlap, the other end of the rope having been made fast to

AN UNPROVOKED ATTACK.

a large tree planted slantingly at some distance. Rope and wheel are buried, and the ground smoothed over with a branch, in order to destroy the human scent. The animal, upon putting its foot into the hole, nooses his leg while the wheel renders release impossible. The five or six hundred pounds weight of tree the rhinoceros will drag after it, until being caught by the trees he becomes a captive doomed to death, for his intelligence is insufficient to suggest to him the cutting of the rope. A hunter, having dismounted and tied his horse, soon espied a rhinoceros charging directly upon the steed, and nothing but a lucky shot prevented the change of an equestrian into a pedestrian. Once, when one of a couple was wounded, the unharmed one returned and, adjusting his pace to that of his companion, walked off in his company. The next day the wounded rhinoceros was found lying dead, while its companion stood guard over the body. Upon receiving bullet wounds it fell repeatedly, but until its strength was utterly exhausted it always struggled to its feet and again charged upon the hunters.

UNGULATES.—HORSES.

The **Tarpan**, or **Wild Horse** (*Equus caballus*), is found throughout the steppes of Asia, and of the Oural Mountain region. These uninviting prairies

RHINOCEROS FIGHT.

are, except when carpeted with green and illuminated by flowering shrubs, awe-inspiring from their seeming want of limits, and excite dread from the wearying sameness and unattractive barrenness which surrounds one on all sides. Think of the ocean as frozen, but with the ice covered with dust, which from time to time is hurled hither and thither by fierce blasts of wind; think of the desolateness of such motionless water, the dreary sameness of the prospect, and the unreasoning terror inspired by mere immensity, and we may form some slight idea of these terrestrial seas. Still even here is animal life, and life so constituted as to flourish where even the imagination of man sinks awe-stricken and exhausted; even here is Divine wisdom displayed in peopling these vast realms with creatures who require no other environment. If we regard the *tarpan* as the original of the modern horse, and these steppes as the cradle of the species, what vistas of antiquity, what forcible suggestion of æons during which the changes were worked which distinguish the *tarpan* from the famous Bedouin steed, or from the blooded stock of Kentucky. What an illustration of that possible change for which modern science accounts by the theory of natural selection!

If, on the other hand, we choose to regard the *tarpan* as finding his ancestors in horses which straggled from the habitations of man or from the herds of wild horses, what an illustration is this return to

PURSUED BY A RHINOCEROS.

a more primitive type of the scientific doctrine of degeneracy as the complement of the doctrine of evolution. Zoology still regards either answer as doubtful, but as has been said, either view leads to the same conclusion in regard to the adaptation of life to its conditions. The *tarpan* lacks the grace of form, the suggestion of facile but vigorous power, and the beauty which has made the horse so favorite an illustration that even the Bible again and again employs the type in completing its wonderful similes.

A SHOT AT CLOSE QUARTERS.

The *tarpan* does not even look like the horse; his stature may be fully as great as that of the useful broncho, but his want of symmetry suggests rather the misplaced body of the sheep than the proper body for even the small horse. His legs are lean and lank instead of exhibiting that tapering from thigh to fetlock which even the sorriest equine specimen possesses. His coat of hair is so coarse, so rough, as to suggest some wool-bearing animal, rather than the hairy apparel of even the least-cared-for specimen of the species horse. Even the woolly horse of the great showman was sleek and well curried in comparison. The *tarpan* replaces the mane, which is as distinctive of the horse as the pig-tail is of the "celestial," by bushy, furzy side-whiskers which extend the whole length of his jaw. The mouth and nostrils, too, are "bearded like a pard," and suggest the organs of a goat; the ears are niggardly in appearance because set far back, and looking like those of some cropped wolf; the forehead

THE FALLEN MONARCH BESET BY HYENAS, JACKALS, VULTURES AND LIONS.

projects, the curves of the body are all replaced by straight lines, so that taken all in all the *tarpan* looks as if he might have come out of a Noah's ark constructed for the amusement of the children of a Titan. Finally there is hardly a variation of color, since the dirty whitish-brown which belongs to most *tarpans*, passes in exceptional cases into nothing but a blacker or a whiter shade. They are gregarious to but a slight degree, since a herd will rarely exceed twenty-five in number, unless the necessity for seeking other lands, or the presence of some great danger happens to unite several herds into a troop. Still the *tarpan* is adapted by structure and by habits to life upon the steppes, for his diminutive stature causes him to require less food, his life in small herds renders it more possible to find subsistence in a land where the deserts are more numerous than the oases; his covering is more suitable for protection in a region whose climate consists simply in violent, sudden and trying changes, while his appearance must be a protection against his easy discovery by the wolves, which prey upon

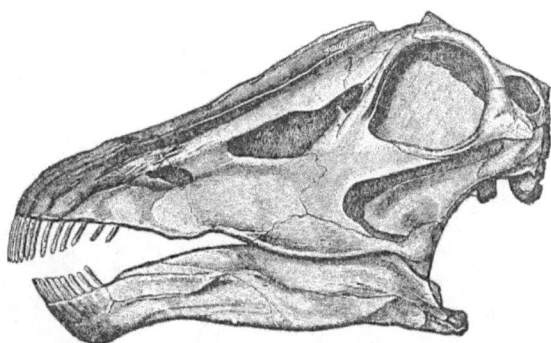

SKULL OF DIPLODOCUS, OR PRIMITIVE HORSE, FROM AMERICAN JURA.

FRONT VIEW.

him and his. So likewise his flight across the steppes enlivens the prospect, while to the barbarians who inhabit the neighboring region, his pursuit as game, or his capture for the service of man, is highly exciting. The flesh of this animal, like that of the horses which now supply the cheaper markets of France, may not be tempting to those who can have the equal of "the roast beef of old England," but to pervert a proverb, tastes differ, and the Cossack would turn in disgust from the less palatable flesh of ox, or deer, or bird.

The cowboy of the period and his Mexican rival are not the only experts in the use of the lasso, for the barbarians of the steppes have from time immemorial practised this art so celebrated since our travellers for pleasure have told and written of the ranches of the West. The *tarpan* is naturally docile and speedily subjects himself

TARPAN, OR WILD HORSE.

to the service of his captors, whether this be to act as the military war-horse of this martial people, the less pampered animal that carries his master in his nomadic life, or the patient beast which returns his fodder in the form of flesh upon which his master shall subsist. In any or all of these functions, the *tarpan* must have the spirit of the horse united to the donkey's patient endurance, hardiness and capability of making a hasty, but sufficient, meal off of anything that can be digested by anything less omnivorous than a shark. That the *tarpan* can answer all these expectations would seem to suggest that his degree of evolution must have had reference to the functions which he was to be called upon to discharge. Though in case of need the Cossack not merely "lives upon his horse," but subsists upon him, he will under other circumstances treat him with the care and kindness which so useful a creature deserves, and which results in a mutual affection such as is familiar

WILD ASSES OF THIBET.

to us in the case of the Arab of the desert. Who has not responded to Caroline Norton's poetical account of the Arab and his steed?

> "My beautiful! my beautiful! that standest meekly by,
>   With thy proudly arched and glossy neck, and dark and fiery eye!
> Fret not to roam the desert now with all thy winged speed,
>   may not mount on thee again, thou'rt sold my Arab steed!
> *   *   *   *   *   *   *   *
> Will they ill-use thee? If I thought—but no, it cannot be,
> Thou art so swift, yet easy curbed; so gentle, yet so free."

Such, too, is the feeling of the Cossack, and the need for "Humane Societies" among more highly civilized men would seem to lend point to the cynicism which defined gratitude as a lively sense of favors yet to be received. Doubtless the Cossack is an inferior man, and the *tarpan* but a sorry specimen of the horse, and yet each teaches a lesson which might lead one to conclude that possibly religious superstition is not in all respects so harmful as "enlightened skepticism and agnosticism." The *tarpan* is considered by the latest authorities not the ancestral horse, but the descendant of domesticated animals which has reverted to a state of nature. The South American horses are known to be the descendants of twelve European horses deserted by colonists in 1537. The horses of Mexico, likewise, are known to have sprung from the few animals brought over by the early Spaniards.

The *wild horse* of the pampas in reverting at least towards the ancestral stock has undergone these changes: An increase of size in head and ears, an enlargement of the joints, a difference in character of coat, and a resumption of dun as a color; any dappling is regarded as indicative of zebraic descent.

The **Ass**, or **Onagra** (*Asinus onager, Asinus sylvestris, Equus onager*), suggests that uncomplaining submission which invites martyrdom and real contentment with the merely useful, which in spite of the boast that "peace hath her conquests," made the satirist declare that homely features should be worn at home. Still, in its primitive condition, before degraded by the treatment of mankind, the ass had all the qualities which we look for in the noblest horse. The *onagra*, or free wild ass of the steppes, like the animal described by Job, is an object calculated to excite sincere admiration: "Who hath sent out the wild ass free? or, who hath loosed the

AFRICAN WILD ASS.

bands of the wild ass? Whose house I have made the wilderness, and the barren land his dwellings. He scorneth the multitude of the city, neither regardeth he the crying of the driver. The range of the mountains is his pasture." Unlike the tarpan, the *onagra* is well proportioned, his legs graceful yet strong, his skin sleek and either a handsome gray or light brown, and his spirit "proud as Lucifer's." When captured, he becomes no mere despised beast of burden, but is the loved companion of his master in his wanderings, and shares with him the dangers and glories of war, the frequent demands of peace, or the care of supporting the household. The *onagra* lives gregariously, for, unlike the tarpan, he is intelligent enough to migrate regularly according to the seasons. The *onagra*, or *koulan*, has generally been called the *wild ass*, but, as in the case of the tarpan, more recent investigations have raised the question whether it is really a wild animal, or a feral one (that is, one which has returned to a state of savagery); and it would seem as if the ancestor of both horse and *ass* must be sought among the fossil forms. The color of the *onagra* is pale red in summer and gray in winter, while the streak along the back is black. Its habitat is on the plains of Persia and Mesopotamia, and the shores of the Indus. It is difficult to capture and is not domesticated except in Bombay. In fleetness the *onagra* exceeds the swiftest horse, and with the advantage of uneven ground, its speed is unapproachable. It lives in troops, and alternates between the mountains and the high plains, according to the season. It is hunted partly because of the difficulty of its capture, and partly because of the delicacy of its flesh, which is no more

DAW AND QUAGGA OF AFRICA.

like that of the horse "than' Hyperion to a satyr, being, by far, superior."

The Hemione, Kiang, Koulan, Half-ass, Mountain Ass, Dgiggitai, Dzigethai (*Equus polyodon*), has, as will be seen, as many names as a Spanish nobleman, but as no one can tell with which one my readers may meet, we give them for the only purpose of a name—identification. It has a brown body, white belly and legs, a short mane, and a mere tuft as an ornament for its tail. Though Asiatic by birth, he is in the modern spirit, not hide-bound by a narrow patriotism, and hence is now making efforts to become naturalized and domesticated in Great Britain. He is speedy, intelligent, willing, like most continental immigrants, to live frugally and

THE ZEBRA.

work industriously for his living. His fur is short, smooth, and in color bay. He neighs like a horse, has no stripes (unless these are inflicted by his captors), has his habitat on the table lands of Thibet, and sometimes reaches in stature as much as fourteen hands. On his native heath he roams about as a member of a small herd of ten or a dozen, and is an exciting object of the chase to his would-be captors.

The **Wild African Ass** (*Equus tæniopus*) is found in Eastern Egypt and Abyssinia, and inhabits alike highland and lowland, the fertile Nile Valley and the arid desert. The *wild ass* of Africa is very beautiful in form and fleet in motion, but as yet has had its claims to notice obscured by the favor shown to the other mammalia. Sir Samuel Baker, whose felicitous and trustworthy accounts of his African experiences have already been referred to, when speaking of the *wild ass*, selects the *equus tæniopus* as the type. The stripes are distinct on the shoulders, and faintly traced on the legs. It is four and three-quarters feet (or fourteen hands) in height; is the best type of the thoroughbred, and is beautifully marked, as its cream color is distinctly tinged with bay, red or dark crimson hues.

The **Quagga** (*Equus quagga*, or *Asinus quagga*,) is said to have derived its name from its peculiar cry. The ground-color is blackish-brown above, and white for belly, hindquarters and legs. The *quagga* wears lateral stripes from its head well back upon its body. It is about four and a half feet high, and about five and a half feet in length. The head and ears are horse-like, and the long, flowing tail is white. It lives in herds, and is found most frequently in company with the ostrich and the gnu. It is peaceful in its habits, but when hunted is quite fearless in its charging. It has been domesticated, but for the most part is hunted for sport or killed for its flesh. Its habitat is southern Africa, and most

VLACKE VARK.

African travellers speak of having enjoyed the sport of shooting the *quagga*.

The **Daw** (*Equus montanus*) differs from the quagga in continuing his striping the full extent of his body. It is strong and muscular, and has been domesticated. Its habitat is the Cape of Good Hope. Like the quagga, it cultivates the society of the ostrich. Its legs and belly are unstriped. Both the quagga and the *daw* are common in zoological gardens and in menageries, and are erroneously called zebras.

The **Zebra**, or **Wild Paard** (*Equus zebra*, *Asinus zebra*), has white as a ground color, but this is marked throughout, except on the belly, with cross-bands of dark brown or pure black. Mane and neck are short, ears long, the form somewhat like that of the wild ass, but the stature greater. It is a mountaineer, very wild, very swift, very shy. It is savage in temperament, and difficult to deal with, even if captured. With one or two possible exceptions, trainers have met with no encouragement from the *zebra*. Zebras and quaggas will wander, wailing about the spot where one of their companions has been killed. Stanley tells of more than one experience in *zebra*-shooting, when the herd would stand by their wounded companions until two or three of their

36

number were killed. The *zebra* will always charge the hunter, and more than one narrow escape has been due to a lucky shot while the enraged creature was making a courageous effort to punish his wanton assailant.

UNGULATES.—SWINE.

The **Hindostan Wild Boar** (*Sus cristatus*) is the animal which figures in the numerous accounts of "pig-sticking in India." It belongs to the jungle,

WILD BOAR BESET BY DOGS.

and its ferocity lends excitement to the hunt. Its color is a brownish yellow, relieved by the whiteness of its beard. Its crested mane has suggested its Latin name.

The Javanese species are called the **Warty Boar** (*Sus verrucosus*) and the **Crowned Wild Boar** (*Sus vittatus*). Celebes has its own variety, *Sus celebensis*, Borneo the **Bearded Boar** (*Sus barbatus*), Japan the **White-mustached Boar** (*Sus leucomystax*), and China the **Chinese Wild Boar** (*Sus indicus*).

The **Japanese Masked Pig** (*Sus pliciceps*) is a large animal which, when domesticated, secretes an enormous amount of fat, and whose face is so furrowed as to suggest the masks which Mardi-gras brings into fashion.

The **New Guinea Pigmy** (*Porcula papuensis*) is the only ungulate represented in the fauna of that country.

The **Indian Pigmy** (*Porcula salvania*) is about a foot and three-quarters in length, although it will weigh from seven to twelve pounds. It is timid and gregarious, and the wandering droves are hunted for the extremely delicate flesh of the animal.

The ancient boar-hunt had all the elements which could excite the imagination or minister to the love of active adventure. The sport was royal and knightly, and this insured a magnificent cavalcade of mighty hunters, with all the pageantry of regal extravagance. The object sought was primarily the exhibition of superior courage and prowess, and this lent seriousness to the amusement. The animal itself was at once a wary and a dangerous foe, and was certain to yield no easy victory. Into the depths of the forest would plunge the sportsmen, while the attendants in a circle beat the bush. Presently the cheerful sounds of baying hounds indicated the discovery of a boar, and then the sport began. Dogs were torn to pieces or sent howling to the rear; attendants had much to do to keep themselves from becoming the pursued instead of the pursuers; and even the hunter was liable to have the boar suddenly

CALYDONIAN BOAR HUNT (ANCIENT PRINT).

swerve in his direction, and with a single cut of his tusks rip open the horse or maim the hunter.

A wild boar was brought to bay and speared, when it charged and received a second spear between the shoulders, and it was charging yet again, when it was killed by the hunter, thus fighting literally to the death.

The **African Bush Hog,** or **Bosch Vark** (*Potamochœrus africanus*), is yet more hog-like in appearance as well as more ferocious looking, more formidable, and more savage. It is a forester, and, like Robin Hood's men, is apt to rush unexpectedly upon the passer-by. Its coloring is either brown, brown and white, or brown and chestnut. Its ox-like head has the cheeks ornamented by protuberances, the eyes encircled by white bristles, and the ears tipped with white, standing more or less erect. Down its back runs a clipped white mane, and it carries its tail, adorned similarly, like a charging buffalo, to which the animal has, in general appearance, some resemblance. His spoor or tracks resemble a capital M, so that those in pursuit of him have no difficulty in distinguishing his trail from that of other animals. The natives catch them in

pits amply supplied with pointed stakes, and take the keenest delight in taking vengeance for the injury which they themselves have suffered. They wear strings of the tusks about their necks, apparently with the same idea that induces the North American Indian to collect scalps. The *bush hog* is about two and a half feet in height, and five in length. His skin is covered with long, harsh bristles, and altogether he is an unpleasant neighbor.

The **Babyrousa** of Malacca (*Porcus babyrusa*) carries four tusks, which protrude above the snout; the pair set in the lower jaw project upward on the sides of the other pair, the upper pair likewise curve upward, passing through the upper lip and turning backward towards the eyes. The function performed by the upper tusks has not as yet been discovered, some naturalists merely supposing that

BABYROUSA.

they may be intended to protect the eyes; as the sow likewise has eyes but is without these tusks, the explanation does not seem at all satisfactory. The great strength and relentless ferocity of the *babyrousa* have made it an esteemed object of the chase, but likewise a terrible antagonist. It frequently attains the size of a yearling calf, and is as much at home in the water as on land. It prefers marshy ground, where it lives in large-sized herds.

The **Vlacke Vark** (*Phacochœrus africanus*) is even more fearfully and wonderfully made than the babyrousa and the bush hog. Its body of dark brown is gray upon the abdomen, and black upon the head, neck and upper part of the back. The tusks are about three-quarters

WILD BOARS DEVOURING A DEER.

of a foot in length and can disembowel a horse. It is not at all backwards about acting on the offensive, but if in flight, it keeps taking observations by raising its head and trying to see over its own back. It is so clever as to destroy the prevailing belief that the hog is naturally and hopelessly stupid. An

African traveller tells of how a *vlacke vark* outwitted the hunter. The pursuit had been continued for some ten miles before the boar was brought to bay, when the hunter turned it and drove it in the direction of his camp. To his surprise the boar showed no unwillingness, so the hunter rode on a little way in advance, and was surprised to find the boar following him, as if it were a domesticated dog. This continued for several miles, until they reached a region full of ant-hills, when the boar suddenly charged backward into one of these, and not only passed from sight, but found some means of escape.

The **Ethiopian Wild Boar** (*Phacochœrus æliani*) is another variety of the same species.

The **Peccary, or South American Wild Boar** (*Dicotyles torquatus*), is smaller than its trans-Atlantic relatives, has fewer teeth, and only a rudimentary tail. It roams about as one of a vast herd, and is quite willing and able to take care of itself. Hunting the *peccary* is full of dangerous adventure, and therefore very inviting to mankind, who find great pleasure in trying their animal strength and cleverness against those of the brute creation. It has happened more than once, however, that the hunter and hunted have changed places, for the *peccary* does not hesitate about "carrying the war into Africa." One unfortunate sportsman, who was not gunning for boar, was forced to climb a tree, and there remain for hours, while a herd of *peccaries* held the fort at the bottom. Having been born lucky, accidental relief came to him, but not until he had become thoroughly cramped, worn-out and famished, so that it was some time before he could take much plea-

PECCARY.

sure in the recollection or care to dwell upon the story. The *peccary* makes up in fearlessness what it lacks in size, and is dreaded by even the fiercest beasts of prey. Its home is the hollow of some great tree, into which the whole family back themselves one at a time. The outermost *peccary* has to do sentinel duty, and should anything happen to him the others successively assume the duties of a guard. Having learned this habit of the *peccary*, the hunter spears or stabs the sentinel, and thus is able to safely massacre a whole family, whose members will appear one at a time.

The **White-lipped Peccary** (*Dicotyles labiatus*) is larger, stouter and more bandy-legged. It is grizzly-black in color, which changes to white on the under parts of the nose and mouth. It is gregarious, and the droves number thousands. It is migratory, and its destruction of crops is equal to that of the locust, while the animal's ferocity makes submission on the part of the farmer almost inevitable, and its ravages are certainly an offset to the fecundity of vegetable life in South America. The *peccaries* would sometimes make short work of the hunter were it not for his agility and the saving presence of a tree. After the hunter is ensconced in a tree, the *peccaries* will stand guard, and in this duty they exhibit a patience which has caused more than one hunter to wish that, for the time being at least, he was frugivorous, and that he might find protection against the irritation of unslaked thirst.

The **Peccary**, or **Tajacu**, is quite as terrible as the wild boar, and being alike irritable and fierce, it does not await attack, but acts on the offensive. One spoken off by Webber, in his "Romance of Natural History," was three feet long and weighed between fifty and sixty pounds. The tusks of the *peccary*, although not protruding, are lancet-like in their keenness. The animal is victor in contests with all the animal kingdom. It takes its rest in deserted burrows or in the hollows of trees, and the whole herd backs in one at a time, the last one in standing guard, like the former species.

The **Taynicate Peccary** is better known, and is larger, fiercer and more troublesome than the former species. It is gregarious, and the herds are of very

HUNTER ATTACKED BY PECCARIES.

great size. It is a good swimmer, and in coloring is a black-brown; the upper jaw is crossed by a white band, which expands so as to cover the lower jaw; its provision in the matter of mane is but slender. The color of the adult animal is a very dark-brown flecked with gray, but the young are handsomely striped with white. It is among the most mischievous of animals, makes long marches in quest of food and so ravages the fields through which it passes as to completely annihilate the growing crops of maize. When even the least alarmed, the *peccary* stops short and gnashes its teeth, somewhat after the manner of an infuriated boar; nor can it be easily put to flight, especially if its numbers be strong. During its marches it swims the broadest streams, unless beset by the Indians, who seize such opportunities to kill large numbers.

The **Hippopotamus** (*Hippopotamus amphibius*) is ugly enough to seem terrible, although it appears to be almost entirely an eater of herbs, and more frequently injures man by destroying his crops than by inflicting any personal damage. To the African the *hippopotamus* is valuable mainly for its ivory, although the flesh is regarded as esculent. An allowance of several pounds of ivory for each tusk specially excites the cupidity of man and leads him to practise dentistry upon a large scale. Africa has become the land for adventure, and the experiences of Livingstone, Baker, Stanley and many another bold explorer are the entertainment of those who widen their horizon by adding to their purely individual surroundings all that others have found in distant countries. African exploration has special charms because the motive has a more permanent value than any that can attach to the simple hunting of tigers in India. Geographical science is the cause in which so many courageous and capable men have enlisted, and to this are subordinated the contests with wild peoples and fierce beasts—for the progress of science, and not simply as a gratification of a taste for adventure, have these men undergone every trial, privation and danger, and thus their adventures are not simply strange and thrilling, but they are ennobled by the motive which prompts them, and the ends in humanity's progress which they are yet to serve.

HIPPOPOTAMUS.

The *hippopotamus* is wholly African and is frugivorous. It is easily irritated and then becomes dangerous, as it never hesitates to begin an attack. Its only vulnerable spots are the eye and behind the ear, the tough hide serving as a protection against the bites of insects, and the attack of any enemies. When wounded they will attack and upset canoes, so that the natives prefer to use none but the smallest, lightest and most speedy dug-outs. The flesh tastes like pork and adds to the incitement caused by the ivory of the tusks. Generally sociable and peaceful among themselves, they sometimes indulge in the most bloodthirsty battles, and most *hippopotamuses* that have been killed or captured bear marks of having at some time suffered from the anger of their kind.

A *hippopotamus* which had been harpooned, charged out of the water on to the land six several times and was driven back by the sand thrown into its eyes, rather than by the countless spears lodged in its yawning throat. The fight between *hippopotamus* and hunters continued fast and furious for three hours, when Sir Samuel Baker terminated the contest by a lucky shot.

An Arab who, in protecting his melon patch tried to drive away a *hippopotamus*, was himself first put to flight and then killed by the bold burglar. Harpooning the *hippopotamus* is a popular method employed by the natives to destroy this huge creature, as will be soon explained, but other means equally effective are resorted to, among which I may mention the use of what is known as the *hippopotamus* dead-fall, made by attaching a large iron spear, heavily weighted and suspended above the path frequented by the animal in his excursions to and from the water. A line is fixed across this path so that

ATTACKED BY HIPPOPOTAMUSES.

when the *hippopotamus* strikes it the spear above is loosed and it falls upon the animal with fatal effect. The *hippopotamus's* irritation at novelties, increased at times by personal wounds, leads it to charge upon boats, and sometimes to wreck these. Sir Samuel Baker tells of a most remarkable contest of this kind where the animal, although repeatedly driven off with fresh bullet-wounds, returned again and again—even going so far as to retire for some hours and then renew the attack. Finally his career was stopped by a bullet, and the boatmen felt a very decided relief.

Stanley in his "Through the Dark Continent" tells of two *hippopotamuses* taking part in a battle between himself and the natives. While the fight was going on, these two creatures deliberately swam out and commenced belligerent operations against his boats, actuated, doubtless, by a mistaken sense of patriotism. The huge creature has many times been known to rise suddenly under a canoe and, seizing it in his ponderous jaws, crush it as if the effort was no greater than the breaking of an egg-shell. The natives frequently harpoon the *hippopotamus*. They will fasten in him a harpoon attached to a rope which has a large floater, or bob, at the other end. When struck the animal goes plunging off, but in spite of his extraordinary ability for remaining under water, the bob betrays his whereabouts. They next fix three ropes so that two of them will make an acute angle, and when thrown over the bob they can not slip. These two ropes are now twisted into one, while the third is held by natives on the opposite side of the stream. The *hippopotamus* is now harnessed, and is gradually pulled towards the bank of the double ropers, until finally, in spite of frequent vicious charges, restrained by the rope on the other bank, he is landed helpless on the shore, and is then speedily dispatched.

The color of the *hippopotamus* is a dark, fleshy-red, marked irregularly by black spots. The young are very fond of riding about on the back of the mother, who bestows upon it the most zealous care, at which time she is so solicitous of its safety that she will viciously attack anything which she may come suddenly upon, even if it be simply a log which meets her sight unexpectedly.

UNGULATES.—CAMELS.

In the eastern steppes are found large troops of the patient-eyed **Camel**—the ship of the desert. In our thoughts of the *camel* as a means of transit we are apt to overlook the other services which it is called upon to render. It supplies food which, even though inferior to what Shakespeare calls "beeves or muttons," is palatable and nutritious; its milk is as excellent and as pleasant to the taste as that of the cow. Its hairy covering is woven into fabrics, from which tents, clothing and coverings are made; or it is twisted into cords which furnish harness and other necessary conveniences; its skin supplies an excellent quality of leather; its refuse is used for fuel; and all these sources of profit are to be added to that patient and long-continued endurance, without which even the existence of the traveller through the deserts would become impossible. And an omniscient Providence has given the *camel* an anatomical and physiological structure which not only adapts it to its environment but which enables it to so serve the needs of the higher creature—man. The *camel* is no longer

HIPPOPOTAMUS TRAP.

unknown to most of us, for the travelling menageries, together with the zoological gardens, have brought it to the acquaintance of most of my readers. Consider the *camel's* feet, and remember that there is in them the beauty of fitness as well as of color and harmonious form. Its foot is divided into two sections and supplied with two toes furnished with a short nail; the foot is elongated, of great strength, and has a stout, horny sole. This structure would be anything but beautiful for the human being; it is, viewed in the abstract, less pleasing than the foot of the deer, or of many a bird; and yet when tried by the standard of mechanical beauty—the test which the engineer applies when speaking of his locomotive or his engine; the test that the mechanic must use in speaking of the processes of his calling; the test used by the mathematician when speaking of a solution—tried by its proper test, what can be more beautiful? The *camel's* pathway is to be over sands which shift and slide away at each movement. If it would move with the security of man upon land, or of the marine beings in the sea, it must be able to fasten a sharp toe in the sand while the flatness of the foot makes even the shifting base a fresh support, and

HUNTING THE HIPPOPOTAMUS.

its horny covering prevents the sand from sifting through. Were we called upon to re-create the world, as many an irritated person longs to do, should we be able to devise so wonderful a mechanism, so entirely suited to the office which it is to fill? But what end is served by the hump or humps? Is not this a malformation, unsightly to the eye and useless for any necessary service? It is the store-house whence the *camel* draws provisions when they cannot be obtained from the dreary waste over which it is passing. It is composed of cells which secrete and retain fat against the day of need. Still again let us regard the *camel's* third peculiarity. Its stomach has not merely the compartments which belong to all the ruminants, or browsing animals, but it is supplied, on either side, with a mass of cells which serve as tanks or reservoirs, and in which is stored water so pure as to be drinkable, and so necessary that the *camel* can exercise the most extraordinary abstemiousness when circumstances require this, and can in case of desperate need more than moisten the parched lips of its owner and thus preserve his reason, if not his very life. But furthermore, the peculiarities of its structure enable the *camel* to contend, to the best advantage, against fierce winds and clouds of sand, while endowing it with the utmost vigor and the greatest swiftness of pace. It is said by the mathematician that there is a ratio beyond which mere accident cannot pass; that while the realism of DeFoe is such as to lend verisimilitude to the experiences of Robinson Crusoe, yet, though each separate experience may have been possible—nay, probable, if you will—the occurrence of such continuous and useful coincidences was simply impossible. Therefore must we not conclude that when we find in the *camel* so many and such manifest adaptations to the life which it must lead, that instead of blind accident, the only reasonable cause must have been the wisdom of the Almighty, even though it be believed by many that he chose to work by means of " natural selection " and " evolution " instead

CAPTURING A MONSTER HIPPOPOTAMUS.

of by a special and instantaneous exercise of his omnipotence? It is not miraculous that an All-Wise and All-Powerful Being should achieve a success not even conceivable by a merely finite mind; but it would indeed be more miraculous than a miracle—more mysterious than a mystery—if a blind force, acting without purpose and without direction, should accomplish, not a single effect, but a continuous and harmonious arrangement and adaptation of organs. The *camel's* strength enables it to carry, without fatigue, a burden of from six hundred to a thousand pounds; its swiftness and power of endurance enable it to pass over, in a single day, from thirty to ninety miles; and both this carrying-power and this ability to conquer time and distance are absolutely essential alike to the *camel*, when in its wild state, and to the human inhabitants of regions where the habitable portions are separated by great seas of sand—a waste more barren than the pathless ocean.

The camel of the steppes is the **Two-humped Camel**, or the **Bactrian Camel** (*Camelus bactrianus*). The *bactrian camel* is generally about seven feet in height; his hair is of a chestnut color, and is short on the body, long upon the neck and fore legs, and woolly on the upper part of the neck, upon the head, and upon the humps. Pause to think of the different adaptation of wool, of short hair, and of long locks, and you will see that the camel is most suitably clothed to withstand the most violent changes of temperature. The gait of the camel is not that of an ambling palfrey, but when one has learned how to ride the animal, he finds the motion no more troublesome than is the deck of a ship at sea to an old tar. As the sailor finds it awkward to walk upon land, and well-nigh impossible to ride on horseback, so those whose experience has been confined to the camel would find themselves like a fish out of water, if they essayed other beasts of burden; and as the sailor contentedly walks the deck, in spite of the heavings of the sea, so does the camel-driver find himself in entire harmony with the movements of his beast of burden. The camel, although much-enduring, is not patient under

THE HIPPOPOTAMUS AND HER YOUNG.

ill-treatment—to hardship he seems to be indifferent. Doubtless we all remember the story of the elephant that revenged himself upon the tailor who pricked him with a needle. Similar stories are told of the camel. For example, it is well-authenticated that a dragoman, or camel-driver, having abused his camel, found it growing more and more intractable, until it suddenly availed itself of an opportunity, and spit into the driver's face the whole of its last meal, reduced to pulp and mixed with water and saliva. The dragoman accepted his punishment and amity at once prevailed between himself and camel.

The **One-humped Camel**, or **Dromedary** (*Camelus arabicus*), shares with the bactrian camel the burdens of life in the desert. Such *dromedaries* as are not condemned to the life of a mere beast of burden, are most carefully trained for the easy and speedy carriage of their riders, and are related to their less fortunate congeners as the thoroughbred race-horse to the merest scrub. Their gait, however, requires the rider to be no mere novice, if we are to accept the reports of travellers who have experimented with this form of riding. Its length is from ten to eleven feet, and its height from seven to eight. Its single hump serves it for a storehouse in which it accumulates strength against the days when forage is scarce. It will eat anything green, so that its owners are compelled to protect it against eating the plant known as "camel poison."

When not on the march it drinks once each day, but if striding across the desert it is watered but once in six days. While carrying from five hundred to six hundred pounds of freight, it will travel regularly twenty-five miles a day. The thoroughbreds, or riding *dromedaries* are called *hygeens*, and will travel from fifty to one hundred and fifty miles a day. The baggage-carrying *dromedary* is an inferior creature and can be purchased at an average price of fifteen dollars.

The **Auchenia**, except for their large heads, long necks and great ears, would look very much like sheep. Their coat is most abundant and valuable.

The **Guanaco** (*Auchenia huanaco*) is from seven to eight feet in length, and four feet in height, and makes its home on the southern Andes. It is gregarious

DROMEDARY AND BACTRIAN CAMEL.

and the herds vary from six to thirty, but never allow the presence of more than one male. Its fleece is a dirty brown. It always carries its tail erect, and defends itself by spitting at its foe; it is a fair swimmer and a swift runner.

The **Llama** (*Auchenia llama*) is used in Bolivia and Peru as a beast of burden, and will carry from one hundred to two hundred pounds travelling from six to twelve miles a day, up and down the mountain; for beasts of burden only the males are used.

The **Vicugna**, or **Vicuna** (*Auchenia vicugna*), is a member of the family whose size is intermediate between that of the alpaca and the llama. The *vicugna* is more shapely, and for coloring is marked with reddish-yellow above, white for the lower portion of the body and for the breast, except on the underpart of the neck and the inside of the legs where ochre prevails. The animal is polygamous and moves about in small herds which all belong to the family of the male. The head of the family is expected to keep watch over the safety of his flock, and to advise them of the approach of danger. The *vicugna* has one quality in common with the turkey: after having been corraled, and when confined by no barrier which it could not leap, it will make no effort to jump, being seemingly dazed by fluttering rags tied to the encircling boundary of rope. The Indians hunt the *vicugna* with a sort of combination of hand-ball and of lasso. It consists of three stones fastened to strings, which, in turn, are fastened to each other. The hunter who is skilful, holds on to one string and its shot or stone, and throws the others so as to tangle up the legs of the *vicugna*, whose

THE LLAMA (*Auchenia llama*).

hide and flesh are both valued. The *vicugna*, as well as the llama and the guanaco, was frequent in ecclesiastical legends of the Peruvian church, and any one not acquainted with Prescott's "Conquest of Peru," will take pleasure in reading what the American historian has to say about the animal.

Among its other enemies, the condor is most dreaded, as the great bird pursues the *vicugna*, and, seizing upon the head, plucks out its eyes, and the blinded animal soon after dashes itself to death among the mountain fastnesses where it is confined.

The **Paco**, or **Alpaca** (*Auchenia pacos*) is the most sheep-like of the Auchenias, and is kept in large flocks on the elevated plains of the Andes, tended by

Peruvian slaves. The wool of the *paco* was introduced into England by Sir Titus Salt, who, at Saltaire, built factories of immense size, and inaugurated a new and immense industry. The wool, as used in Peru, furnished such marvellously fine products as to vindicate the enterprise of Sir Titus Salt. The famous merino sheep yield no such quality of material, and those who like the best of clothing may well recognize the serviceableness of the *paco*. When shorn, the fact that the *paco* belongs to the same family as the camel becomes instantly and almost comically evident to any one. The *paco* is to be found in zoölogical gardens, but visitors may well be cautious of too near approach, as it is given to the very filthy habit of expectorating its food upon any one who may excite its enmity, which a near approach is likely to do. It is kept in flocks, is sheep-like in appearance, and its wool, (which it never sheds) is in great demand for the manufacture of a superior quality of blankets. The fleece is taken every year, and the hair grows to the length of some eight inches. The fossil forms of the suidæ and camels are the *pœbrotherium*, which has the typical teeth of the *auchenia;* the *protolabis* and *procamelus*, which lack the incisor teeth of the camel; and the *pliauchenia* which varies in point of premolars, having three only.

CONDORS ATTACKING VICUGNAS.

UNGULATES.—DEER.

The **Kanchil** (*Tragulus kanchil*) is Malayan in its habitat, and belongs to the smallest sized family of ruminating animals. Its delicate head is pointed, the body arch-like, legs slender, and hoofs small. Its motion is that of a bounding ball rather than walking, running, or leaping. It sleeps resting on its bended knees. It is found also in Java, where the species is named by naturalists, *tragulus javanicus*. It is somewhat less than a foot and a half in length, and the prevailing color of yellow is varied by gray on the throat and brown on the tail, except the under part and the tip, which are white. Three white collars surround its neck, its sides are tinged with red, and the color of its upper parts is shaded with black.

The Sumatran Kanchil (*Tragulus napu*) is larger and more extravagant in the matter of neckties, wearing five of them.

The Ceylon Kanchil (*Tragulus meminna*) is intermediate in size, but is banded and spotted throughout. The animal is solitary in its mode of life. The *kanchil*, first named *tragulus kanchil*, is supreme in smallness of size, as well as superior in quickness of action and in intelligence. It is said when pursued by its enemies to jump up and fasten itself to the branch of a tree, where it will remain feigning death.

WOLVES ATTACKING A CARIBOU.

The **Water Deer** (*Hyæmoschus aquaticus*) is the mammoth of this family. It is found in Sierra Leone, and wears a brown coat, relieved on the throat by five up and down stripes and on the flanks by bands and spots of white.

The **Muntjac** (*Cervulus muntjac*) has its habitat in India and the neighboring islands.

The **Weeping Muntjac** (*Cervulus lachrymans*) belongs to northern China.

The *muntjacs* are only a little more than two feet in height, and their short legs and their bodies are covered with smooth, reddish-yellow hair, white, however, being the color of the throat and belly, and brown that of the face and legs. A species is found in Borneo, which is darker in coloring and smaller in stature. It is called the **Borneo Sambur** (*Cervus equinus*). There is a species in Java (*Cervus hippelaphus*), and the Indian species (*Cervus rucervus*) is notable for the projection of its antlers at right angles.

The **Caspian Deer** (*Cervus caspicus*) is a spotted mountaineer, which changes to a sober brown in the winter.

The **Unarmed Deer** (*Hydropates inermis*) has no antlers. It belongs to China, is a water deer, and lives amidst the rushes.

The **Russian Roe** (*Capreolus pygargus*) has six-tined horns, is in habits and appearances much like the roebuck, and is not gregarious.

A FIGHT TO THE DEATH.

The **Brazilian Marsh Deer** (*Cervus paludosus*) and the **Pampas Deer** (*Cervus campestris*) are frequently mentioned in books of South American travel.

The **Red-coated Coassus** (*Coassus rufus*) is only about two and a quarter feet in height, and it is spiked rather than antlered. Its habitat is South America.

The **Asiatic Sambur**, or **Rusine Deer** (*Cervus aristotelis*), has horns set upon a footstalk, projecting, and forked only at the end. It is large and power-

37

ful, sooty-colored except on the root of the tail and over the eyes, where it becomes tan-colored. It is very fond of the water and lives in low lands. It is vicious and morose. The buck is ornamented with a mane.

CANADIAN LYNX AND MOOSE.

The **Axis, Spotted Hog Deer**, or **Chittra of India** (*Cervus axis*) belongs to the sambur family. It is golden brown in color and has a dark brown back stripe lightened up with a double oblique line of white spots. It is nocturnal, but otherwise has the habits of the fallow deer.

The **Roebuck** (*Capreolus capræa*) is little over two feet in height, but is very quick in its movements, and very powerful. It lives in pairs. It is brown, or brown shot with gray or red, except for the root of the tail, the belly, and on the inside of the legs, which vary from gray to pure white. Its horns have one antler in front and two behind. Its habitat is Europe. Its looks are deceptive, for it has an ugly temper, and is exceedingly dangerous when irritated.

The **Stag**, or **Red Deer** (*Cervus elaphas*), is still to be found in Scotland, although generations of hunting have almost exterminated the species. Its praises have been rehearsed by poet and novelist, and its portrait painted by the artist. It is an expert swimmer and a very fleet runner. It can be domesticated, but its temper

HUNTING THE MOOSE BY NIGHT.

is so uncertain as to keep one constantly in danger from the knife-like hoofs of its front feet. It is gray in winter, but changes with the season until it attains a brownish-red. The males are solitary beings, and their contests are most exciting, equalling in ferocity that of the most desperate bulls. These fights rarely occur except during the mating season, when the sight of one buck by another is a sufficient challenge, and they immediately rush to the encounter.

The **Wapiti, Carolina Stag,** or **American Elk** (*Cervus canadensis*), lives in large or small herds, commanded by an old buck, whose orders are never disputed. This leadership, however, is neither gained nor retained without many a battle with rivals. Occasional instances have been known where two stags, getting their horns inextricably interlocked, have perished of hunger and left the vacant throne to a successor. It is a hard task-master to the does and keeps them in constant and not unnecessary fear, evidently believing in the British doctrine that a man may punish his wife. When wounded

MOOSE (*Cervus alces*).

the *wapiti* at once becomes aggressive and is a dangerous antagonist. It is a swift runner and an expert swimmer, frequently submerging itself to the point of the nose, with the seeming object of escaping the heat and insects. It is valued for its flesh, its skin and its sport-giving qualities. It can be domesticated, but submits to no punishment. It is said that a tame *wapiti* once treed its owner. He was annoyed by its attentions and struck it with his cane, whereupon the *wapiti* at once charged, and the gentleman, while running, happening to fall

ROEBUCK.

between two logs, was kept captive for several hours, while the *wapiti* used its horns on the outside.

In the early settlement of the West, the *wapiti* was very numerous, and afforded both sport and subsistence to the hardy pioneers, but in latter years it has become so scarce that it is seen at rare intervals, and only in the almost inaccessible regions of the extreme northwest, near the British line. A few more years and the species will become extinct, unless a few specimens be preserved in zoological gardens, where, however, it does not seem to thrive.

A hunter tells of an amusing battle which occurred between his companion and an elk that had been excited by a red handkerchief which he had about his neck. It seems that the man was incorrigibly lazy and confined his exercise to the inevitable duties of camp-life, and to rendering both day and night hideous by his attempts at singing. According to his own report, while resting his back against a tree and warbling a ditty for his own entertainment, he was suddenly approached by an elk. He resolved to enter a protest against the chronic complaints of his laziness, so he seized a gun and blazed away at the *wapiti*, which, however, instead of at once succumbing to the invitation for its conversion into venison began a vigorous charge upon the aggressor, and one which, but for the opportune return of the hunter, might have resulted fatally to the vocalist. As it was, the unskilful sportsman had his clothing converted into fringe, and if not dead, still wears marks of the stag's prowess.

RED DEER (*Cervus elaphus*).

The Fallow Deer (*Cervus dama*) has a spotted coat and palmated horns, which branch widely. It is so graceful a pet that it is most commonly to be found in deer parks. The head of the family keeps by himself, or is surrounded only by very few of the most highly-favored. If away some other acts as regent, but he must surrender his authority the instant that his monarch appears or he will be ingloriously pushed aside by the horns of the king. His authority, once obtained, is unquestioned until failing strength tempts the ambition of aspiring courtiers. As a rule the *fallow deer* is reddish, with white spottings and two or three white body lines, but sometimes the animal is brown or black throughout. The venison is specially good, the skin is manufactured into leather, and the horns serve a variety of human needs.

The **Moose** (*Alces malchis*, or *americanus*,) is still found in considerable numbers in Maine. It is about seven feet high, and its large, heavy horns are palmated. It is dark brown, with yellow legs, and corresponds to the elk of northern Europe. It can go over obstacles and through the brush (at which times it lays its head well back on its shoulders), is awkward and clumsy, but fleet and enduring, shy, but capable of the most dreadful attacks with its hoofs and its horns. Its flesh is held in esteem and its horns and skin serve many useful purposes. It will, on the average, weigh some seven hundred pounds, of which its horns claim a twelfth; such a weight brought into sudden contact with a tree of ordinary size will at once uproot it, and, as it were, crush it. Its sense of hearing is

A SAVAGE PURSUIT.

painfully acute, and its watchfulness correspondingly great. The rustle of a leaf, or the snapping of the least twig will be excuse enough to the *moose* for a full hour's listening. Were it not for the winter season, with its provision of moose-yards and deceptive snow-crust, the *moose* might safely defy pursuit. The snow-yards which it constructs are frequently four or five miles in diameter. During the rutting season the Indians imitate the call of the buck, and there always seems to be one to answer the challenge. It can remain under water almost indefinitely, which fact adds to the difficulties of the chase. During the winter, if a *moose* is found outside of a moose-yard, it is pretty sure to be chased and killed by the wolves.

The *moose* is hunted in Maine during the winter, when the hunter, on his snow-shoes, can move rapidly over the crust which gives way beneath the ungulated feet of the *moose*. The *moose* will, on the average, stand sixteen and a half hands (or five feet and a half) high, has an ass-like head, protruding eyes,

FALLOW DEER.

broad ears, long legs and immense, palmated antlers. The chase is always exciting, for the animal goes through the densest brush, without noise or any apparent diminution of its speed, and by mere momentum will press down quite large saplings, brushing them away as if they were made of floss instead of wood. Its senses of sight, scent and hearing are extraordinarily acute, and create the necessity for the greatest caution on the part of the hunter. It is semi-aquatic during the summer months, when it has to protect itself not solely, against the insect which uses the *moose's* flesh for a nest, but also against the numberless species of flies, gnats and mosquitoes which infest its habitat. Its tenacity of life is very

remarkable, as may be illustrated by the fact that a *moose*, which had lost the use of one hind leg, and had had one of its shoulders broken, still managed to keep the hunters chasing it for forty-eight hours before it finally eluded them. Frequently the *moose* will take to the water, and expose only its horns and nostrils. This renders the shooting of them, while in the water, a matter of the nicest skill, and, as dead deer do not float, the enterprise has all the

REINDEER.

elements of interest for tne true sportsman. Compared to this, the chase over the snow, into which the creature cuts at every step, and, still more, the slaughter of the *moose* in the snow-yards which it builds for its protection, is tame.

The **Virginian Deer**, or **Carjacou** (*Cervus virginianus*), is a good swimmer and jumper, a lover of the outskirts of civilization, attached to its feeding-grounds, elegant in appearance and quite numerous in America. Its many-pronged horns bend first backward and then forward, the ends branching out just above the nose. It wears three suits a year: a brown one in winter, a reddish one in the spring, and a blue one in the autumn. The fawns are white-spotted. When hunted, it takes to the water, and has even been known to swim a mile or so out to sea. Sovereignty is not secure when once obtained, for there are constant contests between the bucks. It always jumps into the air when unexpectedly disturbed; but, when aware of the presence of the hunter, it will crouch in the grass until it believes that it sees its opportunity for escape. It can be tamed, but its mischievousness is said to render it a very troublesome pet. The most common species is the American red deer.

The **White-tailed Deer** (*Cervus leucurus*) has a long tail, narrow hoofs, short hair, long, slender feet. It is found on the upper Missouri.

The **Sonora Deer** (*Cervus mexicanus*) dif-

THE MOOSE HUNTERS.

fers from the Virginian or Carjacou deer principally in being smaller.

The **Mule Deer** (*Cervus macrotis*) is larger in size, and distinguished by ears of such proportions as to give it its name. It is found on the Yellowstone.

The **Black-tailed Deer** (*Cervus columbianus*) resembles the mule deer, and is found in California and Oregon. On its forehead it wears an ornament, which resembles a horse-shoe, doubtless carried for luck.

The smallest known deer, the **Dwarf Deer** (*Padua humilis*) is found in Chili, and its antlers, two and a half inches in length, are large, in proportion to the size of the animal.

The **Reindeer** (*Rangifer tarandus*) is about the size of an English stag, but less graceful and symmetrical. The horns are long, slender and round, re-curved, branched with palmated summits. Its color is brown above and white beneath, although the upper parts grow gray with increasing age. The lower part of the neck is drooping, the hoofs large, long and black, and accompanied by secondary hoofs on the hind feet. Julius Cæsar mentions meeting with the *reindeer* in the Hyrcanian forests, but its present habitat is the Arctic polar belt. Its skin makes the warmest of clothing, and its flesh and tongue very delicious food. Its fat is used in making "pemmican," by pouring one-third melted fat over two-thirds pounded meat. The œstrus fly is a great and constant

REINDEER SLEIGH.

annoyance to the *reindeer*, because of its preference for the flesh of the *reindeer* as a depository for its eggs.

The *reindeer* continues to play too important a part in human life, not at once to suggest his appearance to the reader. In winter it is dressed in long, grayish-brown fur, which changes to white on the abdomen, neck, hindquarters and end of nose. The Laplanders own thousands which have been domesticated, and life without the *reindeer* would be unlivable. It will draw a load of from two hundred to three hundred pounds, at the rate of nine miles an hour, and can keep this up for twelve hours a day, as a regular exercise. It is able to sustain life upon a lichen found under the snow, so that it costs nothing for its support. In Kamtchatka the *reindeer* is used as a saddle-horse besides being put to many other useful services.

The **Caribou**, or **American Reindeer** (*Rangifer caribou*), is a large animal, some three and a half feet high, but has never been domesticated. When chased in winter, it will seek a body of water, and sitting on its haunches will slide itself along the ice. The skin is its most valuable possession, so far as man is concerned, although it has a thin layer of fat which, being poured over pounded meat, makes "pemmican." It runs in herds, which vary in size from a dozen to three hundred. It is trapped in two ways: First, a slab of ice is balanced on a pivot, so as to precipitate into a pit any unwary *caribou* that may step upon it; secondly, a herd is driven into a large inclosure, cut up into numerous alleys, each one of which is provided with a noose which strangles the deer as it moves about.

The **European Caribou** is only a different species, whose description does not differ greatly from the above.

The **Barren Ground Caribou** (*Rangifer grœnlandicus*) is smaller, more graceful and larger horned. Its habitat is the American Arctic region.

The **Musk Deer** (*Moschus moschiferus*) is regarded as a primitive or undeveloped species of deer. It is about a foot and three-quarters in height, (measuring from the top of the shoulder), and about three feet in length. It has no horns or antlers, and its reddish-brown hair is sleek and short; its coloring becomes white on the belly. The male has glands on the thighs near the tail and on the posterior ventral part, which secrete the article known to commerce as musk. Although the nature of the use of these glands to the deer is not known, their service to man as a contribution to his pharmacopœia and to his perfumes is unquestionable. A single deer will yield about three drams of musk, and this will sell in first hands for as much as a dollar a sac or pod. We have been told about the antiquity of the Chinese civilization, and one of our own poets has introduced us to the craftiness of the "Heathen Chinee." Of course, with the luxury and increased knowledge of an old civilization comes corruption which must ever be the shadow cast by any virtue. Hence, while it may be regretted, it can hardly surprise us to know that the Chinese have been so clever and so persistent in their adulterations of musk, that a pure article is no longer to be found, except in the possession of the deer itself. Possibly the *musk deer* knows that a price has been set upon its head, for it is exceedingly shy, and though approaching human habitations, it does this like a thief in the night, solely with the intention of stealing sweet potatoes. It is, for the most part, captured only by trapping.

MUSK DEER.

The **Giraffe** (*Camelopardalis giraffa*) is at home in the regions near the sands of Sahara, and when in his native element, is a very symmetrical creature in spite of his tremendous stature. His head resembles that of a camel, but he has a special hair sand-protector for his nostrils, whose safety is further secured by the obliqueness of their setting, and the possession of a muscular cover by which the *giraffe* can close them at will. The beautiful, silken skin of the *giraffe*, and its mild, large, lustrous eyes are certainly objects of beauty. He is an accomplished kicker, and his length and power of limb render him a very dangerous antagonist. A large-sized *giraffe* will measure sixteen or seventeen feet from the top of its head to its fore feet, and the larger part of this is neck. But since he is to graze on nothing smaller than trees and has at the same time to protect himself against enemies who will not disdain to take advantage of any philosophical wool-gathering, it must be conceded that he seems to be built to order for the conditions of his life. His skin makes an excellent leather, and his flesh is edible. He moves about in small

GROUP OF GIRAFFES.

herds and is safe against carnivorous animals, unless these surprise him. He runs slowly on level ground and can easily be overtaken by a horse but he is bullet-proof, except in spots. The natives catch him in pits which they dig in such fashion that when he falls in he will find his belly on a hill, and his legs dangling in the air below. The *giraffe* is easily domesticated, and is interesting from its affectionateness, its active curiosity and its delicacy and fastidiousness in eating.

UNGULATES.—BUFFALO AND OX FAMILY.

The **Auroch, Bonassus,** or **Zubr** (*Bison europea*), is Lithuanian in habitat. It always diffuses an odor of violet musk. Naturally shy and retiring, yet it will fight when irritated, and is a terrible adversary. It is a good swimmer and escapes domestication by its exceeding moroseness and untamableness. It is about six feet in height. The species is substantially extinct, as its habitat is so narrow in extent. It is light-brown in color, has a forehead whose breadth is lengthwise, is bearded, and wears small horns which curve upward.

The **American Buffalo** (*Bison americanus*) has been slaughtered by tourists and visiting foreigners who owned guns, until it is about extinct as a species. The last herd of fourteen known in Colorado were shot by a single party which had crossed the salt water for the sake of having an adventure which would glorify the rest of their lives. It has a remarkable power of storing up water "against a dry spell." Among the many un-sportsman-like methods of hunting the creature, was that of driving a whole herd over the sheerest precipices. When the Union Pacific Railway was built, immense herds of *buffalo* were frequently met with, and travellers, while in the cars, would shoot the animals. Not so very many years ago steamboats on the Mississippi and Missouri were not infrequently compelled to stop and give the right of way to herds of *buffalo* swimming across the river.

NORTH AMERICAN BUFFALO.

The **Gayal** (*Bibos frontalis*) is East Indian, and is named from its forehead. It is a mountaineer, and its stout, large, pointed horns run (excepting for a slight curve at the tips), at right angles to the head, still further increasing the frontal development.

The **Gaur,** or Gour (*Bibos gaurus*), is the mammoth among oxen. It is from six to ten feet in length, has an elevated ridge on its back, varies from deep-brown to black, but the legs are white below the knees. It moves about in herds of from ten to thirty, in which there are but three or four males. It seeks the deepest recesses of the forest, and is held in absolute dread by the fiercest of the carnivora. In addition to always keeping sentinels on guard, they invariably feed standing in a perfect circle, with their heads in the circumference. For some reason it is entirely indifferent to the presence of the elephant, so that a mounted hunter has no difficulty in approaching it. Its lowing and its bellowing both resemble the grunting of the hog. Its two inches of thick skin cover flesh which is specially tender and palatable. It lives in the jungles of India.

The **Anoa Buffalo** belongs to Celebes, is straight-horned, and although small of size is exceedingly fierce. One of them which was penned up with fifteen large stags, killed them all.

The **Banteng** (*Bibos banteng*) belongs to the fauna of Borneo and the vicinity, is grayish-brown, and wears tri-curved horns which end by curving inward. It is often called the *Javan ox*, is strong, fleet and active. The bulls are brown in color, while the cows are bay. The lower legs, inner ears and hips are white, and the hind quarters have white patches. It is from five to six feet in height, is frequently domesticated, and is ridden, driven and used as a beast of burden, rendering all the services which ordinarily fall to the lot of the horse.

The **Indian Zebu** (*Bos indicus*) is a slender-limbed animal, humped on the shoulders, dewlapped, and with a back which, after sloping upward from the shoulders to the haunches, seems suddenly to drop away and vanish. It is not confined to India but is found in China and

THIBETAN YAK AND BIG-HORNED SHEEP.

has been met with on the coast of East Africa. It is readily domesticated, when it displays good temper and intelligence. It is used not merely for heavy draught, but likewise in harness, and while not at all speedy is steady and reliable, being capable of sustaining a gait of five miles an hour for as long as fifteen hours a day. To this species belongs the **Sacred Brahma Bull** for which there is such great reverence that one of them will, without opposition, walk through the market place, push people to the right and left, and help himself to any article that strikes his fancy.

The **African Sauga** (*Bibos africanus*) wears a hump on its shoulders, and the natives from time to time subject it to vivisection that they may enjoy fresh steaks. This treatment may give the animal transient pain, but it seems to cause it no permanent inconvenience.

The **Yak**, or **Grunting Ox** (*Pœphagus grunniens*), is a native of Thibet, where it is a creature of importance whether for domestic service or industrial uses. We hear less of Thibet because by conquest it has become a part of the Chinese Empire, but it is a land full of strangeness and full of nature's wonders. Its Mongolian inhabitants are nomadic in the northern steppes, but agricultural, or engaged in commerce and the industries as they live toward the south or in the cities. To offset the boiling geysers of other countries, Thibet has freezing geysers, although not destitute of the other kind. One might think that the Titans had opened great ice-houses, as solid and massive columns of ice fall in quick succession from the height of the geyser; or he might think as the sunlight is separated into the

THE AFRICAN SAUGA.

prismatic colors that he had stumbled upon the abode of the ice-king. The arable land is confined to the warm valleys, but there abound everywhere rich

THE INDIAN YAK.

mines of the precious and useful metals, and of gems so rich and rare as to render possible the experience of Sinbad the sailor. Animal life abounds and may be represented by the *yak*. This animal, which can be seen in many a zoological garden, would be called a buffalo by those unfamiliar with it. It is very gregarious, so that the immense size of its herds again suggests the American buffalo, before that animal was by ruthless slaughter rendered as scarce as the descendants of the Indians, who lent animation to the life of the early colonists. The hair of the head upon the *yak* is curly or frizzly; its mane is long and thick; its body is covered by short hair, but the legs and flanks have the greatest abundance of soft long hair; its long flowing tail is white. It looks as if some great hairy creature had through some cause grown bald on its back, or as if some prairie fire had burned away the middle of a hairy spread, or some new fashion in the tonsorial art had brought into vogue an inverted pompadour for animals. The *yak* is hunted for its flesh, hair and hide, and the sport has all the excitement of danger, for when excited the bulls are very ferocious. When a herd is attacked the calves are gathered together while the bulls and cows form a solid square around them. But the *yak* is quite as necessary and useful as a domestic animal, and readily adjusts itself to the laborious service of man's daily life. It supplies the family with a plentiful quantity and an excellent quality of pure milk; it is strong and enduring as a beast of burden; and it is able to forage for its own subsistence and to be satisfied with "plain living" even though it may not indulge in "high thinking." There are castes among *yaks*, so that while the beautiful, white-trimmed patrician holds his head high in air, the *common* or *plough yak* recognizes the lack of gorgeous apparel and great stature, and in humility walks with head bent down.

There are two species of the African buffalo, the **Short-Horn Buffalo** (*Bubalus buffelus*), and the **Buffalo of Caffraria** (*Bubalus*

GORED BY A WOUNDED BUFFALO.

caffer); the former is brown in its coloring, the latter black. The **Asiatic Buffalo** (*Bos bubalus*) is the type most commonly known to all but African travellers. The celebrated Livingstone describes methods of hunting the African buffalo, which are even more wantonly destructive than the unsportsman-like warfare which has rendered almost extinct the American bison. The natives are in the habit of digging covered pits, and then beating up the country for miles around to drive first into corrals, and then into the pits, everything that has life, and which does not succeed in breaking through and making its escape. These corrals open into deep pits, in which are planted sharpened stakes, upon which the animals impale themselves. The method of hunting among the Chinese emperors omits the pitfalls, but is equally secure and destructive. They have their soldiers drive the game together, and when the animals have been sufficiently terrified by noise and wounds, the emperor amuses himself by killing such as he is pleased to select, and then the object of the pageant has been accomplished. The *Cape buffalo*, or *buffalo of Caffraria*, is black, as has been said, but a blue-black. Its hide is dense, and its feet are like those of the ox. It will attack anything, and is dangerous because it will lie in wait in the jungle and take advantage of the unwary traveller or hunter. It is more emotional even than the elephant, and sometimes has absolute paroxysms of rage. It was always an enjoyment for those who attended gladiatorial shows to watch the buffalo and the tiger, for the buffalo not merely is combative, but has the greatest antipathy to this ani-

BATTLE BETWEEN A LEOPARD AND A CAPE BUFFALO.

mal. The buffalo is almost uniformly successful, and after tossing its foe, will kneel upon him and crush him, butt him, lick him raw with its tongue, and finally feast upon his blood. A native was pitched into the branches of a tree by a wounded buffalo, and was glad of the exchange from the tossings and tramplings which he had previously undergone. On one occasion, as a hunter was about to dispatch a wounded bull, he was surprised by its peculiar moan which proved to be a successful appeal to its comrades for succor. Tenacity of life is a characteristic of the buffalo, and even when at the distance of a few feet it has been shot full in the head, it has been known to scamper away as if unharmed.

The buffalo has been known to chase the tiger just for sport. A bull having been wounded, its fellows tried to support it on either side, and thus enable their injured companion to escape with his friends. On another occasion a large bull having been desperately wounded, a younger bull came back and devoted itself partly to stimulating the wounded animal to fresh efforts, and partly with attempts to divert the attention of the hunter. The most interesting feature in this story is that the devoted efforts of the young bull were successful, and it was not until many months later that the hunter found the carcase of the bull which he had wounded mortally. The buffalo constantly changes the route by which it goes to and from its watering-place.

ADVENTURE WITH A CAPE BUFFALO.

An amusing experience is related by one of our distinguished African travellers. He had wounded a bull, which charged so savagely that he had to put spurs to his horse, and while riding at full speed the horse put his foot into a hole, while the rider went on as if being fired through a pneumatic tube. Fortunately, an ant-hill was near at hand, and he scrambled to the top, while the buffalo vainly endeavored to imitate his example. The buffalo next established himself as a sentry, being careful not to expose his vulnerable parts. At this stage, the riderless horse was inspired to try co-operation. Attracting the attention of the buffalo, he would keep it charging after him around the hill, until the hunter succeeded, after several trials, in putting an end to the triangular contest. When life was extinct in the buffalo, the horse walked up to the carcase, sniffed at it contemptuously, and then withdrawing a short distance, patiently awaited any further call upon its services. If the buffalo succeeds in catching the hunter, a mad ox is amiable and inoffensive in comparison. Many a native has been carried on the horns of a buffalo, either to his death, or to escape grievously wounded, even if relief should come.

On one occasion a native, after being carried twenty yards upon the "cowcatcher," was thrown into the branches of a tree, where he preferred to pose as a new species of over-ripe fruit, rather than to proceed to "that bourne from which no traveller returns." Henry M. Stanley tells of a fearful experience on the part of one of his native attendants. The native, believing the buffalo to have been rendered helpless, drew near with the intention of using his knife, when the mortally wounded animal made a last herculean effort, and seizing the native, tossed him hither and thither until the last spark of life had fled, and there remained nothing but a mangled corpse.

The buffalo is either hunted by mounted huntsmen, or else by the method of stalking. The shoulder shot is the only one that is effective, and the experienced hunter is always slow to approach the animal after it has fallen. The bulls engage in the most furious contests when seeking the favor of the cows, and the victory belongs to the one which succeeds in pushing the other backward until it relinquishes the conflict and sullenly retreats. When the buffalo strikes a quicksand, it at once loses its fierce courage, and will, in a broken-spirited way, allow itself to be entombed, even when by the slightest effort it might escape. The average weight of a bull is from four hundred to five hundred pounds, so that the wanton destruction which has nearly rendered the animal extinct, is still less pardonable.

Hunting the reeds for buffalo, although "extra-hazardous," is always profitable, as the hunter never fails to find a herd; at one time twenty-nine were thus shot. Two bulls, when fighting, do not remove

BATTLE BETWEEN AMERICAN BISON AND GRIZZLY BEAR.

their horns, but exert all their strength to push each other backward. They strike for the chest, but it is rare that either is unskilful enough not to parry the thrust and to get horns locked with horns. A native disturbed a cow and a calf, and frightened the latter, whereupon the cow made a successful charge, and after striking him in the back and inserting her horns in the man's belt, swung him round and round until the belt gave way. The cow then returned to her calf, licked it affectionately, and the two slowly walked away without paying any further attention to the man. Another less fortunate native was gored from back to breast. Another, while engaged in flight, was overtaken by the buffalo, tossed again and again, and escaped finally through being pitched up into the branches of a tree. Another, while with a

38

party engaged in " pig-sticking," was severely prodded by an intrusive buffalo. A celebrated traveller having gone to bathe was suddenly charged by a buffalo and knocked over the bank. Even this did not satisfy the angry brute, for clambering down to where the traveller lay between two narrow ledges of rocks, it first butted at him ineffectually, and then licked his legs with his rasping tongue, finally desisting solely because the traveller had the self-control to simulate death even while being sand-papered.

The Hon. W. H. Drummond tells of shooting a solitary buffalo bull in the forehead and then hastily climbing a tree. From this point of advantage

A FIGHT IN THE FOREST.

he saw the buffalo charge wildly hither and thither, and finally start off for parts unknown, irritated but unharmed. He followed in its track all day long, and finally came upon it in the jungle, and having taken the risk of entering a great cane-brake escaped by a quick single shot, which happened to touch the buffalo in the right spot. Livingstone and Baker tell of the great skill of the native aggageers, who pursue and hamstring with their swords elephants, buffaloes and other dangerous game.

The Musk Ox, or Musk Sheep (*Ovibos moschatus*), can be successfully hunted only by stalking, but the venison-like taste of its flesh attracts the hunter. It has very acute sight and hearing, and is exceedingly nimble as a leaper. It runs in herds of from ten to twenty, and when pasturing always posts sentinels. It is from eleven and a half to twelve and a half hands in height, and its coat is soft, long, and dark brown in color. Its eyes express gentleness, but during the rutting season it is exceedingly quarrelsome and vindictive, and uses its sharp-pointed horns with great effect. The mothers are very zealous in caring for their young, and while compelled to conceal them, keep a very watchful eye upon any person or animal approaching the place of their concealment.

UNGULATES.—ANTELOPES.

The **Eland of the Steppes** (*Antilope oreas*) wanders in company with a mate or as one of a large band. It equals the horse in stature, and its huge antlers, starting from the head at a right angle, suddenly curve upward and terminate in a broad palm, which finally gives way to deep indentations lying between projecting prongs. As would be supposed, the neck, which must support the weight of these antlers, is short, thick, shaggy and powerful; but if the beauty of fitness is thus attained, the beauty of symmetry is lost through the undue projection of the shoulders and the extravagant length of the fore legs. It is often called the *canna*, is found in Thibet, and is as large as an ox, being six feet in height and even disproportionately stout. It is so unwieldly that it is easy to drive it to the vicinity of camp before shooting it. Another species is found in South Africa, which differs from the *steppe eland* in having straight horns, heavy brisket, is destitute of mane, but has a thick tuft of hair growing from the centre of the forehead. Its flesh is specially pala-

HUNTING THE MUSK-OX.

table in a region where most flesh is dry. It seems to live without any need for water, but it is the opinion of leading naturalists that the *eland* is able to extract its drinking water from its food, and hence while seeming to be an advocate of total abstinence, still manages to secure its supplies. In color it is grayish-brown, and its horns are large and spiral. There is a striped variety, but though belonging to the same region and having substantially the same habits, it is rarely met with. A traveller relates quite an illustrative experience. Having found a herd of *elands* he wounded one of them, but the herd would not desert their unfortunate companion. After driving the herd before him for several miles, he succeeded in separating the wounded one and in driving it in the direction of his camp. At one time he rode into an ant-hill and was dismounted, but speedily resumed the pursuit. Again a lion scared both hunter and *eland*, and gave a certain obliquity to their line of march. Finally the wounded *eland* was driven near the camp, and as a reward for having served as a butcher's delivery-wagon, was then shot. The *eland* is rapidly being exterminated, for zoology to African natives and to African hunters seems to mean only the procuring of some kind of meat to eat. The weight of those killed in Central Africa varies from eight hundred to fifteen hundred pounds.

ELAND OF THE STEPPES.

The **Boschbok** (*Tragelaphus sylvaticus*) belongs to southern Africa, and though numerous is rarely seen and still more rarely captured. It has a white line along the back, succeeded by black, relieved irregularly by white spottings. It is the most suspicious and keen of hearing of all the deer. It is pronounced the finest of African antelopes, and as it always charges, is interesting to the sportsman. On one occasion at least it killed a leopard which had attacked it, and escaped, wearing the leopard's blood upon its horns.

The **Harnessed Antelope** (*Tragelaphus scriptus*) is a striped eland. It belongs to western and southern Africa. It is distinguished by white stripes, arranged like harness, is thickly spotted on the haunches, and has a few spots on the shoulder, which impart a curiously pleasing appearance, resembling breeching, from which the name is derived. The species has been seen by few travellers and may be considered as nearly extinct.

AFRICAN ELAND.

The **Nakong**, or Nzoe (*Tragelaphus spekii*), is an almost aquatic animal, found in the interior of Africa. When swimming it exposes itself very little. It lies in the rushes by day and pastures by night. The hunter burns the rushes, and uses the fire thus kindled to drive the *nakong* into clear water, as well as for enabling him to see the horns which are the only part of the animal above the surface of the water. The *nakong* is smaller than the leche, and is better provided with a paunch than are most antelopes. It is grayish-brown, and its coat is long, sparse and coarse-looking, and is spotted, not striped. It carries small, twisted horns, which are provided with double, winding ridges. It moves with the shambling gallop of a mangy cur dog.

STRIPED ANTELOPE.

The **Nilgau** (*Portax pictus*) is frequently to be met with in zoological gardens. It comes from India, and its grayish-brown coloring is relieved by white patches on the face and legs, and by a full black mane and throat whiskers. Its short horns are almost straight.

The **Abyssinian Beisa** (*Oryx beisa*), though sometimes found on the coast, prefers the desert. It is parti-colored, and while generally inclining to yellow, as the color of its coat, it is white at the extremity of the mouth and nose, the lower ears, the belly and the antlers. Its face bears black triangular marks, and a slanting black line under the eye. It is stout and large in size, has a bushy tail, and its three-foot horns run straight back without spreading; they are annulated on the lower half. Its habitat is Abyssinia.

ABYSSINIAN BEISA.

The **Gemsbok** (*Oryx capensis*) is sometimes called the *kokama*, is nearly four feet in height, and has South Africa as its habitat. It is prevailingly gray in color, but the flanks, hind-quarters, tail and back are black, and it has a black streak across its face. It has an erect short mane, and long, heavy, sharp-pointed horns, which it has been known to use to the discomfiture of the lion. It is supposed by many to require no water, but others more reasonably sup-

HEADS OF AFRICAN ANTELOPES.

WATER BOK (*Kobus ellipsiprymnus*).
BLUEBOK (*Cephalophus grimmius*).

BASTARD GEMSBOK (*Hippotragus leucophæus*).
KALA BOK (*Cephalophus mudequa*).

HARTEBEST (*Bubalis caama*).
BLESBOK (*Kobus leucotis*).

pose that it finds its supplies in moist plants. It is too shy to be stalked, and stays too far from water to be "pitted." Hence it is hunted by mounted men, and the palatableness of the flesh is esteemed a sufficient reward.

The **Oryx Gemsbok** (*Oryx lucoryx*) is grayish, with black and brown markings spread over its body. It is somewhat shorter than the gemsbok, but carries horns three feet in length. Its habitat is northern Africa.

The **Addox, Spotted-nose Antelope, Mahyna,** or **Lyrate-horned Ante-**

GAZELLES (*Antelope dorcas*) AND DEISA (*Oryx beisa*).

lope (*Addox nasomaculatus*), is found in eastern Africa and is distinguished by short, lyre-shaped horns and white nose-markings.

The **Roan Antelope,** or **Bastard Gemsbok** (*Hippotragus equinus*, or *leu-cophœus*), is nearly as large as the eland, and carries long, massive horns of which it makes good use when cornered. It is solitary in its habits and not commonly met with, though not at all scarce among the African fauna. Its hair is unusually long and of a brownish-gray, except on the under parts where it is white. Its mane is long, thick, and bright brown in color. Its erect head, set off by sharply-pointed ears, is supported by a neck whose curl is exceedingly graceful and striking.

WILD DOGS (*Lycaon pictus*) PURSUING A SABLE ANTELOPE (*Oryx leucoryx*).

The **Maharif** (*Hippotragus bakeri*) is mouse-colored, although its face is striped in black and white. Sir Samuel Baker, in his charming narration of his African travels, speaks of securing the horns of a *maharif* which had been killed and eaten by a lion, who seemed willing to divide with mankind if they would accept the indigestible portions of the antelope.

The **Ariel Gazelle** (*Gazella ariel*, or *dama*) is not allowed by many naturalists to constitute a distinct species. The back and upper portions of the

HERD OF AFRICAN ANTELOPES.

body are a dark fawn color, banded with black along its sides, and white upon the abdomen. Its habitat is Arabia and Syria, and its grace and docility has

BLESBOK (*Bubali albifcrus*).

converted it into a household pet. In fleetness it is not approached by the swiftest hunting-dog, and it seems to float rather than run over the ground. This little animal is less than two feet in height, and is hunted out of proportion to its insignificant size. In hunting for sport, falcons are used, and the trained bird by its attacks so confuses the *gazelle* that it uniformly falls a victim. When hunted for its flesh, a large stone-walled corral is built and the animals driven into it. At intervals have been left low walls, on the outer side of which are wide pits or trenches. The *gazelles* attempt to escape over these and fall into the trap prepared for them.

The American world has grown so cosmopolitan in its tastes that there is no quarter of the globe which some of its people have not penetrated—no literature too foreign for it to be unknown to some, at least, among American readers. Firdusi and other Persian poets have not only exercised great influence upon Byron and Moore· and other British poets, but their works have become familiar to many an American reader. Among the most frequent comparisons made by Persian poets are those in which a part is played by the Asiatic gazelle, or jairou, ahu, or dsherin.

The **African Gazelle** (*Gazella dorcas*) is described at length by Sir Samuel Baker, who lays special emphasis upon their adaptation to life in the desert, their beauty, and their strength and symmetry of development. It is unusually fleet, but it is sometimes caught by the hounds because it is ignorant of the fable of the hare and the tortoise. After it has distanced its pursuers it will stop in reckless defiance and exhaust its strength in vainglorious exhibitions of its ability as a leaper. The hounds " sticking strictly to business " finally overtake the *gazelle* when it has exhausted its strength. It is clothed in satin whose color varies from a golden-brown to a mauve; the belly and the legs below the knee are· spotless white. Its head is lighted by the most perfect of oriental eyes and ornamented with black, curved, annulated horns, a full foot in length.

The **Ugogo** (*Gazella granti*) wears fawn color and purple, and its skin looks like watered silk. Its horns converge at the two extremities and are two feet in length. Its habitat is likewise African.

The **Springbok** (*Gazella euchore*) is named from the extraordinary flying-leaps which it makes. It will leap as high as twelve feet into the air—and that without the aid of any springboard. It is a cinnamon-brown above,

IBEX, OR STEINBOCK (*Capra ibex*).

spotless white below, and has these colors separated by a reddish-brown band. It is unusually shy and timid, and especially dislikes the company of human beings. Trustworthy travellers tell us that it carries this dislike so far that it will always leap over any road or pathway that man has trod. The *springbok*, like all the other boks, is liable to have its name spelled in as many different ways as that of Shakespeare, so that the reader need not be troubled if he reads in other works of the *springbock*, *springbuk* or *springbuck*. It is abundant about the Cape of Good Hope, in spite of its frequent slaughter. It is gregarious, and accounts of herds of *springboks* are frequent in all accounts of southern Africa. It has been known to migrate in herds of forty thousands.

The **Pallah**, or **Roovebok** (*Æpyceros melampus*), runs in large herds throughout southern Africa. It is of a bay color, which changes to white on the abdomen and at the base of the tail. Its haunches are ornamented with a black half-moon, and it stands upon hoofs of the deepest jet. Its horns are lyrate, ringed, and nearly two feet in length. The *pallah* has too much self-confidence to be timid, but when a herd decides to put a greater distance between themselves and their enemies, they will solemnly stalk away in single file. The *pallah* is gregarious only in the winter season.

The **Saiga** (*Saiga tartarica*) is an antelope of the steppes which in graceful-ness rivals the gazelle. In appearance, however, it is very like a sheep, so far as its head and face are concerned. It has long tufts of hair beneath the eyes, and corresponding ear fringes. It is quite celebrated for its achievements in the matter of butting and jumping. Its horns are erect, annulated, transparent and yellow, and are never worn by the female. Its nose is no nose, but a snout; its forehead is covered with folds which run crosswise. It moves in large bodies—

SPRINGBOKS (*Antidorcas euchore*).

several thousand being no uncom-mon spectacle. When upon the march it keeps up a large service of scouts and officers, and provides a rear guard to prevent being surprised.

The gazelles when attacked by a panther will arrange themselves in a circle with their heads making the circumference, and invariably succeed in protecting themselves against their dangerous enemy.

The **Prong-horned Antelope** (*Antilocapra americana*) is called likewise the *cabrit, prong-buck, cabree* and *North American springbok*. Its home is on the western plains. Its horns are pronged at the point of curvature, in which respect it is singular among hollow-horned animals. It sheds its horns annually. Its color is a brownish-yellow above and a white below; its brown face grows white on the cheeks, and it wears on its throat a crescent and a triangle both of which are white; white also prevails on the haunches, and sometimes on the tail. It is gregarious only when migrating, or during the rutting season. Its venison is held in high esteem, and while its fleetness might easily insure its safety it falls a victim to the weakness which endangered the life and happiness of Bluebeard's wife and her sister Anne. The hunters find that by merely tying a colored cloth to their guns they can approach quite near to this over-curious antelope. Its habitat is from the Pacific slope to the Missouri river.

The **Reedbuck** (*Cervicapra arundinacea*) is a hooked-horned antelope of southern Africa. It frequents the reeds and moves about in couples. It is about five feet in length and three in height, exclusive of an additional foot for its horns. Its ashen-gray color becomes white on the under parts. Its forays upon the cornfields render its destruction an imperious necessity to the farmer. It is not difficult of approach, as it will lie still in the reeds until the hunter almost stumbles over it, when with a chamois-like whistle it jumps up, gallops a short distance, and then stops

SAIGA ANTELOPE, OF THE STEPPES (*Saiga tartarica*).

to take a fresh observation. This toying with fate is less hazardous than would appear at first sight, for although not invulnerable, it is somewhat rhinoceros-like in its general indifference to bullets. When about to take to flight it always whistles to its mate, so that hunters employ the *reedbuck* whistle as a decoy. It is sometimes called the *umseke* or the *rietbok*.

ANTELOPE.

The **Waterbuck, Photomok, Waterbok,** or **Kobus** (*Kobus ellipsiprymnus*), wanders about South Africa in small herds, and when disturbed takes at once to the water. It is brown except for a white ellipse at the base of the tail. Its horns are about two and a half feet in length, rather inclined to be lyrate, and bent back until they near their extremities when they again bend, but forward. Sir Samuel Baker on one occasion succeeded in bringing to bay upon an island a water antelope and two koodoo bucks, and was fortunate enough to shoot all three. On another occasion he rode down, after an exciting race, a koodoo buck. White is the color of chest, abdomen and eye-orbits; brown prevails on the front legs, while the body is brownish-yellow and the tail black-tufted. The neck is the vulnerable part —if shot elsewhere it is as safe as Achilles, and as tenacious of life as a cat. For example, a water antelope which had been mortally wounded, took to the water where it was seized by a crocodile. In spite of its expiring strength it dragged the crocodile from the water and several yards on the land, and finally escaped from it to die peacefully in the rushes.

The **Leche** (*Kobus leche*) is another species of African waterbok, but, strange to say, it cannot swim, and hence contents itself with wading in the shallows. The *leche* is found feeding in herds positively enormous in size. When the flooding of the flats drives the *leches* to the mounds, the natives with their light canoes chase them from mound to mound, and succeed in spearing great numbers of them.

The **Pookoo** (*Kobus vardoni*) is red in coloring and haunts the immediate vicinity of the rivers.

The **Klipperspringer**, or **Kawdi** (*Nanotragus neotragus*), is a peculiarly attractive antelope belonging to South Africa. Like the chamois, it skims over

WATERBOKS (*Photomok*).

rock and precipice in the most seemingly reckless manner. The natives believe that the *klippersprINger* can, at its will, bring rain, and are therefore in the habit of catching and worrying these creatures during periods of drought. It is about a foot and three-quarters in height. Its coat consists of gray hairs which become brown towards the middle and yellow at the tips.

The **Steinbuck** (*Nanotragus neotragus*) is nocturnal, sleeping in fastnesses during the day and issuing at twilight. The **Madoqua** is the smallest, most fragile-looking and most charming of the antelopes. It is barely more than a foot in height, and most beautifully symmetrical. Its brilliant silvery-gray lightens into white on the under parts and at the base of the tail, and deepens into chestnut along the back. It is frequently spoken of by the great Bruce in his accounts of the Abyssinian country. The *madoqua* is a mountaineer and is not gregarious.

The **Diver Antelope, Duyker, Duykerbok, Impoon** (*Cephalophus mergens*), takes its name from its habit of diving into the brush, and is a solitary little antelope of southern Africa, whose ingenious devices for protecting himself are alike numerous and curious. When approached it will leap over bushes and again dive through them, making so many sharp turns as to make it almost impossible for man or dog to follow it. It will then quietly crawl along under the bushes for quite a distance, when it will jump to its feet and bound away. Even when a shot is had the sportsmen will often lose the game because of its power of carrying without great inconvenience quite a heavy load of shot. It belongs to southern Africa. It is small, gray, and has straight, sharp horns.

ARABIAN GAZELLES OPPOSING A LEOPARD.

It is very tenacious of life and will frequently turn to bay and charge the hunter.

The **Four-horned Antelope**, or **Chouka** (*Tetracerus quadricornus*), is an Indian species. The front horns, situated just above the eyes, are short, the back pair longer; the female wears no horns. The *four-horned antelope* is only about a foot and three-quarters in height, and is bay above and gray beneath.

The **Chamois** (*Rupicapra*) is a wary, mountain antelope, gregarious in its habits. It always keeps a sentinel upon guard, and when danger is announced the whole herd gives one look of inspection before taking to flight. The *chamois* can be domesticated and converted into a household pet. It is brownish-yellow, with a black streak along the back; in winter the color becomes blackish-brown. The face and throat are a yellowish-white, varied by a dark bank forming rings about the eyes, and extending

CHAMOIS.

thence to the corners of the mouth. The horns, doubtless familiar as articles of ornament, are jet-black, sharp-tipped and highly polished; they are about seven inches long. The heights which the *chamois* scales, its grace and agility, and the excitements and perils of the *chamois*-hunter are too often described to be more than alluded to in such a work as THE LIVING WORLD.

The **Gnu, Gnoo, Wildebeest** (*Connochetes gnu*), is a South African animal which looks as if some one in sport had put together parts of the ante-lope, the ox and the horse, or as if the fabulous forms of heathen antiquity had some substra-tum of fact upon which to rest. The head, like that of the buffalo, is armed with wide-spread-ing horns, which first bending down-ward again curve upward and terminate in sharp points; the body and tail are those of the horse, while the an-telope charac-ter pronounces itself most dis-tinctively in the legs and feet. They are gregarious and the herds are large.

CHAMOIS DEFENDING ITS YOUNG.

They frequently associate with the ostrich, the giraffe and the zebra. They are curious, suspicious and capricious. When their attention is caught by any unusual appearance, they will begin pawing, capering and bellowing; will leap high in air, begin to fight each other, and finally take precipitately to flight. It is susceptible of domestication, but is so much favored by the bot-fly as to render it an unwelcome addition to the domestic herd. It is usually dark brown or black, except the mane and tail, which are white.

The **Brindled Gnu** (*Catoblephas gorgon*) has no mane on its chest, is taller, always moves in single file, and is striped with gray.

The **Yakin** (*Budorcina taxicolor*) is found on the steeps of the Himalayas, and exhibits a nose which might properly belong to a sheep, and a tail which seems to have been borrowed from a goat.

The **Blue Gnu** (*Catoblephas taurina*) is still found in Zululand, and the black stripes on the neck and shoulders, its long and flowing mane as well, together with its black tail and symmetrical form, render it quite a sightly creature to behold. Suspicious, timid, curious about strange objects, and exceedingly irritable, their performances, when startled, are indescribably ludicrous. Its curiosity is frequently fatal to the *gnu*, for, at the sight of a strange object, it will gallop about it in constantly decreasing circles, until it brings itself within reach of the hunter who then kills it at his leisure. A *blue gnu* was being harried by two of Sir Samuel Baker's hunting dogs, when it used its ox-like horns and spitted the younger and more venturesome dog. The other dog would run away until it exhausted the patience of its pursuer, when it would turn and begin its work all over again. Baker says that he was so much interested that nothing but presence of mind and a quick and lucky shot made it possible for him to live to tell the tale, for the *gnu* suddenly charged upon the hunter. The *gnu* is a stoic, for on one occasion, in spite of a broken shoulder-blade and a broken hind leg, a *gnu* ran six miles before the hunter succeeded in killing it. A *blue gnu*, or *wildebeest*, was

GNUS.

found, which while fighting had got one of its legs over one of its horns, and was compelled to wander around in a manner at once uncomfortable, awkward and comical, until an African traveller relieved its discomfort by the application of a bullet.

The **Hartebeest**, or **Lecama** (*Bubalus caama*), is about five feet high, and moves about in herds of ten or twelve, which contain but one male. It is grayish-brown in color, with a black streak on the face, a brown patch on the outside of the legs, and a triangular white mark on the haunches. Its thick horns, knotted at the base, become lyre-shaped, and then suddenly curve at right angles. Its habitat is tropical Africa. It is very strong, and so swift that it cannot be run down with dogs. On one occasion a *hartebeest* upset a horse, nearly killed the rider and completely wore out the dogs before it dropped from utter exhaustion.

The **Bubale of Northern Africa** (*Alcephalus bubalis*) is substantially only a variety of the hartebeest.

The **Saisin**, or **Indian Antelope** (*Antilope bezoartica*), runs in herds of about fifty, which, however, contain but one buck. It is so swift as to be hunted successfully only by means of the falcon. It is said to jump twenty-five feet at each bound, and to spring some ten feet above the ground. The young males are promptly expelled from the herd, and at once undertake the task of gathering unto themselves wives from their neighbor's flocks. The young are grayish-brown or black, with white for the abdomen, circles about the eyes, breast and lips. It is frequently called the *black buck*, stands thirty-three inches at the shoulder, and has a length of forty-six inches, exclusive of the tail. In middle life it is fawn above and white below, but as the bucks grow older, they become brown on the neck and head. It wears black vertical shoulder-stripes, and its black horns are four or five-spurred.

ALPINE IBEX (*Capra ibex*).

The **Grysbuck** (*Nanotragus malantis*) is chestnut-red in color, with white hairs scattered over it. The ears are unusually long and the tail correspondingly short. It is an inhabitant of southern Africa.

The **Ourebi** (*Scopophorus ourebi*) will suffer death rather than forsake a favorite locality, and if all of one herd be killed off, others, by some strange law of instinct, will appear to assert the pre-emption claims of the family. It lives in pairs among the long grass. When in flight it bounds into the air from time to time for the purpose of taking observations. When pursued it will run from side to side, bound into the air, and frequently change or even reverse its course. It will go crouching through the tall grass and conceal itself behind an ant-hill, but as soon as the hunter has passed it will leap up and speed away. In color it is tawny above and white beneath. It is found in the region of the Cape of Good Hope, and is handsome, graceful and its flesh is palatable. It prefers treeless plains and the flatter portions of the country.

The **Sable Antelope**, or **Pontaquaine** (*Hippotragus niger*), is shy and crafty, and never approaches the South African villages. It is glossy black in color, which contrasts strikingly with the snow-white of the belly. The horns, some three feet long, sweep back nearly to the haunches. It lives in very small herds and the bucks make the females do sentinel duty, and their vigilance is such that they are very rarely hunted with success.

The **Blue Buck** (*Cephalopus pygmæa*) is only about two feet in length, and but a foot in height. It is dark blue in color, and its two-inch horns are straight and closely ringed. Its color is such a protection that it is seldom killed. It belongs to South Africa. It lives on the hillside, and moves about in herds of ten or twelve. It is remarkably swift and active. Its forehead is tufted.

The **Sassaby**, or **Bastard Hartebeest** (*Damalis lunatus*), is reddish-brown, having a blackish-brown stripe down the middle of the face. It lives near the southern limit of the American tropical zone, and is always a welcome sight to the thirsty traveller, for its presence tells of water near at hand.

HIMALAYAN IBEX.

It is much sought for its flesh, which is held in high esteem. The **Munni, Pied Antelope, White-faced Antelope**, or **Bontebok** (*Damalis pyarga*), is purplish, with blackish-brown on the outside of the legs, and white upon the inside of the legs the haunches and the face. It is larger than the stag, and has a wealth of horns which, annulated and black in color, attain a length of a foot and a quarter. It is found in the region of the Cape of Good Hope.

The **Blaze Buck**, or **Bless Buck** (*Damalis albifrons*), is a species closely connected with the nunni, which also have the blazed face, from which this species takes its name. It is South African in habitat.

The **Natal Bushbuck**, or **Rhoodebok**, or **Red Buck** (*Cephalophus natalensis*), lives a solitary life in the densest parts of the forests. It is about two feet high and wears straight, long-pointed horns. Its deep red color reflects the light and helps to conceal the animal. When alarmed, it keeps moving its large ears and walks, as it were, on tip-toe. If a twig

MOUFLON (*Ovis musimon*).

snaps, it stops at once in whatever attitude it may be, and then again, after this dramatic display, moves on with the same caution. If satisfied of the reality of the danger, it at once bounds away into the forest. It is yel-

39

lowish-red above, gray below, and has a tuft of hair growing out of its forehead.

The **Koodoo** (*Strepsiceros kudu*) is South African, stands about four feet in height, is heavy, not very swift or enduring. Its three feet of horns are much twisted and keeled throughout. It runs in herds of four or five and, though living in the brushwood, finds no inconvenience from its antlers, which it lays back, after the manner of the moose. It is reddish-gray, with streaks of white on back and sides. Its flesh is specially palatable, so that the *koodoo* is frequently hunted. The natives, in imitation of the method of the wild dogs, take turns chasing the *koodoo* at full speed and falling back when exhausted. Such a hunting party includes women who act as vivandieres. It is often called the *nellut*, and the males incline to a bluer-gray than the females.

The antelopes may be conveniently separated, so far as the African fauna is concerned, into those which affect the open country and those which prefer the thickets and the thorn-scrub. To the former belong the ourebi, steinbok, vaal roebok, klipperspringer, and reed buck. To the latter class we must assign the duyker and roebuck. The natives will form hunting parties of as many as eight hundred persons and, making a circle of miles, will gradually close in upon the

MANED GOAT (*Ovis trogelaphus*).

game. The slaughter of antelopes is specially great, but not unfrequently some of the animals will jump sheer of the heads of the hunters. Antelopes, gnus and zebras seem fond of one another's companionship and are generally found pasturing together.

UNGULATES.—SHEEP.

The **Musimon**, or **Corsican Sheep** (*Ovis musimon,*) seems to be the primitive type of the useful, familiar and much-praised domestic sheep. It dresses not in wool, however, but in hair which, short in the warm season, begins to grow wavy as cold weather approaches. It is brown above and white beneath.

The **Asiatic Wild Sheep** (*Ovis orientalis*, or *gmelini,*) is the most graceful of sheep, suggesting the deer just as the saiga antelope suggests the sheep. It is abundant in the salt lake regions of Asia Minor. It varies through the different shades of red to deep-brown, but is white on the abdomen and on the inner parts of the legs.

The **Turkestan Sheep** (*Ovis kerelini*) has been made the subject of a monograph by Severtzolf, who has also discussed another Turkestan species (*Ovis poli*), which has a different coloring but is substantially the same in habits.

The **Argali** (*Ovis ammon*) seems to represent the earliest living type, and is found in the Himalayas, which appear to have been the nursery of the sheep as well as of the human family.

The **Nayaur** (*Ovis hodgsonii*) belongs to the fauna of Thibet, although it is found also in Nepaul. It wears a white collar on its lower neck and carries a short mane.

The Atlas Mountains furnish the **Atlas Maned Sheep** (*Ovis tragelaphus*), which is seemingly an intermediate form between the sheep and the goat, as though, in tribal differentiation, it had endeavored to go two ways at once.

The **Spanish, or Merino Sheep** (*Ovis aries-hispanica*), has, like all that is Spanish, a long and honorable record. It is large, and the ram has great, spiral horns. It has a black face and exhibits a constant disposition to revert to a black color. The wool of these sheep and the skin of their kids are the source of its interest to man.

The **Maned Sheep**, or Goat, is six feet in length of body and three feet in stature. Its curly fleece is not like that of our common species, but is a rusty brown. Its tail is nearly a foot in length and very bushy; its horns are stout and long, and sharply curved over the head; its mane is short, but as a compensation it has long chin-whiskers, and regular valances of hair hang from the shoulders to the feet.

TURKISH SHEEP.

The **Turkish Sheep** (*Ovis aries-steatpyga*) has become quite celebrated as using its tail as a reservoir for the fat which it accumulates.

The **Cape Sheep** are not the celebrated fat-tailed sheep, whose caudal appendage is considered such a table delicacy, but they are prized by the inhabitants alike for their mutton, the fat stored up in the tail and hind-quarters, and for their wool, which is very abundant, soft and warm.

The **Afghan Fat-tailed Sheep** is notable for the size and fatness of its tail and the silki-ness of its coat. The fleece is exported and forms an important article of commerce.

The **Wallachian Sheep** is found in Wallachia and Hungary. Its fleece is long and silken and its large horns spiral.

The **Siberian Argali**, or **Mouflon** (*Caprovis argali*), is about as large as an ox, and has horns four feet long, and nearly a foot and three-quarters around, measured at the base. These horns, rising straight

HEAD OF MERINO SHEEP.

at first, next curve as far as the chin, when they again curve upward. It is a mountaineer and climbs the rocks with a rapidity and ease which are as surprising as unexpected. They are sometimes buried in snowdrifts, when hunters take advantage of their helplessness.

The **Rocky Mountain Big-horn** (*Ovis montana*) runs in small herds over the crags of California. Its flesh is highly esteemed, but the animal is shy and not often captured.

The **Bearded Argali**, or **Aou-dad** (*Ammotragus tragelaphus*), has its habitat in northern Africa, and is a large animal which sports about

MUSK SHEEP.

the rocks and precipices as though it were the smallest of chamois. Its fore legs are ruffled above the knees and have given it the name of the *ruffled mouflon*.

The **Musk Sheep** (*Ovibos moschatus*) is frequently called a musk-ox, and naturalists are not wholly at one as to where the animal should be classed. Its habitat is the North Polar region, but it is rarely found and is best known through fossil forms, which have interest and value for the student of the succession in evolutionary life. It belongs to North America and the Esquimaux call it the *voming noak*. It is long-haired, fleet, irritable and dangerous. The horns of the bull curve downward around the head until they reach the eyes, when they curve upward. Its covering of long, thick hair gives the appearance of great amplitude to a really small body. Its height is about three feet, and its length eight feet. It is amber-colored and during the summer it sheds its long fine coat. The musky odor is always present, but does not, except at certain seasons, impregnate the flesh. It wanders about in herds of not more than twenty or twenty-five

HEAD OF WILD MOUNTAIN GOAT.

and the bull, which plays the role of the head of the family, always keeps sentinels posted. The *musk sheep* have deadly contests with the bulls of other herds and with the bears, and in the latter case the sheep are generally victorious. When angry the *musk sheep* whines like a walrus.

UNGULATES.—GOATS.

The **Ibex, Steinbock, Steinbok, Steinbuck, Steinbuk, or Bouquetin,** (*Capra ibex*), is Alpine in its habitat, and is eminent among the horn-wearing animals. Its color changes from the reddish-brown of summer to the grayish-brown of winter; the belly and the inside of the legs are white or gray; the face and back have a dark stripe running along them. The horns are familiar as the handles of paper-cutters, and are about three feet in length, and full of ridges. It moves about under the guidance of a male who maintains military discipline over his herd of five or ten. He always posts one or more sentinels who, at the least suspicion of danger, whistle for the herd to fly to the heights above.

The **Spanish Goat** (*Capra pyrenaica*) is harmless, lives on the mountain peaks and is rapidly becoming extinct as a species. It is sometimes called the *ibex of the Pyrenees*, and at other times the *tur*.

The **Jemlah Goat**, or **Jharal** (*Hemitragus jemlaicus*), lives on the highest Indian mountain peaks, on the borders of perennial snow. Flocks of about twenty-five, under the guidance of a male, descend during the day to pasture. It is a grayish-fawn in color, has a brown mark on the forehead and the front of the legs, and a dark streak on the back. They wear a long mane falling down on both sides of the neck. The hair is coarse and long, and the horns, wide at the base, keep spreading, until suddenly they narrow and nearly meet in two points.

The **Cashmir Goat** (*Capra falconeri*) has its habitat in Thibet, and its wool or hair is a well-known article of commerce. The animal wears two coats, an outer of long silken hairs and an inner of soft, gray wool; it is the latter which is used in manufacture, and its costliness will be explained when it is understood that seven goats must be despoiled to make but one yard of the woven fabric. Of course, governmental taxes and the charges of commerce greatly increase the cost of the fabric, but then the material is expensive in first hands.

The **Bezoar Goat** (*Capra ægagrus*) belongs to Persia and the Caucasian mountain ranges. Like some valetudinarians it is subject to attacks of stone, and the stones found in the stomach of this animal have quite a legendary history among the superstitious, who regard them as a trustworthy antidote to poisons. Naturalists are not agreed as to the original of the domestic goat, but many of authority assign this honor to the *bezoar*, which certainly has the typical powers of digestion.

WILD MOUNTAIN SHEEP PLUNGING DOWN A PRECIPICE

The **Perbura**, or **Ram Segul**, is an Indian goat, singular through having a smooth, white fur (sparsely intermingled with red), a large dewlap for the male, short ears and no beard.

The **Syrian Goat** has long, pendent ears, which reach nearly to the shoulders.

The **Snake-eating Goat,** or **Markhur,** is found in India and Thibet, and wears unusually large horns, whose convolutions are exceedingly complicated.

The **Rocky Mountain Goat** (*Aplocerus montanus*) is an antelope with small, recurved horns, sheep-like nose, short and furry lower coat, and upper coat long and pendent.

CARNIVORA.—COONS, COATIS, ETC.

The **Raccoon** (*Procyon lotor*) is to those who reside in certain parts of the United States a familiar object. It wears a short, thick, woolly undergarment of gray, and an overcoat of long black and gray hairs. The top of the head is dark brown, and there is a band of the same color across the eyes, and the same bands or rings adorn the bushy tail. It is a common pet, but is so mischievous and destructive as sadly to try the patience of its owner. It is substantially omnivorous, and will drink anything liquid. It has a habit of never eating anything without having first washed it, and this pronounced affectation of cleanliness has given it the name of *lotor*, or the *washer*. Its agility, subtle cleverness, abundance and desirability as an article of food (to those who like its flesh), lead to its being constantly hunted, and few boys in the Southern States have been without the experience of "coon hunting."

The *raccoon* hunt is not dissimilar to the "possum hunt," and is very exciting. Most commonly the *raccoon* is tracked and treed by the dogs, and a fire having been built enclosing the

HUNTING THE MOUNTAIN SHEEP.

tree, some one of the hunters climbs up and dislodges the game, after which the fun begins, and soon grows warm and furious. The least experienced dogs, stimulated by the presence of the rest of the pack and the shouts of the hunters, will rush upon the *raccoon* and seize it by the nearest part of its body. This usually results disastrously for the dogs, as the *raccoon* speedily discourages them by the liberal use of its teeth. The alternations of fortune, the joy in the success of one's favorite dog, and the no less pleasure in the discomfiture of the vaunted pet of some one else, keep one's interest from flagging in spite of the absolute certainty of the result.

The **Raccoon Fox, Ring-tailed Raccoon, Mountain Cat,** or **Cacomixle,** (*Passaris astuta*), is a Mexican animal, dun in color, and wearing a dark cape on its neck; its back has a black stripe, and its tale is ringed with white.

A COATI FAMILY.

The **Crab-eating Raccoon, Agoura, Raton, Maxile** (*Procyon cancrivora*), is possibly no fonder of mollusks and custaceans than is the common *raccoon*, but its opportunities are greater. It is larger than the common *raccoon*, grayish-black mixed with yellow, and its short tail exhibits six black rings. It has great skill in opening oysters, but sometimes miscalculates, and takes a leading role in the play of "The Biter bit." Its habitat is Central America.

The **California Raccoon**, or **Psora** (*Procyon psora*), is taller, and is said to resemble a dog.

The **Black-footed Raccoon** (*Procyon hernandozii*) is found on the Pacific slope, and resembles the crab-eater. It is larger, and, its black foot markings give it its popular name.

The **Coati** have a very small proboscis or a very long snout, which serves many purposes, but which is not used in drinking. They are arboreal and gregarious.

The **Mundeo**, or **Red Coati** (*Nasua solitario*), is red in color except for its black ears and legs, white hair on its jaws, and maroon banded tail. It is a ready climber, and whether ascending or descending always goes head first. It is nocturnal, and its liveliness by night is in marked contrast to its slothfulness by day. When wounded or irritated it is a desperate and formidable fighter.

The **Potto** (*Lemur flavus, Cercoleptes caudivolvulus, Viverra caudivolvula,*) was "sometime a paradox," having been classed with weasels and lemurs. It is widely distributed throughout South America, and hence fairly revels in a multitude of names, of which the reader may meet with the yellow lemur, the honey bear, the yellow macanco, the kinkajou, the guchumbi and the manaviri.

POTTO, OR SPOTTED PARODOXURE.

In color it is dun, banded lightly with a darker color. Its tongue is capable of the most alarming projection and serves all the uses of a miniature proboscis. It uses its tail as an extra paw, both in climbing and in bringing objects near to it. It is so entirely nocturnal, as to be owl-like or bat-like in the light of day. Though fierce in its natural state, it is easily domesticated and becomes a playful and affectionate pet.

The **Wah Chitwa**, or **Panda** (*Ailurus fulgens*), is arboreal and lives near rivers and streams in Nepaul. Its fur is chestnut-colored, darkening on the legs and ribs. The head is fawn-colored, except for a red spot under the eyes and the tail is ringed; its head is short and has a muzzle. It is not often met with, but when seen, rewards the lucky observer.

The **Sloth Bear, Honey Bear, Jungle Bear** (*Melursus labiatus*), belongs to India and is so generously provided with lips as to make these his most

SLOTH BEAR (*Melursus labiatus*).

striking feature. It very early loses its teeth, so that it is more dependent than most bears upon food in a liquid form, such as honey, for instance. It is black in color, except a white tipping on the feet and nose, and a yellowish white mark, resembling a capital V and which is found on the breast. Its hair is long, thick and unkempt-looking. Its queer appearance added to its grotesque movements, its teachableness and gentleness, render it a favorite with the "travelling showmen" of the country where it is found. It regards the ants as a specially delicate article of food and captures them in a very singular manner, by blowing powerfully until he scatters the particles composing the ant-hills and then capturing the ants by successive powerful inhalations. In its attacks upon bee-hives it derives protection from the long hair which falls over its forehead and

eyes. Though, as has been said, amiable and tame, it is quite ferocious in its native state, and as it is some five feet in length and quite stout, it is an adversary to be dreaded.

The **Malayan Sun Bear** (*Helarctos malayanus*) differs from the **Thibetan** species in that it substitutes for the white under jaw a white nose-muff. Its lips are unusually flexible and its tongue susceptible of almost indefinite protrusion. It has little, violet-colored eyes, is long-clawed and muscular out of proportion to its four and a half feet of length. It is readily domesticated,

BROWN BEAR.

when it makes a most amusing and agreeable pet, associating freely with any animal, and displaying no vices and no weaknesses other than a fondness for sweet wines. Its favorite position is that of seating itself upon its hind legs, and when thus made comfortable, it will roll its body about, gesticulate as absurdly as the "dumb orator," thrust out and withdraw its tongue, and altogether serve the uses of a light comedian. It is often called the *bruang*. The Thibetan species loves to bask in the hottest sunshine. It is stoutly built, has a thick neck, large ears and is very active. Its coloring is black throughout, except for a letter Y in white upon its breast, and white upon its lower jaw.

The **Sun Bear**, or **Bruang of Borneo** (*Helarctos eurysipilus*), is not unlike the Malayan species. It replaces the white breast-mark of the latter by orange.

No tree is too tall or too smooth-barked for its climbing, and it is specially fond of the tender top-shoots of the cocoanut palm, of the cocoanuts themselves and of the cocoanut milk. When domesticated, it is very fond of attention, and when it considers itself treated with insufficient consideration, it will refuse to treat further with the persons whom it regards as having hurt its feelings.

The **Spectacled Bear** (*Ursus ornatus*) is found in the Cordilleras. It is a black bear, with yellow semicircles about its eyes.

The **Brown Bear** (*Ursus ornatus*) is found in the mountainous countries of Asia and Europe. When young it wears a white bib or collar, which is dispensed with as the animal grows older. It weighs over seven hundred pounds and does not attain its growth until its twentieth year. It is naturally frugivorous, but after having once tasted blood becomes fond of it, and its union of courage, muscular strength and unappeasable appetite render it very destructive to herds and flocks, after it has once learned to attack them. It is fond of ants and of honey, and never neglects any opportunities which come in its way. The bear is specially equipped for his hibernation, which extends from November to April. In the first place, the fat previously accumulated serves for a support. In the next place, the empty stomach and intestines contract into the smallest' compass. Finally, the entrance to the stomach is blocked up by a mass of leaves and woody substances, called a "tappen," with which the instinct of the bear leads it to provide itself. As the bear re-soles its feet every winter, it has been suggested that his sucking of his paws is not merely an idle habit. The bear not only makes itself a generous, soft bed, but it carpets the floor of its den or cave, after the manner of the Elizabethan nobles. It is easily domesticated, when it becomes playful, affectionate and readily learns tricks, and yet, in its wild state, it is, when attacked, exceedingly savage.

CINNAMON BEAR (*Ursus cinnamoneus*).

A *brown bear* attended mass, although equally unexpected and unwelcome. While the devout worshippers were trying to take their thoughts from earth, a *brown bear* made his way into the sanctuary, and after taking observations retired, but soon returned with the proceeds of a successful hunt, which he proceeded to dispose of amidst surroundings which, however strange to him, seemed to give him much greater satisfaction than his presence caused the paralyzed congregation. Finally the bear withdrew, and after a becoming interval the congregation vanished. An instance is related of a *brown bear* which was used as a guard for the commissary supplies of a regiment. He proved to be entirely trustworthy and captured one or two thievish soldiers, but finally having injured a mischievous boy who was trying to pilfer sweets, he was treated like other benefactors of the human race and put to death, while having his good deeds celebrated in song and story.

A *brown bear* has been known to engage in a game of romps with children too young to feel fear. A mother living on the edge of a forest missed her two young children, and going in search of them found one of them mounted on the back of a *brown bear*, while the other was feeding and hauling around its entertaining playfellow. The bear seemed surprised at the intrusion of the children's mother, but as they left him for her, he seemed to conclude that recess was over and walked away into the depths of the forest. A *brown bear* was once mistaken by a woman for her strayed donkey. The woman approaching, as she supposed in the dark her wandering servant, began to belabor him and the bear was too much astonished to object. Presently the woman discovered her mistake and started off on the double-quick, while the bear retired in the opposite direction.

THE BEAR HUNT.

The *brown bear* not unfrequently extinguishes the camp-fires of the traveller and hunter. Trusting to the thickness of his wrappings, the bear will roll himself into the smouldering fire, scattering the brands, and having thus made clear his title, proceed to possess himself of meat not intended for his enjoyment. On one occasion the hunters, few in number, were forced to take to the trees and watch the feasting of the uninvited guest. The *brown bear* has two families a year and the two sets of cubs are born not far apart in time. The first set is upon the appearance of the second set provided with separate quarters and not permitted to enter the family domicile.

The **Indian White Bear**, the **Syrian Bear**, the **Isabella Bear**, or the **Ritck** (*Ursus isabellinus*, or *syriacus*), is generally considered as forming but one species. At first a gray-brown in color it becomes entirely white upon reaching maturity. Its outer coat consists of curled hair, and on the neck looks like a mane. It is called also the *ritck* and the *dubb*. It is notable for the gentleness of its disposition, and because it is the bear mentioned in the Bible.

The **American Black Bear** (*Ursus americanus*) has offered such inducements through its abundant fur and excellent flesh, that the hunters have greatly diminished its numbers. In the days when slick hair was an essential feature of the toilet, bear's grease was in great demand, but such requirements as still exist are easily met by improvements in the manufacture of hams and pork. Though naturally shy the *black bear* can be dangerous, so he affords ample temptation to the adventurous hunter, and not unfrequently comes off victor. It fights by striking with its front paws, but once that it has its enemy in its power, it uses its teeth. Like the stag, the bear fights for its wives, and many a fearful contest has lacked nothing but a Homer to recite its rapid changes and tragic close. If in good condition the *black bear* hibernates, otherwise it keeps up its foraging during the winter. It brings forth young but once a year when it usually has a promising family of four little gray cubs.

A BEAR WITH HIS SPOILS.

A *black bear* on one occasion had for some time been making free with a farmer's sheep. Keeping on the watch for the bear, the farmer and a visiting sportsman caught the marauder one day just as it was beginning operations and concluded to defer its death until they had taken a short lesson in natural history. The bear entered the pasture and finally succeeded in separating a lamb from the flock. It next proceeded to head it off whenever it endeavored to join its companions. Having finally satisfied the lamb that it was powerless, it proceeded quietly, but steadily, to drive it to the corner of the fence which abutted on the woods. Here a new difficulty seemed to present itself as the lamb could not pass the fence, so the bear picked it up tenderly, pitched it over the fence and then himself climbed over. It now proceeded to drive before it the submissive lamb, and after a while sat down and played with it. This performance was repeated several times until at last the lamb manifesting some weak spirit of revolt was torn to pieces by the bear, when the hunters, having seen the end of the play, now proceeded to shoot the marauder, and to reclaim their rights of property in the lamb.

POLAR BEAR.

The famous **Grizzly Bear** (*Ursus horribilis*, or *ferox*,) is always ready to act on the offensive, and as it seems to suffer no great inconvenience from

ARCTIC BEAR AND WALRUSSES.

wounds not in the head or in the heart, discretion is the better part of valor for any man who is not bent upon hunting the *grizzly*. As before the final attack, the bear sits a moment on its haunches, the hunter must speak then and there or "ever after hold his peace." It would appear that while yet young, the *grizzly bear* is a climber, but as his avoidupois increases, he finds the effort beyond his powers. The name was probably *griesly* at first, for the bear is most frequently of a dull-brown color, although some specimens of gray colored *grizzlies* have been found. The hair of the young bear is. soft, as well as long and abundant. The animal walks with the roll of a tar, and swings its head from side to side like a Chinese manikin. The fore legs are specially muscular, and the extraordinarily long feet are armed with the most cruel claws, several inches in length and working separately. Its tail is ludicrously short, and its head large in proportion. It is really a king among beasts, as

POLAR BEAR AND CUBS ON ICE FLOE.

these fear to approach it even when it is dead, and when alive it does not hesitate to attack and destroy the buffalo. It is tamable when young, very companionable and very amusing, but all this amiability disappears with increased age. Its strength is such that it can drag away the carcase of a full-grown buffalo, and as has been said, it is so much feared by other animals that not even the wolf or the coyote will dare to meddle with game which the *grizzly* may have slain and left, for the time, unguarded.

The *grizzly bear* is said to be frightened by the scent of man, and if not irritated will usually retire. A baby *grizzly* was made a pet on shipboard, and among its other pranks, it deliberately took possession of the pilot's bunk and refused to surrender it to its owner. It formed quite a friendship with a lame antelope, and on one occasion when both animals were allowed to go on shore protected it when it was attacked by a savage dog. This bear is specially subject to ophthalmia, and has been chloroformed and treated successfully by oculists. In one instance the bear was difficult to manage during the first operation, but having found that it procured real relief, it voluntarily submitted to further treatment. A hunter on our western plains shot a deer and concluded to go back to camp and return for the deer later. When he came back he found a *grizzly bear* preparing to carry off his game, and rashly concluded to dispute its possession. He succeeded in wounding and irritating the creature, and as it charged down upon him, a second shot carried away its snout. The bear nevertheless continued its attack, but after inflicting considerable injury was killed. The *grizzly bear* uses its front paws, not for crushing but for holding, and employs its hind paws for lacerating the flesh of its victim. Contrary to popular belief the *grizzly* when young, can "climb a tree," but as stoutness comes with increasing age he has to forego the pastime.

The **Polar Bear, White Bear,** or **Marine Bear** (*Ursus marinus*), has a yellowish-white soft fur, an elongated flat head, a long neck, long legs, large feet whose soles are fleeced, and whose toes are connected by membranes. It visits the land but seldom, as it feasts itself upon dead whales and live

ESQUIMAUX ATTACKED BY A POLAR BEAR.

seals. Its length is from nine to ten feet, and its height about six feet. It is sociable with its kind but ferocious towards other animals and towards human beings. It moves about in small companies and adds to the number of animals which are distinguished by great affection for their young. A very pitiful story of a *polar bear's* dying with its two cubs is as follows: As a bear with two cubs came within range, the two cubs were shot. The mother, forgetful of everything else, tried to revive them. First, she tempted each of them with food; then, with repeated groans, she tried to raise them to their feet; next, she walked away a short distance and plaintively called upon them to follow; returning, she moved them hither and thither, licked their bleeding wounds, and finally, as if realizing the wrong done her, rose erect and gave utterance to a terrible roar of agony and rage,

40

when the hunters put an end to her suffering and their own sympathetic discomfort.

The *polar bear* is active and its senses are unusually delicate. It will perceive at once the vicinity of seal or salmon, and is very successful as a fisherman. It generally captures the salmon and other fish by darting upon them, but it is more ingenious in its seal fishery. It will mark the position of the seal and then swim under water until by coming to the surface it can find itself directly beneath the unsuspecting seal. Its great endurance and tenacity of life render it no mean antagonist of man. Its claws are short, curved slightly, and very strong. The long, sinewy neck supports a small head, whose crown is distinguished from its muzzle by no frontal depression. Its blade feet are long, out of proportion, and its sole is furred with double

NORTH AMERICA OTTER (*Lutra vulgaris*).

reference to warmth and security of its hold upon the ice. The *polar bear*, or *nennook*, often becomes an unwilling and somewhat startled traveller, as the ice-floes float away south and carry him along. Deep-sea fishing is not what he set out for, and hence, after an enforced abstinence his hunger induces him, upon reaching land, to become a depredator, and speedily to lose his life at the hands of the outraged farmer. Whether or not any but the females hibernate is still unknown.

EUROPEAN OTTER (*Lutra vulgaris*).

It prepares its winter quarters on land, under some rock, where it digs a snow cave. The flesh of the *polar bear* is said by Arctic travellers and explorers to be excellent. The female with cubs is specially morose and dangerous.

The *polar bear* when retreating in a hurry will pick up her less fleet cubs and throw them before her. When food is out of reach she will throw things into the water, so that by creating an artificial current or wave, the desired object may be brought within reach.

The **Ingalubi**, or **African Bear**, is a smaller species which inhabits the plains. It does the most unexampled damage to the crops of the farmer, but is no game for the true sportsman, although it will always charge when attacked. When seeking to escape it will frequently reverse itself and back into an ant-hole. When it emerges it comes forth with a somersault, and generally strikes the exact spot where an unwary hunter would be in waiting. When speared it has frequently succeeded in turning upon the native hunter and ripping up his leg.

The **Sea Otter**, or **Kalen** (*Enhydris lutris*), is found on the sea-coasts of the Northern Pacific. It weighs about seventy-five pounds and is much larger than the fresh-water *otter*. It is a successful fisherman, passing its winters on the coast and following up the rivers as the weather grows warm and the fish ascend to spawn. Its glossy fur is generally jet-

OTTERS FISHING.

black, but at times some portion of the animal will be white. It is short-tailed and bandy-legged, but then this is "good form" among otters. Its fur is held in high esteem alike from its warmth, its beauty of coloring, and its scarcity—for the family is neither fecund nor numerous.

The **North American Otter** (*Lutra canadensis*) was formerly abundant in New England, but the species is now almost extinct. It is said by Audubon to coast in the snow or even on the ground, and to do this apparently for the same reason which induces children to slide down hill.

The **European Otter** (*Lutra vulgaris*) is not simply fond of fish, but so fastidious as to select only the best and to eat of these but the choicest portions. It is supple, has a long, broad rudder-like tail, webbed feet, an ability to turn quickly, sharp teeth, and two coats of fur, the under one short, thick, compact, and the outer one composed of long hair. Its coloring is brown with a slight admixture of gray. It prefers finding a hollow. or deserted nest to building a structure of its own. It will fight to the bitter end and is an adversary respected by dog and man. Once it fastens its teeth in anything it will hang on with the tenacity of a bull-terrier. It weighs about twenty pounds, is about three feet in length, and in eating holds its food with its fore paws. Its hind feet are like a seal's, but, though swimming with ease and rapidity, it can run swiftly. Its fur is held in special esteem. It is trained to catch fish for its master, who will at once take them away. The use of the *otter* as a fisherman is most common in China and India.

The **Cape Ratel, Honey Weasel,** or **Honey Ratel** (*Mellivora capensis, Ratelus capensis*), is South African in habitat, and goes dressed in the thickest, furriest, loosest of skin. It is a burrower and manifests quickness only when burrowing. It is a persistent honey-hunter, although entirely dependent upon nests not built in trees. Its covering is an absolute protection against the sting of the bee, as its looseness is against any rough handling on the part of other enemies. In captivity it is very fond of attracting attention, and to secure this end will convert itself into a regular mountebank. The prevailing black color of its body is relieved by gray on the upper part of the head, neck, back and tail, and a stripe on each of its sides.

EUROPEAN BADGER (*Meles taxus*).

The **Indian Ratel** (*Mellivora indicus, Ratelus indicus,*) is regarded by some naturalists as a distinct species, and by others as substantially identical with the **Cape Ratel.** But its habits are different, as it is so ghoulish in its tastes that cemeteries must be protected lest it burrow and disinter the corpses.

The **Badger** (*Meles vulgaris*), though naturally inoffensive, is so worthy an antagonist when forced to close quarters, that " badger baiting " formed a favorite sport during the times when cruelty seemed to be an essential element of enjoyment. It is slow and clumsy in walking, but can out-dig any sexton. In burrowing it uses its nose and hind legs for pushing the earth away. Its home comprises at least three rooms; a living-room, a larder, and a room for refuse. It has a white head with black frame, a grayish-red body, and brown chest, belly, legs and feet. It can be domesticated, and is lest stupid than is generally supposed.

The **American Badger** (*Taxidea americana*) is quite abundant, and while regarded as a separate species, its characteristics have not as yet been fully studied.

. The **Teledu, Skunk of Java, Stinking Badger,** or **Stinkard** (*Mydaus meliceps*), roots in the earth on the elevated table-lands. It digs a circular earth-cave at the foot of some tree and is very scrupulous about its regularity and finish: It lives in pairs, is nocturnal in its habits, and supports life upon a mixed diet of seed, tender roots and insects. If captured before it can use its artillery, it is temporarily inoffensive, and as the flesh is not impregnated it is used for food. When tamed it seems to lose its natural disposition to render itself offensive, so that it is probably only when irritated and full of the idea of self-defence, that it fills the air with an odor whose strength and persistence nothing (human at least) can withstand.

The **Polecat** (*Putorius fœtidus*) is more destructive than a weasel, for not content with making a repast it will destroy every small animal about it. Its habitat is northern Europe.

American Mink, Musk Otter, Water-polecat, or **Vison** (*Lutreola vison, Putorius vison,* or *Vison lutreola*), is found in Europe as well as in North America. Its fur is fine and handsome, but has suffered somewhat in esteem from its frequent use as a counterfeit sable. It is dark-brown as a rule, but frequently is black on the head and white on the jaws. It is a good swimmer and frequents the banks of ponds and streams where it can find aquatic animals which may serve as its food.

AMERICAN BADGERS.

The **Glutton,** or **Wolverine** (*Gulo luscus*), has long been distinguished for its capacity as an eater, and for its great ferocity. Many are the stories which

AMERICAN WOLVERINE (*Gulo borealis*).

grew from the fertile imaginations of the older naturalists, but most of them have been unable to vindicate their claim to consideration. Its habitat is North America, Siberia and northern Europe. In appearance it is not unlike a young bear, brown in color, with its muzzle black, and spotted white on the lower jaw. Where the marten is hunted it robs the traps, and where provisions are stored it robs the owner.

The **Masked Glutton** (*Paguma larvata*) is Chinese in its habitat, is olive in color and takes its name from white face markings, which resemble a mask.

The **Wolverine,** or **Great Weasel** (*Carcajou*), will not touch poisoned food, fall into a trap or get shot by a spring-gun. It will cut the string of the gun and then take the bait. It will steal and hoard up the most worthless articles, even sticks and stones. A hunter once upon returning to his hut found that it had been entirely gutted by a *wolverine*. When surprised it will sit up on its haunches and shade its eyes with one hand while looking directly at the intruder.

The **Tayra** (*Galictis barbara*) has its habitat in tropical America, and is black, with white on the throat and upper chest. Its little brown eyes are very bright and promise the intelligence and alertness which the animal displays. It is a burrower, about the size of a cat, and when domesticated, makes a very amusing pet. It goes about clucking like a hen with a brood of chickens.

The **Grison**, or **South American Glutton** (*Galictis vittata*), is weasel-like and resembles the tayra.

The **Ferret** (*Putorius furo*, or *Mustela furo*,) is well known as a successful rat-catcher. Tame *ferrets* will often escape, but with the approach of cold weather they will return to a captivity which insures them a sufficiency of food and comfortable quarters. The *ferret* though readily domesticated and very useful as the servant of man, has a very ugly and uncertain temper, so that its owner must always be on guard against sudden ebullitions of passion and of savage instincts. The *ferret* when used for hunting rabbits is

GLUTTON, OR WOLVERINE.

muzzled, brought into the proper vicinity, released and again muzzled as soon as it has struck its quarry. It shows the keenness of animal instinct, and a *ferret* which has been muzzled will never afterwards engage in rat-catching.

The **Weasel** (*Putorius vulgaris*, or *Mustela vulgaris*) though but small and slight, is amply able to take care of itself and to worry its enemies. Its little body of about eight inches is covered with bright, red fur which becomes white on the belly. It seems to be absolutely fearless, and to welcome an enemy with the utmost disregard of relative size. A story is told of one which was met by a hunter and shot at, whereupon it coolly sat down on its haunches and observed him, while apparently deliberating whether or not to make an attack. Its persistence in its undertakings, and its ability to squeeze through the smallest crevices, render it so successful a hunter of rats and mice, that farmers have learned that the occasional appropriation of a chicken is none too large a return for its services. It is somewhat fond of birds, and will excel the most active boy in robbing birds' nests. If it once fastens its teeth in an animal it will never relax its grasp until one of the two is dead. A *weasel* and an owl contended for the contents of a hen's nest. At first the owl was the quicker in reaching the spot, but at the second attempt was attacked by the *weasel* which, though carried into mid-air, held on to the owl till it was killed and dropped to the ground. Although the hare is so much fleeter than the *weasel*, it seems to be paralyzed by the mere sight of its "fell foe," and hence falls an easy prey.

A *weasel* has been known to make an unprovoked attack upon the fetlock of a horse, and to hold on until it had been killed. On one occasion at least *weasels* refused to let a boy pass, for gathering together they chased him back the way he came. A hunter falling asleep under a tree was attacked by a band of *weasels*, and escaped simply because his thick clothing protected him while he ran. A kite swooped down upon a *weasel* and bore it aloft, but it soon discovered its mistake, for though it had come to prey it remained to be preyed upon. A *weasel* brought a frog to church, and disregarding the reproving looks of the congregation, turned one of the aisles into a refectory. The *weasel*,

TIBETH AND TAYRA.

when domesticated, is full of the most intense and futile curiosity, and takes a lively interest in every object, action and proceeding. It is said, although not authoritatively, that the *weasel* seeks quarrels with snakes, and manages first to kill and then to eat them.

The **Ermine,** or Stoat (*Putorius erminea*), has passed into song and story as well as into commerce, because of its furry coat, which, though red in summer, becomes white in winter. It is found on both continents, is about three-quarters of a foot in length, and wears a bushy, black-tipped tail. Like the skunk, the *ermine* can teach one to "learn too late that" *stoats* "deceive." The *stoat* delights in carnage, and is a great rat-killer.

The **Bridled Weasel** (*Putorius frœnatus*) belongs to the Pacific slope.

The **Skunk** (*Mephitis americana*), also called *pole cat*, is a common creature throughout America, and is one of the most destructive thieves with which farmers have to contend, stealing, as it does, both eggs and chickens. It is rather prettily marked with white and black stripes, has a bushy tail, and though small in size it has the power of emitting an odor so penetrating that no amount of washing or disinfecting can subdue it.

A very strange but well-authenticated story is told about a *skunk's* holding up a stage full of passengers. As the stage approached it deliberately opened its batteries, and so astonished the travellers that in their consternation they awaited the punishment which his impish majesty inflicted upon them. Even so distinguished a naturalist as Audubon made his acquaintance with the *skunk* by personal experiment.

The **Martens** (*Mustelæ*) are abundant in the more northern portions of both hemispheres. Their fur robes consist of long, soft hair, whose color is uniform, and when brought into the marts of mankind are called sables—the fur which has become so famous through the pens of clever writers.

The **Pennant Marten, Black Fox**, or **Black Cat** (*Mustela pennanti*), is about four feet in length, of which more than a quarter is claimed by the tail. In spite of its stoutness it is an agile climber, and nests high up in some tree, where it can find a hole in which to house

ERMINE AND SABLE MARTEN.

its young—generally twins, but sometimes four in number. Mice and frogs are its chief articles of diet, but it will on occasion feed upon other species of its own family.

The **American Sable**, or **Pine Marten** (*Mustela americana*), is about as large as a cat, but not so tall, as it indulges in great economy in the matter of legs. Like the gallants who wore vest over vest, or the circus-rider who divests himself of garment after garment, the *American sable* wears a triple garment, of which the outermost covering consists of long, glossy hair; the innermost, of short, soft wool; the middle one, of hair, short and kinky, but of greater length than that of the covering below, and of shorter than that of the covering above. Like the cuckoo, it quite

PINE MARTEN.

frequently takes possession of nests belonging to other animals, such as the squirrel, when to prevent any further question of rightful ownership it converts the squirrel into a repast. It will, at times, build its nest underground or use the natural little caves made by the rocks. It is not confined to pine forests, nor does it show any preference for these, so that its name probably arose from the accident of its having first been found there; it seems to require only the neighborhood of water. It does no damage to the farmer, though where meat is stored in pits it is likely to prove a trespasser.

The **Sable of Asia** (*Mustela zibellina*) furnishes the most valuable sable— that fur among furs. The principal hunting ground is Siberia, and involves great hardship, though this counts for nothing where human life is so cheap, and the caprices of the wealthy and powerful are so absolute. The animals are caught by trapping, and as it requires more than one skin to make a single muff, the costliness and regal magnificence of

SABLE AND MARTEN IN THEIR WINTER ROBES.

cloaks made of this material may well be conceived.

The **European Pine Marten**, or **Sweet Marten** (*Mustela martes*), loves to live in the pine forests. Its body is about a foot and a half in length, and is set off by a long, bushy tail. Its brown coat is quite handsome, but the animal is not numerous enough to tempt the cupidity of man. Its habitat is in the northern portion of each continent.

The **Beech Marten, White-throated Marten,** or **Stone Marten** (*Mustela foina*), is probably a distinct species, although regarded by some as merely a variety of the pine marten. It is white-throated and white-breasted, and is given to hanging about farm-yards and dwellings as a simple means of gratifying its appetite for birds, eggs and poultry.

CARNIVORA.—DOGS, FOXES.

The **Otocyon**, or **Big-eared Fox** (*Otocyon lalandii*), is gray, with black for the color of its legs and its tail. Its habitat is South Africa, where its erect furred ears, so large as to exceed its head in size, make it a noticeable creature.

The **Coast Fox**, or **Short-tailed Fox** (*Vulpes littoralis*), is

PENNANT MARTEN.

small in size, and colored black, changing to cinnamon-brown on the fore-legs and sides of the neck.

The **Gray Fox** (*Vulpes*, or *urocyon cinereo-argentatus*,) is quite common in the United States.

The **Caama**, or Asse (*Vulpes caama* or *zaarensis*), is hunted in South Africa for its skin. It, like the fennec, is extravagantly fond of eggs, which it opens by pushing them before it until they strike against a stone or some other hard substance. It has been driven northward by persistent hunting,

SILVER FOX.

FENNECS (*Vulpes zerda*).

and while very scarce around the Cape, is in no immediate danger of becoming extinct.

The **Fennec** (*Vulpes zerda*, or *Megatolis fennec*) is a graceful little African creature, whose cream-colored or fawn-colored body is set off by a bushy tail, whose baseand tip are black. Its slender body, sharp nose and large, pointed ears give it a resemblance to the fox, while its blue eyes peeping forth from its long full-bearded face, result in a very cunning appearance. It is valued as a pet because of its grace, its timidity, and its mere semblance of a bark; but it is held in no less esteem since it furnishes one of the most desirable of furs. It is found from the Cape of Good Hope as far north as Tunis. It is specially fond of ostrich eggs, whose shells it breaks by rolling them against the rocks. It can climb trees, but lives in shallow burrows, which it makes in the sand. The excellence of its fur, and the small size of its robe render a garment of *fennec* skin rare, costly and desirable.

OTOCYON.

The **Corsac** (*Vulpes corsac*) is celebrated as having been the favorite pet of the courtiers of Charles IX. of France. It belongs to Central Asia, and is no larger than a house-cat. The **Bengal Dog** (*Vulpes bengalensis*) is a nearly related species, as is also the American **Swift Fox** (*Vulpes velox*). The African Fox (*Vulpes niloticus*), in the north, and (*Vulpes adusta*) in the south, will be found mentioned in books of travel. The **Cross Fox** (*Vulpes fulvus* or *decussatus*) is so-called because of a dark stripe on its shoulders. It is common in so well-settled a State as New York.

The Silver Fox, or Black Fox (*Vulpes argentatus*), is another variety of the red fox (*Vulpes fulvus*). The Silver Fox, of the Western States, is the American Swift Fox. The Long-tailed Fox, Large Red Fox, or Prairie Fox, (*Vulpes macrourus*, or *utah*), is the largest of the fox family, being nearly three feet in length. It has a sharp muzzle, long-pointed ears, and its haunts are on the western plains of the United States.

The Blue Arctic Fox, Terrienniak, or Perzi (*Vulpes lagopus*), becomes entirely white in the winter season, when its silken fur is an object of interest to man. Its habitat is the North Polar region, and as in passing from its summer color of blue to its winter mantle of white, it furnishes various other shades; it is called also the *stone fox*, the *pied fox*, the *white fox*, and the *sooty fox*. It is stated that it has the power of mimicry, and that it uses it in securing birds for food. It lives in small towns of twenty or thirty inhabitants, and in spite of its appearance of alertness and intelligence is very easily trapped or shot.

BLUE ARCTIC FOX.

The European Fox (*Vulpes vulgaris*) is the hero of fox-hunting stories, and from his cunning and endurance he is well calculated to sustain the role. This is the fox of the Romans, the celebrated Reynard of fable and story, and the one whose brush has so long been the coveted prize of the British hunter. Its range is the whole of Europe, Asia Minor and eastward to Thibet. An English fox-hunt is a spirited affair, and throws into excitement the whole community. The curvetting of the horses, the struggling of the hounds in their leashes, the broken country over which the hunters must ride, the cunning of the fox itself, and the glory of triumphing over one's fellows, all unite to fan the excitement into a blaze, and to cause the successful hunter to believe that he is as great a conqueror as the celebrated military men of yore, but since fox-hunting has been introduced at the East, and because fox-hunting has always been a well-known sport in the South, we refer our readers to the sporting papers.

The *fox* fully deserves its reputation for cunning. Knowing that it is endangered by the strong scent which is one of the crosses which it has to bear, it makes every effort to use "art as nature to advantage dressed." It will eagerly avail itself of the presence of any perfumed shrub; it will often, when pursued, usurp the burrows of another *fox* so that by temporary occupancy it may divert the pursuit and share its undesired adventures with another of its

AMERICAN FOX.

kind; it will take to the water to destroy its trail; it will leap high in air and over wide distances that it may effect the same end. So, too, when hunting hares and rabbits, whose fleetness exceeds its own, it will display a cleverness which lends reasonableness to the various fables in which the *fox* has played the hero. On one occasion an observing hunter gave himself up to watching the proceedings of a *fox*. Reynard first came down and reconnoitered a field in which a great many hares were feeding. He then carefully inspected the possible means of exit, and finally selected one hole under the fence as a good location for a hunter. He next burrowed to a slight depth, and patiently anticipated the time when the unsuspicious hares having feasted, should return to their warren for rest. After about an hour the hares began to straggle through the other openings, but without any effect upon Reynard except that of making his eyes sparkle and his tail quiver. Finally two hares came through the opening which Reynard was guarding, and allowing the first one to pass, he sprang upon the second and was bearing it away in triumph when the unseen hunter brought its life to a close, and added to his increased knowledge of natural history, a *fox* and hare both in prime condition. A tame *fox* found a curious way of turning its curse of muskiness into a blessing in disguise. Finding by experiment that its presence was especially offensive to the cats, it would await the time when they received their supplies of milk, and then intrude upon their company until, when they withdrew in disgust, it would devote itself to enjoying the repast provided for his exclusive neighbors. Learning wisdom from experience, it began to brush against the milkmaid as she returned from the milking, and having thoroughly impregnated the milk, would succeed to its

FOX CARRYING HIS PREY.

possession. On more than one occasion, when closely pursued, a *fox* has been known to vanish as if into thin air. One such clever strategist used always to disappear over a most precipitous cliff, and it was not until after many a fruitless pursuit that a hunter-spy discovered that the *fox* had found a shelving rock just below the top of the cliff, and a passage leading to the top of the ground which

enabled him to circumvent hound and hunter, while these were gazing stupidly
into the darkness of the chasm. Another astute *fox* used to be hunted frequently and always vanished at one corner of the field. His success became a
by-word and, a reproach to every fox-hunter in the district, but no effort of
theirs seemed able to discover the means by which the *fox* "dematerialized"
itself. In the end one of the hunters who had sacrificed his sport to play the
spy, was after innumerable failures fully rewarded upon reaching the corner of
the field. Reynard did not, as has been supposed, run along the topmost rail of
the fence, but clambering up on it leaped many feet into the hollow of a tree,

A FOX FAMILY.

so distant that any suggestion of its affording a means of escape had been finally
dismissed. Unfortunately what is sport for the hunter is death for the *fox*, so
that the much less astute men proceeded to kill Reynard despite his cleverness.
The *fox* displays the greatest cunning in outwitting those who would trap
him, burrowing below the trap and thus springing it before he trifles with the
bait. Unfortunately for the *fox*, the hunters have learned to set their. traps
upside down, so that as a dead *fox* never returns it will be some time before the
foxes learn of the scheme arranged for their capture. In dealing with spring-
guns the *fox* has exhibited equal fertility of resources, for it will either burrow
at right angles to the bait, or gnaw the cord near the trigger, and then safely
feast upon the bait.

The *fox*, like some other animals, has been known to fish successfully, using his tail as bait and line for the greedy, obstinate and tenacious crabs. *Foxes* generally go about in pairs and avail themselves of the benefit to be derived from co-operative effort. Such was the case when two gullies down the side of a precipitous and inaccessible hill were used by the rabbits and hares as a pathway to the valley below. Two *foxes* stationed themselves one at the end of one of the gullies and the other at some distance above. The former chased the rabbits up the gully and the latter sprang from its concealment and captured

JACKAL.

one of them. On another occasion the hunting *fox* allowed its victim to escape, whereupon the other *fox* fell upon its awkward assistant and punished it soundly. The *foxes* are very considerate of their young, but the latter are very ungrateful and, like spoiled children, forget past benefits in present desires. On one occasion a naturalist undertook to satisfy himself in regard to stories about the family discords of the *fox*. Concealing himself near its retreat, he was finally rewarded by seeing the *fox* appear with a goose which it had stolen, and lay it down on the ground while it went to call its children to dinner. The man managed to secure the goose, and when the happy family appeared, the mother hunted high and low for its marketing, and failing to find it, was set upon by her cubs and submitted to being torn to pieces; a good story, but altogether improbable.

The Indian Jackal (*Canis aureus*) is an animal whose nightly howling is as troublesome as his voracious appetite and depredations upon

WOLF AND YOUNG.

the flocks and poultry. It quite frequently dances attendance upon the lion and possesses itself of the remains of the lion's feasts. Still its patient cunning quite as often induces the *jackal* to hunt on its own account, in the vicinity of farms, and it most commonly gets its share of the farm produce. It also hunts after the tiger, relying upon getting possession of all that the tiger cannot eat. It has been known, when chased by hunting-dogs, to summon its fellows to its aid and to disable or kill the dogs. An example of the *jackal's* cunning is furnished by the account of one which, when hotly pursued, selected as its place of concealment, not the jungle, but the most thickly settled portion of a village.

COYOTE.

The **Black-backed Jackal, Duwa,** or **Cape Jackal** (*Vulpes mesomelas*), is mottled in white and black, and renders night hideous by its ear-splitting cries. It shares the taste of the natives for the fat-tailed sheep, but instead of killing the sheep, it bites off its tail. It is very cunning, and its devices might serve as a study for those who have frequent occasion to extricate themselves from difficulties.

The *jackal* is harmless, and indeed it is often serviceable to the hunter, as a trustworthy indication of the vicinity of the lion.

Jackals have been known to surround a hare concealed in the crevices of a rock, and then to depute some of their number to drive it out into the ambuscade. After having killed it and hidden it in the bushes, the *jackals* come forth to see whether any more powerful animal is likely to despoil them, and during this reconnoitering they swagger around with an assumed air of indifference. If an enemy is in sight, they have been known to pick up cocoanuts or any other convenient object and to pretend to carry them away as if they were the special object of their quest. If, however, the coast is clear, the *jackals* return to the bush and enjoy their

GRAY WOLF.

feast. Packs of *jackals* have in India been known to station themselves along the edge of a jungle skirting a pond, and when a deer had rendered itself heavy by large draughts of water, to head it off at each new attempt to enter the jungle, and to keep it moving about the pond until it dropped through sheer exhaustion.

CARNIVORA.—WOLVES.

The **Maikong** (*Canis cancrivorus*) is a crab-eating fox-wolf, found in Guiana, where it runs in small packs.

The **Guara** (*Canis jubatus*), or **Hyena Fox-wolf**, is found in South America. It is a yellowish-red, is solitary, about five feet long, wears a black and red mane, a white spot under the head, and is very fierce. It is called the *aguara culpea.*

The **Wolf** (*Canis lupus*) is known in all countries and has earned a reputation which would be better if one stopped to reflect that he lives the life for which he was created.

The **Indian Wolf** (*Canis pallipes*) resembles the jackal. The **Wolf** of the **Pyrenees** is black. The **French Wolf** is brown. The **Russian Wolf** wears whiskers on throat and cheek, and the **Italian Wolf** is red.

The **Tanate** (*Canis procyonoides*) belongs to China and Japan, and is often called the *raccoon dog.*

The **Gray American Wolf** (*Canis occidentalis*) resembles the European wolf, although its size is less. It varies in coloring and form in different parts of the country, and is valuable as illustrating this doctrine of the creation of varieties from species (not classes from orders).

WOLVES ON THE HUNT.

The **Coyote**, or **Prairie Wolf**, or **Cajote** (*Canis latrans*), is a familiar object on the plains of western North America, and his howl, sufficiently blood-curdling in itself, is always taken up without break by at least two or three others in succession. Like the rest of the wolf family he finds his hunger unappeasable by food, and hence he is always in quest of more, like Oliver Twist at Do-the-Boys Hall. His sneaking persistency and his guardian care over the hunter and his supplies, render him an object of detestation so intense as to prevent the recognition of his excellences, when viewed as abstract qualities.

The *wolf* displays the greatest suspicion of traps and spring-guns, and is very clever in dealing with such interruptions to his enjoyments. He will approach the gun back of the barrel, and, having gnawed the cord enjoy the bait which he has converted into a harmless provision for a meal. So, too, he will seize the set lines of the fishermen, drag them away until the fish has been landed, and then, returning to the other end of the line, enjoy at his leisure the returns for his cleverness.

Two *wolves* when hunting once displayed very great cleverness. One of them concealed itself in a ditch while the other made a wide circuit, and getting behind a herd of antelopes drove them towards the ditch where one of them fell a victim to its unsuspected foe. Darwin authenticates the statement that *wolves*, like men, will station a guard to the leeward of a herd of deer, while the rest of the pack goes to the windward and drives the herd into the jaws of death.

HYENA HOUNDS PURSUING A BEISA.

AFRICAN WILD DOG.

The buffalo is at times pursued by a pack of *wolves*, which will finally manage to hamstring him. On the western plains a hunter found a tired buffalo watched by two *wolves*, who, attacking him only when he lay down, or else merely feigning attack, were about to see him fall a victim. It is said that *wolves* when found in a trap will unresistingly submit to any treatment, and on one occasion when a woman fell into a pit, she found that a similar fate had happened to a *wolf*, and that it was as harmless as a fawning cur. That the *wolf* should always be hungry is not a matter of his own choice, and his tireless persisting is an admirable quality, although to us it may seem like persistence in ill doing. The thrilling adventures with *wolves* have been many, but none, perhaps, has surpassed in interest that of the woman who escaped by sacrificing her children, one at a time, and who, upon arrival at the village, was punished for murder, though it is difficult to understand why, since she could not save them.

41

The **Wild Dog, Hyena Dog, Painted Hyena,** or **Wilde Hund** (*Lycaon venaticus*, or *pictus*), is very abundant in southern Africa, and is generally regarded as the bridge between the hyenas and the dogs. It is not inclined to attack man but prefers to hunt its own game or to devour carcases of animals killed by others. It is smaller in stature than the hyena, is thin, has bristling hair of red or brown (changing in spots to white or black, and to gray upon the bushy tail), has large, straight, white-tufted ears, broad, short muzzle and fang-like teeth. It is nocturnal, hunts in large packs, which give no sound while pursuing their prey, and is altogether a gruesome creature. It has been domesticated, when it exhibited the most utter antipathy to dogs born in captivity. It sometimes happens that a pack of *wild dogs* and one of hyenas will meet and quarrel over the possession of a carcase.

JACKALS ROBBING A GRAVE.

The *wild dog* sometimes finds the wolves enjoying a feast upon the prey they have run down, or it may be upon a carcase which the wolves have discovered. Then there will ensue the most bitter quarrelling and the most animated fighting, which will terminate only when the prey has been reduced to fleshless bones and the combatants have exhausted their strength.

The *wild dog* runs like hounds in a pack, but unlike the hound gives no audible expression to its emotions. It has a great variety of names, such as the *warabo*, the *durwa*, the *wilde hund*, the *painted dog*, the *painted hyena*, and the *hunting-dog hyena*.

A hunter had a somewhat startling experience from suddenly stumbling into the midst of a sleeping pack, which rose about him like spectres or hobgobblins. In order to induce them to withdraw, and thus render escape possible for himself, he shot

HYENA HOUND (*Lycaon pictus*).

two of the most seemingly eminent, and upon this hint the pack sought other lodgings.

The skin of the *wild dog* is found to be superior to all other material for gun-covers, so that since the invasion of Africa by Caucasian hunters, with their armament of small artillery, the *wild dog* has a new danger to fear.

The Warragal, or Dingo (*Canis dingo*), is Australian and very harmful to those engaged in sheepraising. Forming themselves into packs they seem to district

AUSTRALIAN DINGO (*Canis dingo*).

the country, and each pack to strictly confine itself to its own allotment. It is brownish-red or reddish-brown, sharp-muzzled, has a bushy tail, small, crafty-looking eyes and short pointed ears rising straight from its head. It is tenacious of life and a perfect Spartan in the endurance of pain. It relies upon cunning rather than upon strength, but if forced to fight it becomes a very ugly adversary. When domesticated it retains its unpleasant savage habits, and is liable at any moment to attack its owner.

The Anaponda Greyhound is colored black, tawny or white; sometimes it resembles the deer hound and at other times the greyhound. Its ears are erect, its coat short and smooth, its speed unusually great, and it runs entirely by sight. The

DINGO.

wild dogs are sometimes trained as a pack of hunting-hounds. When thus domesticated the wild dog exhibits the most unconquerable aversion to dogs born in captivity. Their method of running down their prey may be illustrated by the example of wild dogs hunting the buffalo. The dogs pursued the buffalo as a pack, and whenever he charged discreetly withdrew. In order to save their own strength, while exhausting that of the buffalo, the dogs would keep up their long, wolf-like run, and from time to time depute two of their number to run ahead of the pack and increase the speed of the buffalo, and when these forerunners became exhausted they would drop back while others took their place. Finally the buffalo sank from exhaustion and the pack of dogs fell upon him and tore him to pieces. The wild dog of Australia, or dingo, frequently hunts the kangaroo in a similar manner.

HYENA HOUNDS BAYING A LEOPARD.

The **Indian Dhole**, or **Kohlsun** (*Cyon dukhuensis*), is confined to a limited district of British India. It lives in the jungle, is very shy, and but rarely seen. It hunts successfully every animal but the elephant, the rhinoceros and the leopard — the last being secure because the *dhole* can not climb. It is of a dark-mahogany color, about the size of a greyhound, gregarious, and harmless to man unless attacked.

WILD DOG (*Icticyon venaticus*).

The **Buansuah**, or **Wild Dog of Nepaul** (*Cyon primigenius*, or *primævus*), resembles the Indian dhole, except for the fact that it bays as it chases its prey. It is often domesticated and is found specially useful in hunting the wild boar. .

The **Cape Hyena**, or **Brown Hyena** (*Hyæna brunni*, or *brunnea*), is named from its habitat or from the color of its hair, which is long, running backward from the head. Around the cheeks and chin the hair is white or gray, producing the effect of very long side-whiskers.

The **Spotted Hyena**, or **Tiger-wolf** (*Crocuta maculata*), is larger and more dangerous. It utters a cry which has given it the name of the *laughing hyena*, and when engaged in uttering this crazy sound it dances about and contorts itself. It has been trapped so often that a piece of rope causes paroxysms of fear. Some of these creatures have learned how to steal small children, which they prefer to other animal food. Its habitat is South Africa. Its taste for human flesh is supposed to have been culti-vated by the heathen custom of

STRIPED HYENA (*Proteles lalandii*).

exposing to beasts of prey the dead bodies of relatives. The hyena adds to its other disabilities as a pet a most nauseating odor. It is stated upon reason-able authority that the lion will chastise the hyena by biting off one or more of its legs.

SPOTTED HYENA (*H. crocuta*).

The **Striped Hyena** (*Hyæna stri-ata*) is a striped creature whose size and ferocity make him a dreaded neighbor by the inhabitants of the Cape of Good Hope. The hyena family is not in any wise attractive, and possibly we shall look for no greater cause for our repul-sion than to the picture of *striped hyenas* snarling and quarrelling over a deserted camel. In color it is a grayish-brown striped with black. Like the rest of its family it is very cowardly, but this does not prevent its paying nightly visits to the village cemeteries and unearthing the bodies for a ghoulish feast.

The spotted hyena tears off the ud-ders of the cow which it attacks, and, secure of having inflicted a mortal wound, postpones till the close of the day's hunting any feasting upon the rest of the animal. Architecture in southern Africa is yet in a somewhat primitive condition, and no one knows it better than the spotted hyena. Hence it not unfrequently occurs that he comes like a perverted Santa Claus through the roof, not to fill the children's stockings, but to empty them of their possible owners. The carrying off of young children is so frequent as to excite no more surprise than a bar-room fight on the frontiers, and when the hyena does not succeed in carrying off the child it will frequently take such liberal bites of its flesh that it might as well have had the remainder. On one occasion when a spotted hyena had secured an infant and escaped from the village, it for some reason dropped it unharmed before retreating in the presence of some unexplained terror. It is a vampire, which, though not sucking the blood, tears to pieces and mangles the faces of any native who can be approached by stealth. The hyena never attacks unless its prey is running away—except, of course, in such raids as have been described—and when intending to prey upon the cattle of the ranchmen or traveller, seeks first to induce them to run away. Livingstone, in his "Zambezi," tells of how he managed to induce a hyena to commit suicide. He hung a piece of meat just high enough up on a tree to make the hyena jump from the ground in order to reach the bait. He then planted a short spear in the ground, and the scheme working successfully the hyena impaled himself.

The **Asiatic Civet**, or Zibeth (*Viverra zibetha*), is whiter than the African civet and its tail is dark-ringed. It is very gentle when domesticated and quietly takes a place in the household.

The **Tangalung** (*Viverra tangalunga*) has blacker markings, and along the back is distinctly black. It has a short muzzle and finds its habitat in Sumatra.

The **Malayan Weasel**,

STRIPED HYENAS OF SAHARA.

or **Rasse** (*Viverra malacensis*), is grayish-brown, with eight parallel lines of dark spots. It furnishes the favorite and most common perfume of Java.

The **Linsang, Sawtooth** or **Delundung** (*Prinodon gracilis*), is found in Malabar and Java, is beautiful in coloring and so shapely as to have been named *the graceful*. It secretes no civet and has no pouches. Its ground-color is gray, but it carries four saddles of brown. From the flanks to the cheeks extend two dark bands.

The **African Civet** (*Viverra civetta*) is an object of desire on account of the civet contained in its pouch. It is Abyssinian in habitat, weasel-like in nature and black and white in coloring. Its lips and eyes are fringed with white and the ears are tipped with the same color. Its mane extends from the head to the end of its tail and can be erected at pleasure. Its greatest value is as a producer of civet which at certain seasons is scraped from its double, abdominal pouch.

The **Common,** or **Blotched Genet** (*Genetta vulgaris*), is found in the greatest abundance in South Africa, although not unknown in central Europe. Dark patches are scattered over a yellowish-gray coat and its tail is banded in black and white. It is a climber, and its claws are retractile.

The **Senegal Genet,** or **Pale Genet** (*Genetta senegalensis*), is whiter and is striped broadly. It is domesticated and used like the common cat.

The **Banded Mungous** (*Mungos fasciatus*) belongs in Java. It is of the size of a large rat and the front part of its body is somewhat rat-like. It is very bad-tempered even with its own kind,

AFRICAN CIVET (*Viverra civetta*).

but being an active climber is frequently caged as a pet. It is blackish , in color, with lines across the back. It keeps up a constant chattering, somewhat like a crow.

The **Mungous of Java,** or **Javanese Ichneumon** (*Herpestes javanicus*), is sometimes called the *garangan.* The chestnut, which is its predominant color, passes into fawn on the head and the under parts of the body. It can inflate its body by inhalation, and is said to do this so as to induce snakes to coil about its body, whereupon it collapses, escapes and kills the snake.

BLOTCHED GENET (*Viverra genetta*).

The **Suricate** or **Zenick** (*Suricata tetradactyla,* or *Rhyzæna tetradactyla*), is frugivorous as well as carnivorous. It is about a foot long, gray brown shot with yellow, and barred above with darker brown. It is a burrower, for which its long, strong claws fit it admirably. It is readily domesticated, when it proves itself to be alike intelligent and useful.

The **Mampalon** (*Cynogale bennettii*) has its habitat in Borneo. It has a long nose or snout, which is plentifully furnished with long mustaches. It lives in the vicinity of streams, and in habits does not differ from other ichneumons.

The **Nandine,** or **Double-spotted Ichneumon,** is dark-brown, and its tail is ringed with black.

The **Binturong** (*Arctitis binturong*) has long, dull, coarse hair, except upon the head and around the ears, where it changes to gray. It carries a long, thick, bushy tail, and altogether is one of the largest species of the ichneumon. Its tail is prehensile, and is used in climbing. It is about two feet and a half long, and is irritable, dull, and rather sluggish.

The **Crab-eating Ichneumon, or Urva** (*Herpestes cancrivorus*), has narrow bands of white from its mouth along its cheeks to the base of the neck.

LUWACK OR MALABAR CIVET-CAT.

The **Common Ichneumon, or Pharaoh's Rat** (*Herpestes ichneumon*), is very useful in the land of Isis and Osiris, where rats and mice, serpents and crocodiles disport themselves; large numbers of which it destroys, including the most poisonous species of snakes. It was domesticated among the Egyptians, and frequently appears in their symbolic art. It is a foot and three-quarters in length, with a tail a foot and a half long. Like the civet it has a pouch, in which a liquid is secreted, but the use of this, either to the animal or to mankind, is yet to be ascertained. It is a grizzled-brown in color and in appearance suggests the ferret.

The **Indian Moongus, or Indian Ichneumon** (*Herpestes griseus*), has hair of a mixed gray and black, is smaller than the common ichneumon, and is frequently enrolled as a member of the household.

The **Nyula** is marked according to a basket pattern, which is never lost, however delicate the markings become. It frequents the densest thickets in the African fever districts, and is very wary, graceful and handsome.

The **Ruddy Ichneumon, Pencilled Ichneumon, or Meerkat** (*Cenictis penicillata*, or *levaillantii*), is tawny, with brown paws. Its habitat is South Africa.

The **Kusimanse, or Mangue** is a Western African plantigrade. It is chocolate-colored, but if the fur is disturbed it shows as a whitish-yellow. Its nose resembles a short proboscis.

FOUSSA.

The **Malabar Civet-cat, Coffee Rat, or Luwack** (*Parodoxurus typus*, or *hemaphroditus*), is yellowish-black, but from the nature of its hair is somewhat chameleon-hued. Its sides have rows of spots, and its shoulders bear spots irregularly. It is plantigrade, and has a tail which, though not used as prehensile, is frequently curled up into a coil. In captivity, at least, its habits are slothful, although it is qualified for the most rapid and daring feats of climbing.

The **Java Linsang**, or **Musang** (*Paradoxurus fasciatus*), though useful in destroying vermin, is unfortunately fond of the coffee plant, and has been known to descend to the robbing of poultry yards.

The **Hemigale**, or **Half-weasel** (*Hemigalea hardwickii*), is grayish-brown, relieved by six or seven saddles of dark-brown. Its nose is black, and a black line extends from it, on each side of the face to the ear, and also encircles the eyes.

The **Foussa** (*Cryptoprocta ferox*) has Madagascar for its habitat. It is light-brown, inclining to red in its color, has large, round ears and stout claws. It looks gentle and inoffensive, but in this case at least appearances are deceitful, for it is alike ferocious and dangerous. Its whole life is but a thirst for blood.

CARNIVORA.—LIONS, TIGERS, ETC.

The **Lion** (*Felis leo*) has suffered alike from undue exaltation as the king of beasts, and from the ebb that always follows the tidal wave. It would seem to depend very much upon the temperament and the mood of the individual *lion* whether he is to be regarded as worthy of sovereignty, or whether, as frequently happens in monarchies, he is simply the lineal successor of more worthy ancestors. It will be seen from the illustrative anecdotes in THE SAVAGE WORLD, that travellers have had various experiences, so that there is sufficient authority for our preserving the traditions of our youth, or for our rising superior to these, and regarding them as the merest nursery tales.

The **South African Lion** (*Leo capensis*) will naturally receive the most attention, since India has become part of the British Empire, and travellers are directing their footsteps towards the " Dark Continent." In color it is tawny-yellow, growing lighter on the under parts. The ears and tip of the tail are black, and the male possesses a mane. The *lion* when full-sized stands about four feet in height and is ten or eleven feet in length. The *lioness* is smaller in size, but fiercer in disposition. With the multiplication of museums and zoological gardens, this product of African soil has become familiar to the average person, so that our readers will doubtless have sufficient opportunity for examining the animal from the standpoint of the naturalist.

There are reasons why the accessibility of the lion in zoölogical gardens does not furnish the means for personal examination, inasmuch as the lion itself is too uninterested in the progress of science to co-operate with the student. Hence, although the lion's tongue is well worth examining, it is safe to say that most of us will be satisfied to take the word of African travellers. The tongue of the lion, even more than that of the cat, is covered with numberless minute cone-like papillæ, so curved as to lie with the points towards the throat. These serve to strip the flesh from animals, while the tongue of a dog is entirely smooth, since it is not used for any such purpose. So, too, the claws of the lion, and of the cat family in general, are retractile, and when not in use are withdrawn within the padding of the foot; when needed, the tendons above relax and those below contract. The lion does not grind its food, but tears it into strips, and then swallows the pieces, which are macerated in the stomach.

The Lion, so long credited with being the "king of beasts," has, in our democratic days, been shorn of much of his glory, for science deals but hardly with mistaken ideas, even though these have the support of antiquity and even the attractiveness of Æsop's Fables. Livingstone and other African travellers say that the *lion* is only a cowardly thief in the night, and that it can be put to flight by the barking of dogs, the shouts of persons, a bright light, or even by a stout whip fearlessly applied. He will fight only in extremity and even then will not "fight to a finish." But let this last characteristic count in maintaining his reputation for magnanimity, even though his ferocious courage is to be but a myth. It seems that a *lion* becomes a man-eater when he no longer has the teeth required for mangling the flocks and wild animals. Entertaining as Buffon is, his reputation as a trustworthy naturalist has been wrecked by the larger opportunities and increased information of the scientists who have succeeded him. *Lions*, like men, form hunting-parties and co-operate in the capture of game, which they drive in constantly narrowing circles. Dr. Livingstone asserts that it is hardly possible to distinguish between the roar of a *lion* and the cry of an ostrich. The *lion* seems to stand a good chance of becoming extinct, for he is no longer found in Asia, Asia Minor, Greece or Persia, and has almost disappeared from Hindostan. The Indian lion is smaller in stature than the African lion, has a shorter and thinner mane, and is rep-

TAKE CARE!

resented by the *Bengal lion*, the *Arabian* or *Persian lion*, and the marvellous *lion* of Goojerat. The African, likewise, represents three species: the yellow-brown, full-maned *Barbary lion*, the light-yellow *lion of Senegal*, and the *Cape lion* which is either brown or yellow in its color. The maximum size of the *lion* is about eight feet, with an allowance of some four feet more for the tail.

Sir Samuel Baker tells of finding a *lion* within thirty yards of his camp. He baited it with a wounded buffalo and thus tracked it to the jungle. Coming suddenly upon the animal, he shot it from the small distance of three yards, and as it bounded away, was greatly astonished to find its place taken by a *lioness* which escaped while Baker's attendants were making up their minds whether to hand their master a loaded gun or to seek safety in flight. The next day, the *lion* was found in a dying condition. At one time Baker's camp was invaded by a *lion* which, to effect an entrance, had to break through a dense and high thorn-picket. It was finally driven away by firebrands and met its death while roaring in rage about the camp. After an unsuccessful all-day's *lion* hunt, Baker suddenly came upon one but ten yards distant. He shot it in the spine and it rolled over and over, roaring and tearing the ground. In the morning Baker found the *lion* still unable to get away and but ten minutes' walk from the camp. Even then, by a supreme effort, it rose to its feet and showed fight. It was found to weigh five hundred and fifty pounds, and required the efforts of eight men to load it upon a camel.

A man and a boy were sitting with their legs hanging over a bank when a *lion* approached beneath, and seizing the man's legs dragged him down and made off with him—which act serves to illustrate the extreme boldness the

KING OF THE JUNGLE REALM.

king of beasts sometimes exhibits in contradistinction with the cowardice with which he has been charged.

One trustworthy traveller tells of riding up to a *lion* in the spirit of mere bravado, and by shooting his pistol above its head, driving the *lion* away as if it were a patient ass, or a much-abused ox. Naturally he came to the conclusion that the *lion* is neither ferocious nor noble. On the other hand, Livingstone, having shot one of several *lions* which were attacking the herds, and believing that the creature was powerless because in the throes of death, had an experience more thrilling than desirable. He approached the seemingly-helpless brute, when it suddenly sprang upon him and treated him as a well-trained terrier does a victimized rat or cat. The doctor found that his sole defence lay in simulating the limpness and unresistingness of death until an attendant could safely get a shot at the *lion* without imperilling his master. When this happened, the *lion* deserted Livingstone and bit through the thigh of his fresh assailant, and a second attendant coming to the rescue was seized by the shoulder and saved from being horribly maimed and mangled only by the wounds of the *lion* opportunely proving fatal at that very moment. Dr. Livingstone found that he had received eleven bites in his arm and was, there-

LION SEIZING A BUFFALO.

fore, well-qualified to report from personal experience as well as with the weight of his profession as a missionary, upon the effects of the *lion's* bite. He says that the *lion's* bite is poisonous and resembles a gunshot wound. Parker Gillmore, whose books always have interest, tells of finding a *lion* and *lioness* temporarily occupying a native's hut while the owner was absent, and how he himself retired quickly, noiselessly, precipitately and without desire to disturb the fierce beasts in their peaceful slumbers and dreams which, doubtless, dwelt upon victories yet to be won. Upon another occasion he came suddenly upon a sleeping *lion*, and while his majesty, aroused by the intrusion, was taking a preparatory yawn before collecting his energies, Gillmore shot and mortally wounded it. At another time a hunter, while stalking a rhinoceros, suddenly discovered that a *lion* was stalking him, so that he was compelled to change his game. The *lion* is so fearful of a trap that it will not attack horses or cattle while these are tied. Hence it first tries by its roarings and odor to get up a stampede, and if successful, then pursues and captures its prey. A very striking instance of this wariness of the *lion* was furnished in the case of a horse which, having thrown its rider and run away, was caught by the bridle on the limb of a tree. Two days afterwards, the hunters found the horse uninjured, although it had had the constant companionship of quite a party of *lions*. Three *lions* were once watched while they hunted a wounded buffalo, which they finally succeeded in "pulling down," biting always at the withers.

The hunters appeared as the *lions* were beginning their feast, and shooting one of the *lions* a second betook itself to precipitate flight, while the third one persisted in continuing the feast until it fell a victim to such a lack of prudence. On the other hand, the *lion* is not unfrequently killed by the buffalo. On one occasion a *lion* attempted to secure a buffalo calf, but the cow protected it both valiantly and successfully, terminating the contest by tearing the *lion* to pieces with her horns. The *lion* has a more than decent respect for the elephant and rhinoceros, and leaves any attacks upon them to other foes. In some parts of Africa the man-eating *lion* is regarded (not sardonically) as the walking tenement

LION ATTACKING A GIRAFFE.

of the souls of chiefs deceased. This unfortunate belief adds yet more to the paralysis which affects the natives when a man-eater is ravaging their villages. The terrible destruction of life in southern Africa is shown by the Government Reports for a single year. There were lost, in a single year, forty-six persons by elephants, eight hundred and nineteen by lions, two hundred and ten by leopards, eighty-five by bears, five hundred and sixty-four by wolves and twenty-four by hyenas. On the other hand, the slaughter of animals was as follows: Thirty-two elephants, fifteen hundred and seventy-nine lions, thirty-five hundred and fifty-nine leopards, thirteen hundred and seventy-four bears, forty-nine hundred and twenty-four wolves and fourteen hundred and seventeen hyenas.

The **Tiger** (*Felis tigris*) is Asiatic in its habitat, and is there monarch of all he surveys, unless it be the buffalo and the elephant. The Asiatic lion is inferior to the African lion, and hence the disputes about the relative claims of the lion and the *tiger* arise from taking a partial case. The *tiger* has been so often exhibited in menageries, and is so constant a member of zoological gardens, that it is wholly unnecessary to dwell upon its beautiful coloring or general appearance. In passing, however, it may be well to challenge the attention to the wonderful suitableness of the tiger's coloring for concealment in the jungle. The noiselessness of its tread, and the crouching in which it indulges when it wishes to escape observation, are further evidences of its having been provided with facilities and instincts suitable for the life appointed to it.

ROME UNDER NERO. - FEEDING CHRISTIANS TO THE LIONS.

The natives capture the *tiger* by the use of *prauss,* a kind of bird-lime. The *tiger,* in approaching the bait, has to walk over leaves which have been liberally sprinkled with *prauss,* and as the leaves stick to his feet, the *tiger* becomes more and more impotent. Lying down in his rage, he becomes yet more dressed in this shirt of Nessus, until finally he is at the mercy of the hunter. Another device is the bow-and-arrow trap, so arranged that the *tiger,* in seizing the bait, shoots himself fatally.

When the native hunters make a game drive it will not unfrequently happen that enclosed within their circle will be several *tigers,* and these will charge most viciously at the human fence, sometimes escaping by killing those who stand in their way, sometimes succeeding in leaping over their heads, sometimes falling before a shower of spears, and at other times driven back in surly rage before the tempest of weapons. Many of the savage rulers from time to time dispatch their armies to hunt in the most approved martial order. *Tigers* have learned that this soldiery is somewhat timid in its enforced obedience, and not unfrequently they will jump over the heads of the vanguard, and clear themselves a path to safety.

A *tigress* charged upon the elephant and hunter and ran them off several times. Having fired into her ambush the hunter supposed her to be dead, but on his approach she rushed forth and seized the driver of the elephant by his foot. Firing again the only response the hunter found was a spring of the *tigress* upon the trunk and jaw of the elephant. The agitated elephant fell upon its knees and dug up the sand with its tusks while the *tigress* safely ensconced between continued its work of destruction. The sudden pose of the elephant affected the hunter like the unexpected stumbling of a horse, and· he found himself burrowing into the sand without any expectation of digging deeply enough and speedily enough to defeat an undesired attack in the rear. The *tigress*, however, either lacked hind-sight or was well enough off as it was, for when· the hunter, failing to find safety downward, concluded to climb upward,

THE GREAT JUNGLE CAT.

he found himself an unnoticed spectator of the battle between the *tigress* and the elephant—if battle that can be called—in which the elephant received and the *tigress* inflicted all the punishment. Not wishing to intrude, the hunter withdrew, until having rejoined his panic-stricken escort, he was carried to meet an elephant which as yet had not participated in the hunt. Thus re-mounted the hunter returned to the scene of conflict, and having again shot the *tigress*, she deserted her position of advantage and threw herself upon the thigh of the elephant (the defeated one making a prompt retreat). Finally the *tigress*, between the appeals of the elephant and of the ammunition of the hunter, was induced to take a reluctant farewell. When found dead the next day, she demonstrated the fact that her body contained eleven bullets, and that any one of the wounds would have proved fatal.

One man-eating *tiger* devastated thirteen villages, controlled fifteen square miles of territory and killed seven hundred and twenty-seven persons. As was remarked in the case of the lion, the *tiger* hunts of India have been so frequent and so often described that it seems unnecessary in a work such as THE SAVAGE WORLD to repeat what would prove as wearisome as a twice-told tale.

The **Leopard** (*Felis leopardus*) has white spottings in the centre of imperfect black rings. It is often called by the names *ingwe* and *onguirara.*

ROYAL BENGAL TIGER AND YOUNG.

The African chiefs are very fond of wearing fezzes made from the tail of the *leopard*. On one occasion a *leopard* invaded a camp and, possessing himself of a goat, carried it away with him. Paul du Chaillu, whose statements, received at first with incredulity, seem now to be receiving the respect which follows confirmation, tells how he was waked from a sound sleep by a *leopard*, but which did not offer to do him any violence beyond reminding him that his surroundings were such as demanded constant vigilance.

The **Ngulule,**. or **Maned Leopard,** is larger but more cowardly, and its color is gray with dirty spottings.

The **Small Ingwe** is the smallest and fiercest of the leopard species. Its

TIGER STALKING HIS PREY.

spots are very black and its coat strikingly handsome. An African traveller tells of finding two cubs at play and of watching for a long while their kitten-like

A MAN EATER.

rompings until one cub starting for the jungle, he shot the other. On another occasion he had a most exciting chase and a most desperate and prolonged contest with a *small ingwe.* The party had been reduced to four persons and two

42

unusually powerful hounds when the *ingwe* was cornered in a cairn and wounded. In spite of four shots, each one of which told, it at once charged

TIGERS BEFORE THEIR LAIR.

upon the hunters, but was after awhile diverted by the persistent dogs. It finally

HUNTERS ATTACKED BY LEOPARDS.

escaped from them and took refuge under a large rock. One of the hunters approached the opening and, having fired, fell back, while before the smoke had cleared away the *ingwe* leaped forth like a projectile from a catapult, but while still in mid-air a shot by another hunter ended its life. The *ingwe* scorns to feast upon dead game.

A leopard has frequently been known to make off with an ox by making the ox serve as an unwilling steed bestrode by an unwelcome rider. It will alight on the ox's shoulders, seize it with its teeth in front of the withers, dig one of its fore claws into the neck of the ox and the other into its back, and thus be borne along until its huge prey falls exhausted and the leopard can complete his feast at his own pleasure. On one occasion a hunter's dogs found a leopard prowling about the camp and made an attack upon it, when it coolly seized one of the dogs and carried it away as a dietary experiment.

WILD CHETAH.

The African Leopard (*Felis pardus*) is a beautiful creature, if, as has been claimed, beauty is a mere matter of the skin. Its ground color is fawn inclining to yellow, growing lighter in shade on the under parts. Upon this groundwork is scattered a profusion of dark or black spots which appear to advantage from the contrast. It lives in

AFRICAN LEOPARD.

the trees, not amidst them, and hence has at times been called the tree tiger. The *leopard*, or panther, carries its cubs as a cat does her kittens, and it is an amusing sight in a menagerie to see the mother pick up a cub in her teeth and carry it to some other part of the cage, as exhibiting a characteristic of the cat species, of which the *leopard* is a member.

The Chetah, Hunting Leopard, Hunting Cat or Youze (*Gueparda*, or *Cynolocrus jubata*), is found alike in Africa and in Asia. It is taller than a leopard,

INGWE.

but is not specially strong-limbed. It cannot climb, and is neither swift nor possessed of great endurance, but it makes up for these shortcomings by its stealthiness and craft. In Asia the *chetah* is what the falcon and hawk were to the knights of old, and is successfully used for the capture of deer. It is readily tamed, and its gentle, even disposition renders it a very acceptable pet. Its skin is like that of the leopard, although the hair is much rougher.

LEOPARD SECURING ITS PREY.

In Africa the *chetah* has not been domesticated, and can be seen from time to time wandering about in search of its prey. It will creep up on the antelope or deer until within easy reach, and then leap upon it, and never leave it while the antelope is alive. When the *chetah* is used as the assistant of man in the hunt, it is hooded until a herd of deer is sighted, when it is turned loose and will speedily make a successful attack. As soon as the *chetah* has sprung upon the game the hunters rush up and tempt it away by a platter full of blood. The hood is then replaced, and if the hunt is to be continued, the same process is repeated. As a pet, the *chetah* can safely be permitted to wander about like a house dog, and will play with the children as if it were a kitten, and even at the zoölogical garden it is a favorite with keeper and visitors.

TRAINED CHETAH.

The **Jaguar**, or **American Tiger** (*Felis onca*), is nearly as large as the tiger and quite as fierce. His sinewy body is marked with long black stripes, except when, as in the case of thighs and legs, he exchanges the stripes for spots. His head is round and large, his legs short and stout, and on the belly his color passes into a pale yellow. He prefers the jungle and the marsh, for while dependent for the most part upon the herb-eating animals, he never objects to a mess of fish which he quite frequently secures for himself. As a rule the *jaguar* will not attack man, but when he does he is an enemy to be dreaded. It is four or five feet long, with an additional allowance of two feet for the tail, and it stands about two feet from the ground. It is yellow colored with roundish black spotted figures lying in parallel lines.

JAGUAR WATCHING FOR PREY.

MEXICAN JAGUAR.

Its white belly sets off the rest of its coloring, and renders its skin very beautiful. It is very stoutly built and its immense head is furnished with jaws which seem capable of indefinite expansion. It is an expert climber and leads the monkey tribe many a merry dance. Its appetite is voracious, and no viand is treated contemptuously. It is surprisingly noiseless and stealthy, and always makes its attack from behind. Its patience is unwearying, and it will for days stalk the traveller hoping for an unwary moment. The adventures ascribed by poets to the lion and tiger are even more appropriate for the *jaguar*, with whom it is an ordinary event to throw itself on the neck of a giraffe, zebra or antelope, and drive it at full speed to its death. It is properly a tropical American animal, but has worked its way as far north as Texas. When it feels an appetite for fish, it baits the finny tribe with its saliva, and as they rise to the surface, kills them and draws them in with its paws. It is very fond of turtles, which it instinctively turns on their backs. As it catches all that it sees, the natives have found it profitable to merely glean in its pathway. The natives kill it with spears, or capture it by lassooing. Its most dreaded enemy is the boa constrictor which is said to get the *jaguar* occasionally in its fatal folds.

JAGUAR FISHING.

The Puma (*Felis*, or *leopardus concolor*,) is found throughout the two Americas from the extreme south to the extreme north. It goes by different names in different localities—the *panther*, *American lion*, *cougar*, *painter*, and possibly one or two others. Many and

SOUTH AMERICAN PUMA.

thrilling are the stories of which it forms the subject, and numerous and frequent have been adventures to which they imparted the liveliest interest. It attains the length of nearly five feet, and the stature, measured from the shoulder, is over two feet. Stealthy as would be expected from its relationship to the cat, it utters neither roar nor growl to warn its foe. It is brownish-red or reddish-gray in color, an apparent provision for its better concealment. In South America it is said to deceive its prey by imitating the sound made by the deer. The *puma* abounds in California, and thence eastward for quite a distance, so that stories true and apocryphal reach us often since mining and ranching, and pleasure seeking, have made communication so frequent between the "Wild West" and the more thickly-settled East. The *puma* and the grizzly bear seem to be natural-born enemies, and many are the evidences of contests in which the *puma* seems always to be the victor It is asserted without contradiction that the grizzlies, being unable to cope successfully with the *pumas*, take

THE PUMA, OR CALIFORNIA LION.

their revenge by killing every *puma* cub that they find unprotected, as if a second Herod in the grizzly kingdom had ordered a new slaughter of the innocents, or as if a boy took his revenge upon the younger brother of a formidable antagonist.

The **Ounce** (*Felis irbis*) is equal to the leopard in size and also resembles it in beautiful colorings as in habits. It is found throughout the central region of Asia especially at very high altitudes, which it prefers, being proof against cold. It will not attack man, though its chief food is both wild and domestic animals.

TEXAS PUMA.

The **Serval, Tiger Cat,** or **Bush Cat** (*Felis serval*, or *leopardus serval*), has the general appearance of a lynx. It is golden-gray above and white below, dark-spotted, and sometimes striped. It is about a foot and a half in height, and inclusive of its tail, about three feet long. Its fur is a valued and not uncommon article of commerce. It is found in South Africa, and seems to be quite amiable in its disposition.

The **Yagouarondi** (*Leopardus*, or *Felis yagouarondi*,) is of a blackish-brown, sometimes tinged with white or black. Its habitat is Guiana, and it is about the size of the common cat.

The **Marbled Tiger Cat** (*Felis*, or *Leopardus marmorata*,) is a mottled brownish-gray, and is found in Malacca.

The **Ocelot** (*Felis*, or *Leopardus pardalis*,) belongs to tropical America. Inclusive of the tail it is about four feet in length, and about a foot and a half high. The fawn color that furnishes the ground work is banded and edged with black, and head, tail and neck are spotted, or patched with dark color. At the base of its black ear it has a singular white spot. The skin is so beautiful as to have high value to those who indulge in rugs made from the skins of wild animals. Although cat-like in size, it has the strength, agility and instincts of the leopard family. The disposition varies with the individual, one being sociable and playful, another surly and savage.

OUNCE.

The **Painted Ocelot** (*Felis picta*) is deeper hued and more distinctly marked than the common ocelot. It is white on the throat, which has distinct black lines upon it.

The **Eyra**, or **Eyra Cat** (*Felis cyra*), has its habitat in Texas, but is now so nearly extinct as to be very rarely seen, and its habits are little known.

The **Clouded, Tortoise Shell Tiger, or Rimau-dahan** (*Felis macrocelus*, or *Leopardus macrocelus*), is marked as if in imitation of all the felidæ. The ground color is gray, shaded with brown, and having two black bands running the entire length. The hair is glossy, silken and long, and the markings, though irregular in form, are very beautiful to the eye.

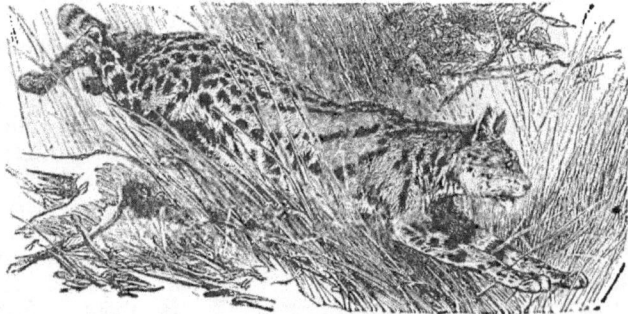

SERVAL.

The **Small Tortoise Shell Tiger** (*Leopardus macroceloides*) differs little but in size.

The **Mitis**, or **Chati** (*Leopardus*, or *Felis mitis*), is smaller and paler. Like the weasel it can get through the smallest holes, and like the monkey it can climb anything. It is very destructive to poultry, which it holds in an esteem above all other kinds of food.

The **Pampas Cat**, or **Jungle Cat** (*Felis pajeros*), is brown striped and sandy colored, is very long haired, and belongs to Buenos Ayres.

The **Caffre Cat** (*Felis caffra*) is South African, and its two feet of length is gray-brown with black stripes; its tail is long and bushy.

MARBLED CAT (*Felis marmorata*).

RIMAU-DAHAN, OR TORTOISE SHELL TIGER.

The **Nepaul Tiger Cat** (*Felis nepalensis*) has black bands and spots arranged lengthwise.

The **Kubouk** (*Felis javensis*) is the species found in Java.

The **Mirvini** (*Felis moormennis*) is a mountaineer of Nepaul.

The **Wogati** (*Felis viverrina*) is found in India.

The **Balu** (*Felis sumatrana*) belongs to Sumatra.

The **Chinese Tiger Cat**, or **Maou** (*Felis chinensis*), is a yellowish-gray species fairly common in

EUROPEAN WILD CAT.

EGYPTIAN CAT.

the neighborhood of Formosa, though its range is throughout southern China. It has a length of body of about two feet. Occasional specimens are found in the Philippines.

The **Egyptian Cat** (*Felis caligulata*) resembles the pampas cat, but is not so bright in its coloring. To the Egyptian the cat has always been a sacred object, and upon its death was piously embalmed.

The **European Wild Cat** (*Felis catus*) is grayish-yellow with tiger-like markings. In length it is about two feet, exclusive of the tail. It is still found in Scotland, but though formerly abundant in England, it has now disappeared from that country.

The **Chaus**, or **Jungle Cat** (*Felis chaus, Chaus lyricus*), is tawny, with indistinct stripes. It wears two coats of hair, and thus seems to be heavier than it is. It is found alike in Asia and Africa, generally about bogs.

The **Texas Wild Cat** (*Lynx maculatus*) is found also in California.

The **Red Cat** (*Lynx lasciatus*), found in the far west, is distinguished by its color, and by the abundance of its soft fur.

The **Caracal**, **Ciagosh**, or **Black-eared Lynx** (*Felis caracal*), is found both in Asia and in

TEXAS WILD CAT.

Africa. Its color is brown, inclining to red above and lighter below. The under parts are spotted, and white hair appears stretching from the tip of the upper lip to the end of the chin. It is about the size of a small dog, and has the most marvellous power of leaping and climbing. It

WILD CAT STALKING A FAWN.

mostly acts as a dependent of the lion, but sometimes hunts on its own account.

The **European Lynx** (*Felis virgatus*) is found from northern Spain to Scandinavia, and also in Asia. The summer dress of the *European lynx* is a dark-gray, intermingled with red, but this becomes a grizzly-white with the

WILD CAT OF THE WESTERN STATES.

EUROPEAN LYNX.

approach of winter. It is spotted and patched in darker color. It is about three feet in length, exclusive of its tail, and is hunted for its fur.

The **Southern**, or **Pardine Lynx** (*Lynx pardinus*), is marked more like a leopard and flourishes in Spain and Portugal.

The **Canada Lynx** (*Felis canadensis*) wears its hair longer and substitutes black for red with an intermixture of gray. It runs in leaps, lighting on all four feet at once. It is harmless and easily killed by a blow, notwithstanding the fact that it is the largest of the lynxes.

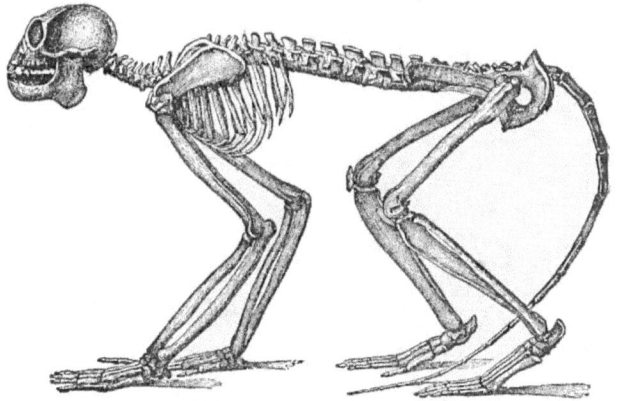

GIANT APE (EXTINCT) (*Mesopithecus pentelici*).

The **Booted Lynx** (*Felis*, or *lynx caligatus*,) wears long, black hair on the sides and back of its legs, so that its appearance has suggested the popular name. It is a small animal and lives in Asia and Africa.

PRIMATES.

The highest of the mammals, excluding man, are the *Quadrumana*, or four-handed animals, including the various species of monkeys, baboons and apes and lemurs. These animals come nearer than others to an actual resemblance to man, and yet, when examined with reference to structure, and not simply with regard to superficial resemblances, they are as easily separable as the lowest of animal organizations. The following scheme may simplify matters for our readers :

QUADRUMANA, OR FOUR-HANDED ANIMALS.

HANDS, NOT PAWS OR FEET: Thumbs generally, but not always opposable.
TEETH: Four incisors above and below; molars like those of a human being and five above and five below (American monkeys have six instead of five); canines project from the least degree noticeable to very tusk-like protuberances.

NAILS: Flat and stretched out.
CENTRE of gravity always such as to prevent a perfectly upright position when moving.
NOSTRILS, anterior in Old World species (down nose): lateral in American (flat nose).
CALLOSITIES and pouches absent in American monkeys.
HIGHEST—Chimpanzee: Skull changes from human shape to that of the baboon. Can walk erect, use knife and fork, do domestic service of a chambermaid, build kitchen fires, unfurl sails, turn capstans, heave anchors, learn rights and lefts of shoes and gloves.

CHIMPANZEE EATING WITH A SPOON.

SECOND HIGHEST—Gorilla: Quadrupedal; elongated arms.
THIRD HIGHEST—Orang-outang.
FOURTH HIGHEST—Gibbon: Callosities like monkeys; pouched and tailless like apes.

The **Primates** (highest class animals) are distinguished from other orders by mental peculiarities, as well as by anatomical structure, but the naturalists by profession prefer to confine their attention to physical organization alone. It is the most recent usage to make man a class by himself, so that for the ends sought by THE SAVAGE WORLD, *primates* will be confined to the *four-handed animals*, or *quadrumana*. The anatomical peculiarities are opposable thumbs, shoulder-blades, nails rather than claws, orbits of the head enclosed in bone, temporary and permanent sets of teeth, at least two mammæ, and posterior part of the brain distinctly developed. As would be expected by those who have followed the successive steps in THE SAVAGE WORLD, all of these characteristics are not always present in every animal and family; for the very fact that one class seems to shade off into another, and that we find animals classed variously because of their contradictions of organism, seems to us but another evidence of the soundness of the doctrine of evolution as a method for working, and the only reasonable explanation of those reversions known to occur among animals, as for example in the case of the wild horse of the pampas. The *primates* are separated into the *lemurs* and the *anthropoids* (or human-like animals), and the two classes are distinguished by structure, distribution, appearance and habits. The *lemurs* have free communication between the orbits of the face and the cavities of the temples; at least one nail is clawed and the others are quite flat, and the hands and feet are large. The habitat of most of the *lemurs* is Madagascar, though one family is African. The appearance of these animals can be deferred until the various species pass in review; and any discussion of their habits will properly find its place when the species are considered.

Considering generally the primates, or the monkey family, we are struck with many curious characteristics which serve to render them anomalous, if we choose to regard them as occupying even a remote relation to man. We find them provided with four hands, by which we might conclude, by analogies, that they were capable of exercising the double function of hands or feet, which is the case; and yet while this is true, they are incapable of using their hinder limbs except in conjunction with the anterior arms, and they are thus forced to walk on all fours. Many species of the monkey family may rise to a sitting posture when resting, or even walk upon the posterior limbs, but they can never assume an erect position. A dog may be taught to walk on its hind legs with quite as much ease and gracefulness as the chimpanzee, orang or gorilla, and, so far as naturalness goes, one is quite as easy when upon two limbs as the other. Again, we are forced to observe the marked difference between the hand of the monkey and that of man. In the former we may perceive a wonderful adaptability, the four fingers, united to an opposable thumb, giving the power of firmly grasping or of picking up the smallest object; but yet it is a hand only by the courtesy of exaggeration, for it has all the appearance of a paw, as well adapted for locomotion as for grasping.

The nose, characteristic of the monkey family, is not always the same, there being three distinct types, viz.: the flat, pointed and proboscis. In all these are beheld the symbolism of the brute, also in the smallness of the eyes and the narrow line that separates them. Hair is not necessarily a distinguishing characteristic of the brute, since considerable growths have developed upon man, especially under circumstances of long-continued exposure to severity of climate, as nature is ever so regardful that she exerts herself to counteract every unfavorable condition. Nor is the presence of a tail the special mark of the brute, since man possesses a rudimentary appendage which physiologists designate as the coccyx. But aside from these possible similarities, observable only to those scientists who make a study of structure, the points at variance between man and monkey are not only distinct, even besides the highest attribute, the power to reason, but in them we see the measureless distance that separates man from the anthropoid.

The **Anthropoids** are marked by greater development of the posterior lobe of the brain; by bony partitions between the facial orbits and temporal cavities; by a human form, and by pectoral mammæ, always two in number. Much unnecessary irritation and uneasiness has been caused by the rash speculations and the unsupported inferences of men writing about natural history, rather than confining themselves, like the most eminent scientists, to gathering data and refraining strictly from analogical reasoning. Scientists, such as Saint George Mivart and Darwin, find ample employment for their abilities in studying nature and endeavoring to understand her scrolls, instead of treating these as palimpsests for their own glosses. In the matter of structure, as well as in that of intelligence, there is a wider gulf between the *anthropoid* and man, than

MARMOSET (*Hapale iachus*).

between the various classes of the rest of the animal world, so that the study of the monkey tribe is quite as likely to lead away from "the missing link" as to hold out any promise of finding it.

Still, with reference to the succession of forms becoming more highly organized as the earth became fitted for their supremacy, the *anthropoids* make a fitting peroration to the story of Genetic development. The lowest forms of the *anthropoids*—the *marmosets*—are found, as would be expected by such of our readers as have accepted our own conclusions, amidst the rank luxuriance of vegetable life which prevails in tropical America. We cannot do better than to cite, as we have done in other of our books, Sir Thomas Buckle's characterization of the Brazilian forests, and of similar

INDRI.

riotness of vegetable life in other lands and

iu other times. Amid this pomp and splendor of nature, no place is left for man. He is reduced to insignificance by the majesty with which he is surrounded. The forces that oppose him are so formidable that he has never been able to make headway against them, never able to rally against their accumulated pressure.

The **Babakoto** (*Lichanotis*, or *Indris brevicaudatus*, or *mitratus*,) is slim but tall, has the sharp muzzle and the head of a dog, is furred to the very tips of its toes, excepting its face, and its soft, somewhat wool-like coat is a curious mixture of black and white—a study, as the British artist would call it. Its black, erect ears rise out of a white ground which, beginning at the back of the head, extends over part of the neck, forms a boa or small collar round the front of the neck, and spreads out on the throat, reaching the under jaw. Its sides, outer surface of the hind legs and inner surface of the fore legs likewise are white or·gray. Its habitat is Madagascar, but its history has not been fully studied.

The **Lemurs** (*Lemures*) occupy the debatable ground between the primates and preceding species. The appearance of the head is fox-like, and its long muzzle would suggest any family but that of the monkey.

The **Diadem Lemur** (*Propithecus diadema*) has a white, tiara-like band on its forehead. The rest of its body is soot-colored above and white beneath, except that the hindquarters and hind legs are brown; the tip of the tail is white and gold, like the furniture of Louis Quatorze. Its head is roundish, its body is about a foot and three-quarters in length, and is supplemented by a tail but little shorter.

The **Woolly Lemur, Madagascar Indri**, or **Wahi** (*Indris*, or *Lichanotis laniger*), is a foot in length and

CROWNED LEMUR.

carries a tail nearly as long as its body. It is brown in color, with a white stripe on·each thigh. Its black face is lighted up by disproportionately large, greenish-gray eyes. Its muzzle is long and pointed, hound-like in appearance, and projects from a small, round head, further ornamented with erect, black-haired ears, and a white pad on the head between. Its cry resembles the wail of a baby, and hence the animal is frequently called the *old man of the woods*.

The **Crowned Lemur** (*Propithecus coronatus*) does not differ greatly from the diadem lemur, except in the arrangement of the hair upon the top of its head, which forms a crown, and in its habits it resembles the indri, from which, however, it is distinguished by the possession of a tail.

The **Ruffled Lemur** (*Lemur macaco*) has its habitat in Madagascar and the neighboring islands, is as large as a cat, has dilatable eyes, and presents

RUFFLED LEMUR.

very striking contrasts of black and white in its long, silken hair. Its black face is set off by a ruff of white whiskers, and its long, full tail terminates in quite a white plume. It is a good runner, jumper and climber. It is shy and timid, but when forced to defend itself, displays no want of courage or lack of effective teeth. It belongs to the class of monkeys which form philharmonic societies, and hold a nightly sængerfest, at which each struggles to outdo the others in the prolonged utterance of the deepest and most sepulchral tones; evidently they believe in conquering the thorough bass and basso profundo before venturing upon the parts which lend liveliness and variety to the body of the song.

The **Maki-Macoa, or Ring-tailed Lemur** (*Lemur macoa*, or *catta*), is a graceful little monkey, which has none of the repulsiveness of features which belongs to apes. It is a cindery-gray in color, the cheeks and throat being quite white, and its extraordinarily long tail being regularly ringed in black. Its little round ears are almost concealed by hair, and its fur is thick and soft. It is nocturnal in its habits, is docile, and readily adapts itself to captivity. Its length is somewhat less than a foot, with the addition of an eight or nine-inch tail. In captivity it is gentle and pranksome, and is specially addicted to weaving itself and its kind into a constantly changing living pattern. Its bones are so peculiar that the stretching out of the arm at once closes the hands.

1. MAKI-MACOA. 2. WHITE-MANTLED MAKI.

The **White-mantled Maki, or White-fronted Lemur** (*Lemur albifrons*), varies the white, which is its prevailing color, by black shoulders, muzzle and tail.

The **Red Maki** (*Lemur ruber*) is satisfied to use white for his neck, head and for the ends of his feet; for the rest he indulges in nothing less than the brightest of red. The makis are represented by many species, each of which indulges in a different coloring. The hair of the *red maki* is more woolly than that of the other species.

The **Mongoz** (*Lemur mongoz*) has its muzzle white, and the Collared **Lemur** (*Lemur collaris*) carries a brownish-black muzzle.

The **Gray Lemur** (*Hapalemur griseus*) is about two and a half feet in length, equally divided between body and tail.

The **Red Lemur** (*Hapalemur simus*) belongs to the bamboo forests, is reddish, and its teeth fit beside each other like a knife blade shutting into the handle. It is sometimes called the *broad-nosed lemur*.

The **Short-tailed Lemur** (*Lepilemur mustelinus*) is varied in color, the prevailing red becoming gray on the under parts, white on the throat, and brown on the tail, which, though not exactly short, is but two-thirds the length of the body.

MONGOOS, OR INDIAN ICHNEUMON (*Lemur mongoz*).

The **Slender Loris** (*Loris*, or *Stenops gracilis*,) has only a rudimentary tail; has a muzzle which terminates very unexpectedly and abruptly; is very delicate in its structure, and is about three-quarters of a foot in length. Its habitat is Ceylon and the neighborhood; its color is gray above and white beneath; and its large, round eyes are made more weird by the dark hair of their orbits and the white streak which runs down the nose.

SLENDER LORIS.

The **Slow-paced Loris** (*Stenops tardigradus*) is much larger than the slender loris; its fur is more wool-like and lacks the reddish tinge to be found in the preceding species. The back of the head, the ears and the orbits of the eyes are of a dark brown or chestnut color.

Its circulatory system is peculiar because the usual arrangement seems to be reversed, as the reservoirs are located in the shoulders and thighs, and from them the network of veins and arteries is spread out. In captivity, at least, it is by preference carnivorous, and it is specially noticeable for its slothfulness.

43

There is a variety of the *slow-paced loris* in Java (*Loris*, or *Nycticebus javanicus*) and, like its family, it has the first finger on hands and feet turned backward.

The **Potto**, or **Aposora** (*Perodicticus potto*), is a loris whose habitat is West African. Its tail, though exceedingly short, is still a real tail; its index finger is more than rudimentary, and of its three-quarters of a foot of length the body claims two-thirds. Its color is a grayish-chestnut, growing lighter on the under parts; its muzzle is short and sprinkled with scattering white hairs, and it is regarded as sacred by the natives.

The **Angwatibo**, or **Bear Macaque** (*Arctocebus calabarensis*), is the potto of Calabar. It has but an incipient tail, and has its broad. cat-like ears ridged. Its body is less than a foot in length, and is covered by a short, thick woolly coat, brown dappled with gray. Its second fingers are rudimentary, but its thumbs are opposable. It is stoutly and symmetrically built, and its round eyes, although not unduly large, are made prominent by its long, slender, cylindrical muzzle, enhanced also by the brevity of its ears.

BEAR MACAQUE (*Arctocebus calabarensis*).

The **Brown Mouse Lemur** (*Cheirogaleus milii*) is frugivorous and nocturnal. It is but a little more than half a foot in length, a grayish-brown, becoming whitish on the under parts; has a pointed muzzle, prominent eyes, a liberal allowance in the matter of ears, and a tail which might belong to a rat, and which is of no service as an extra hand. It walks like a quadruped, but when eating it behaves like a squirrel. In common with all the Madagascar *lemurs* it sleeps through the winter, supporting life for that very brief period by an inactive digestion and by the fat accumulated in its body.

The **Dwarf Lemur** (*Microcebus smithii*) is the smallest of the *lemurs*, and is arboreal, feeding upon plants and insects. It belongs to the fauna of Madagascar.

The **Thick-tailed Galago** (*Galago crassicaudatus*) carries a tail of a foot and a third, while the body which is to wag it is but about a foot. It is arboreal and nocturnal, and is found in the forests of Mozambique. It is unduly fond of palm wine, a beverage said by Sir Samuel Baker to be harmless to man, and drinking to excess from the supplies furnished by the natives recovers its ordinary faculties to find itself a captive.

The **Little Galago**, or **Madagascar Rat** (*Galago demioffi*), resembles a rat in size and in coloring. Its large ears are perfectly translucent, and its eyes are full and lustrous.

The **Moholi** (*Galago moholi*) is a larger species whose coloring is gray above, white beneath, golden on the legs, and brown at the extremity of the tail. It builds a nest-like home in the fork of some tree, and during the day sleeps folded up in its ears. It leaps from bough to bough, is insectivorous, and belongs to the South African fauna.

The **Tarsier** (*Tarsius spectrum*) belongs to Borneo and its vicinity, and is distinguished by the great length of its hands and feet. Its hands have palms which are springy and cushion-like, and the unnatural size of its great, owl-like eyes compensates for the shortness and smallness of its erect ears. Its tail is hairless except for a brush. Its fur or wool is short but thick and abundant, and its color is a mixture of olive, gray and brown. Its motions resemble the hoppings of the frog. It is sometimes called the *podji* and the *banca tarsier*.

The **Aye-Aye** (*Chiromys madagascarensis*) is like Polonius' animal, built upon several quite distinct plans. In appearance it resembles the galago, while its dentition might lead one to suppose that it was one of the rodents. Its teeth are incisors, deeply set in sockets and acutely pointed. Its coloring is gray beneath and reddish-brown above; the cheeks and throat are gray and the feet black. It has large ears which are destitute of hair, and drags after it an exceedingly long and bushy black tail. It, like the other lemurs, sleeps by day and forages by night. Except for its bird-like nails or claws, it would bear some resemblance to the raccoon. It is called by the natives of

GALAGO (*Galago moholi*).

Madagascar the *handed mouse*. Its three feet of length is so distributed as to allow fully one-half to the tail, and its brown eyes, tinged with yellow, dilate only at night.

The **Flying Lemur**, or **Flying Colugo** (*Galeopithecus volitans*), takes its name from a striking resemblance to the flying fox bat. A hair-covered membrane covers its arms and legs as far as the elbows and knees, and when the creature spreads out its limbs this membrane serves all the uses of a parachute. The *flying lemur* necessarily has always to fly downward, and when he wishes to ascend he folds away the membrane and uses his feet and claws. The head is rat-like.

The **White-tufted Marmoset** (*Hapale vulgaris*) has a black head and white ear tufts.

The **White-fronted Marmoset** (*Hapale humeralifer*) is a mixed-brown and white, but is distinctly white on the fore part of its body.

The Barba **Variegated Marmoset** (*Hapale chrysoleucos*) adds a red tail and red hands and feet to the white which colors its body. It is small-sized, has pencilled ears and a long, silken coat, and makes a very acceptable pet.

The **Black-headed Marmoset** (*Hapale penicillata*) has a white spot on its forehead, and uses the same color for the mouth or muzzle; its tail is annulated in black of a darker shade. It belongs to the *ouitilis* or *sagouins* and is quite common in monkey cages.

The **Pigmy Marmoset** (*Hapale pigmaeus*) is untufted and about half a foot in length. It is a brown Brazilian species, notable mainly for its lack of size.

TARSIER.

The **Black-tailed Marmoset**, or **Mico** (*Hapale melamuurus*) uses even a darker shade for its tail than prevails in the coloring of the body; the thighs in front are white, and it is banded with the same color across its loins. Its habitat is Brazil and the States to the West.

The **White Mico** (*Hapale leucopus*) has its extremities and its fore arms white. It is found in Columbia and is about a foot in length, with a tail which is longer than its body.

The **Albino Marmoset** (*Hapale argenteus*) is the *albino* variety of the black-tailed marmoset.

The **Marmoset** (*Hapale*) has long, soft white and reddish-yellow fur, striped distinctly with black. Its long, large, white tail is ringed with black, and its bead-black face is set off by long wings or tufts of white hairs which stand erect above its ears. In confinement it is active and affectionate, but not very wide in the limits of its intelligence. It is insectivorous, and is particularly fond of cockroaches. The *marmoset* species are the smallest of the monkey family, and are all peculiar to South America.

The **Tamarin** (*Midas ursulus*) has no tuft, and its lips are not white. Its bushy tail is much longer than its body, the two together being something over three feet. Its color is black, changing here and there to a gray-brown or a red-brown.

The **New Granada Marmoset**, or **Geoffroy's Marmoset** (*Jacchus*, or *Midas geoffroii*), is whitish-brown, tuftless, short-haired, and has chestnut colorings on its neck and at the root of the tail, which become black at its tip.

The **Two-colored Marmoset** (*Midas bicolor*) is grayish-brown, changing to white in the front. Its tail is black, really adding a third color.

The **Mustached Marmoset** (*Midas mystax*) is a Peruvian species whose lip is adorned with long white hairs.

The **Red-bellied Marmoset** (*Midas labiatus*) is more conspicuous from its ventral coloring than for its mustaches. It belongs to the country of the Amazon.

The **Capped Marmoset** (*Midas pileatus*) wears a cap of gold, and the **Rufus Marmoset** (*Midas rufiventer*) seems to unite the peculiarities of the red-bellied and the *capped marmosets*. Its ventral color is distinctly red, and the head and neck are marked by triangles of red or gold.

The **White-cheeked Marmoset** (*Midas leucogenys*) has white triangular cheek markings.

AYE-AYE.

The **Yellow-headed Marmoset** (*Midas flavifrons*) has the top of its head a brownish-yellow.

The **Pinche** (*Hapale œdipus*) is white on the throat, chest, belly and arms; brownish-gray on the body and has reddish markings on the shoulders. The long, thick, white hair which, starting from the centre of the head just above the eyes, widens out at the top of the head and falls upon the neck and cheeks, is a very striking peculiarity and looks as if it were some temporary and accidental ornament of the monkey. It carries a long, well-rounded tail, which beginning with a brown color soon deepens into deeper black. Its vocal achievements resemble the squeaking of a mouse rather than the sounds usually made by the monkey tribe.

The **Lion Monkey, Marakina**, or **Silky Monkey**, has its face so shrouded in hair, and the shape of its head so leonine, that its popular names are more than usually descriptive. Its hair is a golden-lustred chestnut, darkening on its feet and on its forehead. It is the most cleanly of animals, and any dirtying of its fur at once depresses its spirits. It is very active and very timid, and is domesticated mainly because of its gentleness and beauty.

The **Marakina** (*Jacchus,* or *Midas rosalia,*) belongs to the region of the Amazon, and wears its hair brushed back from its forehead, like a typical professor of music. Its mane and its ruff are likewise quite conspicuous.

The **Black Marakina** (*Jacchus,* or *Midas chrysomelas,*) is a differently colored variety of the *marakina.* It is about six or eight inches in length, and very playful.

MARMOSETTES.

The **Brazilian Night Monkey** (*Nyctipithecus trivirgatus*) is nocturnal, and somewhat of a howler. Its large eyes are surrounded with hair, like those of an owl. It passes the day in a sleep so sound that even handling it does no more than color its dreams of the night to come. It is quite small, but exceedingly quick and active. It is strictly one-wifed, and is never found except alone or in the company of its own immediate family. It secretes itself during the daytime in the hollows of trees, where for the most part it is undisturbed. Its color is a silvery gray, lined with brown on the ridge of the back, and orange-hued on the breast, belly and inner limbs. It takes its scientific name partly from its nocturnal habits and partly from three black lines, two of which bind its forehead, while the third runs from the top of the head in its centre to the base of the eyes. Its ears are so small that it has been called the *earless monkey;* its other names are the *douroucouli*, *aotes* and *three-striped monkey*. Its voice is variable, as at times it produces a miniature roar like that of the jaguar, at another time it spits, and mews, and hisses like a cat, and still again barks shrilly like a little puppy.

The food of the *night monkey* is almost exclusively of an animal nature, and it may with propriety be classed among the insectivora, though it occasionally captures small birds when they are nesting. As soon as twilight begins to gather over the always deep shades of its forest home, the *night monkey* revives from its lethargy and soon after issues forth on its nightly rounds, full of extraordinary activity. The large eyes beam like coals of fire and the listless limbs become instinct with an astonishing activity. Its agility now becomes such that it not only leaps with wonderful animation from tree to tree but contrives to catch the swiftest-winged insects that make nocturnal excursions through the woods. It is generally very restless during the night, but will pause on some high perch for a few moments and watch for prey. So marvellously acute is its vision that if a winged insect comes within

NIGHT MONKEY.

reach in a trice, and with a movement the eye is not quick enough to follow, it strikes out and with such precision that the prey is invariably secured.

In making its quest for birds, of which it does not appear to be overly fond, the *night monkey* uses no small amount of cunning. Having discovered its nest, it takes a position always below and gradually crawls upward, observing the utmost care in order not to startle the bird that may be sitting thereon. Having gained the required closeness to the nest, the monkey raises its forehand slowly above the prey and then suddenly seizes the bird, if she chance to be upon the nest, of which he is not certain until he makes the attack. The males are not only monogamous, but exhibit great devotion to their mates and young, being rarely seen separated by more than a few yards, and are continually manifesting their affection for one another by the most loving caresses.

The **Roaring Monkey**, or **Bellowing Monkey** (*Nyctipithecus vociferans*), though small and harmless, has a deep voice, which to the stranger suggests the roar of the lion or jaguar, or the suppressed bellow of the buffalo. Paraguay is its habitat.

ROARING MONKEY.

The **Lemur-like Monkey**, or **Spectral Monkey** (*Nyctipithecus lemurinus*), belongs to New Granada, where it steals forth at night and utters constant cries as it wanders about in search of food.

The **Cat-like Monkey** (*Nyctipithecus felinus*, or *commersonii*) is found in the Brazilian forests. It is sometimes called the *vitoe*.

The **Gray Callithrix** (*Callithrix donocophilus*) is quite common in menageries and is worthy of no special description.

The **Black-footed Callithrix** (*Callithrix nigrifons*) has the glossy hair of the family, and is distinguishable only by the coloring of its feet.

The **Black-handed Callithrix** (*Callithrix melanochir*) wears only a single coat of short hair or fur.

The **Brazilian Squirrel Monkey** (*Callithrix*, or *Chrysothryx ustus*), has a burnt-brown color.

The **Bolivian Squirrel Monkey** (*Callithrix*, or *Chrysothrix entomophagus*), is named from its fondness for a diet of insects.

The **Rubicund Saki** (*Brachyurus rubicundus*) is found in the upper Amazon, where its various shades of red suggest an affliction like Bardolph's, but one which has not been confined to the nose. The red of its body becomes a pronounced vermilion in the face; its cocoanut-shaped head is entirely bald, its body and legs are stout and like those of a man; its mouth is drawn down with an expression of the most pronounced disgust, and its tail consists of an exceedingly hairy ball.

BRAZILIAN TITI.

The **Parti-colored Sajou** (*Callithrix sciurea*) is a Peruvian monkey, whose reddish-gray coat is marked with chocolate-colored spots. It carries its young on its back, and does not appear to be at all embarrassed in its movement. When asleep it rolls itself up into a round ball. From its cry it has been called the *singing monkey*, and the *ventriloquist;* by the natives it is called *ouapoussa*, or *ouappo*.

The **Death's-head Monkey** (*Chrysothrix*, or *Callithrix scieureus*,) is black, with a tinge of gray; its ears and face are white, and the mouth and nose a blue-black. Its tail is disproportionately long, and when the monkey is at rest he generally circles it about his middle so that he looks as if framed in a hoop of dead evergreens.

The **Brazilian Titi** (*Callithrix personata*, or *sciurea*,) is named by naturalists from its beautiful hair, which is white on the ears, gray on the under parts, black at the extremity of the tail, golden on the legs and light olive on the body. The *titi* is small in size, exceedingly active and affectionate, and always wears an innocent, pleading expression, which is heightened by its ready use of its tear-glands; its brown eyes generally have that appearance of suppressed tears which is always so pathetic. Its most marked peculiarity, however, is its resemblance to the deaf, through its unflagging interest in watching the lips of any one who is talking. Its tail is not prehensile, but is flexible without being muscular. The *titi*, from its activity, has been called the *squirrel monkey*, and it may not be unserviceable to mention that it is sometimes called the *saimara*, and sometimes the *tee-tee*.

BEARDED SAKI.

BLACK HOWLER.

The **Cuxio, Black Cuxio, Bearded Cuxio, or Saki** (*Pithecia*, or *Brachyurus satanas*), is cowled and heavily whiskered and bearded in black, and is vainer of its natural ornament than even the Diana monkey. It is solitary in its habits, or at most lives only with its immediate family. It is quite fierce and its unusually powerful jaws and teeth can do the greatest execution. It sleeps in the daytime and prowls about at night. Its long hairs are gray at their points of insertion and for some part of their length, and brown the rest of the way, so that, as the fur is stirred, the animal may assume any hue which can be drawn from these component colors. Its well-rounded head is covered by long, black hair scrupulously parted in the middle.

The White-headed Saki (*Pithecia leucocephala*) is graceful in form and unusually symmetrical in its proportions.

URSINE HOWLERS.

From the top of its head, which is the deepest black, a partial cowl, coming off into whiskers and beard extends, and by its excessive whiteness forms a startling contrast. The lower part of the throat and chin is orange-colored and entirely destitute of hair. It is said to live principally upon wild bees, although it sweetens its repast with their honey. Its other names are the *fox-tailed monkey* and the *black yarke*.

The **Cacajao** (*Mycetes caraya*) has a flattish, black head and an extremely short, docked tail. It is brownish-yellow in color, with a change to black for the head and front legs. Its three inches of tail seem a weak ending to its two feet of stout body. It is scarce, even in its habitat on the Rio Negro. It

HAIRY HOWLING MONKEY.

is easily domesticated, but always remains shy and timid. The reader may find it called, in books of travel, the *caruiri*, the *hideous monkey*, the *chucuzo*, or the *chucuto*.

The **White Acari** (*Brachyurus calvus*, or *Ouarcharia calva*,) is another species, considered a dainty by the natives, and very much resembling the preceding species.

The **Paranacu** (*Pithecia monachus*, or *hirsuta*) is a Peruvian form, which wears a rough, thick, gray coat. Its bushy tail is somewhat less than twice as long as the body, and in common with the other species of its family, it looks as though it had taken a vow never to let a barber touch its hair or whiskers.

The **White Paranacu** (*Pithecia albicans*) differs only in color and is most notable, perhaps, for illustrating the suddenness with which one passes from a district occupied by one variety of monkeys to another, whose occupants are altogether different, at least in their garb.

CATCHING MONKEYS BY MEANS OF SUGARED COCOANUTS.

The **Red Paranacu** (*Pithecia rufiventer*) is red below and its upper coat is a blackish-gray, ringed with yellow. It sports a golden-hued mustache.

The **Black Howler** (*Mycetes caraya*) belongs to the lower coast of western South America; its forehead is ornamented with hair, which projects, and imparts a bold, uninviting and tousled appearance. Like the rest of its family it is prodigally endowed with sounding-boards, having in addition to the largely-developed hyoid bone, two pairs of resonators. Its hair is coarse and seems to purchase its great length at the expense of having its under parts almost entirely naked. It is conscientious in the development of what teachers of elocution term the "calling voice," and as it is not solitary in its habits it manages to make a most deafening noise.

The **Golden Howler** (*Mycetes auratus*) is red and yellow in its coloring.

The **Ursine Howler** (*Mycetes ursinus*) has a body about three feet in length and a tail equally as long. It is a golden-red in color and from its long bared head and throat proceed sounds so piercing as never to be forgotten after once having been heard.

The **Hairy Howler** (*Mycetes villosus*) is a South American vocalist, whose hairy coat is more than ordinarily abundant.

The **Gray-footed Howler** (*Mycetes barbatus*) and the **Yellow-footed Howler** is the same species discusssed below under head of preaching monkey.

The **Alqualte**, or **Araguato** (*Mycetes seniculus*) is a large-sized South American monkey, whose shaggy head and face suggests that he shaves only on the upper lip and chin. Its stentorian voice is increased in volume by a special development of the hyoid bone, and it by no means allows its talents to rust for want of usage. It is gregarious and the troops are very large in number, so that when they raise their evening song the woods and hills and rocks resound. It is stated that their voices can be heard at the distance of more than a mile, and, in spite of their imitation of every known cry and call, that the sound is so discordant that even persons who are not unusually nervous feel as if their noise was unbearable. These monkeys seem to yell and cry in concert and under the direction of a recognized orchestral leader. They are hunted for their flesh, which, though somewhat tasteless, is esteemed in a country where, although game abounds, man as well as animal, is always hungry. The natives fill a large nut with sugar, and the monkey being unable to withdraw its hand when closed, and unwilling to

PREACHING MONKEYS.

lose its treasure-trove, is as easily taken as a cat shod in nut-shells.

The **South American Howling Monkey** (*Mycetes strumineus*) has a body of about sixteen inches, which is covered by long, shiny, dirty white hair, while its beard and whiskers are tawny. It is believed that its ceaseless howling is a performance designed to add in some mysterious way to the creature's means for self-protection.

The **Preaching Monkey** (*Mycetes beelzebub*) is about the same size as a fox and is covered with long, glossy, black hair. It has sparkling black eyes, a circular throat-beard, small, round ears, and a very long tail. It is found in Brazil and Guiana, where its oratorical performances are an unvarying feature' of each day's existence. The leader will ascend to the topmost branch of the tree, await the congregating of the other monkeys upon the lower branches, and when satisfied he opens the ceremonies by a continuous howl. The preacher next waves his hand and the congregation takes up the refrain and chants a response. When this has been continued to the end of the ritual the leader waves his hand for silence and proceeds to pronounce a benediction.

The **Common Capuchin** (*Cebus apella*) is small and playful and is frequently kept as a pet. Its head suggests that of a pug dog and all but the eyes and muzzle is furnished with an enormous coat of hair, which, on the forehead, cheeks and neck, changes from a grizzly-brown to a gray or white.

The **Capped Sapajou** (*Cebus capillatus*) has the hair of its head long, and running in all directions, so that it resembles the odd worsted caps worn by young misses. It is very playful, though uneven in temper, and in captivity seems to take special interest in mechanical contrivances, which it invariably applies in a manner novel, but entirely satisfactory to itself.

The **Weeper**, or **Sai** (*Cebus chrysophus*), is a medium-sized South American monkey, whose constant mood is Niobe-like, "all tears."

The **Whitish Sai** (*Cebus hypoleucus*) is differently colored, as is also the **Olive Sai** (*Cebus olivaceus*).

The **Curled Capuchin** (*Cebus vellerosus*) is marked by its curly or wool-like hair.

The **Tufted Capuchin** (*Cebus cirrhifer*) is a Brazilian species, notable for its tuft. It is this family, though not so frequently this species, which furnishes the serious, hard-work-

CAPUCHIN MONK MONKEY.

ing, melancholy companions of the itinerant organ-grinder. The *capuchins* are small and playful. It readily makes friends, playmates and allies of all other animals.

The **Caparras** or **Negro Monkey** (*Lagothrix humboldtii*), is a less noisy howler, found along the Oronoco. Its tail is prehensile, its hair close, thick and soft, and it uses its intelligence to devise schemes for pilfering and for gratifying its well-developed gluttonous instincts.

The **Spider Miriki** (*Eriodes arachnoides*) forms an intermediate species between the spider monkeys and the howling monkeys. It is more spider-like than the mono, but otherwise does not differ except to the anatomical naturalist.

The **Mono**, or **Miriki** (*Brachyteles*, or *Eriodes hypoxanthus*), is a brown-colored, short, thick-furred monkey, which becomes gray on the under parts and in its whiskers and mustache. It is quite large in size.

The **Black Chameck** (*Ateles ater*) is Brazilian and distinguished by its black face. Domesticated *chamecks* have been known to make friends of dogs and to use them for horseback exercise.

The **Grizzled Coaita** (*Ateles gridescens*) has the characteristics and habitat of the other *coaitas* and is known by its grizzled coat. The **Hooded Coaita** (*Ateles cucullatus*) wears a hairy hood on its forehead. The **Coaita of Nicaraugua** (*Ateles albifrons*) has a white forehead. The **Red-bellied Coaita** (*Ateles rufiventer*) belongs to Columbia, is smaller and has a flesh-colored face, while the black of its body is in marked contrast with the red of its under parts. The **Hairy Coaita** (*Ateles vellerosus*) belongs to Vera Cruz, lives at high altitudes, and each of its long hairs seems to have a direction of its own.

The **Marimonda** (*Ateles belzebuth*), of Guiana and Central America, is another of the spider monkeys. It is quite small and slender, its head is diminutive and its tail is almost whip-like. Its color is black, becoming lighter on the under parts and frequently relieved by chestnut color on its sides. When resting, it throws its arms back of its head. It is easily tamed and makes so amiable a pet, as to be a great favorite.

The **Spider Monkey** or **Coita** (*Ateles paniscus*), is another South American form, and it is so wholly adapted to life in the trees as to be almost helpless on the ground. It uses its tail not simply for prehensile purposes, but likewise as antenna, and when in motion curls it over its body, so that it projects in front of its head; he is said to use his tail, as the ant-eater does its tongue, and to insert it into the nests of birds, extracting the eggs and conveying them, one by one, to its mouth. It is small-sized in body, long and spider-like in its hairy legs, and the ductility of its tail adds to its general resemblance

NEGRO MONKEY.

to a black, hairy spider, as it sprawls about in its climbing. The male, at least in captivity, is quite playful, while its mate yields a somewhat reluctant assent to its overtures for a romp. Although averse to terrestrial locomotion, it will, when necessity compels, walk off on its hind feet, using its tail as a reversed rudder, or balancing-pole. Although exceedingly active when so disposed, it is exceedingly fond of its *dolce far niente*, and numbers of them, almost equal to the leaves on the tree, will suspend themselves by their tails and rock themselves for hours at a time, until the inexperienced observer would suppose that he had not only found an unknown species of tree, but that he had met with the phenomenon of a spectral breeze which agitated the leaves of the tree, without being otherwise perceptible. This is the monkey whose success in transforming himself into a suspension bridge has so often formed the subject of popular description and illustration. One of them will fasten his tail to a bough, and then a series of monkeys will tie themselves to it and to each other until a sufficiently long chain has been formed. Next, they use their powerful muscles to give them the necessary swing, and finally the endmost. monkey will be able to seize a tree on the opposite bank. The bridge now being formed, the troop of unoccupied monkeys run along it, and when the last one is safely over, the monkey on the wrong side of the river relaxes his hold and the swing is started from the other end.

The tail of the coaita is about two feet in length, or twice as long as the body. The face of this monkey is copper-colored, and its black hair grows to great length on the shoulders, thighs and legs. Like some of the monkeys already described, the coaita is always ready to assault the traveller or hunter with missiles and insulting remarks, using its best efforts to convey the idea that it is by nature in open hostility to man, or, at least, prefers his absence to his company.

The **Chameck** (*Ateles chameck*) is found in Brazil and, like other American monkeys, illustrates, by its difference from the forms of the old world, that constant and singular adaptation to environment which no theory of accident will account for. It has perfectly black, long, thick hair, which is inclined to kink. Its length is about a foot and three-quarters and its tail is fully two feet long. The last foot or less of the tail is hairless, and as the member possesses the greatest flexibility, and is controlled by numerous powerful muscles, it substantially renders the *chameck* and the other species belonging to his family, five-handed. It will frequently suspend itself wholly by its tail, and then swing its body, which has a rat-like subtleness, to the next branch or tree. Its nailless thumb, like the thumb of other monkeys, cannot be opposed to the fingers. In captivity, it is

SPIDER COAITA.

docile, amiable and playful, and is susceptible of a high degree of cultivation. Nor does it appear to be so capricious of temper as most monkeys, for, while delighting in any kind of sport, it does not become spitefully tricky even under abuse. When subjected to ill treatment it manifests great grief, repairing to some corner, where it spends a long while in dolorous exhibitions that will excite the pity of any warm-hearted person. Instances are on record where the *chameck* has actually grieved itself to death. Unlike most of the monkey species, the *chameck* does not possess the posterior callosities, but is provided with long hair on the hinder quarters instead. The nostrils, too, are peculiar, in that they open from the side, whereas in all other species the opening is from beneath.

The **Hebe**, or **Tartarin** (*Cynocephalus hamadryas*), is from four to five feet in height when erect, and infests the mountainous regions of Abyssinia and Arabia. It is covered with long, shaggy hair, except on the legs where it is specially short. Its face is long and looks like an unwashed human skin. It moves about in large companies, the old males bringing up the rear. It is the *tot*, *tota*, or *thoth* of the ancient Egyptians, and it is often found as a mummy.

The **Papion**, or **Papio** (*Cynocephalus papio*), is of various shades of red, inclining, however, to a brownish-yellow. On the march the young lead the van and reconnoitre, the females occupy the security of a central position, and the old males bring up the rear, while officers are appointed to see that no one straggles; these packs will number more than a hundred. It, like other baboons, is most commonly captured by the use of drugged beer. The *papio* is the common baboon, and is quite a familiar sight in Guinea, where it will stroll about the streets of the towns like a sailor taking a holiday on shore. When domesticated it has been taught to drink mugs of beer,

MANDRILL.

and more reluctantly to smoke a pipe, which protruded between its fawn-colored whiskers. Its face, hands and ears are hairless and black, but its eyelids are as white as an albinos.

The **Mandrill**, or **Hobgoblin** (*Ateles maimon*), belongs to Guinea, and its immense size and forbidding appearance are but indications of its ferocious and malicious character. Its head looks like that of a hornless buffalo, and terminates in an enormous snout and a wide, thick-lipped mouth. The extremity of the snout is a bright red, and its ridges are marked by lines of blue, azure, purple and scarlet. It wears a yellow, Shakespearean beard, its hairless ears are blue, its under parts are gray, and the upper parts olive-colored tinged with brown. All of these colors are most pronounced in hue and look as if the creature had robbed the goddess of the rainbow of her palette and brushes. Its cheek-bones are ridged into pouches and add to the ugliness of the creature —at least as tried by the laws of Greek æsthetics. Like all the monkey tribe, the *mandrill* is capricious and liable to sudden fits of anger, but unlike most of the monkeys, he is unforgiving and vindictive after the immediate cause for his anger has passed away. It forms with its kind a sort of combination of outlaws, and its strength, cunning and fearlessness render it a very much dreaded neighbor. It will, in large bands, enter a village when the hunters are absent and help itself with the utmost disregard for the old men, women and children. In the woods it is frugivorous and insectivorous, but in captivity it eats almost

BABOONS (*Cynocephalus babuin*).

anything. It is teachable and learns willingly to drink beer and spirits, and more reluctantly to smoke tobacco. It wears its stump of a tail pointed over its back and makes as much use of it in gesticulation as a sidewalk politician

GROUP OF BABOONS.

does of his hand and index finger.

The **Drill** (*Cynocephalus*, or *Papio leucophæus*), is smaller than the mandrill, its cheek-pouches are smaller, its colors are duller, with a predilection for green, and its hands and feet are copper-colored.

The **Gelada** (*Cynocephalus gelada*) belongs to Abyssinia, is maned, brown except on the fore legs and the feet, where it is black, and with a head ornamented like that of the macaques. It walks on all fours and sits in the way peculiar to the most pronounced monkey as distinguished from a human being. It climbs trees and rocks with the utmost facility, and as the able-bodied creature goes about in large-sized crowds,

it is more than abundantly able to deal with any adversary except a bullet. Legs, nails, jaws, and teeth are all brought into play when they engage in battle,

44

and woe betide the unlucky hound, or the unwary hunter whom it once gets into its power, for it seems to unite the offensive armaments of the kangaroo, the bear, the lion and the elephant.

The **Baboons** are the ugliest, the most ill-tempered, the fiercest, and the most repulsive of the monkey kingdom, so that it is no injustice to them that in popular language they have been used as the symbol of extreme ugliness, and thoroughly awkward and disagreeable behavior. In all ways they represent anarchy as against good order, and flourish best where civilized man can flourish least. The muzzle looks as if truncated, or suddenly chopped off, and the nostrils are located in its very extremity. The *baboon* is as filthy and unseemly in its habits as in its appearance, so that altogether it is not exposed to the danger of having its head turned by becoming a popular favorite.

The **Ursine Baboon, Imfena,** or **Chacma** (*Cynocephalus porcarius*), has a front like that of a bear, is as large as a full-grown wolf, and more than a match for any number of ordinary dogs. It is the most expert of plunderers, and sending two or three of its number into an orchard the rest will fall in line and the fruits stolen are passed on from hand to hand, until when enough has been gathered together, all will retire amicably to enjoy the plunder. They consider the laborer worthy of his hire, and hence make no objection to any quiet bites in which the line of battle may indulge while the fruit is passing through their hands. It is readily domesticated and is employed by the natives in hunting roots and in find-

CRESTED BABOON, OR BLACK MACAQUE.

ing concealed supplies of water when the drought has been unusually severe. It is when tame very playful, and seems specially to delight in teasing any one whom it can frighten. In common with many monkeys it is fond of drinking strange mixtures and seems to have a natural appetite for ink. Many an amusing incident has happened to those who have had the *chacma* as a pet. When young it seems to be as playful and as light-hearted as a young child; it will pet puppies and other small animals, but handles them with the same disregard manifested by many a miss in her treatment of cats and dolls. If the animal is too persistent in its objections to such rough usage, the *chacma* will fling it away in disgust. The *chacma* is supposed under ordinary circumstances to fill out two score years of age. The hunter's dogs once discovered a baboon sitting in solitary grandeur upon an ant-heap. Driving him to bay, he seized the foremost dog and rent it into pieces. The hunter then threw his spear which stuck fast in a tree; whereupon the baboon plucked it out and hurled it back at the hunter. The second spear struck the baboon on the arm, and, as he was about to try an imitation, a sudden mortal wound, inflicted by another hunter, put an end to the adventure.

The **Sphinx Baboon** (*Cynocephalus sphinx*) is a frequent denizen of the menagerie and of the zoological garden. It is native to western Africa, and is unusually docile and intelligent for a baboon.

The **Babouin**, or **Babuin** (*Cynocephalus babuin*), is as yet but little known.

The **Crested Baboon** (*Cynopithecus niger*) is an inhabitant of the Philippine Islands, and differs from the baboon proper by possessing but a rudimentary tail, and a crested head surmounted by a face whose flat nose and bristling brows, quite as much as its black color, justify our calling it not only " niger," but " nigger." It is gregarious, and as a band descends from the trees the sight might well terrify one, who if not like Ajax "afraid in the dark" is capable of being startled by the dusky apparitions, even though in size they do not exceed that of a bull-terrier. These creatures are very destructive to the fruit raisers· and gardeners, whose products these marauders descend upon in the night and sometimes destroy whole fields of ripening fruit or vegetables.

MACAQUES.

The **Dog-monkeys** (*Cynopitheci*) are tailless, have round, rimmed ears, and an elongated face. The Celebean species wear a head-dress strongly suggestive of the feathers assigned to the Indian of the books, and has its face fringed with long hair. It is mild but lively in its disposition.

The **Macaque** (*Macacus*) is an oriental baboon or monkey, and is quite abundant in India and the East Indian Islands.

The **Common Macaque** (*Macacus cynomolgus*) has a round head from which the face protrudes, and this, in turn, is distinguished by the prominence claimed by the eye-brows. Its tail is twice as long as in the satyr, but it compensates for this superiority by lacking the tuft. Some varieties of this species lack the tail, or have it in merely a rudimentary form. It is hardy, and does not suffer from confinement, so that it can generally be found in zoological gardens, and its appearance is thus rendered familiar.

The **Tailless Macaque** (*Macacus niger*) lives upon the rocks and feeds upon mollusks, crustaceans, insects or vegetable life.

The **Tailed Macaque** is a denizen of the forest. While quite young the macaque is gentle and teachable, but after a few years its disposition seems to undergo a change, caused, perhaps, by its trials and disappointments.

HAMADRYAS (*Cynocephalus hamadryas*).

The **Munga**, or **Bonneted Macaque** (*Macaque sinicus*) is a long-tailed species, clad in brownish-green, with a whitish shirt front. It parts the hair on the forehead in the middle, but brushes back the rest of its capillary adornment. Its habitat is Ceylon, but it is frequently imported, and is relatively hardy and able to endure a change of climate. Its face is flesh-colored, its body olive and gray above and white beneath. It is larger than the green monkeys, has a much shorter tail, is more muscular, and the callosities or hardened spots on its hind legs are quite decided. The *munga* belongs to the sacred monkeys of India, and it may be that the ready attention and obedience with which he meets at home will explain the sullenness, and surliness, and spitefulness which the specimens occupying our monkey cages generally exhibit. In India, the temple of the monkeys is inhabited mostly by this species, and hundreds of them meet the visitor in expectation of the propitiatory offerings which he is required to bring for them as well as for the priests whose lives are devoted to their service.

The **Bunder Monkey**, or **Rhesus** (*Macacus rhesus*), is another sacred monkey of India, the natives in some districts going so far as to pay tithes to it.

It is a short, stout-bodied animal, with a mere excuse for a tail, and callosities specially prominent because of their colors. Green, orange and yellow contend with each other in coloring the hair, though the back is brown, and the arms dun-colored beneath. It is naturally free from shyness, and possessed of the most insatiable inquisitiveness, and these qualities seem to be developed by the veneration in which the natives hold it.

It can leap many feet, and that even though encumbered by its young, and overloaded with the results of successful expropriation. It is mischievous and pranksome, and like some other species does not hesitate to hurl missiles as well as objurgations. One traveller tells of the liberties taken by a pack of *bunder monkeys.* He says that while in camp he was suddenly informed that a pack of *bunders* had taken possession of the surrounding trees, whence they made constant descents, as one coveted object after another attracted their attention. The loss of turbans, spears and other ordinary possessions had not greatly disturbed the natives, but when the *bunders* began amusing themselves with the horses and cattle, the natives thought it high time to communicate with "the central office" before a stampede should have been initiated. The traveller taking his gun, shot one *bunder* as an example, and the lesson, though a surprise, proved to be sufficient. Still, the hunter was compelled to withstand on the one hand the superstitious fears of the natives, and on the other the pathetic spectacle of the wounded *bunder* coming directly to him for medical aid, which proved useless. On another occasion several officers were put to death by the natives for killing a *bunder.*

THE GELADA.

An ingenious scheme of a European farmer or planter is to be found "in the books." Knowing that he did not dare kill the *bunders*, and not being willing to raise crops for their benefit solely, he caught quite a number of the young ones, and having covered them with a mixture of syrup and tartar emetic, set them free and started them to rejoin the rest of the pack. The whole tribe now engaged in licking up so well-tasting a repast, and when the aftermath came it at once and forever deserted a neighborhood where the cooks poisoned the food. The *bunder* is jealous, envious, spiteful and malevolent, and sight-seers about our monkey cages will do well to identify him, and remain upon the most distant terms of acquaintance. Another peculiarity in the appearance of the *bunder*, is the looseness of its skin, and its bagginess about the throat.

The **Gibraltar Magot** (*Innus ecaudatus*) is really only a naturalized citizen of Gibraltar, as its real habitat is the Barbary States. It prefers to pass its life on the rocks, where it moves about in large packs, seemingly under the guidance and rule of a single, absolute monarch. The *magot* is very powerful, active, nimble, quick-sensed and ingenious. Like most monkeys, it dreads the panther and its relatives, but is more irritated ·than frightened if approached openly. It is about three feet or less in length, gray in color, eyes sunk under prominent brows, which set off a roundish, heavy, dog-like head, which is supported by a short, thick neck: altogether its appearance is fierce and formidable. If captured when young it displays great intelligence, and readily learns new tricks; but as it grows older, and its captivity becomes more grievous, it is apt to sink into a lethargy, or at least exhibit all the signs of a broken spirit, and of a heart bowed down with woe. In a state of freedom it is frugivorous and insectivorous, but in captivity it becomes omnivorous. When eating, it carefully examines each separate insect or article of food, and adds to the grimaces and contortions of other monkeys that of sucking in its cheeks. It moves about on all fours, but rests in the same attitude as that of a person occupying a chair, and sleeps either lying at length on its side or sitting on its haunches, with its head reposing between its hind legs; its tail is so rudimentary as to be almost a mere symbol.

The **Black Macaque** (*Cynocephalus niger*) belongs to the Philippine Islands, and bears some faint resemblance to the magot. It is, however, much larger; its prominent eyebrows extend like a rubber band beyond the eyes, the face is elongated so that it looks like that of the true ape, and its head is ornamented by a crest of long hair which, though noticeably erect, keeps pointing backward. It is stout and muscular, but differs from the other species in appearance rather than in habits and characteristics.

The **Bruh,** or **Pig-tailed Macaque** (*Macacus nemestrinus*), belongs in habitat to Sumatra, and even among the many ingenious species of monkey, distinguishes himself by his cleverness. It is trained by the natives to gather cocoanuts from the lofty palms, and to discriminate with the greatest nicety the choicest fruit of the tree. It is of medium size, mainly fawn-colored, with browner shades on the top of the back and head, on the sides to a small extent, and on the uppermost side of the tail, which is short and curved like that of a pig. It is exceedingly cunning, inquisitive and mischievous and will concoct the most subtle schemes for coming into possession of any article of wearing apparel, which it destroys after having amused itself in its examination.

The **Wanderoo** (*Silenus veter*) is classed with monkeys by some, and with baboons by other naturalists; it resembles the baboon in having a bushy tail, and for our purposes may be regarded as a convenient form of transition from the monkeys. It is called by the natives the *neelbunder*, and it is spoken of by travellers as the *bunder*. Its head and face are surrounded by long, bushy hair, looking like a combination of a judge's wig and an enormous pair of false whiskers; this capillary ornament is entirely white, so that it lends the most comical appearance of undisturbed dignity and seriousness to the monkey's melancholy countenance. It avoids the habitations of man, and seems to have selected the calling of some mediæval ascetic. In captivity it is very capricious, but grows increasingly ill-tempered as it gets older.

The **Gelada** (*Cynocephalus gelada*) is an Abyssinian creature whose singularly profuse hairy adornment distinguishes it above all the macaque species. Its prevailing color is brown, with lighter hue on the crown. Its manner of progression is on all fours, having a swaggering walk and moving in a gallop when running. When sitting it seems almost enveloped in a mantle of very long coarse hair and presents a rather forbidding appearance, notwithstanding its rather benign countenance. The mantle is confined to the neck and shoulder, the hind quarters and limbs being covered with short hair. In size it is equal to the chacma.

The **Tailed Apes** (*Cercopitheci*) furnish most of the useful servants employed by the itinerant organ-grinder, whose ubiquity seems unlimited, and whose apathy and narrowness of musical range are wearing upon any but children, who disregard the less intelligent man, in their enthusiastic interest in the appearance and antics of the monkey.

The **Green Monkey** (*Cercopithecus sabæus*) has hairs in which the colors black, yellow and blue alternate so frequently as to produce the effect of green. It is liberally endowed in the matter of cheek-pouches, which it uses as larders for storing away food until it is needed for consumption.

The better-fed American or European traveller finds

GREEN MONKEYS.

roast monkey a rather tasteless dish. The deep melancholy which the faces of monkeys always express, has rendered sportsmen disinclined to shoot an animal whose flesh they do not care for, and whose pleading eyes, pathetic looks and dying moans seem to them too human to be agreeable as a recollection. The natives, like their wild dogs, wolves, hyenas and jackals, are always hungry, and rudely thrust away all sentimentality when the question is one of appetite, and the *green monkey* accordingly serves them as a popular article of diet.

The **Variegated Monkey** (*Cercopithecus mona*) is sometimes imported, and manifests all the cunning and fondness for imitation which "is the badge of all his tribe." It has the same habits and habitat as the green monkeys, but the green of its coat is varied by white, maroon and gray.

The **Red Monkey** (*Cercopithecus ruber*) is larger, and the red, which is its prevailing color, becomes cream-colored on the legs. It belongs to the Senegal fauna. It is gregarious and combative, following the hunter by leaping from tree to tree, and hurling at him every available missile while it scolds away as a further means of offence, or as a relief to its highly-wrought feelings. It is not an unfrequent denizen of our monkey cages, and is exceedingly active and ingenious, though its humor is distinctly spiced with malevolence. It is specially resentful of mimicry or ridicule, and treasures its wrongs in a memory remarkable for its tenacity. On one occasion at least, it was so offended at a small boy, who thoughtlessly undertook to imitate its gestures and chattering, that whenever afterward he approached the cage it had a paroxysm of fruitless fury, until after about a year, it found an opportunity of seizing his hat and tearing it to shreds with the most marked manifestations of anger and contempt.

The **Diana** (*Cercopithecus diana*) is specially notable from wearing a long, pointed, cavalier-like beard of white, and like the unusually long-bearded among men, it is very vain of its hirsute appendage, and devotes much time to caressing it and keeping it spotless and undefiled. Did it live where it could know of and purchase a mustache cup, it would be quick to procure one, for whenever it drinks it is most particular about pushing back and protecting its much-prized beard and whiskers. Its forehead is marked by a white crescent, whence probably its being named after the goddess of the moon. Its coloring is a rich, deep-chestnut above, separated distinctly and sharply from the bright orange color below, by a band of the purest white. The eyes and legs are gray, and the hands and feet are quite black. These colors are each so pronounced and the hair is so glossy as to fit the *diana monkey* to enter into rivalry with the most gorgeous of the feathered tribe. It is four or five feet in length, and the body and tail divide this length about equally. It is easily domesticated, but in spite of its scrupulous cleanliness (in which respect it finds but little emulation among the monkeys), it has not as yet ceased to be rare in our collections. Its habitat is the Congo and vicinity.

The **Sooty Mangabey** (*Cercocebus fuliginosus*) likewise belongs to western Africa. It is less irritable than most monkeys, is easy to tame, and makes an amusing, even though mischievous, pet. It is only about a foot and a half in length, and might in a London fog easily be mistaken for the stunted and shrivelled children who pass their lives in removing soot from the chimneys of their more prosperous fellow-beings. The *sooty mangabey* is specially addicted to wearing the most pronounced and constant grin, and adds to its peculiarities by curling its tail along its back toward its head whenever it goes forth for a promenade on foot. It is frequently to be found in monkey cages, and its passion for being noticed will soon call attention to it, while its apparent bonelessness enables it to perform feats and to indulge in contortions whose reward is none too great if these excite the enthusiastic admiration of the observer. It is fond not merely of "keeping itself before the public," but of any glittering objects, such as jewelry, and will exhibit much cunning in its efforts to secure the coveted object. It is black in color, but pink callosities gleam from the midst of the black fur. It is treated by other species of monkeys as though its venerable appearance was but the outward symbol of qualities entitling it to the most profound respect and esteem of the whole monkey kingdom. It fills its cheek-pouches before it breaks its fast, seemingly with reference to unforeseen and unwelcome interruptions.

The **Burmah Macaque** (*Macacus arctoides*) has a red face and a dark-brown coating. The **Thibetan Macaque** (*Macacus thibetanus*) is short-tailed and woolly-coated; and the **Ochre Macaque** (*Macacus ochreatus*) is an ash-colored, short-tailed species.

The **Black Hocheur** (*Cercopithecus melanogenys*) has its nose ornamented with white hairs; and the **Samango** (*Cercopithecus samango*) has its nasal ornament of the same color as the body.

The **Black-templed Hocheur** (*Cercopithecus erxlebenii*) has black streaks on its temples.

The **White-collared Mangabey** (*Cercocebus collaris*) differs only in marking from the sooty mangabey.

The **Japanese Red-faced Macaque** (*Macacus speciosus*) has the characteristics, but not the marking of the other *macaques*.

The **Collared White-eyelidded Monkey** (*Cercocebus æthiops*) belongs to western Africa, and differs from the sooty mangabey in having white mustaches, a white neck-tie, and brown as the color of the upper part of its head.

The **Hocheur** (*Cercopithecus nictitans*) belongs to Guinea, and its large, white nose makes a singular contrast to the olive-spotted black, which prevails throughout the rest of its coloring, excepting only its white or gray side-whiskers.

The **Little White-nosed Monkey** (*Cercopithecus petaurista*) has a body less than a foot in length, black hands and feet, and grayish-brown for its general coloring. Like the hocheur it belongs to the fauna of Guinea.

The **White-throated Monkey** (*Cercopithecus albogularis*) belongs to Madagascar, and is distinguished by the marking of its throat.

The **Grivet** (*Cercopithecus grivet*) is dark-green, with legs and tail inclining to gray, while the ears and soles of the feet are a violet-black. It sports side-whiskers. It is found in Abyssinia, and the way in which its canine teeth protrude renders it noticeable amidst " the wilderness of monkeys" to be found in Africa, where it is often called the *tota*. It is medium sized, and quite active in its movements and habits.

The **Beautiful-haired Monkey** (*Callithrix*) belongs to Senegal, where it is frequently domesticated, and whence it has often been exported. It belongs to the green monkeys. Its under parts are white, the outside of its legs and thighs gray, and its whiskers the most golden yellow.

The **Vervet** (*Cercopithecus pygerrythrus*), like the callithrix, belongs to the fauna of Senegal, where the trees will be fairly alive with immense troops of this bright-eyed little monkey. Its canine teeth project, and its coloring is various—most frequently resembling the grivet, with the exception that it is likely to be dun-colored on the head, throat and breast. It resents the presence of man, at whom it will continue to scold and throw small branches, even though it sees its companions being shot down at its side. The *vervet* is frequently to be seen in the monkey cages of our menageries and zoological gardens, and is one of the most active and amusing of the monkey tribe. This monkey, like some others, is hunted by the natives for his flesh, which they hold in high esteem.

The **Maned Monkey**, or **Colobus** (*Colobus*), is African in its habitat, and its long, silky, valuable covering renders it exceedingly handsome—for a monkey. It takes its name from having only rudimentary thumbs on its fore-paws.

The **Bear Colobus** (*Colobus ursinus*) is small in size and prevailingly black in color; its head, cheeks and chin are surrounded by long, white hair, which reappears on the thighs, and on the tail, which, white throughout its whole length, terminates in an abundant tuft.

The **Black Colobus** (*Colobus satanas*) is dressed throughout in glossy black, and its black crest and whiskers add to the uncanniness which has suggested its being named from its impish or demon-like appearance; it wears a long, thick, pendent, yellow mane.

The **Angola Colobus** (*Colobus angolensis*) is, like the rest of its family, very handsomely apparelled. It is clothed in a brilliant, short, black fur, and beginning with a narrow line at the base of the forehead, short, white hair covers the cheeks, neck and fore part of the body, and after passing the shoulders, grows into long, pendant, silken fringe, hanging from the sides. The tail terminates in a large, silvery plume. The gleam of this silvery sheen as the animal leaps from bough to bough is exceedingly beautiful.

The **Zanzibar Colobus** (*Colobus kirkii*) is white beneath, ruddy above, and black on the back of the neck and on its legs. Its hair seems to form a fur cap for the top of its head.

The **Bay-colored Colobus** (*Colobus ferrugineus*) belongs to Sierre Leone, and is distinguished only for its color.

PARTI-COLORED COLOBUS (*Colobus guerza*).

The **Parti-colored Monkey**, or **Guerza** (*Colobus guerza*), is Abyssinian in habitat, and the black of its upper parts suddenly changes on the sides to an equally pronounced white; the cheeks, forehead, and the tuft of the tail likewise are furnished with a white framework, or are themselves suddenly changed to white.

The **Kahau**, or **Proboscis Monkey** (*Presbytes laureatus*) is Bornean, and has a wealth of nose, which resembles a proboscis, from which characteristic the animal takes its name. It is gregarious and very noisy, exchanging the ordinary chattering for a most unmistakable howling. It is of a rich chestnut color,

PROBOSCIS MONKEY.

which on the under parts of the body, on the shoulders and on the face becomes a true golden. The nostrils do not, as in man, run the length of the nose, but seem to be mere slits at its extremity. It is said that in making its immense leaps from branch to branch, and from tree to tree, that the *kahau* always holds up one hand as a protection to what the poet called "a most unlikely feature, but mine own, sir."

The **Proboscis Monkey** (*Semnopithecus nasalis,* or *larvatus,*) belongs to Borneo, lives in large troops in the trees which skirt the banks of the rivers, resembles a shrivelled, bowed, long-nosed, little old man or woman, and is sacred in the eyes of the natives. Its noisy outcries, malignant disposition and fondness for irritating mischief, seem to add a fresh illustration to the truth that the uncivilized animal nature is perfectly unfit for the government of self or of others.

The **Proboscis Monkey** (*Nasalis larvatus*) is only about two feet in length, and the tail claims "the larger half" of this. Its color is a dark chestnut, but its face markings are blue and red. It is frugivorous and peculiar to the fauna of Borneo.

The **Hoonuman,** or **Entellus** (*Presbyter entellus,* or *semnopithecus*), exceeds the simpai and the negro monkey in size, being about eight feet in length, equally distributed between the body and the tail. It is at first gray in color, with brown lines on back and loins, but with increasing age changes to

GIBBON (*Hylobates agilis*) AND HOONUMAN MONKEYS (*Semnopithecus entellus*).

black. The *hoonuman* is the sacred monkey of India, and like the Sacred Brahma Bull has learned that he has "the freedom of the city," and like the bull uses his opportunities to their fullest extent. He mingles freely with the natives, not that he values social intercourse with them, but simply because they furnish opportunities for his amusement or profit. He is fond of shopping, and it is always understood that his purchases can be charged only as an offset to the faults and follies of poor humanity. His demands are rendered none the less exorbitant by the fact that for him to have a handful requires a provision fourfold as great as for a human being. The banyan-tree, so familiar an illustration in school geographies, is a favorite resort of the *hoonuman* or *entellus,* and his numbers seem a sufficient excuse for its branches bending to the earth, and having taken root to furnish yet further support to the parent tree. No one is without enemies, so in spite of the veneration accorded by man to the *hoonuman,* the serpent—man's earliest and most irreconcilable foe—is specially harmful and inimical to the *hoonuman*. It is not, however, an unresisted tyranny, for whenever a *hoonuman* finds a snake asleep, it at once seizes it by the back of the neck, drags it to the ground, and puts it to death with the utmost refinements of torture, slowly grinding off its head against the edge of some sharp stone or rock; nay, more than this, after death has befallen the snake, the *hoonuman* treats its lifeless remains with contempt and tosses them to make sport for its young. It will follow tiger-hunters, and is frequently useful in pointing out the hiding-place of their game. It has learned that it is perfectly safe in the presence of man, for its position as a sacred being protects it as fully as the law in the most highly civilized communities guards against the commission of murder. Still further, the doctrine of the transmigration of souls, so common in some of the ancient philosophers, and popularly known through stories of the Egyptians, adds to the sacredness of its person.

GIBBON (*Hylobates lar*).

The **Lungoor** (*Semnopithecus schistaceus*) is a sacred monkey, belonging to the mountain districts of Nepaul.

The **Negro Monkey, Moor,** or **Budeng** (*Semnopithecus maurus*), is jet-black, and its long silken hair furnishes most of the monkey-fur which from time to time receives the imprimatur of fashion. It is difficult to tame, and not very agreeable or amusing when domesticated. It ranges the woods of Java in companies of forty or more, and keeps up the most constant and noisiest chattering. Some of the family secrete in their stomachs a substance called bezoar and an essential part of an Oriental pharmacopœia. It is sometimes called the *bezoar monkey* from this fact. When young it is reddish-yellow and gradually changes to black.

The **Gibbon** (*Hylobates*) is considerable in size, having a stature of quite three feet. Its head is small and round, muzzle short, face pleasant in expression. It is wrapped in dark fur which is relieved in part by white. Its arms and hands are unusually long and out of proportion to the body. It is frugivorous, gentle, intelligent and marvellously supple. It is readily domesticated, in which state it is very affectionate.

The **Mourning Gibbon** (*Hylobates funereus*) is black with a change to ashen gray on the outside of its arms and legs, which are covered with long hair.

POSITION OF THE GIBBON WHEN STANDING.

The **Silver Gibbon** (*Hylobates leuciscus*) is silver-gray in color, changing to black on the face and the palms of the hands, and to white in the bushy hair and whiskers which cover its head, cheeks and neck. Its height rarely exceeds three feet.

The **Cinder Gibbon** (*Hylobates cinereus*) belongs to Java and takes its name from the color of its hair. It is gentle, affectionate and easily domesticated.

The **Simpai** (*Presbytes metalophors*) is long-tailed, its arms are delicate and well-proportioned, and hands slender, and is most distinctly four-footed in its locomotion. In color it is a light chestnut-brown intermixed with golden tints, changing to gray on the under parts. Its form is slight and graceful, and its foot and three-quarters of length is increased by some three feet of tail. The black hair on its head, cheeks and neck resembles a fur hood and has given rise to its popular name of the *black-crested monkey*. Its habitat is Sumatra, whose fauna receive such frequent mention.

The **Wou-wou** (*Hylobates agilis*) is of fair stature—about four feet—and but for the excessively long hands and fingers its body would be not unlike that of a human being. There is a great variety in the heads of persons, but even one who is not a craniologist will at once perceive the wide gulf which separates them from the patterns used by the monkey tribe. The cocoanut-shaped head of the *wou-wou* has a low forehead, from which it retreats until it reaches a peak or table-land. The eyebrows project, its nose is broad and flat, its mouth a long, thin slit, and its chin quite short. Its long, fine hair is chocolate-colored, the back and thighs tending towards yellow. The face of the male is dark blue and is furnished with white whiskers and a white band above the eyes. Sumatra is the habitat of the *wou-wou*, where it is often called the *ungaputi* and the *active gibbon*.

The **Siamang** (*Hylobates syndactylus*) is Sumatran in its habitat, and is covered with short, black hair, except upon the upper breast, which is bare. It takes its scientific name from the fact that a membrane connects the upper joints of the first and second fingers or toes on the hind feet; the hands are narrow and the fingers slender. It wears two throat-pouches which, when excited, it puffs out with air. In the early morning and the early evening the *siamangs* gather in great crowds and engage in a shouting and crying tournament. The *siamang* is harmless, indeed almost unable to inflict harm. In the family economy, the males are entrusted to the care of the father, and the females to that of the mother, and, were the young as much inspired as Cowper, they would lack no opportunity for the celebration of their parents' scrubbing powers and unpleasant fondness for being assured of the cleanliness of their young children. The *siamang* is tailless, and when domesticated, affectionate and playful, although very mischievous. Those that have been in captivity display the greatest sensitiveness to any attempts at ridicule, feeling, like most professional humorists, that their own province was being invaded.

The **Apes** (*Simiadæ*) are without the callosities which ornament the hinder parts of other monkeys; they are tailless, and have no cheek-pouches or plumpers.

Of the *apes*, the **Cynocephali** or **Dog Headed** variety have the

ORANG-OUTANG.

muzzle of a dog, and the callosities on the thighs are brilliantly blue or red. They are quadrupedal in their locomotion, stand, when erect, from two and a half to three feet, and are characterized by ferocity, audacity, malignancy, love of mischief, brutality and viciousness. Of course, this is as they appear to mankind, and doubtless they do not echo Burns's refrain.

At one period they fairly swarmed in Sweden, and the trees are said to have been covered with them. In Arabia they rob the natives of all kinds of fruits, vegetables and nuts, and keep posted the most vigilant sentinels. At the Cape of Good Hope they are so fierce that one species has earned the appellation of *man-tiger*. The wounds which they inflict are deep and dangerous, and emphasize the fierceness which they uniformly display. The Gibraltar species is tailless.

HUNTING THE ORANG-OUTANG.

The **Orang-Outang** (*Simia satyrus*) is one of the animals whose name can so easily be misspelled, as philology seems to delight in making constantly fresh experiments in attempting to express in English the gibberish of a savage tongue. Whether or not the *orang-outang* is the original of the satyrs celebrated in mythology, and thence passing into the imaginative literature of every people, he certainly can be used to give bodily substance to these poetical fictions. Asia is the habitat of the *orang-outang*, and he especially affects Sumatra and Borneo. The inhabitants of Borneo call the *orang-outang*, *mias*, *mias-pappan*, or *pappan*. Like the gorilla, it is solitary in its habits, and seems to find pleasure in meditating upon subjects which increase its gloominess and sullenness. It is almost helpless on the ground, as it is compelled to walk on its hands and on the outside edge of its feet. But once in the trees, its length and muscularity of fore-arm enables it to swing along with a rapidity, grace and ease truly surprising. Indeed, its capture is possible only by cutting down trees, so that the *orang-outang* is finally confined to a single one, which is then felled. The color of this creature is reddish-brown, the face blue and red-bearded, the hair long and abundant on breast, arms, face and

HAND AND FOOT OF ORANG-OUTANG.

HAND AND FOOT OF CHIMPANZEE.

back. The *orang-outang* is about half a foot shorter than the gorilla, and wears the singular air-pouches which belong to the monkeys. This receptacle, however, is not used for respiration; it is fingered like a glove and serves some use not yet definitely known, although it has been suggested that it may increase his levity when swinging through the air. The *orang-outang*, when young, is easily tamed and is quite an amusing pet, but should it not meet an early death from consumption, and failure to become acclimated, it grows intractable and savage with increasing years. In its natural state it is frugivorous, but in captivity learns to be omnivorous, and in particular grows passionately fond of spirituous and fermented liquors. In the woods it shows great fondness for cocoanuts whose shells it is able to crush. This same power, partly mus-

BORNEANS CAPTURING AN ORANG-OUTANG.

cular and partly mental, renders it an adept at opening any bottles which it may find while in a state of captivity.

The *orang-outang* is found in Borneo, Java and Sumatra, where it is commonly called the *wild man of the woods*. Its height is about four feet, its arms particularly muscular, and so long that when standing erect, it can touch the ground with them, and legs correspondingly short and weak. His success as an acrobat might lend support to the belief that he was the original trapezist. The hair is coarse and reddish, and thick except about the fore parts. The face is blue and for the most part naked; the eye-brows are bushy and prominent; the palms of the hands are hairless, and the creature sports mustache, chin-beard and side-whiskers. The ears are small, the muzzle long and thin, the nose flat, the lips remarkably extensible, and the creature indulges in what among

45

corpulent men is called a "bay window." As the animal grows older the forehead becomes depressed and the creature atrophied. The head, which inclines forward, is set on a thick short neck, which in turn is re-enforced by a pouch which extends beneath the arm-pits. He makes a bed every night, and when the weather is inclement, he erects a temporary roof. It is timid and inoffensive, but if provoked too far is a dangerous enemy, although it never uses its teeth in the conflict. When pursued it utters the most dolorous cries, but to which the natives, who are fond of his flesh and use his skin for helmets and caps, are entirely deaf.

The **Mias-kassar** is a smaller, slighter, less dangerous species of the orang-outang, but is found in the same localities.

The **Chimpanzee** (*Pithecus troglodytes*) rankes highest among the quadrumana. It is called *engecko* by the natives, (whence the familiar name *jocko*,) and sometimes the *quimpezee* (whence *chimpanzee*). An African traveller found a native chief, who, after having feasted upon the flesh, had from superstitious reasons made a garden, whose rare plants consisted of the skulls of one hundred

FAMILY OF CHIMPANZEES.

and eighteen *chimpanzees*.

The **Black Chimpanzee** (*Troglodytes niger*) belongs to Guinea. Its face is larger and flatter than that of the orang-outang; its ears are large and shaped like those of a human being; the head, shoulders and back are covered with long black hair; its legs have calves, and the *chimpanzee* can easily walk on its hind legs, though the position is awkward. It is tailless and pouchless, and wears its hair parted in the middle. Its hairless face is copper-colored and wrinkled, its nails are black, and the palms of its hands and its fingers are copper in color. Its eyes are small, sunken, and hazel in color; its height is about four feet; its cranium is depressed, and its forehead not entitled to be called even retreating since it is a mere ridge-like projection. As it grows older it becomes less and less amiable, and its physical degeneration is quite marked. It lives, or at least congregates, in groups and is much given to throwing stones at the travellers, or using the branches as missiles. For the most part it is frugivorous, but to a very small extent it is insectivorous. It interweaves leaves and branches for the sake of furnishing itself with a comfortable bed. The natives regard it as a man dumb through choice lest it share their curse of labor.

The **Bald Chimpanzee** (*Troglodytes calves*) adds to the luxury of its home the most elaborated leafy screens.

The **Chimpanzee** has the same habitat as the gorilla, to which it seems to be nearly allied. Its color, like that of the gorilla, is black, but it ornaments its nose with a few white hairs, and which give its face some resemblance to that of the Chinese. Its muzzle projects at the expense of its nose which is unusually flattened. Unlike the gorilla, the *chimpanzee*

POSITIONS OF THE ORANG-OUTANG WHEN WALKING.

is social and gregarious instead of being sullen and solitary, and as it is widest awake at night and much given to exhibitions of its vocal powers, the silence of the forest is often broken by its cries. The *chimpanzee* is not as a preference arboreal although climbing is not difficult to it. It prefers to find caverns or openings in the rocks where it builds huts for the occupancy of its family, but the male refuses to enter, keeping guard on its roof as if he were fearful of the stifling atmosphere of his tenement house, and prefers a continuous existence in the open air. The *chimpanzee* in captivity has always proved amiable and docile, but soon succumbs to consumption, which is the prevailing disease among the monkey tribes. The *chimpanzee*, unlike the gorilla, can act in co-operation with his kind, and it seems no mere accident that this power should have been restricted to the species which is certain not to combine for injury to mankind. The *chimpanzees* always keep sentinels posted, which are never untrue to their trust, but utter a warning cry at the least appearance of danger. Like the gorilla, the *chimpanzee* fears none of the animal kingdom except the leopard, for which it has the greatest aversion (possibly because it is arboreal), and when one is pursued by hunters, it will follow from tree to tree scolding and vilifying its enemy in a manner exhibiting the greatest excitement and fury.

The **Gorilla** (*Troglodytes gorilla*) has its habitat in the west of Africa, is the largest of the *simiadæ*, and for many reasons the most interesting to the naturalist, and the most exciting to the popular imagination. For much more than two thousand years explorers and navigators had told of a wild man of the woods, but it is only recently that this singular creature has been identified. The immense size, relatively erect position, and imitative habits of the *gorilla* may well account for his having been mistaken for a wild man by those in the midst of constant dangers, and who naturally endeavored to relate every new object to their ordinary and familiar experiences. The myths of the ancient Phœnicians, Greeks, Romans, Egyptians and Carthaginians were not mere fabrications, but mistaken and imaginative accounts of actual experiences. These myths were first actual beliefs, then popular superstitions, then the material for poetic illustration, next the scoff of those who believed their knowledge to be exact and final, and at length the subject of scientific investigation, as supplying the confused notions of times past in regard to actual existences and happenings in the world of nature. The *gorilla* has not as yet become possessed of the spirit of the scientific movement, so that he does nothing to decrease the difficulties in the way of making a study of him and his habits. His habitat is limited in extent, and so distant from our great commercial and literary centres as to multiply the obstacles in the path of the enthusiastic naturalist. Then, again, the *gorilla*, without being in the least timid, prefers to dwell afar from the haunts of man, selecting the deepest parts of the jungle

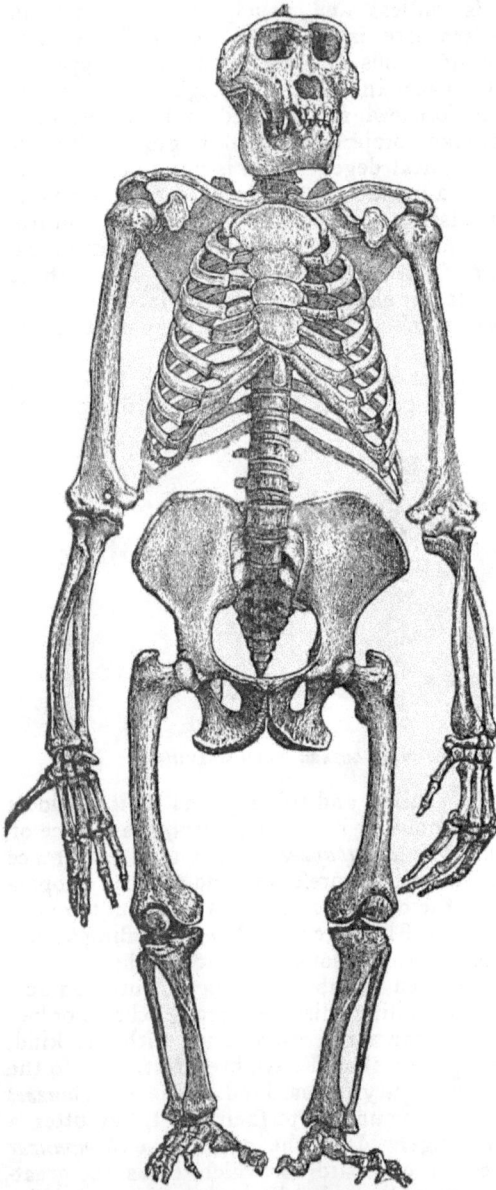
SKELETON OF THE GORILLA.

that, like the American poet, he may find "a solitude where none intrudes."
Furthermore, he is so British in his political economy as to resist violently

MALE GORILLA.

MALE GORILLA ALARMED.

and somewhat effectually any invasion of his personal rights, among which he
seems greatly to esteem the sanctity of
his domestic privacy. His breadth of
shoulder, powerful muscles, long arms,
immense and well-furnished jaws, together with his pugnacity, fearlessness
and tenacity of life, qualify him extremely well for resisting effectually any
ordinary attack, and suggest to all but
the devoted naturalist, or the irrepressible
sportsman, that it is more expedient to
hunt other animals than to make game
of the *gorilla*. It is no uncommon experience for a native to suddenly find
a hand thrust forth from the branches

POSITION OF GORILLA WHILE RETREATING.

of a tree and himself in clutches which will never relax until he be killed. The *gorilla* has but a small brain, and would seem to enjoy but a limited range of intelligence; but within his limits of cunning and muscularity he is unapproachable. His imitative powers are not complete, for though he will collect ivory, following an example of the natives, he has not sense enough to lay it down, but will frequently exhaust his strength by aimlessly carrying about a heavy and useless load. It will stir up a deserted camp-fire and sit by its light and warmth, but it never occurs to him to put on fresh fuel. It will build huts in imitation of the work of man, but it will then stay on the outside. It will carry about its sick child, but if the child dies it will never think of laying it down, or of burying it. It will gather its plunder into bundles after the manner of the farmer, but if it makes these too heavy, it will never think of a remedy.

The *gorilla* is in color black, or grizzly black, wearing a coat whose hair is somewhat more than two inches in length, and which on the arms and legs runs down from shoulder to elbow, and up the rest of its length. Though its eyes are naturally brown they become very green under the effects of excitement and add to an

FEMALE GORILLA WITH YOUNG.

appearance which is exceedingly dreadful and repulsive. Its paws and fingers are large-sized and enormous in their muscular power, and the hind ones have the use of the thumb very much like the hands of a human being. In size it is nearly three feet across the shoulders, more than five feet tall when erect, and the body, excluding the legs, is about two and a quarter feet in length. The *gorilla* is not gregarious in his habits, but usually solitary; like the criminal classes among mankind, whom the *gorilla* resembles in malignity, brutality and entire surrender to passion and appetite, he does not know enough or has not faith enough to enter into combination with his fellows.

The *gorilla* is believed to have been referred to by Hanno and other ancient writers, whose allusions were unintelligible or ridiculous until Du Chaillu re-discovered the animal. Its habitat is lower Guinea, or the Gaboon district of Africa, and its fierceness, savageness, extraordinary strength, and its constant effort to shun as well as to resist human companionship, has prevented that continuous study of its woodland habits which might simplify the discussions which turn upon the *gorilla* as the missing link.

In stature it varies from four to six feet, and is distinguished by the length and width of its face, the development of its lower jaw, and the possession of an elevated orbital arch. Its nose is flat, its eyes sunken, its ears small, and its mouth large. The neck is thick and short, the shoulders exceedingly broad, the chest massive and sounding like a deep base drum when the gorilla, in his wrath, strikes his fists upon it. Its belly is large and expansive. It is sometimes brownish-black in color, though sometimes an iron-gray, while the palms of the hands and the naked face are of a deep black. The male sleeps on the ground, lying on its back; the female takes her rest in the trees.

MALE GORILLA COVERING THE RETREAT OF HIS YOUNG.

It uses both paws and feet for walking, is swift, frugivorous, nomadic and solitary, except being devoted in its family relations, keeping close to the female and young, so that occasionally as many as half a dozen may be seen together.

Its hearing is unusually acute; its chest is its "tendon Achilles," and the hunter who fires and misses is generally lost, for with a single blow the creature can annihilate a man. When fatally wounded it utters a dying cry of anguish whose humaneness is so great as to haunt the hunter.

Of monkeys as a class many interesting characteristics may be predicated, and endless stories possessing interest have been told. For example, monkeys have been detected keeping the flies off from their sleeping young. So, too, their social attachments can be illustrated by the fact that fifty monkeys pursued a hunter who had killed one of their companions, and scolded and plead until he allowed them to carry back the lifeless body. Towards each other they are sympathetic and most attentive and gentle in their ministrations to the sick. On one occasion at least, when a monkey was bitten by an ill-natured baboon, it was immediately coddled by a monkey of a wholly different species. One of two monkeys on shipboard fell overboard, when the other first held out its hand, and finding this ineffective caught up a rope and threw one end of it to his drowning campanion. A hunter having fired into a band of monkeys to still their noisy chattering, fatally wounded one of their number, which at once descended to the ground, and holding its hands went directly to the hunter, as by its reproachful looks to induce him to repair injuries so unprovoked. Unsatisfied curiosity is very trying to the monkey. One of them failing to discover anything behind a mirror into which it was looking, dashed it to pieces and then repeated the action with each fragment large enough to renew his curiosity and re-awaken his anger. A monkey which had an ulcerated tooth refused to take an anæsthetic but cheerfully submitted to the dentistry. Many species of monkeys manifest the greatest interest in mechanical devices, and much skill in the application of their principles. They have, when in confinement, shown an acquaintance with the principle of the lever, screw and wedge.

APPEARANCE OF A YOUNG GORILLA WHEN WALKING ON ITS HIND LEGS.

By the many manifestations of an intelligence certainly superior to that of other animals, no less than the human appearance which many of the higher primates exhibit, we have come to regard the monkey family as next to our own, though there is a gulf between the two infinitely wider than that which separates the other orders of animal life. For this reason, in following the ascending series of animate creation, we are compelled now to halt for want of a bridge over which to pass to another sequential order. Man stands alone, isolated from all other species, and "The Story of Man" is therefore reserved for a work which I have prepared with much care, to prove that, like others of God's creatures, since his fall he has developed from a very low condition to the attainment of such intellectual powers as now distinguish him.